COUPLED CFD-DEM MODELING

COUPLED CFD-DEM MODELING

FORMULATION, IMPLEMENTATION AND APPLICATION TO MULTIPHASE FLOWS

Hamid Reza Norouzi
Reza Zarghami
Rahmat Sotudeh-Gharebagh
Navid Mostoufi
School of Chemical Engineering, College of Engineering, University of Tehran, Tehran, Iran

WILEY

This edition first published 2016

First Edition published in 2016

Registered Office
John Wiley & Sons, Ltd, The Atrium, Southern Gate, Chichester, West Sussex, PO19 8SQ, United Kingdom

For details of our global editorial offices, for customer services and for information about how to apply for permission to reuse the copyright material in this book please see our website at www.wiley.com.

Library of Congress Cataloging-in-Publication Data

Names: Norouzi, Hamid Reza, 1985– author. | Zarghami, Reza, 1973– author. | Sotudeh-Gharebagh, Rahmat, 1964– author. | Mostoufi, Navid, author.
Title: Coupled CFD-DEM modeling : formulation, implementation and application to multiphase flows / Hamid Reza Norouzi, Reza Zarghami, Rahmat Sotudeh-Gharebagh, Navid Mostoufi.
Description: Chichester, UK ; Hoboken, NJ : John Wiley & Sons, 2016. | Includes bibliographical references and index.
Identifiers: LCCN 2016025074 | ISBN 9781119005131 (cloth) | ISBN 9781119005292 (epub)
Subjects: LCSH: Computational fluid dynamics–Mathematical models. | Discrete element method–Mathematical models.
Classification: LCC TA357.5.D37 N67 2016 | DDC 620.1/064015111–dc23
LC record available at https://lccn.loc.gov/2016025074

A catalogue record for this book is available from the British Library.

Cover (background) image: Gettyimages/KTSDESIGN/SCIENCE PHOTO LIBRARY

Set in 10/12pt Times by SPi Global, Pondicherry, India

1 2016

To our families, for their continuous support and understanding.

Contents

Rahmat Sotudeh-Gharebagh is currently Full Professor of Chemical Engineering at the University of Tehran (email: sotudeh@ut.ac.ir). He has taught process modeling and simulation, transport phenomena, and fluidization courses for over 17 years. His research interests include computer-aided process design and simulation, and fluidization. He holds a B.Eng degree in chemical engineering from Iran's Sharif University of Technology, plus an M.Sc. and a Ph.D. in Fluidization from Canada's Ecole Polytechnique de Montréal. He has been a Visiting Professor at Qatar University. Professor Sotudeh has more than 250 publications in major international journals and conferences, plus four books and three book chapters. He is the Founder and Editor-in-Chief of Chemical Product and Process Modeling (www.degruyter.com/view/j/cppm) and winner of two prestigious awards, University of Tehran's International Award, 2015 and Allameh award from Iran National Elite Foundation, 2015.

Navid Mostoufi is currently Full Professor of Chemical Engineering at the University of Tehran (email: mostoufi@ut.ac.ir). He has taught advanced mathematics and fluid mechanics courses for over 16 years. His research interests include process modeling, simulation and optimization, and fluidization. He holds B.Eng and M.Sc. degrees in chemical engineering from Iran's University of Tehran, plus a Ph.D. in Fluidization from Canada's Ecole Polytechnique de Montréal. He has been a Visiting Professor at METU, Turkey. Professor Mostoufi has more than 270 publications in major international journals and conferences, plus five books and four book chapters. He is the co-author of the textbook *Numerical Methods for Chemical Engineers with MATLAB Applications*, published by Prentice Hall PTR in 1999. He is the Founder and Editor-in-Chief of Chemical Product and Process Modeling (www.degruyter.com/view/j/cppm) and winner of University of Tehran's International Award, 2015. He was also the University of Tehran's distinguished researcher, 2013.

Preface

This book provides an up-to-date description of the formulation, implementation, and applications of combined Computational Fluid Dynamics and Discrete Element Method (CFD-DEM) modeling. It is an integrated text that deals with theoretical and practical concepts of CFD-DEM, its numerical implementation accompanied by a numerical code and industrial applications. In the DEM part, different contact force models for spherical and non-spherical particles, as well as free-shape bodies, are discussed, along with their applicability and limitations. In the second part, couplings between solid and fluid equations for momentum, energy, and mass for particles and fluid are described and implementation of external forces on particles in multiphase flows is presented.

Over the years, many excellent books have been published dealing with various aspects of CFD. The level of sophistication of these books varies from academic to complex industrial systems and this is the main reason why we started this book with thorough treatment of the DEM. The distinctive feature of this book is its emphasis on coupled CFD-DEM (momentum, energy, and mass) as compared with books written on CFD or DEM alone. In addition, hands-on numerical codes are also delivered with the book in order to be used by readers as is, or modified as desired.

CFD-DEM has found wide range of applications in nearly all systems dealing with solids in various fields of science, engineering and technology such as chemical, food, pharmaceutical, biochemical, mechanical, energy, material, and mineral engineering. However, the prime concern of this book is to provide a more comprehensive treatment of DEM and CFD-DEM in chemical and process engineering with applications in granular and multiphase flow systems. In these systems, the DEM is commonly used for analysis of granular flow, including solid mixers, hoppers and silos, and CFD-DEM for fluid-solid flows, such as fluidized beds and conveyers, spouted beds, coal combustors, and solid incinerators.

Experimentations on multiphase flow systems are of vital importance in research and engineering. Nevertheless, they are lengthy, cost intensive, tedious, and challenging. We are unable to conduct experiments on the micro- and meso-scales in many cases and this is why most industrial processes fail at the early stages of development, design and operation. With the constant evolution of efficient computational tools, we now can analyze these issues and provide solutions. This book also helps the reader to acquire a better insight into these complex systems. With the diffusion of computational skills in industry and academia, we see the future of computation in process engineering rather promising. This would allow the better utilization of existing computational knowledge along with limited experimentation efforts.

The content of this book has gelled over the last 10 years through the collaborative research efforts of the authors on the subject. The book is primarily intended to serve students, scientists, and practitioners in process, chemical, mechanical, and metallurgical engineering. However, other engineers, consultants, and scientists concerned with various aspects of multiphase flow systems may also find it useful. Scientists and graduate students who want to learn and excel in DEM and CFD-DEM would find this book helpful. The content of this book can be used in a graduate course on advanced modeling and simulation in chemical engineering or as a complementary book to other engineering areas.

The authors acknowledge the contribution of many colleagues, former, and current students from the University of Tehran. Special thanks is extended to Dr. Zahra Mansourpour, Sedigheh Karimi, Bahram Haddani-Sisakht, and Shahab Golshan who have greatly contributed to some of important results presented in this book. We also express our gratitude to Mohammad Amin Hassani, Yasaman Norouzi, Mohammad Foroughi-Dahr, Mahsa Okhovat, Maryam Karimi, Maryam Sanaie-Moghadam, Hanieh Sotudeh-Gharebagh, Dr. Jaber Shabanian, Dr. Ebrahim Alizadeh, Dr. Rouzbeh Jafari, and Mr. Christian Jordan for their help extended to us during the completion of the book. Iranian National Science Foundation (INSF) is acknowledged for supporting our research efforts in the multiphase flow laboratory, and process design and simulation research center where experimentations on and simulations of multiphase flow processes are the main concern.

Finally, we should emphasize that much remains to be done in this area and the utilization of CFD-DEM is expected to be increased rather than diminished. Adapting CFD-DEM to new areas will undoubtedly keep scientists and engineers busy for a long time. We can only hope that we have provided a useful base from which to start. The authors hope that this book would serve the industry and academia in the coming years. No human attempt is flawless, including this book. With your help, shortcomings and mistakes can be remedied and corrected. You are kindly requested to send your comments and corrections to mostoufi@ut.ac.ir.

Hamid Reza Norouzi
Reza Zarghami
Rahmat Sotudeh-Gharebagh
Navid Mostoufi
June 2016, University of Tehran, Tehran, Iran

1

Introduction

Industry demand for efficient and faster computational tools has facilitated the development of Computational Fluid Dynamics (CFD). This has allowed the utilization of CFD as a specialized tool to solve mass, momentum, energy, and species conservation equations. Advances in computer technology have now changed the entire frame of CFD modeling, allowing it to be a tool for engineers and scientist to carry out design, simulation, and optimization of various processes. Today, the application of CFD not only covers the conventional engineering fields, such as chemical and mechanical engineering, but is also widely extended to multidisciplinary areas, such as environment and healthcare.

With CFD, fluid and solid particulate phases can be modeled by the "Eulerian–Eulerian" approach, which is a way of looking at the motion of fluid and particles from a continuum point of view. This hypothesis may be true for fluids but it may bring less accurate results when considering solid particles as a continuum. In order to properly model particle motion, the Discrete Element Method (DEM) has been developed, in which the motion of individual particles is tracked in space and time using the Lagrangian approach. This approach is complementary to the Eulerian approach for modeling multiphase flows and is referred to as the "Eulerian–Lagrangian" approach, detailed in the following sections.

Multiphase flows exist in many industrial applications such as gas or liquid fluidized bed reactors, fluidized bed dryers, spotted beds, three-phase gas-liquid-solid fluidized beds, pneumatic conveying of solids, and so on. A detailed knowledge of these flows is crucial for design, scale-up, optimization, and troubleshooting of such processes. Although this may be achieved by experimental techniques, modeling can be considered as an alternative tool for exploring different aspects of multiphase flows. Modeling enables us to understand different phenomena occurring in these processes, to perform sensitivity analysis on different input parameters and to test different

---

Coupled CFD-DEM Modeling: Formulation, Implementation and Application to Multiphase Flows, First Edition.
Hamid Reza Norouzi, Reza Zarghami, Rahmat Sotudeh-Gharebagh and Navid Mostoufi.

Companion website: www.wiley.com/go/norouzi/CFD-DEM

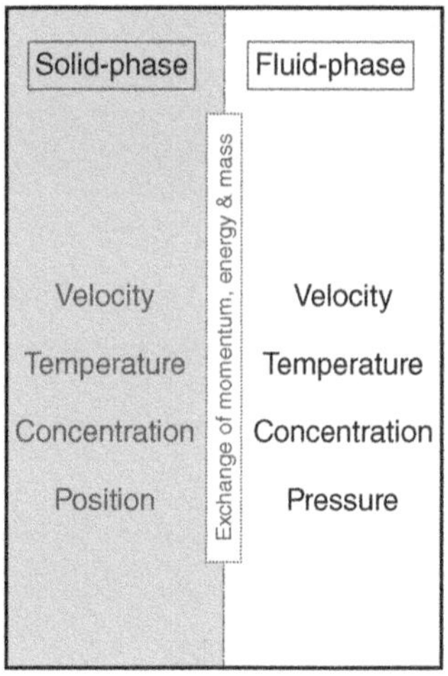

Figure 1.1 Momentum, energy, and mass transport between solid and fluid phases

configurations and operational conditions at lower expense compared with experimental methods. In the following, we discuss the overall view of the modeling of granular and multiphase flows.

1.1 Multiphase Coupling

Phase coupling, in terms of momentum, energy, and mass, is a basic concept in the description of any multiphase flow. The coupling can occur through exchange of momentum, energy, and mass among phases as shown in Figure 1.1. In principle, fluid-particulate properties can be described by position, velocity, size, temperature, and species concentration of fluid and/or particle. While the phenomenological description of multiphase flow can be applied to classify flow characteristics, it also can be used to determine appropriate numerical formulations. In various modes of coupling, depending on the contribution of phases and phenomena, different coupling schemes can be adapted. This may allow independent treatment of phases or simultaneous integration of momentum, heat, and mass exchanges between phases. In general, modeling complexity increases as more effects associated with time and length scales are included in the simulation.

1.2 Modeling Approaches

Real systems are rather complex in nature and modeling allows analysis and simulation of these systems to be conducted more accurately. Depending on the length scales considered for fluid and particle systems, various combinations of modeling scales can be suggested. These are classified as micro-, meso-, and macro-scale models. In a micro-scale model, trajectories of individual particles are calculated through the equation of particle motion and the fluid length scale is the same as the particle size or even smaller. At the same time, instantaneous flow field around individual particles is calculated. In the meso-scale model, both solid and fluid phases are considered as interpenetrating continua. The conservation equations are solved

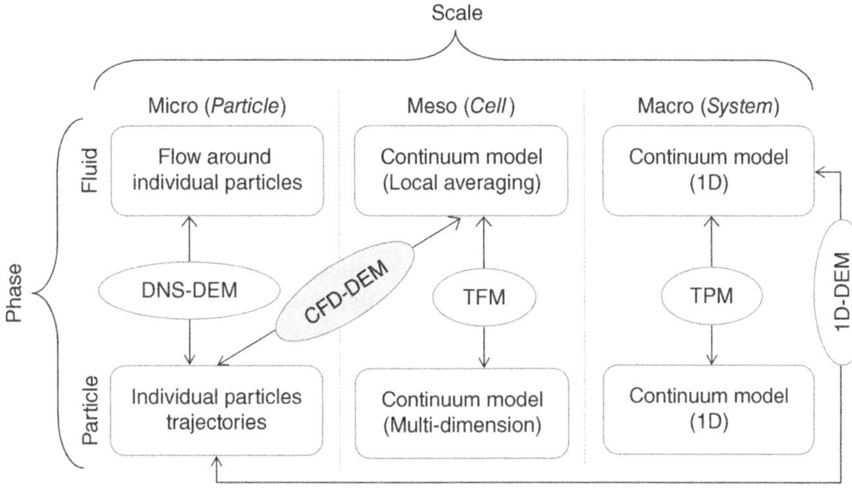

Figure 1.2 Modeling scales in fluid–particulate systems

over a mesh of cells. The size of the cells is small enough to capture main features of the flow, like bubble motions and clusters, and large enough (essentially larger than the size of individual particles) to allow averaging of properties (porosity, interactions, etc.) over the cells. Anderson and Jackson [1] first presented this formulation for fluid-particulate systems. In the macro-scale model, the fluid length scale is in the order of the flow field. This means that motions of the fluid and the assemblage of particles are treated in one dimension based on overall quantities [2]. It is also possible to develop some intermediate models in which the length scales of fluid and solid phases are different. For example, the length scale of solid phase can be kept at the micro-scale while changing the length scale of fluid phase to meso or macro. Under these conditions, the affective interactions in the larger scale can be calculated by averaging the information in the smaller scale.

In multi-scale modeling, the smaller scale model takes into account various interactions (i.e., fluid–particle, particle–particle) in detail. These interaction details can be used with some assumptions and averaging to develop closure laws for calculating the effective interactions (e.g., drag force) in the larger scale model [3]. This allows capture of the essential information needed on the larger scale. Alternatively, calculation of effective interactions can be performed through the local experimental data, if available. Combination of fluid/particle motion with different modeling scales can provide different modeling approaches, as sketched in Figure 1.2 and detailed here:

1. *Micro approach (fluid–micro, particle–micro)*: In this approach, the fluid flow around particles is estimated by the Navier–Stokes equation. Since the forces acting on particles are calculated by integrating stresses on the surface of the particle, the empirical correlation for drag and lift forces are not required. This approach is used in cases where particle inertial force is relatively small (e.g., liquid–particle flow) or the fluid lubricating effect on particles is rather significant (e.g., dense-phase liquid–particle flow). A typical example of such an approach, shown in Figure 1.2, is the direct numerical simulation–discrete element method (DNS-DEM).

2. *Meso approach (fluid–meso, particle–meso)*: In this approach, which is shown in Figure 1.2 and is referred to as the two-fluid model (TFM), in addition to the real fluid, the assemblage of particles is also considered to be the second continuum phase. The flow field is divided into a number of small cells to capture motions of both phases, provided that the cell size is larger than the particle size. The two continuous phases are modeled by applying laws of momentum and mass conservations in each fluid cell, leading to averaged Navier–Stokes and continuity equations. Capability of the TFM in capturing the solid phase motion greatly depends on the closure laws used for this phase. These closure laws always involve some simplifications or are obtained by semi-empirical correlations. While this approach is preferred in commercial packages for its computational simplicity, its effectiveness depends on the constitutive equations and is not easily applicable to all flow conditions. The TFM has been successfully utilized to obtain the flow behavior of various non-reacting and reacting multiphase flows in laboratory, pilot, and industrial scales.
3. *Macro approach (fluid–macro, particle–macro)*: This approach provides a one-dimensional (1D) description of gas-particle flows [4]. The main output of such a model is the pressure drop, which is considered as the sum of pressure drops due to flow of fluid and particles. Usually, a formula for the single phase flow, such as Darcy–Weisbach equation, is used for the fluid pressure drop and that of particles is balanced with the fluid drag formula from the momentum balance. This approach would also allow the calculation of averaged flow properties by empirical correlations that are essential in design and analysis of industrial processes. A typical example of such approach, shown in Figure 1.2, is the two-phase model (TPM) in fluidization. In this model, conservation equations are written for bubbles and emulsion, both having the length scale of the system in a fluidized bed.
4. *Macro-micro approach (fluid–macro, particle–micro)*: In this approach, shown in Figure 1.2 by 1D-DEM, the fluid forces acting on particles are calculated from empirical correlations (e.g., drag and lift) while translational and rotational motions of particles are described based on Newton's and Euler's second laws. At very low concentration of particles, effect of particles on the fluid motion can be neglected. However, at higher concentrations, closure laws should be modified to account for the closeness of surrounding particles. Generally, in this approach the flow field, which is considered to change in one dimension, is not divided into cells and additional pressure drop is taken into account to reflect the effect of particles on the fluid motion.
5. *Meso-micro approach (fluid–meso, particle–micro)*: In this approach, referred to as CFD-DEM and shown in Figure 1.2, the flow field is divided into cells with a size larger than the particle size but still less than the flow field. Effect of motion of particles on the flow of fluid is considered by the volume fraction of each phase and momentum exchange through the drag force. This approach is the focus of this book and is explained in detail in the following sections.

Let's consider an example for illustrating the abovementioned approaches for modeling. Various modeling approaches including TPM, TFM, CFD-DEM, and DNS-DEM for a gas-fluidized bed are shown in Figure 1.3. The macro approach TPM, which is a one-dimensional model, is mostly used in industrial applications for long term simulations. For multi-dimensional modeling, the TFM can be used to predict the characteristics of fluidized beds at the meso-scale. The increased accuracy of the model is obtained at the expense of more computational costs and simulations are restricted to shorter periods. Handling the solid particles at a micro scale and fluid at meso

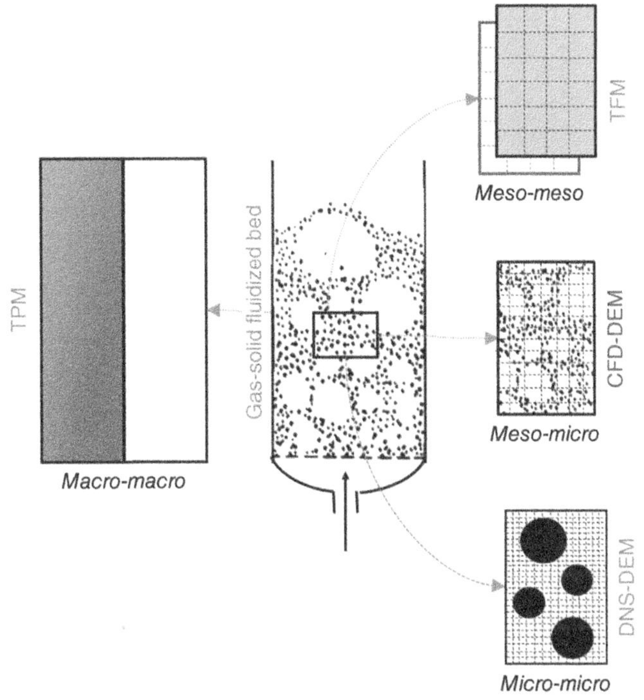

Figure 1.3 Different modeling approaches for fluid–solid systems

scales is carried out in the CFD-DEM approach. While this increases the computational cost, it provides results with a higher resolution when compared with the TFM. If a higher resolution is needed for the fluid phase, DNS-DEM is the choice of modeling considering the fact that it needs higher computational effort. It should be mentioned that applying CFD-DEM, and especially DNS-DEM, is mostly limited to lab scale units.

In a gas-solid fluidized bed, various structures (micro, meso, and macro) coexist with different scales, as shown in Figure 1.4. Single particles and individual particles in clusters are typical examples of micro scale phenomena, while small bubbles and clusters are considered meso structures. Large bubbles, as well as the whole reactor, are at macro scale. However, it is important to mention that we are not limited to use the scales of these phenomenological structures in the modeling. Nevertheless, modeling with a finer scale would provide characteristics of that structure and larger ones while coarser scale modeling would provide only an averaged description of finer structures. For example, if using the CFD-DEM model, characteristics of particles, clusters, and bubbles can be captured, while in the TFM only characteristics of clusters and bubbles can be obtained and individual particles cannot be observed.

1.3 Modeling with DEM

DEM is a type of modeling tool through which the dynamics of a system comprising of a large number of distinct bodies with arbitrary shapes are studied. In granular flows, these distinct bodies are solid particles. Particles may interact with each other through their contact area or

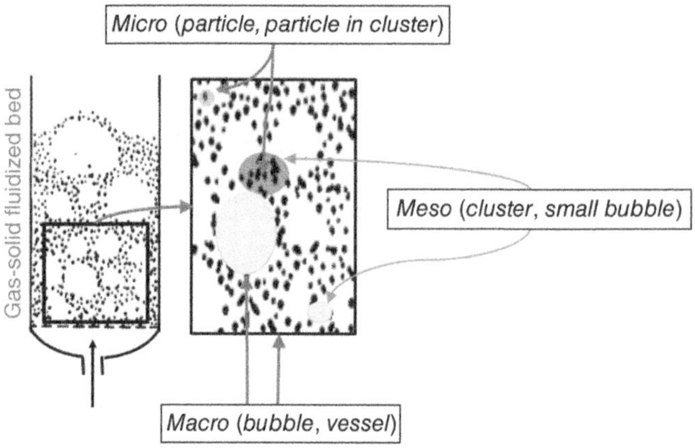

Figure 1.4 Different scales in a fluid–solid system

interparticle effects. Particles are assumed to be either rigid or deformable, leading to two different formulations of their collision, that is, hard-sphere/event-driven and soft-sphere/time-driven. In the soft-sphere formulation, which is the main focus of this book, particles are allowed to overlap and their contact lasts for a certain period. This allows a particle to be in contact with more than one particle at a time. This formulation is suitable for the motion of particles in both dense and dilute phases at quasi-static and dynamic conditions. Translational and rotational motions of a particle are tracked by integrating Newton's and Euler's second laws of motions, respectively. In addition to contact forces between particles, inclusion of other forces, like interparticle and fluid–particle interaction (if the fluid effect is significant), can be performed by introducing proper terms in the equation of motion.

The contact force is calculated according to normal and tangential overlaps of particles (or particle and wall) using a set of force-displacement expressions combined with friction laws. Many force-displacement models, such as linear and non-linear visco-elastic, elasto-plastic, and visco-plastic models have been developed for calculating the contact force between particles depending on the material properties and operating conditions. Particles are surrounded by walls as system boundaries. Among different methods, walls can be introduced to the model by decomposing the actual geometry into triangular or quadrilateral elements. This seems to be the flexible and rather general method to deal with simple to complex geometries in DEM simulations.

All motion and force equations associated with the DEM should be solved using proper integration methods. The explicit integration of these equations, in contrast to the implicit integration, increases the flexibility of the DEM simulation, even though it requires adapting a small time step for integration. A small time step and existence of a large number of interacting particles in the DEM simulation demands a huge computational resource. Without implementing efficient numerical algorithms and parallelization, this modeling approach is restricted to short-time simulations with the number of particles not exceeding 10^5. Nowadays, the DEM is a powerful technique, allowing scientists and engineers to analyze rather complex systems for which analysis and understanding are not possible by current experimental techniques. In Chapters 2–5 of this book, essential components related to formulation, implementation, and sample application of DEM can be found in detail.

1.4 CFD-DEM Modeling

CFD coupled with DEM is a computational approach used to model fluid–particle systems. In the CFD-DEM, the fluid phase is assumed as a continuum and its meso-scale motion is described by the volume averaged Navier–Stokes equation, while the micro-scale motion of the solid phase is described by Newton's and Euler's second laws. Forces acting on a particle are gravity, contact between colliding particles, fluid–particle interaction, and interparticle forces. Normally, there are thousands or millions of distinct bodies in the system for which the equation of motion should be simultaneously solved along with fluid equations over fluid cells. These equations are usually presented for phenomena occurring with different length scales in the system. The flow field of the CFD-DEM is shown in Figure 1.5, which demonstrates the relation between micro- and meso-scales for modeling of a gas-solid system. The coupling between fluid and particulate phases is performed through the local porosity and the mutual fluid–particle interaction forces. Comprehensive reviews of CFD-DEM technique and its applications have been published [5–7].

In the CFD-DEM, fluid–particle interaction for the fluid phase is based on total interactions per unit volume of the fluid cell (data exchange from the micro- to meso-scale), while the fluid–particle force for the solid phase is based on the force acting on each individual particle (data exchange from meso- to micro-scale). When equations of these phases are coupled together, at each time step, the DEM calculates the position and velocity data for each individual particle and the CFD provides the fluid flow profiles over fluid cells for the next time step. The resulting profiles allow calculation of the fluid–particle interaction acting on particles and in fluid cells. A fully explicit scheme is assumed for coupling,

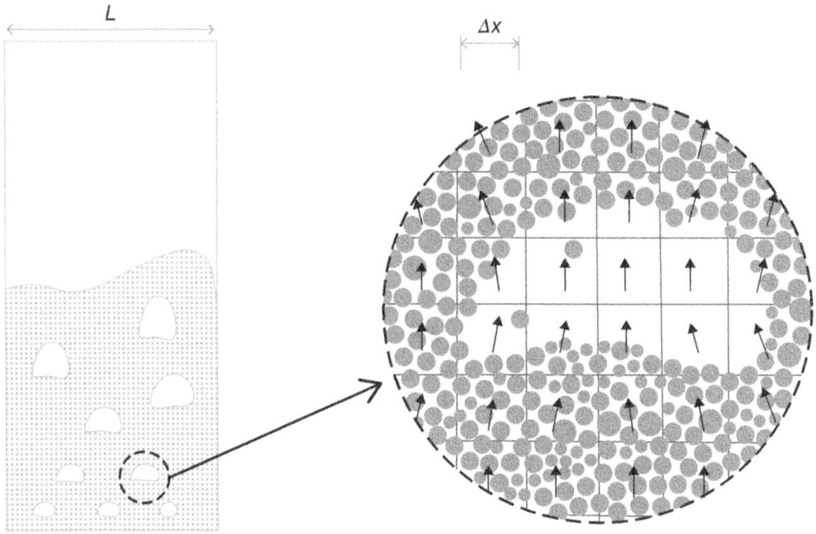

Figure 1.5 A typical bubbling gas–solid fluidized bed. Averaged fluid fields (pressure, velocity, porosity, etc.) are calculated in a fluid cell at the meso-scale. The cells are small enough to capture all properties of the flow (e.g., bubbles) and large enough to contain a number of particles to ensures proper averaging. Vectors show the fluid velocity interpolated on the center of fluid cells

meaning that the coupling terms are calculated with the information at the current time step. Then, equations of motion of both phases are solved to advance the calculations one step at a time.

Outputs of the CFD-DEM are velocity, pressure, temperature, and species concentration fields of the fluid phase as well as contact time, position, velocity, temperature, and interparticle forces of particles at any moment of the simulation. Obviously, these variables are not easily accessible via experimental measurement. Unique capabilities of the CFD-DEM enable one to deeply investigate overall and local quantities of the whole system based on information generated for individual phases. These quantities include mixing and segregation rates, formation, growth, coalescence, and breakage of bubbles, pressure drop, solid flow rate, agglomeration and aggregation of particles, residence time distribution, and concentration and temperature profiles.

The CFD-DEM simulation is computationally expensive and has two major limitations; time step of integration and the size of problem. The first limitation is related to the integration time step for equations of motion of both phases. For the particle phase, integration is performed using the explicit scheme. This is essential for a general purpose DEM code. This integration scheme, however, imposes a stability condition on the particle time step in a system with physical contacts. As a rough estimate, the particle time step should be a fraction (at least 1/10th) of the natural frequency of the linear elastic mass-spring system or a fraction of critical time step determined by Rayleigh analysis (see Chapter 3). Based on this condition, the particle time step generally falls in the range of 10^{-7} to 10^{-4} s. The stability condition for the fluid phase is expressed by the Courant number. The Courant number should be kept to less than 0.5 to ensure a converged solution of coupled pressure-velocity fluid phase equations, even if the solver is implicit [8]. In practice, the fluid time step falls in the range of 10^{-5} to 10^{-2}s based on this condition.

These conditions are applied when dealing with the numerical solution of equations of fluid and particulate phases separately. When solving equations of motion of the coupled fluid–particle flow, the interphase momentum exchange terms in these equations impose new conditions on particle and fluid time steps. The condition for the particle time step is imposed by the particle response time defined as the time required for acceleration of a particle to the fluid velocity due to the drag force [9]. Also, the condition for the fluid time step is imposed by the fluid response time defined as the time required for the acceleration of fluid to the particle velocity due to the drag force in the fluid cell [9]. Ranges of particle and fluid response times are 10^{-6}–10^{-1} and 10^{-8}–10^{-1} s, respectively. To ensure stability of solution and prevent lagged integration of equations of motion of phases, particle and fluid time steps should be less than these response times. In practice, the fluid time step is chosen equal or larger than the particle time step in typical fluid-solid systems. This difference leads to developing different coupling strategies (treatment of momentum exchange between phases) that have different stability behaviors as detailed in [10].

The second limitation in the CFD-DEM simulation is size of the problem. The size refers to the number of fluid cells and number of particles in a simulation. Basically, a CFD-DEM numerical code is developed to be executed in either sequential or parallel mode. With the sequential mode, and based on the current computational capacities, a few seconds of real-time simulation of a fluid–particle flow with less than 10^5 particles and cells is feasible while long simulations are not feasible in this mode. Therefore, implementation of efficient numerical algorithms and parallelization is a *must* for any CFD-DEM code.

Table 1.1 Open source and commercial software packages available for CFD-DEM simulation (accessed on November 2015)

Package	Parallel	Type	Description
LIGGGHTS®	MPI and CUDA®	Open Source	DEM code in C++ on the LAMMPS platform www.cfdem.com/liggghts-open-source-discrete-element-method-particle-simulation-code
PFC™	Multi-threaded	Licensed	DEM package, with a built-in fluid coupling package www.itascacg.com/software/pfc
EDEM®	Shared memory	Licensed	DEM package, can be coupled to Ansys Fluent® www.dem-solutions.com/software/edem-cfd-co-simulation/
CFDEM®	MPI	Open source	CFD-DEM code in C++, OpenFOAM® for CFD and LIGHHHTS for DEM www.cfdem.com/cfd-dem
Yade	OpenMP®	Open source	In C++ for DEM simulations https://yade-dem.org/doc/
SAMADII©	CUDA®	Licensed	DEM for single and multiple GPUs http://166.104.13.30/metariver/Technology.htm
MFIX-DEM	MPI-OpenMP®	Open source	CFD-DEM code in FORTRAN for fluid-solid flows like fluidized beds https://mfix.netl.doe.gov/mfix/

Parallelization falls into two categories: coarse-grained distributed-memory parallelization and fine-grained shared-memory or multi-threaded parallelization. The distributed-memory parallelization gives an almost unlimited number of processing units and memory space while it suffers highly from communication and inter-node data transfer latencies, which make it suitable for coarse-grained computations like CFD simulations. However, the parallelization and load balancing of the code for a dynamic DEM is a crucial task, which still needs more sophisticated algorithms to be implemented. The shared-memory parallelization suits DEM-based codes, since it benefits from loop level load distribution among processors. However, the number of processing units and memory space provided by the shared-memory machines are limited, which restrict the size of problem in some cases.

Accordingly, a unique and robust parallelization model cannot be suggested for a CFD-DEM code and mixed solutions have been used so far [11–13] to benefit from advantages of both parallelization models. This would lead to various configurations for software packages. Current open source and commercial CFD-DEM software packages are listed in Table 1.1. They use different parallelization models, ranging from distributed memory in Message Passing Interface (MPI) to shared memory in OpenMP®. Some packages employ advanced platforms such as compute unified device architecture (CUDA), which requires graphical processing units (GPUs) for computation (can be categorized as the shared-memory parallelization).

Nowadays, software companies have been developing ready-to-use tools for users with a basic knowledge of CFD-DEM. They regularly update their packages with new models and integrate them with mathematical tools and allow users to develop their own computational models within the software. However, troubleshooting, customizing, and extending of these packages need one to properly understand the basis of CFD-DEM as provided in this book (or in similar books coming in the future).

1.5 Applications

CFD-DEM modeling and simulation can be applied to many multiphase flows in scientific and engineering applications. This can be included in initial stages of process and product design. While these flows have been the subject of numerous experimental investigations, their CFD-DEM simulation can provide more detailed knowledge in different time and length scales to capture all features of such flows. The CFD-DEM has been widely used for analyzing the behavior of particles in fluid systems such as gas fluidization with cohesive and non-cohesive particles, dense phase and pneumatic conveying, drying, granulation, coating, and blending, segregation, agglomeration and clumping, caking, tribo-charging, and particle dispersion in solid and fluid. Industrial processes also can be successfully analyzed by this technique. Examples of such processes are gas cyclones, filtration, solid-liquid mixing, spray coating, drying, cooling, heating, and handling of powders. Moreover, complex processes such as flow in a circulating fluidized bed with application to coal combustion and incineration, fluid catalytic cracking (FCC) units for cracking high-molecular-weight hydrocarbon fractions of crude oil, metallurgical processes (e.g., blast furnaces), chemical reactors, off-gas scrubbing, industrial dust recycling, sedimentation and erosion, trickle beds, and so on, can be properly handled with this numerical technique.

Among the many applications, we chose to investigate chemical engineering examples such as packing behavior of particles, discharge of particles from a hopper, and hydrodynamics of fluidized beds (i.e., bubble behavior, gas and solid flows, and particle mixing). In addition, heat transfer, mass transfer, and chemical reactions are common phenomena in chemical engineering processes such as combustors, incinerators, fluidized bed reactors, dryers, granulators, and coaters. With the wide applications mentioned here, we can see that the future of CFD-DEM modeling and simulation is promising. CFD-DEM modeling would mimic chemical engineering applications as close to reality as possible and expand its application to new challenging problems while reducing the cost of experimentation to a great extent. Research institutions and companies are now integrating CFD-DEM into the early stages of their research and product development. This can largely decrease the development costs, increase the productivity, and shorten the marketing time. With powerful computers and knowledgeable users, modeling and simulation of more complex geometries will become believable in the near future.

1.6 Scope and Overall Plan

The focus of this book is to deal with granular and multiphase flows. The book is organized into two interconnected parts. Part I deals with the flow of solid in the absence of fluid in four chapters and Part II treats the flow of solid in the presence of fluid within three chapters. Although there are excellent books on CFD and DEM alone, limited number of books can be found on CFD-DEM coupling basics. The book chapters contain the most recent theoretical developments in this advanced modeling approach as well as the corresponding numerical implementation. It is worth mentioning that, due to the availability of more books on CFD, this book starts with DEM with more focus on the coupling of CFD-DEM models and corresponding applications.

A large number of references are cited and discussed, covering the essence of most published literature on CFD-DEM. A DEM code in FORTRAN standard accompanies the book to help graduate students and researchers in solving the formulations addressed, since it takes a long time to develop such a long code. This code is mature enough to be directly used (with no modification) in most related applications. However, it also can be partially modified to suit other applications not discussed in this book. Methods of setting up codes for different applications are also presented. It is worth mentioning that the elements of CFD-DEM model described in this book have been assessed by different investigators in terms of overall quantity. It is not our intention to verify the quality and the correctness of their approaches and we leave this to readers. In this book, there are eight chapters in two parts on various relevant topics with the following general descriptions:

Chapter 1 gives a brief introduction to topics related to multi-scale modeling, DEM, and CFD-DEM. A few industrial applications, among many applications that can be handled by this efficient technique, are described in brief. Organization of different chapters of the book and online content forms the last section of this chapter.

Part I

Chapter 2 begins with an overview of DEM for granular flow and is followed by the basic formulation of hard-sphere and soft-sphere frameworks. Since the focus of this book is the soft-sphere framework, the rest of the chapter is devoted to common models for evaluating contact forces and torques among spherical particles in detail. Boundary (walls) and initial conditions are also covered in this chapter. In addition, the calculation procedures and sequences are detailed wherever deemed necessary.

Chapter 3 provides an in-depth treatment of the numerical implementation and algorithms of DEM. Contact search algorithms, integration methods, as well as wall treatment are addressed. Dealing with a large number of equations needs choosing efficient numerical algorithms and parallelization techniques as addressed in this chapter. In order to help the reader to understand the implementation properly, the codes are partially given in this chapter, which eases understanding of the techniques presented.

Chapter 4 describes the implementation of non-spherical particles in DEM. The kinematics and dynamics of non-spherical rigid body are presented first. Superellipsoid and multi-sphere methods, as two main approaches for shape representation of non-spherical particles in a DEM simulation, are also discussed in detail.

Chapter 5 provides applications of the DEM code accompanied with this book to granular flow as formulated and implemented in previous chapters. Packing of particles as a common practice in the powder industry, flow pattern in hoppers, mixing operation and design, screw conveying, and film coating are among the application presented in this chapter with details.

Part II

Chapter 6 covers the CFD-DEM formulation and coupling of momentum, energy, and mass equations. Different coupling strategies and interaction parameters among phases and fluid flow field are also discussed in this chapter with details. This is followed by detailing the solid phase flow, interphase coupling, coupling framework, fluid volume fraction, and mapping from the Eulerian to Lagrangian domains. The heat and mass transfer formulations with and without chemical reactions for both solid and fluid phases as well as their coupling strategies are also presented.

Chapter 7 summarizes the most practical applications related to fluidization, spouting, and pneumatic conveying from literature with some figures and facts generated by our in-house CFD-DEM code developed with formulations and implement approach presented in this book. In subsequent sections, non-isothermal and reactive flows are presented in brief in order to cover the essence of these flows. The chapter ends with miscellaneous applications related to fluid–solid interaction systems practiced in chemical industries.

Chapter 8 briefly introduces interparticle forces and external fields in particulate systems. The governing equations of the system as well as interparticle forces such as van der Waals, liquid bridge, and electrostatic and effect of external fields such as electric, magnetic, vibration, and acoustic are discussed due to their importance in the CFD-DEM modeling. The chapter ends with some relevant applications.

1.7 Online Content

The supplementary online content associated with this book contains the numerical code, animations, and simulation setups for case studies and applications addressed in the book. The content was prepared as the assembly of (i) numerical FORTRAN code that accompanies this book, (ii) animations related to problems solved in Chapters 5 and 7, and (iii) the code files and guides to setup most of the simulations in Chapter 5 using the numerical code in order to support the reader's hands-on simulation experience. The online materials are available from the Wiley website at: www.wiley.com/go/norouzi/CFD-DEM.

References

[1] Anderson, T.B. and Jackson, R. (1967) Fluid mechanical description of fluidized beds. Equations of motion. *Industrial and Engineering Chemistry Fundamentals*, **6**(4), 527–539.

[2] Tsuji, Y. (2007) Multi-scale modeling of dense phase gas–particle flow. *Chemical Engineering Science*, **62**(13), 3410–3418.

[3] Van der Hoef, M., Ye, M., van Sint Annaland, M., Andrews, A., Sundaresan, S., and Kuipers, J. (2006) Multiscale modeling of gas-fluidized beds. *Advances in Chemical Engineering*, **31**, 65–149.

[4] Klinzing, G.E., Rizk, F., Marcus, R., and Leung, L. (2011) *Pneumatic Conveying of Solids: A Theoretical and Practical Approach*, Vol. **8**, Springer Science & Business Media.

[5] Deen, N., Annaland, M.V.S., Van der Hoef, M., and Kuipers, J. (2007) Review of discrete particle modeling of fluidized beds. *Chemical Engineering Science*, **62**(1), 28–44.

[6] Zhu, H., Zhou, Z., Yang, R., and Yu, A. (2008) Discrete particle simulation of particulate systems: a review of major applications and findings. *Chemical Engineering Science*, **63**(23), 5728–5770.

[7] Zhu, H., Zhou, Z., Yang, R., and Yu, A. (2007) Discrete particle simulation of particulate systems: theoretical developments. *Chemical Engineering Science*, **62**(13), 3378–3396.

[8] Jasak, H. (1996) Error analysis and estimation for the finite volume method with applications to fluid flows. PhD thesis. Imperial College of Science, Technology and Medicine.
[9] Crowe, C.T. (2005) *Multiphase Flow Handbook*, CRC Press.
[10] Radl, S., Capa Gonzales, B., Goniva, C., and Pirker S. (2014) State of the art in mapping schemes for dilute and dense euler-lagrange simulations. 10th International Conference on CFD in Oil & Gas, Metallurgical and Process Industries, Trondheim, Norway.
[11] Yakubov, S., Cankurt, B., Abdel-Maksoud, M., and Rung, T. (2013) Hybrid MPI/OpenMP parallelization of an Euler–Lagrange approach to cavitation modelling. *Computers & Fluids*, **80**, 365–371.
[12] Amritkar, A., Deb, S., and Tafti, D. (2014) Efficient parallel CFD-DEM simulations using OpenMP. *Journal of Computational Physics*, **256**, 501–519.
[13] Liu, H., Tafti, D.K., and Li, T. (2014) Hybrid parallelism in MFIX CFD-DEM using OpenMP. *Powder Technology*, **259**, 22–29.

Part I
DEM

2

DEM Formulation

The discrete element method (DEM) refers to a type of modeling approach in which the dynamic/time transient behavior of a system comprising of a large number of distinct bodies with arbitrary shapes is studied. These bodies may be considered as either deformable or rigid, which continuously come into contact with each other and rebound. The translational and rotational motions of each body are formulated in the model and are highly dependent on type and strength of interactions of the bodies with the surrounding medium through contact and interparticle effects. The bodies may interact with each other through their contact points/areas. Depending on the operating conditions of the system, interparticle interactions, such as van der Waals, liquid bridge, and electrostatic forces, may exist between distinct bodies even if they are not in contact. In multiphase flows, the fluid interacts with the distinct bodies in the system. The interaction between each body and the fluid is mutual. During the motion of bodies, they interact with stationary or moving walls. Moving walls increases the kinetic energy of bodies leading to convective movement.

In the DEM, the bodies are considered either deformable or rigid. The choice of deformable or rigid formulation depends on operating conditions and available computational resources. These two are mostly referred to as soft-sphere and hard-sphere formulations, although these formulations are not specific to spheres. In general, the soft-sphere formulation covers more practical granular and multiphase flows than the hard-sphere formulation. For a dilute system, the hard-sphere formulation is computationally effective, although soft-sphere formulation is also applicable. This is due to the fact that the contact time of each body is much shorter than the mean time between successive collisions and hence each contact can be considered instantaneous and pairwise. In a dense system, for which long contact times and multiple contacts exist between particles, the soft-sphere formulation is more appropriate.

Coupled CFD-DEM Modeling: Formulation, Implementation and Application to Multiphase Flows, First Edition.
Hamid Reza Norouzi, Reza Zarghami, Rahmat Sotudeh-Gharebagh and Navid Mostoufi.

Companion website: www.wiley.com/go/norouzi/CFD-DEM

Discrete element formulations detailed here are mainly based on the soft-sphere, considering the following justifications. First, it covers most of granular and multiphase flows with and without interparticle forces. Second, modeling time-progressive phenomena, like solid-bridge formation, liquid-bridge progression, agglomeration, caking, and particle breakage, are straightforward in the soft-sphere formulation. Third, the soft-sphere approach can be modified and extended to suit the hard-sphere approach. Fourth, the rapid development in computational resources and numerical algorithms make it possible to simulate a system containing hundreds of thousands of distinct bodies in a reasonable time.

In this chapter, both hard- and soft-sphere formulations are covered for spherical particles. These are followed by introducing common models for evaluating contact forces and torques. In addition, the ways we can treat different boundary conditions/walls in the model formulation are also covered here. The rest of the chapter is devoted to boundary (walls) and initial conditions. In addition, the calculation procedures and sequences are detailed wherever deemed necessary. The DEM formulation for non-spherical particles is covered in Chapter 4.

2.1 Hard-Sphere

The soft-sphere formulation is based on the collision forces that are established during a contact between bodies. The collision force is calculated based on the overlap between colliding particles. This overlap, and hence the collision force between bodies, changes in the course of a collision. Then, the Newton's second law of motion is applied to balance the acceleration and the forces acting on it (see Section 2.2). Integration of the Newton's second law gives new velocity and position of each body in each time step. This is known as the force-based/soft-sphere/time-driven formulation in the DEM. As we will discuss in Section 2.2, the time step for numerical integration and tracking bodies should be so small that each contact is processed in several time intervals. Moreover, the procedure of finding new collisions and relaxing old collisions should be applied in each time step.

Let us consider a situation in which the collisions between bodies are not very long and frequent. In this system, each body travels freely (based on external forces and translational velocities) until the next collision happens, similar to what we have in the dynamics of granular gases. Thus, new states (position and velocity) of bodies are not evaluated until the next collision/event happens. The post collision velocities of colliding bodies are calculated based on pre-collision velocities via establishing the momentum conservation law. This is the collision-based/hard-sphere/event-driven formulation in the DEM. From the computational point of view, this formulation, in some cases, is more efficient than the force-based formulation since the velocity of bodies is updated whenever a collision/event happens.

This formulation implies that the mean time in which a body freely travels is much more than the duration of the contact. In this case, we can confidently assume that contacts between bodies are pair-wise and there are no multiple contacts. Moreover, we further assume that each collision between bodies is established with no overlap and lasts for an extremely short time ($t_{col} = 0$). The hard-sphere formulation works well for describing the dynamics of dilute systems where the number density of particles is low and the previously mentioned assumptions are held. The speed gain of the calculation is mostly due to elimination of redundant calculation of the system state (velocity and position of bodies) between successive events. Although its robustness has been proven for dilute systems, it has been successfully implemented also for dense systems. This approach was first introduced by Campbell and Brennen [1] to simulate a granular system and then was used to simulate particles flow in different systems like rotating drums [2, 3]. The hard-sphere DEM was combined by computational fluid dynamics (abbreviated as CFD-DEM) to study bubble and slug formation in a gas-solid fluidized bed [4]. The hard-sphere CFD-DEM model was used to study size segregation in fluidized beds [5–7], fluidization characteristics at high pressure [8, 9], fluidization regimes in spout-fluid beds [10], and granulation in spout-fluid beds [11].

As we discussed before, most theories and implementations in this book are based on the soft-sphere formulation. We bring the hard-sphere formulation here mainly for the sake of comparison between different aspects of these modeling approaches. In addition, this gives the reader a better perspective to enable choosing between these two types of formulations for each application. Thus, we just present the headlines of the hard-sphere formulation to give a general overview of it. The interested reader is referred to other references [12–14] for more detail on theoretical aspects as well as numerical implementation of this approach.

2.1.1 Equation of Motion

The velocity change of each particle with mass m_i during its free motion (between successive collisions) in the system is individually tracked by integrating the Newton's second law of motion.

$$m_i \frac{d\vec{v}_i}{dt} = m_i\vec{g} + \vec{f}_i^{\,f-p} \tag{2.1}$$

The first integration of Equation 2.1 yields the new velocity of the particle and the second integration yields its new position. The velocity of particle changes due to external forces, like gravitational force $(m_i\vec{g})$ and total fluid–particle interaction force $(\vec{f}_i^{\,f-p})$. We considered the fluid-particle interaction force in this equation in order to extend the basic equation of motion to multiphase flows where fluid–particle interactions are important. There are different types of fluid–particle interactions in a multiphase flow such as, buoyancy, drag, and lift forces. We will discuss them in detail in Chapter 6. The equation of motion for the hard-sphere formulation is very similar to that for the soft-sphere formulation. However, there are some differences that will be explained in Section 2.2 after we present the soft-sphere formulation. Note that this equation gives the velocity change of particle i during the free motion of the particle. At the moment of collision between particle i and another particle j, which is assumed to last for an infinitely short time, the hard-sphere collisional model is used to calculate the post-collision velocities (both translational and rotational) based on the pre-collision velocities. Therefore, Equation 2.1 is not valid for the moment of collision between particles. After the collision, the post-collision translational velocity will be used as the initial condition for Equation 2.1 until particle i comes into contact with another particle. In the following section, we present the main framework of the hard-sphere collision model.

2.1.2 Collision Model

Consider two particles with radii R_i and R_j, position vectors $\vec{x}_i$ and $\vec{x}_j$, and masses m_i and m_j. They are traveling with translational velocities $\vec{v}_i^{\,0}$ and $\vec{v}_j^{\,0}$, and rotating with angular velocities $\vec{\omega}_i^{\,0}$ and $\vec{\omega}_j^{\,0}$. We consider that the collision is impulsive and any external force, like attractive, drag and so on, are zero during the collision. This is true since the basic assumption of the hard-sphere formulation is the instantaneous collision. The relations between pre-collision and post-collision translational and rotational velocities are established by impulse equations as follows:

$$m_i \vec{v}_i^1 = m_i \vec{v}_j^0 + \vec{J} \tag{2.2}$$

$$m_j \vec{v}_j^1 = m_j \vec{v}_j^0 - \vec{J} \tag{2.3}$$

$$I_i \vec{\omega}_i^1 = I_i \vec{\omega}_i^0 + R_i \vec{n}_{ij} \times \vec{J} \tag{2.4}$$

$$I_j \vec{\omega}_j^1 = I_j \vec{\omega}_j^0 + R_j \vec{n}_{ij} \times \vec{J} \tag{2.5}$$

$$\vec{J} = J_n \vec{n}_{ij} + J_t \vec{t}_{ij} \tag{2.6}$$

where superscripts 0 and 1 represent particle properties before and after the collision, respectively; $\vec{n}_{ij}$ is the unit vector pointing from particle i to particle j and is perpendicular to the contact plane; $\vec{t}_{ij}$ is the tangential unit vector at the contact point that lies on the contact plane and I_i is the moment of inertia of the particle.

$$\vec{n}_{ij} = \frac{\vec{x}_j - \vec{x}_i}{\left|\vec{x}_j - \vec{x}_i\right|} \tag{2.7}$$

$$\vec{t}_{ij} = \frac{\vec{v}_{ij}^t}{\left|\vec{v}_{ij}^t\right|} \tag{2.8}$$

$$I_i = \frac{2}{5} m_i R_i^2 \tag{2.9}$$

There are three parameters that are used to determine the impulse force $\vec{J}$ in these Equations 2.2–2.9. These parameters are coefficients of restitution in normal and tangential directions and dynamic friction coefficient. Once the impulse force is known, the post-collision translational and angular velocities can be calculated. The impulse force represents the mutual repulsion of colliding particles due to their elasticity (see Figure 2.1).

The relative velocity between colliding particles at the contact point is calculated as follows:

$$\vec{v}_{ij} = \vec{v}_i - \vec{v}_j + \left(R_i \vec{\omega}_i + R_j \vec{\omega}_j\right) \times \vec{n}_{ij} \tag{2.10}$$

$$\vec{v}_{ij}^n = \left(\vec{v}_{ij} \cdot \vec{n}_{ij}\right) \vec{n}_{ij} \tag{2.11}$$

$$\vec{v}_{ij}^t = \vec{v}_{ij} - \vec{v}_{ij}^n = -\vec{n}_{ij} \times \left(\vec{n}_{ij} \times \vec{v}_{ij}\right) \tag{2.12}$$

The coefficient of restitution is a measure to show that how much kinetic energy is lost during an inelastic contact between two particles. It is simply defined as the ratio of post- and

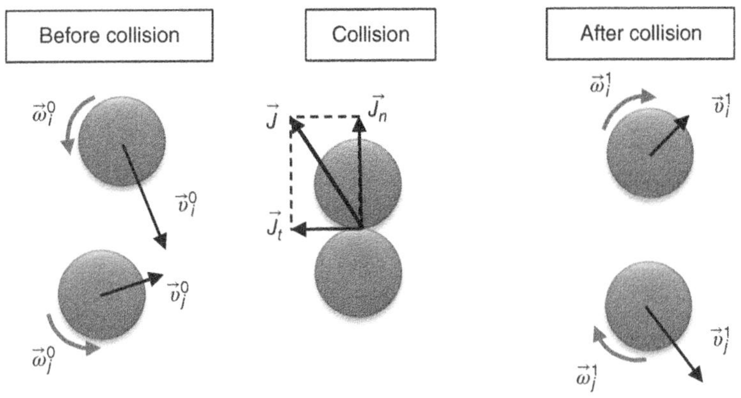

Figure 2.1 Successive steps in an instantaneous collision between two spherical particles in the hard-sphere formulation. $\vec{v}_i$ and $\vec{v}_j$ are translational velocities, $\vec{\omega}_i$ and $\vec{\omega}_j$ are rotational velocities, $\vec{J}_n$ and $\vec{J}_t$ are normal and tangential impulse forces. Superscripts 0 and 1 refer to pre- and post-collision velocities, respectively

pre-collision velocities. The coefficients of restitution in normal direction, e_n, and tangential direction, e_t, are defined by the corresponding velocities in the contact plane.

$$\vec{n}_{ij} \cdot \vec{v}_{ij}^1 = -e_n \left(\vec{n}_{ij} . \vec{v}_{ij}^0 \right), \quad 0 \le e_n \le 1 \tag{2.13}$$

$$\vec{t}_{ij} \cdot \vec{v}_{ij}^1 = -e_t \left(\vec{t}_{ij} . \vec{v}_{ij}^0 \right), \quad -1 \le e_t \le 1 \tag{2.14}$$

The normal component of the impulse force is related to the normal coefficient of restitution and pre-collision relative velocity by:

$$J_n = -m_{eff} \left(1 + e_n\right)\left(\vec{n}_{ij} . \vec{v}_{ij}^0 \right) \tag{2.15}$$

$$m_{eff} = \frac{m_i m_j}{m_i + m_j} \tag{2.16}$$

Calculation of the tangential component of impulse force needs more attention since there are some circumstances that lead to sliding or sticking collisions. When the tangential component of the relative velocity is high in comparison with the normal velocity, sliding occurs throughout the collision. In this situation, the tangential component of impulse force is obtained by Coulomb's friction law:

$$J_t = -\mu J_n \quad \text{for } \mu J_n < \frac{2}{7}\left(1 + e_t\right) m_{eff} \left(\vec{t}_{ij} . \vec{v}_{ij}^0 \right) \tag{2.17}$$

In other circumstances, when the normal relative velocity is high in comparison to the tangential relative velocity, sticking collision occurs. Therefore, the tangential component of impulse force is obtained by:

$$J_t = -\frac{2}{7}\left(1 + e_t\right) m_{eff} \left(\vec{t}_{ij} . \vec{v}_{ij}^0 \right) \quad \text{for } \mu J_n \ge \frac{2}{7}\left(1 + e_t\right) m_{eff} \left(\vec{t}_{ij} . \vec{v}_{ij}^0 \right) \tag{2.18}$$

The normal coefficient of restitution varies between 0 and 1. It was experimentally observed that it decreases as the impact velocity increases. We can assume two following extreme cases. When the impact velocity approaches zero, the normal coefficient of restitution becomes unity, meaning no loss of the kinetic energy during the contact. On the other hand, when the impact velocity approaches infinity, the coefficient of restitution approaches zero, meaning that all the kinetic energy is lost during the contact. However, the restitution coefficient is not only a function of impact velocity but also a function of particle properties and the type of particle deformation (elastic or plastic). The same is true for the tangential coefficient of restitution. It is a function of impact velocity, material properties, and contact conditions.

2.1.3 *Interparticle Forces*

Interparticle forces are mostly referred to as non-contact interactions between bodies that are separated by a distance of *d*. These forces may be attractive or repulsive. Generally, the attractive or repulsive interparticle force is a continuous function of separation distance of bodies. For example, in a simplified situation, the attractive van der Waals force between two bodies is inversely related to square of the separation distance [15]:

$$f_{vdW} \propto \frac{1}{d^2} \tag{2.19}$$

Interparticle forces exist between particles during their movements and each particle interacts with all other surrounding particles. The physics of interparticle forces does not allow us to directly incorporate them into the hard-sphere formulation since this formulation is based on instantaneous and binary events. However, Weber *et al.* [16] and Weber and Hrenya [17] introduced the square-well method to handle granular flows with interparticle cohesive forces in the hard-sphere formulation. We introduce this methodology here to give a good picture to the reader of how this formulation can be further extended to a similar system in which interparticle forces play a key role. For example, dilute phase electrification of gas-particle flow in pneumatic conveying of particles [18] can be modeled by event-based/hard-sphere DEM. Charge transfer and induction occur in particle–particle and particle-wall collisions/events [19, 20]. At the same time, charged particles experience electrostatic force from neighboring charged particles that can be represented by the square-well method. The interested reader is also referred to Kosinski and Hoffmann [21] in which the hard-sphere formulation is extended to include inter-particle interactions in an agglomerating gas-particle pipe flow.

The square-well potential is shown in Figure 2.2. Whenever the separation distance is longer than a threshold distance d_0, the cohesive potential is zero. Once the separation distance is shorter than d_0, the potential energy is equal to D_{sw} and an instantaneous, binary cohesive collision occurs. Similar to inelastic collisions in the hard-sphere formulation, which leads to a reduction in the kinetic energy of colliding particles, the cohesive collision leads to either an increase or a decrease in the kinetic energy of colliding particles. Therefore, the same analogy between inelastic and cohesive collisions can be considered. When particles are approaching each other and a cohesive collision occurs, the impulse force of this collision is calculated as follows:

$$J_{n,coh} = -m_{eff}\left(\vec{n}_{ij}.\vec{v}_{ij}^{\,c} + \sqrt{\frac{2D_{sw}}{m_{eff}} + \left(\vec{n}_{ij}.\vec{v}_{ij}^{\,c}\right)^2}\right) \tag{2.20}$$

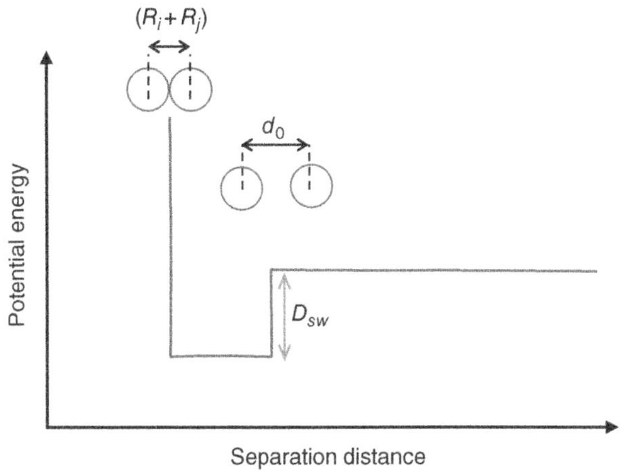

Figure 2.2 Square-well potential used for simulating instantaneous and cohesive forces between particles. When the separation distance is longer than d_0, no cohesive interaction exists between particles. Once the separation distance is d_0, a cohesive collision occurs and thereafter particles undergo an elastic collision

This equation clearly shows that the kinetic energy of particles, which are exposed to cohesive interactions, is increased. In this equation, $\vec{v}_{ij}^{c}$ is the relative velocity of particles before the cohesive collision. We used a new symbol for the relative velocity here to distinguish it from $\vec{v}_{ij}^{0}$. During the approach of two cohesive particles to one another, they first experience a cohesive collision and then an inelastic collision. The translational velocities of particles with cohesive collisions can be obtained from Equations 2.20, 2.2, and 2.3. For an inelastic collision, Equations 2.2, 2.3, and 2.15 should be used.

After the impact, the rebound step starts. Two particles may escape from or adhere to one another. The rebound impulse force is calculated by the following equation if the value inside the square root is positive:

$$J_{n,reb} = -m_{eff}\left(\vec{n}_{ij}.\vec{v}_{ij}^{1} + \sqrt{\left(\vec{n}_{ij}.\vec{v}_{ij}^{1}\right)^{2} - \frac{2D_{sw}}{m_{eff}}}\right) \tag{2.21}$$

In this case, two particles do not adhere and may escape depending on their normal relative velocity. This equation shows that the kinetic energy of particles decreases – due to the cohesive force – when they depart. However, when the value inside the square root is negative, this means that the cohesive force is large compared to the normal impulse of colliding particles, thus, particles adhere to one another. The impulse force is then calculated from:

$$J_{n,adh} = -2m_{eff}\left(\vec{n}_{ij}.\vec{v}_{ij}^{1}\right) \tag{2.22}$$

The magnitude of relative velocity does not change, while its direction changes, meaning that there is no loss of kinetic energy. The procedure described here can be easily implemented into a hard-sphere code and mimics the behavior of multiphase flow with attractive interactions between particles.

2.2 Soft-Sphere

Contrary to the hard-sphere formulation with binary and instantaneous collisions between bodies, interactions between bodies are progressive and multiple contacts are allowed in the discrete soft-sphere formulation. The interaction between bodies is treated as a dynamic process with a local equilibrium for internal forces of colliding bodies. We establish the soft-sphere formulation based on the following assumptions in this book:

1. Bodies are deformable, but the deformation is reversible and they retain their original shapes after the contact is released.
2. The interaction between contacting bodies is via the contact area. The overlap between bodies is very small compared to the size of bodies.
3. This overlap progressively changes with time during a contact and its maximum is a function of physical properties and impact velocity of bodies.
4. The magnitude of contact force between colliding particles is obtained from force-displacement laws. The force-displacement is a function of overlap, relative velocity, the contact history, and shape and properties of bodies.

It should be noted that non-collisional interactions, like interparticle forces will be covered in Chapter 8. We first develop the soft-sphere formulation for spherical particles. This helps us to understand the fundamentals of collision dynamics easier. There are many occasions that a granular flow or the solid phase of a multiphase flow can be reasonably estimated by spherical particles. We then extend this formulation to non-spherical particles to include the shape effect in the DEM simulation.

The dynamic characteristic of the whole system is numerically estimated by an iterative time-integration of equations of motion for each individual particle. We assume that velocities and accelerations (linear and angular) are constant during each time step. By integrating, the new positions and velocities of particles, and hence their overlaps, are calculated. Based on the new overlaps and the collision history of particles,[1] the collision forces acting on particles are calculated. Thereafter, the linear and angular accelerations of particles are calculated and next iteration starts with integrating equations of motion.

Such an iterative solution is applicable when the integration time step is chosen so small that within a time step, a disturbance wave propagates a distance not very far from the vicinity of the particle. That is, the disturbance is only experienced by neighboring particles. In this way, the forces acting on a particle are only determined by its interactions with surrounding particles that are in contact with it. The speed at which a disturbance wave can propagate through an assembly of solid bodies depends on the physical properties of bodies. Therefore, having the size of individual bodies and propagation speed, we can calculate the time that the disturbance passes over a body. Accordingly, choosing a time step smaller than this value will satisfy the assumption of explicit time integration. We will come back to this issue in Chapter 3 and discuss it in more detail.

[1] Parameters that should be saved for each collision depend on the type of force-displacement laws used to describe contact interactions between bodies. The minimum number of parameters that should be saved is the tangential displacement of collision point.

The use of explicit scheme for the numerical implementation of equations of motion of bodies has some advantages. First, both linear and nonlinear contact force models can be used for particle interaction without taking care of how to solve a large set of equations (the same integration scheme can be used for both linear and nonlinear models). Second, this assumption implies that each contact between bodies is processed in several time steps. This enables us to insert some additional sub-models to account for other time-dependent processes into the model. For example, the time-dependent solid bridging model proposed by Seville *et al.* [22] was implemented into the soft-sphere framework to study the agglomeration of polyethylene particles in gas-solid fluidized beds [23]. Based on this model, the neck growth rate and solid bridging depends on the collision duration and radius of contact area, which are readily available in the soft-sphere formulation.

2.2.1 Equations of Motion

Granular or multiphase flows consist of a large number of solid particles. These particles may have a size distribution ranging from a few micrometers to centimeters. The dynamics of the granular material, or the solid phase in multiphase flow, is governed by the Newton's second law of motion for the center of mass of each particle and by the Euler's second law of motion for the angular momentum change. The main governing equations for N spherical particles in the system are as follows[2] [13]:

$$m_i \frac{d\vec{v}_i}{dt} = m_i \frac{d^2\vec{x}_i}{dt^2} = \vec{f}_i\left(\vec{x}_j, \vec{v}_j, \vec{\varphi}_j, \vec{\omega}_j\right) \quad for\ j = 1, \ldots, N \tag{2.23a}$$

$$I_i \frac{d\vec{\omega}_i}{dt} = I_i \frac{d^2\vec{\varphi}_i}{dt^2} = \vec{M}_i\left(\vec{x}_j, \vec{v}_j, \vec{\varphi}_j, \vec{\omega}_j\right) \quad for\ j = 1, \ldots, N \tag{2.23b}$$

where $\vec{f}_i$ and $\vec{M}_i$ are sum of different forces and torques that act on particle i, respectively. They are complex functions of variables such as particle position $\vec{x}_j$, angular position $\vec{\varphi}_j$, translational velocity of center of mass $\vec{v}_j$, and rotational velocity around center of mass $\vec{\omega}_j$. We call these four variables *state variables* of the particle. We used subscript j for the input parameters of forces and torques functions to denote the fact that not only the state variables of particle i but also the state variables of other particles in the system affect the resultant interactions.

Equations 2.23a and 2.23b show that a set of $2N$ nonlinear differential equations should be solved to obtain the dynamics of a granular flow. Analytical solution of these equations is impossible and a numerical solution should be obtained instead. In each time step, forces and torques acting on particles using the state variable of current time step are first calculated and then new state variables of the next time step are obtained by integrating Equations 2.23a and 2.23b.[3] This is the core part of the DEM simulation of granular flows: first, calculating the interaction forces on each particle and then, integrating the equations of motion to obtain new state of the system.

---

[2] The Equation of angular momentum (Euler's equations) is a bit different for non-spherical particles. This will be discussed later in Chapter 4.

[3] This integration scheme is a one-step explicit integration method. We will introduce two-step explicit integration methods later in Chapter 3. Note that implicit integration schemes are also available for discrete-element method simulations.

The equations of motion introduced here are in the general form and are not suitable for applications in this book. Therefore, we rewrite them in the following form that is more applicable to DEM simulations:

$$m_i \frac{d\vec{v}_i}{dt} = m_i \frac{d^2\vec{x}_i}{dt^2} = \sum_{j \in CL_i} \vec{f}_{ij}^{\,p-p} + \vec{f}_i^{\,f-p} + \vec{f}_i^{\,ext} \tag{2.24a}$$

$$I_i \frac{d\vec{\omega}_i}{dt} = I_i \frac{d^2\vec{\varphi}_i}{dt^2} = \sum_{j \in CL_i} \left(\vec{M}_{ij}^{t} + \vec{M}_{ij}^{r}\right) \tag{2.24b}$$

The first term on the right-hand side of Equation 2.24a is the sum of particle–particle interaction forces acting on particle i. These forces may be collisional forces or interparticle forces, like electrostatic or van der Waals forces. The summation is performed on all particles that are in the contact list of particle i, CL_i. The second term stands for all fluid-particle interaction forces. This is mostly applicable to multiphase flows in which forces acting on particle by the fluid are of the same order of other interaction forces.[4] Thus, in the absence of continuous fluid or where the fluid effects are not important, this term is assumed to be zero. The fluid–particle interaction forces will be discussed in detail in Chapter 6. The third term represents all external forces acting on particle i due to uniform or non-uniform external fields. For example, we can name the gravitational force due to existence of the uniform gravitational field of earth or the force acting on a magnetic particle in an electromagnetic field.

We should distinguish between different contacts that may exist in the contact list of particle i in Equation 2.24a. There are two types of contacts between particles: physical and non-physical. The physical contact refers to a condition that the surfaces of two particles touch each other while the non-physical contact refers to a condition that particles do not necessarily touch but still interact with each other. The collision force between two particles with physical contact is calculated according to the force-displacement laws – which are covered in this chapter – and the interaction force between two particles with non-physical contacts according to interparticle interactions – like the van der Waals and electrostatic forces that are covered in Chapter 8. It is clear that when there is no particle in the contact list of particle i, the particle–particle interaction term is zero.

In Equation 2.24b, the first term in the summation represents *tangential torque*, $\vec{M}_{ij}^{t}$, produced by particle–particle collision. Since the particle–particle collision force acts on the contact point (surface of the particle), it causes a torque. This is the origin of rotation of particles. $\vec{M}_{ij}^{r}$ stands for rolling friction, which is also another torque acting on particle i. Since it always opposes the rotation of particle, it is also known as the *rolling resistant torque*. We can conclude from Equation 2.24b that, as long as the particle is not in physical contact with other particles, the angular velocity of particle does not change.

[4] The fluid-particle forces are considered mutual in the CFD-DEM. That is, the force acting on the particle is counter-balanced by the particle-fluid force with the same magnitude and in the opposite direction (see Chapter 6).

Basic equations of the soft-sphere DEM and the iterative solution of the equations of motion require us to deal with the following issues to implement the framework of DEM into a numerical code:

- *Particle-particle interaction forces*: We present different models for collisional forces in this chapter and interparticle forces in Chapter 8. We also cover models for non-spherical particles in Chapter 4.
- *Initial and boundary conditions*: Considering the appropriate initial and boundary conditions to obtain a specific solution of the governing differential equations is necessary for discrete element simulations as covered in this chapter. We also covered the numerical implementation of walls in Chapter 3.
- *Contact detection*: Updating the contact list of all particles in each time step, the extent of overlap and direction of overlap between colliding bodies are very important part of a DEM simulation that will be covered in detail in Chapter 3.
- *Integration*: The dynamic behavior of the system is obtained by integrating Equations 2.24a and 2.24b. Since there are sharp changes in the interaction force acting on particles, an accurate and efficient method for integration is required. Different integration schemes will be presented in Chapter 3.

2.3 Force-Displacement Laws

When there is a physical contact between any two particles, force-displacement laws are used to calculate collision properties as a function of normal and tangential overlaps, physical properties, and collision history of the colliding particles. A comprehensive model that accurately computes the collision forces needs complicated contact mechanics and its implementation is really hard and time consuming [24]. Since the common granular or multiphase flow consists of more than thousands of particles, there are numerous collisions for which collision forces should be evaluated in each time step. This is not very practical for our purposes. Therefore, simplified models, called force-displacement laws, are utilized to reduce the computational effort and keeping the accuracy at a level acceptable for engineering applications.

There are many force-displacement laws developed and applied to molecular dynamics and DEM. They can be classified into two main categories: continuous potential force-displacement models (which are mainly used in molecular dynamics [25]) and non-continuous force-displacement models (which are widely used in the DEM simulations of granular flow). The force model in continuous potential force-displacement models is similar to the Lenard–Jones potential equation. Kruggel-Emden *et al.* [26] discussed different aspects of continuous and non-continuous laws and their applicability for discrete element simulations. They concluded that continuous models have non-zero force at the beginning and the end of collision, which is physically unrealistic in granular flows. Non-continuous force-displacement models have zero repulsive force if particles are not in physical contact, which is more realistic in granular materials. These types of models can be categorized into different groups. We present linear viscoelastic, nonlinear viscoelastic, and elastic perfectly plastic models in this text. The variety of proposed models does not allow us to cover them all in this book. Thus, more popular ones are presented and discussed here. More details on force-displacement models and their characteristics can be found elsewhere [26–33].

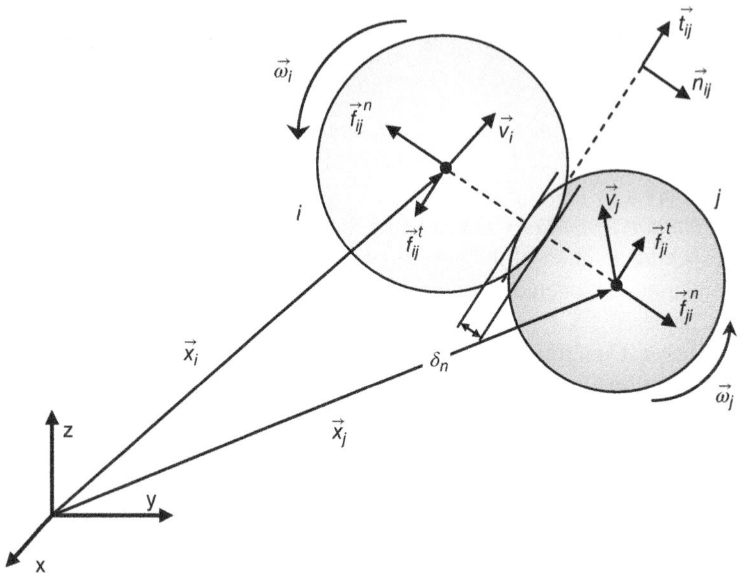

Figure 2.3 Schematic illustration of colliding particles i and j and forces acting on particles as a result of collision

Figure 2.3 shows two particles i and j with position vectors $\vec{x}_i$, $\vec{x}_j$, and radii R_i and R_j, which are in physical contact. Normal overlap of two particles is defined by:

$$\delta_n = R_i + R_j - \left|\vec{x}_j - \vec{x}_i\right| \tag{2.25}$$

If $\delta_n > 0$, there is a physical contact between particles. The collision force between the two particles is comprised of normal and tangential collision forces represented as $\vec{f}_{ij}^n$ and $\vec{f}_{ij}^t$, respectively.

$$\vec{f}_{ij}^c = \vec{f}_{ij}^n + \vec{f}_{ij}^t \tag{2.26}$$

The normal vector is a vector points from the center of particle i to the center of particle j and is defined as:

$$\vec{n}_{ij} = \frac{\vec{x}_j - \vec{x}_i}{\left|\vec{x}_j - \vec{x}_i\right|} \tag{2.27}$$

The relative velocity between colliding particles at the contact point and its normal and tangential components are defined as:

$$\vec{v}_{ij} = \vec{v}_i - \vec{v}_j + \left(R_i\vec{\omega}_i + R_j\vec{\omega}_j\right) \times \vec{n}_{ij} \tag{2.28}$$

$$\vec{v}_{ij}^n = \left(\vec{v}_{ij}.\vec{n}_{ij}\right)\vec{n}_{ij} \tag{2.29}$$

$$\vec{v}_{ij}^t = \vec{v}_{ij} - \vec{v}_{ij}^n \tag{2.30}$$

Having the tangential velocity at contact point, the tangential overlap is then calculated as:

$$\vec{t}_{ij} = \frac{\vec{v}^{\,t}_{ij}}{\left|\vec{v}^{\,t}_{ij}\right|} \tag{2.31}$$

Calculation of the tangential overlap is a bit different from the normal overlap. The tangential overlap should be calculated from the tangential relative velocity ($v_{rt} = \vec{v}_{ij}.\vec{t}_{ij}$) of particles at the contact point. When a new physical contact happens at time t_0 between two particles, the tangential overlap is zero and the tangential overlap can be calculated by:

$$\delta_t = \int_{t_0}^{t} v_{rt}\,dt \tag{2.32}$$

Obviously, the step-wise time integration of the state of the system does not allow us to perform an analytical integration. Instead, we can calculate the tangential overlap using a discrete approach. At each time step, by calculating the incremental tangential overlap ($v_{rt}\Delta t_p$) and adding it to the tangential overlap of previous time step, $\delta_{t,0}$, the new tangential overlap is obtained from:

$$\delta_t \cong \delta_{t,0} + v_{rt}\Delta t_p \tag{2.33}$$

Since each contact is processed for several time steps, this approach can be a good estimate of analytical integration in Equation 2.32. The tangential overlap of each collision should be saved at each time step, since it should be used in Equation 2.33 to calculate new tangential overlap in the upcoming time steps.

2.3.1 Linear Viscoelastic Model

One of the most commonly used viscoelastic force-displacement laws is the linear spring-dashpot (LSD) model introduced by Cundall and Strack [34]. This model is widely used due to its ease of implementation in numerical codes and its applicability to the case of multiple collisions and non-spherical particles. It is most often seen in literature [35–38] that a small value for the spring stiffness (spring constant) is selected, which makes it possible to use larger time step for integration and accelerating calculations. However, it affects the maximum overlap and the magnitude of repulsive contact force between particles which may lead to unrealistic results. This matter will be covered in detail in Chapter 3.

In the LSD model, the collision force in the normal direction consists of two terms: elastic and viscous forces (see Figure 2.4). The elastic force, $\vec{f}^{\,n}_{el}$, is calculated by Hook's law, which relates the normal elastic force to the normal overlap by the proportionality factor of spring stiffness and conserves the kinetic energy of collision. In contrast, the viscous force, $\vec{f}^{\,n}_{diss}$, is proportional to the relative velocity of particles and dissipates the kinetic energy of collision. Accordingly, the collision force in normal direction is calculated as follows:

$$\vec{f}^{\,n}_{ij} = \vec{f}^{\,n}_{el} + \vec{f}^{\,n}_{diss} = -\left(k_n\delta_n\right)\vec{n}_{ij} - \left(\eta_n v_{rn}\right)\vec{n}_{ij} \tag{2.34}$$

$$v_{rn} = \vec{v}_{ij}.\vec{n}_{ij} \tag{2.35}$$

where k_n is the normal spring stiffness of linear spring, η_n is the normal linear velocity proportional dapper (normal damping coefficient) and v_{rn} is the relative velocity in normal direction.

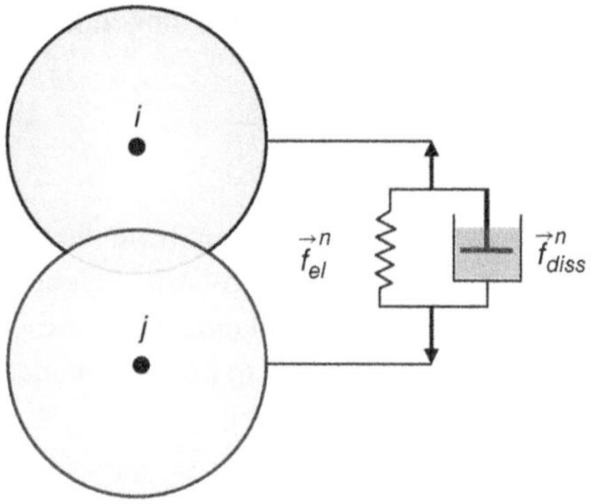

Figure 2.4 Schematic of a viscoelastic collision model in normal direction. The elastic force is repulsive and conserves the kinetic energy of collision while the viscous force resists the motion and dissipates kinetic energy of the collision

The equation of motion in normal direction (Equation 2.24a) in the absence of external and particle–fluid interactions for two spheres and considering linear viscoelastic collision between them leads to the following second-order ordinary differential equation:

$$\frac{d^2\delta_n}{dt^2} + 2\psi\frac{d\delta_n}{dt} + \kappa_0^2\delta_n = 0 \tag{2.36}$$

$$\psi = \frac{\eta_n}{2m_{eff}} \tag{2.37}$$

$$\kappa_0^2 = \frac{k_m}{m_{eff}} \tag{2.38}$$

where κ_0 is the frequency of un-damped harmonic oscillator and ψ is the reduced damping coefficient. Equation 2.36 can be solved analytically with initial conditions $\delta_n(0) = 0$ and $d\delta_n/dt(0) = v_{rn,imp}$ from which the following solutions are obtained:

$$\delta_n(t) = \frac{v_{rn,imp}}{w}e^{-\psi t}sin(wt) \tag{2.39}$$

$$\frac{d\delta_n}{dt} = \frac{v_{rn,imp}}{w}e^{-\psi t}\left(wcos(wt) - \psi sin(wt)\right) \tag{2.40}$$

with $w = \sqrt{\kappa_0^2 - \psi^2}$. The collision duration, t_{col}, is obtained by solving Equation 2.39 with $\delta_n = 0$ and the normal rebound velocity $v_{rn,reb}$ (just after the collision is released) by putting t_{col} into Equation 2.40.

$$t_{col} = \frac{\pi}{\sqrt{\kappa_0^2 - \psi^2}} \tag{2.41}$$

$$v_{rn,reb} = -v_{rn,imp}e^{-\psi t_{col}} \tag{2.42}$$

As we discussed before, the normal coefficient of restitution, e_n, is a measure of the fraction of kinetic energy recovered during a collision. Based on this definition, we can obtain a relation between normal spring stiffness and normal damping coefficient via normal restitution coefficient for the LSD model:

$$e_n = -\frac{v_{rn,reb}}{v_{rn,imp}} = \exp\left(-\pi \frac{\psi}{\sqrt{\kappa_0^2 - \psi^2}}\right) \tag{2.43}$$

Substituting the values of ψ and κ_0^2 into Equation 2.43 and rearranging it gives the following relation for normal damping coefficient:

$$\eta_n = \frac{-2\ln e_n \sqrt{m_{eff} k_n}}{\sqrt{\left(\ln e_n\right)^2 + \pi^2}} \tag{2.44}$$

In practice, the coefficient of restitution is measured experimentally and reported as a basic property of colliding particles rather than the normal damping coefficient. In these experiments, collision of a particle with a wall is recorded by a very high speed digital camera and the impact and rebound velocities are precisely determined by analyzing the captured images [39–43]. Alternatively, the coefficient of restitution can be determined in free fall experiments by the following equation:

$$e_n = \sqrt{\frac{h_1}{h_0}} \tag{2.45}$$

where h_0 is the height from which the particle is initially dropped and h_1 is the maximum height after rebound. Equations 2.41 and 2.43 show that neither collision time nor coefficient of restitution are a function of impact velocity, while the experimental observations reveals an opposite trend [30, 44].

Let's examine the impact of a sphere with a rigid wall with the LSD model. Properties of the sphere are listed in Table 2.1. We use the following equation to calculate the normal spring stiffness [31]:

$$k_n = 1.2024\left(m^{1/2} E_{eff}^2 R_{eff} v_{rn,imp}\right)^{2/5} \tag{2.46}$$

$$E_{eff} = \left(\frac{1-\upsilon_i^2}{E_i} + \frac{1-\upsilon_j^2}{E_j}\right)^{-1} \tag{2.47}$$

where E_{eff} is the effective Young's modulus. E and υ are the Young's modulus and Poisson's ratio of the particle, respectively.

Using DEM code, we performed normal impacts between a sphere and a wall with various normal coefficients of restitution. The dimensionless normal collision force is plotted versus dimensionless normal overlap in Figure 2.5. The dimensionless normal collision force was calculated by dividing the normal collision force by maximum normal force of the elastic collision and the dimensionless normal overlap was calculated by dividing the normal overlap

Table 2.1 Physical properties of the sphere used in normal and oblique collisions with wall

Property	Value	Property	Value
Radius (mm)	1	Impact velocity (m/s)	0.1–2
Density (kg/m^3)	7850	Initial rotational velocity (rad/s)	0
Young's modulus (GPa)	200	Normal spring stiffness[a] (MN/m)	6.5
Shear modulus (GPa)	77	Tangential spring stiffness (MN/m)	5.4
Poisson's ratio	0.3	Tangential to normal stiffness ratio	0.824
Dynamic friction factor	0.3	Normal coefficient of restitution	0.5–1.0
Shear viscosity (η_1/E) (s)	10^{-6}	Bulk viscosity (η_2/η_1)	0.022

[a] The value of normal spring stiffness is obtained for normal impact velocity of 1 m/s. This value is different for other impact velocities based on Equation 2.46.

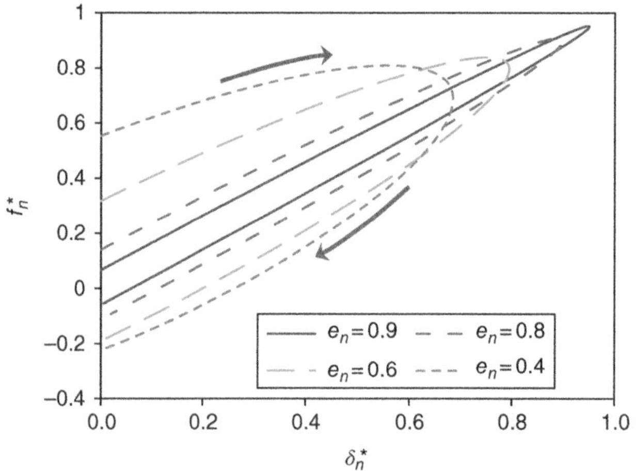

Figure 2.5 Dimensionless normal force versus dimensionless normal overlap in the linear spring-dashpot model for various normal coefficients of restitutions. f_n^* is the ratio of normal collision force to maximum elastic normal force and δ_n^* is the ratio of normal overlap to maximum normal overlap of an elastic collision

by maximum normal overlap of an elastic collision. As can be seen in this figure, every normal viscoelastic collision undergoes two phases. In the loading phase, normal overlap and normal force increase until they reach their maximum values. In the unloading phase, the particle moves in the opposite direction and rebounds from the wall. The normal forces in loading and unloading steps are not equal at the same overlap. This is due to the existence of the viscous term in the model that dissipates the kinetic energy. In fact, the area enclosed between these two curves shows the amount of dissipated energy. The maximum dimensionless overlap decreases as the normal damping coefficient increases (smaller normal coefficient of restitution). Although a non-zero force at the start and end of collision is physically unrealistic for a non-adhering collision, the LSD model shows this behavior that is caused by the viscous term in the model. This term is in fact proportional to the relative velocity that is not zero at the start and the end of the collision. The non-zero forces become greater in collisions with greater e_n. This is a very important disadvantage of the LSD model and should be avoided in discrete element simulations. Kruggel-Emden *et al.* [26] extended the linear model to remedy these problems and proposed a four-parameter linear model. One needs experimental data of particle impacts to obtain the parameters of this model. These parameters are not reported widely in literature. Moreover, the extended linear model of Kruggel-Emden *et al.* [26] has some limitations when applied to a real system in which multiple contacts occur. Later in this chapter, we will show that nonlinear force-displacement laws do not have such a problem at the beginning and end of a collision.

Analogous to the normal direction, the tangential collision force is defined by the following equation, assuming no micro-slip in the contact area of colliding particles with constant tangential spring stiffness:

$$\vec{f}_{ij}^{t} = \vec{f}_{el}^{t} + \vec{f}_{diss}^{t} = -(k_t\delta_t)\vec{t}_{ij} - (\eta_t v_{rt})\vec{t}_{ij} \tag{2.48}$$

where k_t and η_t are tangential spring constant and damping coefficient, respectively. This equation is valid when the Coulomb's criterion is not violated. If $|\vec{f}_{ij}^{t}| \geq \mu|\vec{f}_{ij}^{n}|$, the Coulomb's criterion is violated, gross sliding occurs and the tangential force is given by the model of Cundall and Strack [34].

$$\vec{f}_{ij}^{t} = -\mu\vec{f}_{ij}^{n}\,\mathrm{sgn}(\delta_t)\vec{t}_{ij} \tag{2.49}$$

where μ stands for the coefficient of dynamic friction between particles i and j. This method of applying Coulomb's law was first used by Cundall and Strack [34] and it is the most commonly used method among researchers.

There are some variations of the Coulomb's law that differ from the original method. The method of Maw *et al.* [45], which was used by Tsuji *et al.* [46], is based on limiting tangential deformation as well as tangential force. According to this method, the tangential deformation is limited by the following equation when gross sliding occurs:

$$\delta_t = \mathrm{sgn}(\delta_t)\frac{\mu|\vec{f}_{ij}^{n}|}{k_t} \tag{2.50}$$

The tangential force is then calculated by Equation 2.48. These two approaches were compared for different oblique particle-wall collisions and it was shown that the limited tangential deformation produce different force–displacement curves during different collision regimes from non-limited tangential deformation [47]. Kruggel-Emden *et al.* [27] combined different linear and nonlinear contact models with limited and non-limited tangential displacements and tested the results against experimental data in terms of tangential coefficient of restitution, particle rotation, rebound angle, and final tangential velocity. They concluded that the limited tangential displacement can provide better estimation of experimental data.

The tangential LSD model needs a value for damping coefficient. This coefficient can only be determined from the comparison of simulation contact results with experimental data.

An analytical relation similar to Equation 2.44 was presented by Deen *et al.* [14] that needs the tangential coefficient of restitution, β_0:

$$\eta_t = \begin{cases} \dfrac{-2\ln\beta_0\sqrt{\dfrac{2}{7}m_{eff}k_t}}{\sqrt{\pi^2+\ln\beta_0^{\,2}}} & for\ \beta_0 \neq 0 \\ 2\sqrt{\dfrac{2}{7}m_{eff}k_t} & for\ \beta_0 = 0 \end{cases} \tag{2.51}$$

β_0 should be obtained by experimental measurements.

Discussion on the LSD model in this section finishes with analyzing elastic oblique particle-wall collisions at different collision regimes. The frictional behavior of an oblique collision depends on the impact angle θ_{imp} and physical properties of the particle. There are three different collision regimes for oblique particle-wall collisions. To determine the collision regime, two dimensionless parameters are introduced[5] [45, 47], which fully relate the collision conditions to physical properties:

$$\kappa = \frac{k_t}{k_n} \tag{2.52}$$

$$\psi_{imp} = \frac{\kappa}{\mu}\tan\left(\theta_{imp}\right) \tag{2.53}$$

where κ is the ratio of tangential spring stiffness to normal spring stiffness and ψ_{imp} is the dimensionless impact parameter.

We performed some numerical elastic oblique particle-wall collision tests with different collision angles and the results are shown in Figure 2.6. The physical properties used in these tests are listed in Table 2.1. There are three curves in each regime. The dashed curve shows the limit imposed by the Coulomb's law. The dotted curve is the tangential force without friction and the solid curve corresponds to actual tangential force during the collision. In the first collision regime, which occurs at small collision angles, the condition $\psi_{imp} \leq 1$ is held. The second regime occurs for $1 < \psi_{imp} \leq 7\kappa - 1$ and the third regime for $\psi_{imp} > 7\kappa - 1$.

Figure 2.6a shows the tangential force during the oblique collision with $\theta_{imp} = 5°$. In this situation, $\psi_{imp} = 0.24$ that is less than 1 and we are in the first regime in which the normal component of relative velocity is much larger than its tangential component. The collision starts with the elastic mechanism in both normal and tangential directions until the Coulomb's criterion is violated and gross sliding occurs. Thereafter, the tangential force is governed by the Coulomb's law until the end of the collision. In fact, the collision starts with the sticking condition in the growing contact area and micro-slip condition in the annulus of contact area. When the normal force reduces during the rebounding of the particle, the tangential

[5] It should be noted that the criteria proposed for determining different collision regimes are valid for elastic collisions and the transition points between different regimes changes with introduction of dissipation mechanism in the model. As noted by Zheng *et al.* [48], the micro-slip zone of a viscoelastic sphere develops more rapidly than that of an elastic sphere which means that the transition between regimes should be a bit different.

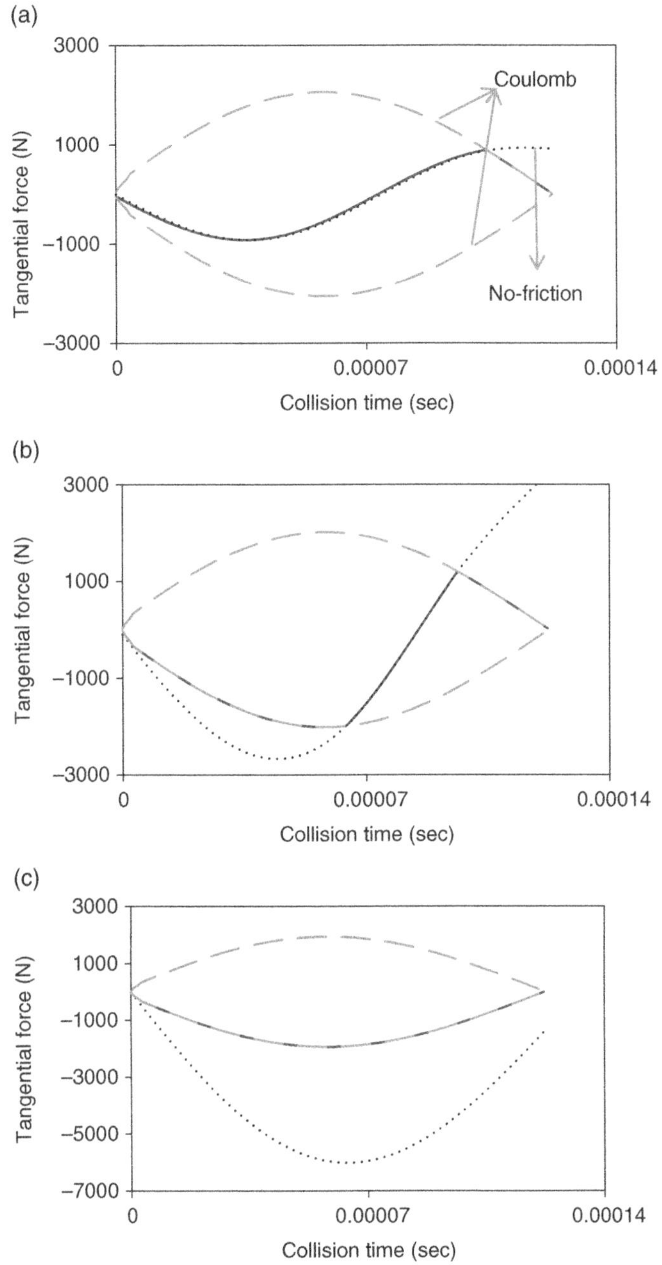

Figure 2.6 Different elastic oblique collision regimes in the LSD model. The collision angle is (a) 5° (first regime), (b) 25° (second regime), and (c) 65° (third regime)

elastic recovery causes the annulus of micro-slip to develop rapidly throughout the whole contact area and gross sliding occurs.

Figure 2.6b shows the tangential force during the oblique collision with $\theta_{imp} = 25°$ ($\psi_{imp} = 1.3$), that corresponds to the second regime. The collision starts with gross sliding with the Coulomb's law mechanism. At a point during the collision, the sticking area spreads from the center of contact to its perimeter, which turns the tangential force into the elastic mechanism. Then, during the rebound phase (unloading phase) due to tangential recovery, the micro-slip zone develops and gross sliding occurs until the end of the collision.

At high collision angles, for example, $\theta_{imp} = 65°$, the third collision regime happens, which is demonstrated in Figure 2.6c. In this condition, the tangential velocity is comparable with the normal velocity and the collision entirely occurs in the gross sliding mechanism. If we perform the same simulation test for the oblique impact of a viscoelastic sphere with wall ($e_n < 1$), six different collision regimes can be observed due to non-zero normal forces at start and end of collision in the viscoelastic LSD model. These regimes are discussed by Kruggel-Emden *et al.* [27].

Let's compare the oblique impact of spherical particles in hard-sphere and soft-sphere formulations. For large collision angles, gross sliding occurs throughout the collision for which the tangential impulse force is given by the Coulomb's law (Equation 2.17). The same behavior can be observed in the soft-sphere formulation. However, for intermediate and small collision angles, the mechanism of collision changes during the collision in the soft-sphere formulation. In contrast, in the hard-sphere formulation, the collision mechanism is determined before collision and the assumption of instantaneous collision does not allow switching between sticking and gross sliding mechanisms.

The linear model is very simple and its analytical solution is also available. This makes it very attractive in calculating collision parameters. Implementation of the linear model into a numerical code is straightforward. It can be used for situations with multiple collisions and non-spherical particles with no or minor modifications. Due to these reasons, it is one of the most popular models used in the DEM. However, as we mentioned before, there are some unrealistic physical behavior associated with the linear viscoelastic model like:

- Contact time and coefficient of restitution do not change with the impact velocity.
- There is a non-zero force at the start and the end of the collision in viscoelastic models.
- The tangential collision is almost independent of the normal force (except the situation when the Coulomb's law is used). The tangential stiffness is constant and independent of normal overlap and collision history.

It should be noted that the linear model should be used with caution, especially in cases in which particle–particle interactions are dominant in the flow of granular material. This does not mean that using the LSD model should be avoided for DEM simulations, as we can find many examples in which this model is accurate enough to match experimental results [49–52]. However, there are some cases that the LSD model results are less accurate in comparison with nonlinear force-displacement models [53].

2.3.2 *Nonlinear Viscoelastic Models*

Experimental investigations and finite element analysis (FEA) show that the linear relationship between normal overlap and normal elastic collision force is not a good representative of force-displacement model for contact between two spherical particles. Thus, we present nonlinear viscoelastic models here if higher accuracy for calculating collision forces is required. The model for collision between two elastic spherical particles in the normal direction was first proposed by Heinrich Hertz in 1882 [54]. Later, the theory of nonlinear contact force model was further improved and extended to contact between particles with adhesion [24]. In conjunction with the theory of Hertz, the Mindlin and Deresiewicz (MD) theory [55] is also used for elastic tangential contacts between particles. Based on the MD theory, the force-displacement relationship completely depends on the full loading history and instantaneous rates of changes in normal and tangential displacements. Later, Maw *et al.* [45] presented a discrete procedure in which the contact area was considered circular and discretized into a set of co-centric annuluses. Then, evolutions of normal and tangential forces were solved numerically for each of these discretized nodes. This enabled them to calculate the tangential force without the need to memorize the full contact history [28, 29, 56].

Although the approach of Maw *et al.* [45] gives a very accurate tangential collision force and covers all collision regimes, implementing them into a real DEM simulation is not practical. There are numerous bodies in a common discrete element simulation, which implies hundreds to millions of collisions in each time step. In addition, these models are developed for particles with smooth surface, while such an ideal particle does not exist in real applications. Thus, applying an accurate model is not somehow reasonable. Various simplified models based on this theory, which can be used in discrete-element simulations, have been developed [46, 57–60]. We do not cover them all here and just mention the most commonly used ones. The interested reader is referred to the literature for more information [26–31, 33, 47, 61, 62].

2.3.2.1 Normal Direction

Like the LSD model, the nonlinear viscoelastic contact force model consists of two components, elastic and viscous forces. The theory of Hertz accurately describes the elastic behavior of collision of spherical particles. The energy loss during a contact leads to different velocity of particles before and after the collision. By introducing a damper, similar to what considered in the linear viscoelastic model, the energy dissipation is introduced in this model. All nonlinear normal viscoelastic force-displacement models use the Hertz theory to describe the elastic behavior of collision. The only difference among these models is the viscous force. Thus, the Hertzian elastic model is the first presented here, followed by nonlinear viscoelastic models.

Consider an elastic sphere which is in contact with a frictionless rigid wall as shown in Figure 2.7. The particle deforms as a result of normal force, F_l, and interacts with the wall via a circular contact area with radius a. On the contact area, an axisymmetric normal pressure as a function of radial coordinate of contact area, r_{ca}, is produced as follows:

$$P(r) = P_{max}\left(1-\left(\frac{r_{ca}}{a}\right)^2\right)^{1/2} \tag{2.54}$$

In this equation, it is assumed that the radius of contact area is much smaller than the radius of particle. The maximum pressure is produced at the center of contact area where the maximum deformation occurs and is given by:

$$P_{max} = \frac{3F_l}{2\pi a^2} = \left(\frac{6F_l E_{eff}^2}{\pi^3 R_{eff}^2}\right)^{1/3} \tag{2.55}$$

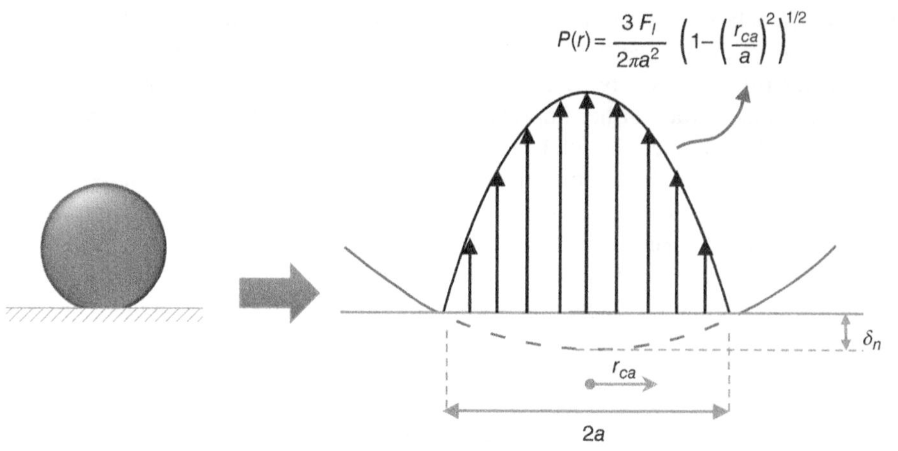

Figure 2.7 The normal pressure distribution in the contact area of an elastic sphere as a function of contact radius

The radius of contact area is a function of normal force, F_l, effective radius and Young's modulus and is calculated by:

$$a = \left(\frac{3}{4}\frac{F_l R_{eff}}{E_{eff}}\right)^{1/3} \tag{2.56}$$

Also, the normal deformation is obtained by:

$$\delta_n = \frac{a^2}{R_{eff}} = \left(\frac{9}{16}\frac{F_l^2}{R_{eff}E_{eff}^2}\right)^{1/3} \tag{2.57}$$

If we integrate the pressure over the contact area, we would obtain the following equation that calculates the normal force produced during an elastic collision between the sphere and rigid wall:

$$f_{Hertz} = \frac{4}{3}E_{eff}\sqrt{R_{eff}}\,\delta_n^{3/2} \tag{2.58}$$

The effective radius is expressed as:

$$R_{eff} = \left(\frac{1}{R_i} + \frac{1}{R_j}\right)^{-1} \tag{2.59}$$

These equations are formulated for contacts between two particles i and j. We can generalize these equations for the contact between a sphere and a rigid wall. Take sphere as object i and the wall as j in Equations 2.58 and 2.59. In this case, the curvature of the wall is considered as zero (infinite radius or $R_j = \infty$), thus, the effective radius becomes equal to the particle radius.

Moreover, the Young's modulus of the wall can be set to infinity ($E_j \sim \infty$) and the term corresponding to the Young's modulus of the wall is ignored in Equation 2.47.

Zhang and Vu-Quoc [32, 62] performed a nonlinear FEA on the dynamics of the collision between an elastic sphere and a rigid wall and compared their results with the values obtained from Equation 2.58. They found that Hertz's theory is valid for an elastic contact and the energy dissipation caused by the elastic wave propagation inside the body of sphere can be neglected for most materials (quasi-static manner) that are involved in practical applications of granular flows.

The collision duration between the elastic sphere with a normal impact velocity $v_{rn,imp}$ and density ρ_i and the rigid wall is calculated from [63]:

$$t_{col} = 2.94\left(\frac{5}{4}\frac{\pi\,\rho_i}{E_{eff}}\right)^{2/5}\frac{R_{eff}}{v_{rn,imp}^{1/5}} = 2.86\left(\frac{m_{eff}^2}{R_{eff}E_{eff}^2 v_{rn,imp}}\right)^{1/5} \tag{2.60}$$

This equation also applies to contacts between two spheres. In this case, the normal relative velocity between the two spheres should be considered as the impact velocity. Based on Equation 2.60, the contact duration for the nonlinear Hertzian model is an inverse function of the impact velocity, while it is not a function of velocity in the LSD model.

In all proposed nonlinear viscoelastic models, the elastic force is given by Equation 2.58 and the only difference between these models is the treatment of viscous (dissipative) force [46, 48, 64, 65]. The first nonlinear model that we present here is the one proposed by Kuwabara and Kono [65]. We name it the KKn model for convenience (the letter n at the end refers to "normal"). Normal collision force in the KKn model is calculated from:

$$\vec{f}_{ij}^{\,n} = \vec{f}_{el}^{\,n} + \vec{f}_{diss}^{\,n} = \left(-\frac{4}{3}E_{eff}\sqrt{R_{eff}}\,\delta_n^{3/2}\right)\vec{n}_{ij} - \left(\bar{\eta}_n\delta_n^{1/2}v_{rn}\right)\vec{n}_{ij} \tag{2.61}$$

where $\bar{\eta}_n$ is damping coefficient for the KKn model. Considering the same normal coefficient of restitution, value and dimension of $\bar{\eta}_n$ are different from value and dimension of the damping coefficient utilized in the LSD model. An analytical expression was also presented for $\bar{\eta}_n$ by Kuwabara and Kono [65], which was a function of Poisson's ratio, Young's modulus, coefficients of shear, and bulk deformation viscosities (material properties). Later, the same expression for normal collision force was obtained by Brilliantov *et al.* [66] with a different functionality for $\bar{\eta}_n$. It is not very common to calculate $\bar{\eta}_n$ using material properties of particles and in discrete element simulations, the value of $\bar{\eta}_n$ is adjusted to fit the experimental results of the normal coefficient of restitution.

Tsuji *et al.* [46] proposed a nonlinear force model similar to the KKn model but with a slightly different viscous term. We name this model the TTIn model. The TTIn model gives Equation 2.62 for the normal collision force:

$$\vec{f}_{ij}^{\,n} = \vec{f}_{el}^{\,n} + \vec{f}_{diss}^{\,n} = \left(-\frac{4}{3}E_{eff}\sqrt{R_{eff}}\,\delta_n^{3/2}\right)\vec{n}_{ij} - \left(\tilde{\eta}_n\delta_n^{1/4}v_{rn}\right)\vec{n}_{ij} \tag{2.62}$$

where $\tilde{\eta}_n$ is damping coefficient of the TTIn model. It can be seen that the viscous forces in both models are functions of normal deformation and normal relative velocity of particles. However, the viscous force is proportional to $\delta_n^{1/2}$ in the KKn model while it is proportional to $\delta_n^{1/4}$ in the TTIn model. This leads to different force-displacement curves and values for damping coefficients for these two models.

The equation of motion for a particle-wall contact with the TTIn model in the absence of external forces leads to an ordinary differential equation similar to Equation 2.36, which can be solved by proper initial conditions. However, the analytical solution is not possible. To derive a relation that calculates the damping coefficient as a function normal restitution coefficient and particle properties, we follow the procedure suggested by Malone and Xu [67]. Since the normal coefficient of restitution is not a function of impact velocity, we can derive a relation for damping coefficient of the TTIn model similar to that we obtained for the LSD model. The following equation can be considered for the dissipative force in terms of Young's modulus, normal displacement, and normal relative velocity.

$$f_{diss}^{n} = \alpha\left(\tilde{k}_{Hertz} m_{eff}\right)^{1/2} \delta_n^{1/4} v_{rn} \tag{2.63}$$

where α is an empirical parameter that is a function of e_n. k_{hertz} is the nonlinear Hertzian-spring stiffness, which is defined as:

$$\tilde{k}_{Hertz} = 4/3 E_{eff}\sqrt{R_{eff}} \tag{2.64}$$

Comparison between Equation 2.63 and the dissipative term of Equation 2.62 yields the following relation for the damping coefficient of the TTIn model:

$$\tilde{\eta}_n = \alpha\sqrt{\tilde{k}_{Hertz} m_{eff}}\,;\alpha = f\left(e_n\right) \tag{2.65}$$

We performed normal particle-wall impacts using the TTIn model with different values of $\tilde{\eta}_n$. The integration of equations of motion was performed by the fourth order Runge–Kutta method with $\Delta t = 10^{-9}$ seconds. Then, the corresponding coefficient of normal restitution was calculated using impact and rebound velocities. Different particle sizes and different Young's modulus (10^5–10^{10} Pa) were applied in calculations and 80 data pairs of $\left(e_n, \tilde{\eta}_n\right)$ were obtained. We performed a nonlinear regression on this data and obtained the following correlation:

$$\tilde{\eta}_n = \frac{-2.2664 \ln e_n \sqrt{m_{eff}\tilde{k}_{Hertz}}}{\sqrt{\left(\ln e_n\right)^2 + 10.1354}} \tag{2.66}$$

Since the coefficient of normal restitution in the KKn model is also a function of collision velocity, we cannot establish such a relationship for the damping coefficient of the KKn model.

Recently, Zheng *et al.* [48] extended the KKn model and proposed a relationship for normal collision force. This model is called the ZZYn model hereafter. They performed a set of finite-element method analyses for the viscoelastic contacts between a sphere and a plate and proposed a "semi-analytical" model for the normal collision force as follows:

$$\vec{f}_{ij}^{n} = \vec{f}_{el}^{n} + \vec{f}_{diss}^{n} = \left(-\frac{4}{3}E_{eff}\sqrt{R_{eff}}\,\delta_n^{3/2}\right)\vec{n}_{ij} - C_n\left(\hat{\eta}_n\left(R_{eff}\delta_n\right)^{1/2} v_{rn}\right)\vec{n}_{ij} \tag{2.67}$$

where C_n is the modification factor of viscous force in the ZZYn model. They introduced this modification factor to obtain a more similar force-displacement curve to that resulted from the finite-element method analysis. C_n reads as:

$$C_n = \left(0.8 + 26\,\upsilon_i^3\right)\left(\frac{\eta_2}{\eta_1}\right)^{-0.5}\left(\frac{\delta_n}{R}\right)^{0.04} \quad \text{for} \begin{cases} 0 \le \dfrac{\delta_n}{R} \le 0.01 \\ 0 \le \dfrac{\eta_2}{\eta_1} < 4 \\ 0.1 \le \upsilon_i \le 0.45 \end{cases} \tag{2.68}$$

The damping coefficient $\hat{\eta}_n$ is expressed as:

$$\hat{\eta}_n = 2\frac{E_{eff}}{E_i}\left(1 - 2\upsilon_i\right)\left(1 + \upsilon_i\right)\left(2\eta_2 + \frac{\eta_1}{3}\right) \tag{2.69}$$

In these equations, η_1 and η_2 are coefficients of shear and bulk deformation viscosities, respectively. It can be seen that the valid range for Equation 2.68 allows us to use it for many practical applications.

We performed a set of normal particle-wall impacts with TTIn, KKn, and ZZYn models. The dimensionless force-displacement results are shown in Figure 2.8. Note that the results of the ZZYn model are not included in this figure, since the KKn and the ZZYn models have similar damping coefficients. Physical properties of the particle are listed in Table 2.1. We did not report the exact values of damping coefficients and only reported the normal coefficient of restitution in each test. This enables us to compare the results of nonlinear models with the LSD model. The maximum overlap of the particle decreases and the distance between loading and unloading curves becomes broader with increasing the damping coefficient. A very important behavior of these models is the zero collision force at the start and the end of the collision, which is an improvement over the LSD model (see Figure 2.5).

Once we perform normal particle-wall collisions, we can also calculate the duration of collision and the normal coefficient of restitution. The collision duration as a function of impact velocity and damping coefficient for KKn, TTIn, and ZZYn models are shown in Figure 2.9. The normal coefficient of restitution obtained from the finite-element method simulations, according to Zheng *et al.* [48], are also reported in this figure. Figure 2.9a shows that in all models, the collision duration decreases as the impact velocity increases. The same behavior is also reported in the normal particle-wall impact experiments. Moreover, the results produce a straight line in the logarithmic scale with a slope of –0.2. This is in agreement with the collision duration equation proposed by Timoshenko and Goodier [63] for an elastic particle-wall collision as we presented in Equation 2.60. This is also another improvement over the LSD model in which the collision duration is not a function of impact velocity.

The normal coefficient of restitution versus impact velocity and damping coefficient is plotted in Figure 2.9b. The damping coefficient for the TTIn model was obtained from Equation 2.66 at $e_n = 0.56$. The damping coefficient for the ZZYn model was calculated by Equations 2.68 and 2.69 according to the physical properties listed in Table 2.1. Also, the normal damping coefficient of the KKn model, $\bar{\eta}_n$, was adjusted to obtain the best fit and was found to be $5300\,\mathrm{kg \cdot s^{-1} m^{1/2}}$. As can be seen in this figure, the TTIn model predicts a constant normal coefficient of restitution for various

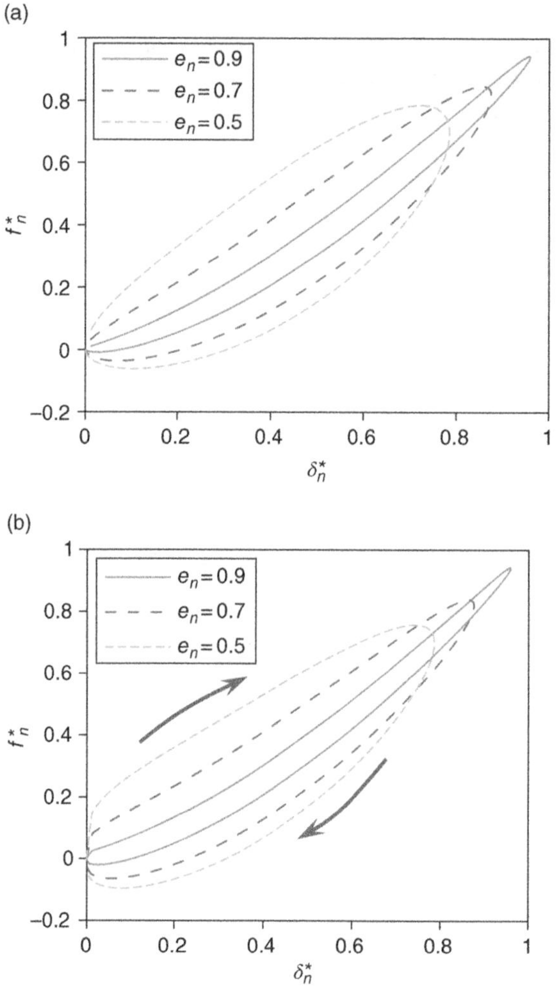

Figure 2.8 Dimensionless normal force-displacement curves for (a) KKn model and (b) TTIn model. f_n^* is the ratio of normal collision force to maximum elastic normal force (Hertz's theory) and δ_n^* is the ratio of normal overlap to maximum normal overlap of an elastic collision

impact velocities while KKn and ZZYn models predict lower e_n at higher impact velocities. Both KKn and ZZYn produce very close values for the coefficient of normal restitution. When the required material properties of the ZZYn model are available, using the ZZYn model is preferable since there is no need to adjust normal damping coefficient against the experimental data of e_n.

2.3.2.2 Tangential Direction

Tsuji *et al.* [46] derived a very simplified version of the MD theory for no-slip condition and constant normal force on the contact surface. They did not account for the effect of viscous force in the tangential direction. Their model reads as:

$$\vec{f}_{ij}^{t} = \vec{f}_{el}^{t} = -8G_{eff}\sqrt{R_{eff}}\,\delta_n^{1/2}\delta_t\vec{t}_{ij} \tag{2.70}$$

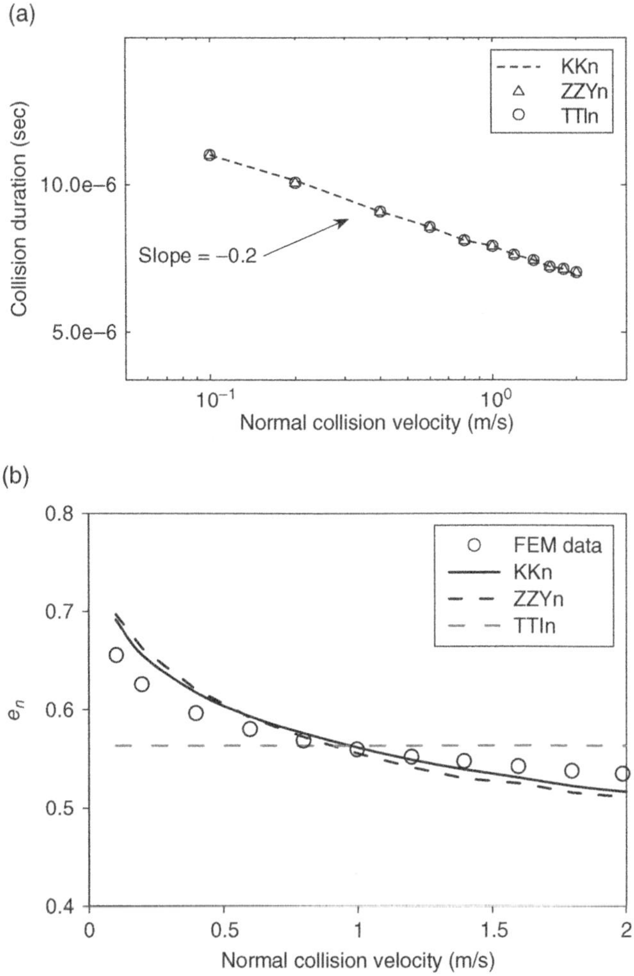

Figure 2.9 Predictions of different nonlinear viscoelastic models (a) collision duration and (b) normal coefficient of restitution, versus normal impact velocity. The FEM data are collected from [65]

where G_{eff} is effective shear modulus and is defined as:

$$G_{eff} = \left(\frac{2-\upsilon_i}{G_i} + \frac{2-\upsilon_j}{G_j} \right)^{-1} \tag{2.71}$$

Unlike the linear model, the nonlinear spring stiffness ($\tilde{k}_t = 8G_{eff}\sqrt{R_{eff}\delta_n^{1/2}}$) is continuously changing during a collision because the normal displacement is not constant. Since the real simulations involve micro-slip conditions and varying normal force at the contact surface, the use of such a model for calculating the tangential force does not lead to accurate results. Di Renzo and Di Maio [29] suggested that the tangential force obtained by the model of Tsuji *et al.* [46] (Equation 2.70) should be multiplied by a correction factor to account for the variable normal force and micro-slip at contact area. They used the integral form of the MD theory and showed that at both large and small impact angels, the tangential force calculated from the MD theory is a fraction of $\frac{2}{3}$ the anticipated value from the model of Tsuji *et al.* [46]. Therefore, they suggested the following relation for elastic tangential force:

$$\vec{f}_{ij}^{t} = \vec{f}_{el}^{t} = -\frac{16}{3} G_{eff} \sqrt{R_{eff}} \delta_n^{1/2} \delta_t \vec{t}_{ij} \tag{2.72}$$

We hereafter call this model DDt model (the letter "t" at the end refers to tangential). This model does not consider the dissipation of kinetic energy in the tangential direction. However, this dissipation term can be incorporated by a damping force, which is proportional to the velocity, similar to that is used in the linear model for the tangential direction. Coulomb's law of friction may be applied to provide gross sliding condition and limiting the tangential force by applying Equation 2.49. As discussed for the LSD model, an alternative way of applying Coulomb's law of friction is to limit the tangential displacement (Equation 2.50). In this way, better results were obtained [27]. The methodology is similar to that explained for the linear model.

Langston *et al.* [58, 68] presented a simplified model based on the MD theory for a contact with constant normal force. We hereafter call this model the LTH (Longston, Tuzun, and Heyes) model. This model, which is widely used [59, 60, 69], is given as:

$$\vec{f}_{el}^{t} = -\mathrm{sgn}(\delta_t)\mu \left|\vec{f}_{ij}^{n}\right| \left[1 - \left(1 - \frac{\min\left(\left|\delta_t\right|, \delta_{t,\,max}\right)}{\delta_{t,\,max}}\right)^{3/2}\right] \vec{t}_{ij} \tag{2.73}$$

$$\vec{f}_{diss}^{t} = -\tilde{\eta}_t \left(6 m_{eff} \mu \left|\vec{f}_{ij}^{n}\right| \frac{\sqrt{1 - \min\left(\left|\delta_t\right|, \delta_{t,\,max}\right)/\delta_{t,\,max}}}{\delta_{t,\,max}}\right)^{1/2} v_{rt} \vec{t}_{ij} \tag{2.74}$$

where $\tilde{\eta}_t$ is the tangential damping coefficient and $\delta_{t,max}$ is the displacement at which sliding starts and is defined as:

$$\delta_{t,\,max} = \mu \frac{2 - \upsilon}{2(1 - \upsilon)} \delta_n \tag{2.75}$$

Note that the use of min $(|\delta_t|, \delta_{t,max})$ ensures us that gross sliding occurs when the tangential displacement is greater than $\delta_{t,max}$.

Zheng *et al.* [48] extended the model of Dintwa *et al.* [70] and proposed the following equation, which we refer to as the ZZYt model hereafter, for the viscous force in the tangential direction:

$$\vec{f}_{diss}^{t} = -\mathrm{C}_t \frac{\eta_1}{2 G_i \delta_{t,max}} \left(1.5 \mu \left|\vec{f}_{el}^{n}\right| \sqrt{1 - \min\left(\left|\delta_t\right|, \delta_{t,max}\right)/\delta_{t,max}}\right) \vec{v}_{ij}^{t} \tag{2.76}$$

where C_t is the correction factor that is expressed as:

$$C_t = 1 - \frac{0.4\eta_1 \left|\vec{v}_{ij}^{\,t}\right|}{2G_i\delta_{t,max}} \tag{2.77}$$

The elastic force in the tangential direction is obtained from Equation 2.73. In derivation of this model, they performed a set of finite-element simulations for elastic and viscoelastic particle-wall collisions and showed that the micro-slip region on the contact surface of a viscoelastic sphere develops faster than that on the contact surface of an elastic sphere. As a result, when the viscoelastic sphere has no-stick zone and undergoes gross-sliding, the elastic sphere has both stick and slip zones. Thus, to account for such an effect (which was not considered in previous similar models), they modified the tangential force-displacement by introducing a semi-empirical correcting factor into the original model.

2.3.2.3 Calculation Procedure

A very important part of a DEM simulation is to correctly implement the force-displacement laws into a code. Although the goal of this chapter is to describe the theoretical principles of the DEM formulation rather than the numerical implementation aspects, we present the procedure of calculating the tangential force here to emphasize the importance of correct implementation of the model into a numerical code. The tangential force models presented here assume a constant normal force during the collision, while it varies in real cases. Thus, utilizing such models for real cases needs additional numerical treatments as described by Zheng *et al.* [48].

A key point in the calculation of tangential force at time $t+\Delta t_p$ is the variation of normal force from time t to time $t+\Delta t_p$. Therefore, we must rescale the tangential overlap at time t, to a new value that corresponds to the normal force[6] at time $t+\Delta t_p$ and then add $\Delta\delta_t$ to the rescaled tangential overlap for obtaining the new tangential overlap at time $t+\Delta t_p$. Thereafter, the tangential force can be calculated from corresponding equations. Figure 2.10 shows the step-by-step procedure for rescaling the tangential overlap and force for the LTH model based on Equations 2.73 and 2.74 and for the ZZYt model based on Equations 2.73 and 2.76.

2.3.3 Comparison of Viscoelastic Force-Displacement Models

To compare the behavior of different viscoelastic force-displacement models, we performed a set of oblique elastic impacts between a sphere and a flat wall. To obtain a better judgment for the accuracy of different models, results of different models were compared with the results of experiments at the same operating conditions. It is useful to analyze the dynamics of oblique impact in terms of rigid body dynamics, which is a simplified view of real oblique impacts with small deformation. Thus, before we proceed to the comparison, the dynamics of an oblique impact of a rigid sphere with a rigid wall is presented.

[6] There is a specific $F_t-\delta_t$ curve for each value of constant normal force. The tangential stiffness changes by changing the normal force. This leads to different $F_t-\delta_t$ curves for different values of normal forces.

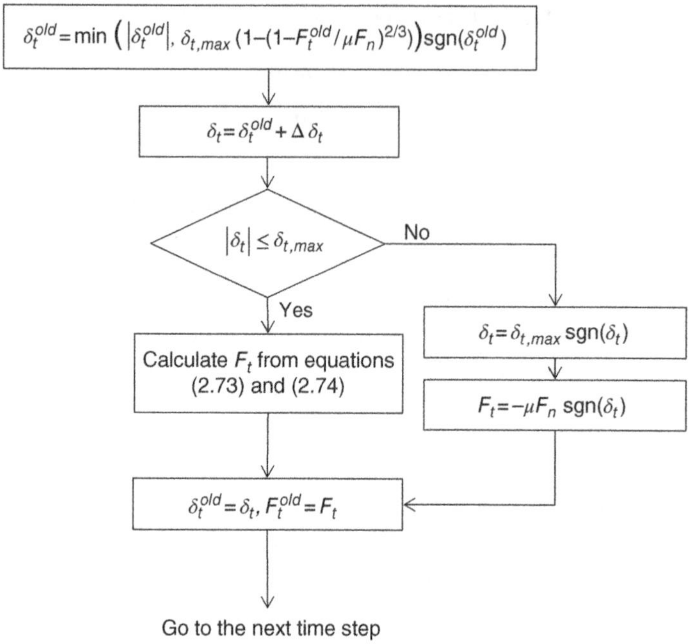

Figure 2.10 The procedure for calculating tangential forces based on LTH and ZZYt models

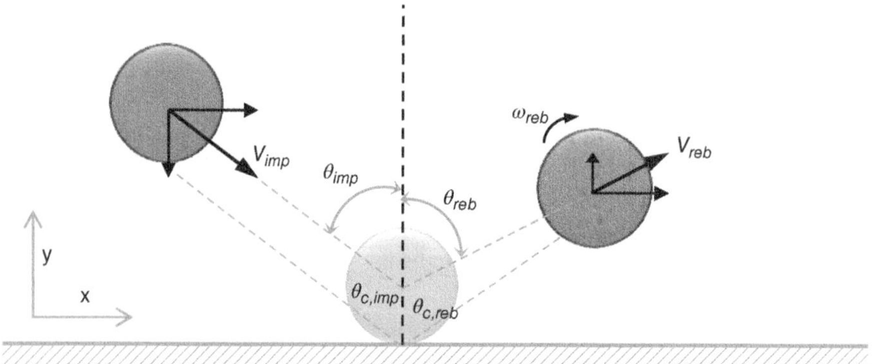

Figure 2.11 Schematic of an oblique impact of a sphere with a flat wall. The rotation speed is zero before the collision. The impact angle may range from 0 to 90° and different collision regimes may occur. At low impact angles (near normal) sticking regime occurs and at high impact angles (near horizontal) sliding regime occurs

The schematic of an oblique impact of a rigid sphere is illustrated in Figure 2.11. The impulse forces in the normal and tangential directions are obtained by applying the Newton's second law:

$$J_n = m\left(V_{n,reb} - V_{n,imp}\right) \tag{2.78}$$

$$J_t = m\left(V_{t,reb} - V_{t,imp}\right) \tag{2.79}$$

The rotational impulse is defined by:

$$J_\omega = I\left(\omega_{reb} - \omega_{imp}\right) \tag{2.80}$$

The rotational impulse is related to tangential impulse by the following equation:

$$J_\omega = RJ_t \tag{2.81}$$

With $\omega_{imp} = 0$ and the definition of I by Equation 2.9, the rotational velocity can be obtained in terms of tangential impact and rebound velocities:

$$\omega_{reb} = \frac{5\left(V_{t,reb} - V_{t,imp}\right)}{2R} \tag{2.82}$$

For avoiding confusion, we redefine normal and tangential coefficients of restitution by notations used in this section:

$$e_n = -\frac{V_{n,reb}}{V_{n,imp}} \tag{2.83}$$

$$e_t = \frac{V_{t,reb}}{V_{t,imp}} \tag{2.84}$$

The rebound tangential velocity at the collision point between the sphere and the plane is:

$$V_{tc,reb} = V_{t,reb} + R\omega_{reb} \tag{2.85}$$

Substitution of Equations 2.82 and 2.84 into Equation 2.85 and rearranging yield the following equation for tangential coefficient of restitution:

$$e_t = \frac{5}{7} + \frac{2}{7}\frac{V_{tc,reb}}{V_{t,imp}} \tag{2.86}$$

As we described in Section 2.1, there are two possibilities for a rigid body collision. At low impact angle, where the tangential impulse is low, the sticking regime occurs. In this condition, $V_{tc,reb} = 0$ and Equation 2.86 gives $e_t = 5/7$. The other possible behavior is the sliding regime that occurs at high impact angles. In this condition, the Coulomb's friction law can be applied:

$$J_t = -\mu J_n \tag{2.87}$$

By substituting Equations 2.78 and 2.79 into Equation 2.87 and using the definitions in Equations 2.83 and 2.84, we obtain the following equation for e_t in the sliding regime:

$$e_t = 1 - \frac{\mu\left(1 + e_n\right)}{\tan\left(\theta_{imp}\right)} \tag{2.88}$$

where θ_{imp} is the impact angle and is evaluated from:

$$\tan\left(\theta_{imp}\right) = -\frac{V_{t,imp}}{V_{n,imp}} \tag{2.89}$$

Another important parameter that is considered here is the relationship between impact and rebound angles at the collision point. We can define the rebound angle at the collision point by the following equation:

$$\tan\left(\theta_{c,reb}\right)=\frac{V_{tc,reb}}{V_{nc,reb}}=\frac{V_{tc,reb}}{V_{n,reb}} \quad (2.90)$$

where $V_{nc,reb}$ is rebound normal velocity at the collision point that is equal to the rebound normal velocity $V_{n,reb}$. Here, we follow the work of Maw *et al.* [45] who showed that for an elastic contact, the ratio of tangential and normal spring stiffnesses, κ, and the friction factor are effective parameters on the mechanism of oblique collision (see Section 2.3.1). We define dimensionless impact parameter ψ_{imp} and dimensionless rebound parameter $\psi_{c,reb}$:

$$\psi_{imp}=\frac{\kappa}{\mu}\tan\left(\theta_{imp}\right) \quad (2.91)$$

$$\psi_{c,reb}=\frac{\kappa}{\mu}\tan\left(\theta_{c,reb}\right) \quad (2.92)$$

We chose the experimental results by Kharaz *et al.* [40] to compare the accuracy of the models in elastic oblique collisions. The experiments were carried out with 5-mm aluminum oxide spheres impacting a thick soda-lime glass anvil under different impact angles ranging from near normal to near glancing incidence. The initial rotational speed of the sphere was kept zero and the impact velocity was 3.9 m/s. Other properties of these impacts are presented in Table 2.2.

Figure 2.12 shows the variation of tangential coefficient of restitution as a function of impact angle. The value of tangential coefficient of restitution starts from 0.75 at very low impact angles and reduces to the minimum value of 0.6 at impact angle of 20°. Then, it gradually increases with increasing the impact angle and approaches the unity at very high impact angles. The tangential coefficient of restitution obtained from both linear and nonlinear models are also plotted in this figure. All models follow a trend similar to the experiments. However, the LSD model with limited tangential displacement and the Hertz-LTH model show the best agreement with experimental results. The Hertz-DDt model with limited tangential

Table 2.2 Properties of sphere and wall used in elastic oblique tests

Property	Value	Property	Value
Radius (mm)	2.5	Impact velocity (m/s)	3.9
Density (kg/m^3)	3950	Initial rotational velocity (rad/s)	0
Young's modulus of sphere (GPa)	380	Young's modulus of wall (GPa)	70
Shear modulus of sphere (GPa)	154	Shear modulus of wall (GPa)	28
Poisson's ratio of sphere	0.23	Poisson's ratio of wall	0.25
Dynamic friction	0.092	Normal spring stiffness[a] (MN/m)	157
		Tangential spring stiffness (MN/m)	136

[a]The value of normal spring stiffness is obtained for normal impact velocity of 3.9 m/s. This value is different for other impact angles.

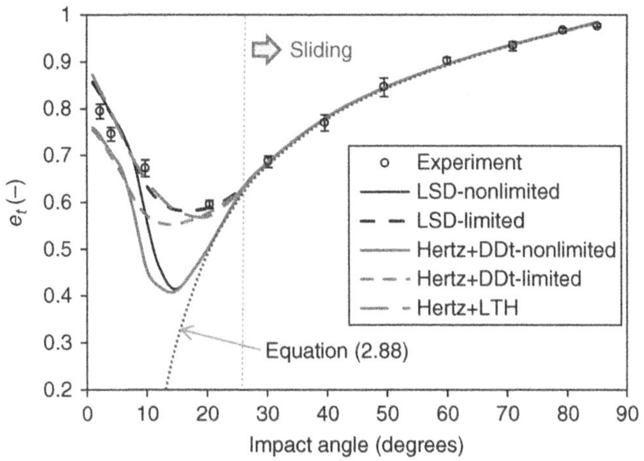

Figure 2.12 Tangential coefficient of restitution as a function of impact angle for elastic particle–wall collisions. The dashed curve shows Equation (2.88) corresponding to the sliding regime. Experimental data are collected from Kharaz *et al.* [40]

displacement shows a rough agreement with experimental results. It should be noted that the tangential restitution coefficient of an elastic sphere exactly matches the equation obtained for the rigid collision with sliding. This shows that at the sliding regime, for $\theta_{imp} > 28°$, the dynamics of an elastic collision is similar to the dynamics of the rigid body collision. Although the collisions are elastic, the tangential coefficient is not unity in all impact angles. This shows that a significant amount of initial kinetic energy is converted into the rotational kinetic energy and is lost as a result of friction at the collision point.

Figure 2.13a shows the variation of dimensionless rebound parameter at contact point as a function of dimensionless impact parameter. Three different oblique impact regimes are also marked in this figure. The LSD model with limited tangential displacement, the Hertz-DDt model with limited tangential displacement, and the Hertz-LTH model show good agreement with the experimental data in all three regimes. The LSD model and the Hertz-DDt with non-limited tangential overlap deviate from the experimental data in the second regime in which a sequence of sliding, sticking, and sliding mechanism occurs. The rebound angle as a function of impact angle is plotted in Figure 2.13b. Again, the LSD model and the Hertz-DDt model with limited tangential displacement agree well with the experimental data in the whole range of contact angle while the LSD model and the Hertz-DDt model with non-limited tangential displacement fail at intermediate impact angles.

2.3.4 *Elastic Perfectly Plastic Models*

When two particles composed of elastic material collide in normal direction, the kinetic energy of collision dissipates due to stress wave propagation throughout the particles. However, the amount of energy dissipation due to this mechanism depends on the material properties as well as the impact velocity. In the case of elastic deformation, the particle shape is preserved after the collision. The other mechanism for dissipation of kinetic energy of a normal collision is plastic

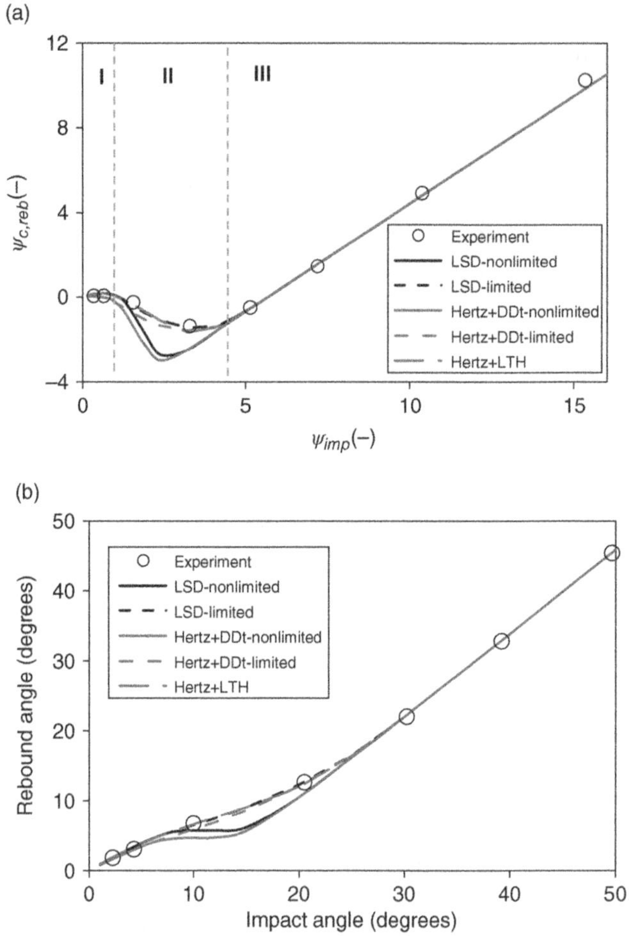

Figure 2.13 (a) Dimensionless contact parameter as a function of dimensionless rebound parameter at contact point and (b) rebound angle as a function of impact angle. Experimental data are collected from Kharaz *et al.* [40]

deformation. Plastic deformation occurs when the impact velocity is high enough to initiate it. When a high velocity collision occurs, the kinetic energy of collision is converted into three forms: elastic strain energy, plastic strain energy, and propagating waves through the body. The plastic strain energy is used to plastically deform the particle and the elastic strain is stored in the colliding particles. The plastic deformation is irreversible, hence, the plastic strain energy is not conserved after the collision is released. The elastic strain energy is the only recoverable part that is converted back into the kinetic energy after the rebound. It has been shown that most of the energy loss is due to the plastic deformation rather than the wave propagation [32].

Figure 2.14 shows typical force–displacement curves for collision of an elastic sphere with elastic perfectly plastic substrates at different impact velocities. These results are obtained by FEA simulations. The whole collision is divided into two phases: the loading phase (in which compression occurs) and the unloading phase (in which restitution occurs).

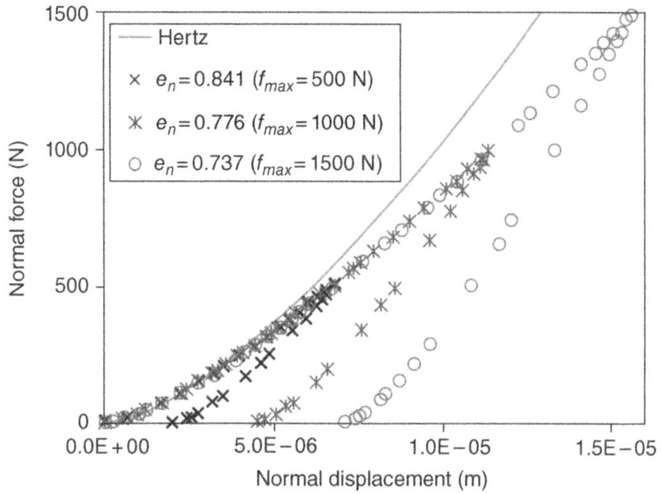

Figure 2.14 Typical force-displacement curves for normal elastic perfectly plastic collisions at different maximum normal force. Results are based on finite-element analysis. Adapted from [62]. As the normal applied force becomes larger, more energy is dissipated in the plastic deformation of sphere

As can be seen in this figure, the loading curve initially follows the Hertz's theory, indicating the elastic behavior of the collision. However, for higher overlaps, the force–displacement curve diverges from the Hertz's theory. At the center of the contact area, where the contact pressure is maximum value, the micro-yield of material is reached and the plastic deformation starts from the center of the contact area. Therefore, the collision continues with plastic deformation at the center of contact area and with elastic deformation at the perimeter of the plastic region. The pressure at which the plastic deformation starts is proportional to the yield stress of the particle, for example, based on Davies [71], it is 1.587 times the yield stress of particle. Unloading phase starts when the maximum overlap is reached. The unloading curve does not follow the loading curve due to plastic deformation in loading phase. Moreover, loading curves are the same at different velocities while the slope of unloading curve of a high impact velocity is higher than that of a low impact velocity. This means that the fraction of kinetic energy dissipated increases with increasing the impact velocity [32, 62].

To account for the plastic deformation and to obtain a similar behavior for force–displacement curves, a number of force-displacement laws based on the elastic perfectly plastic collisions have been developed [32, 33, 72–74]. The model developed by Walton and Braun [74] is based on a semi-latched linear spring force-displacement model (hereafter, WB-I model). Figure 2.15 shows a schematic of the WB-I collision model between particles i and j. In the loading phase, the spring with low stiffness is active while in the unloading phase, the spring with high stiffness governs the interaction force between colliding particles. In the WB-I model, linear force-displacement behaviors are considered for loading and unloading phases with different linear springs. The normal collision force is expressed as:

$$\vec{f}_{ij}^{\,n} = \begin{cases} -\left(k_l \delta_n\right)\vec{n}_{ij} & v_{rn} \geq 0 \\ -k_{ul}\left(\delta_n - \delta_{n,0}\right)\vec{n}_{ij} & v_{rn} < 0 \end{cases} \tag{2.93}$$

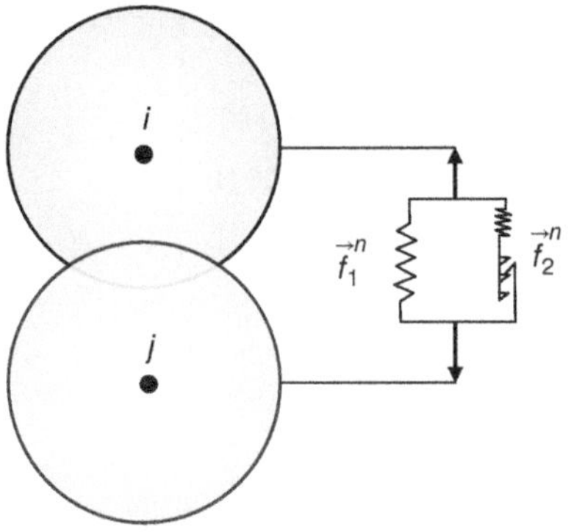

Figure 2.15 Schematic of a force-displacement law with different spring stiffnesses in loading and unloading phases corresponding to the model proposed by Walton and Braun [74]

where k_l and k_{ul} are loading and unloading spring stiffnesses, respectively, and $\delta_{n,0}$ is the residual overlap after completing unloading phase. Since different values of spring stiffness are considered for loading and unloading periods, a part of the kinetic energy is dissipated during each collision. The coefficient of normal restitution is defined as:

$$e_n = \sqrt{\frac{k_l}{k_{ul}}} \tag{2.94}$$

Values of k_l and k_{ul} are constant in the WB-I model, which results in a constant normal coefficient of restitution. The collision duration, given next, is a function of particle properties and does not change with the impact velocity:

$$t_{col} = \frac{\pi}{2}\sqrt{\frac{m_{eff}}{k_l}}\left(1+e_n\right) \tag{2.95}$$

The following relationship gives the residual overlap in terms of impact velocity, $v_{rn,imp}$, effective mass, and loading and unloading stiffnesses:

$$\delta_{n,0} = \frac{v_{rn,imp}\sqrt{m_{eff}}\left(k_{ul}-k_l\right)}{k_{ul}\sqrt{k_l}} \tag{2.96}$$

This model gives a constant coefficient of restitution and a constant collision time regardless of the impact velocity. To overcome this problem, Walton and Braun [74] modified their original model (WB-II hereafter) by introducing an adjustable parameter and relating the unloading stiffness to k_l and f_{max}:

$$k_{ul} = k_l + sf_{max} \tag{2.97}$$

where s is an adjustable parameter and f_{max} is the maximum normal force achieved prior to the unloading phase that is given by:

$$f_{max} = k_l\delta_{n,\,max} = k_l\sqrt{\frac{m_{eff}}{k_l}}v_{rn,\,imp} \tag{2.98}$$

Since the maximum normal force is directly related to $v_{rn,imp}$, Equation 2.97 linearly relates the unloading spring stiffness to $v_{rn,imp}$. Substituting Equations 2.97 and 2.98 into Equation 2.94 leads the following relationship for the normal coefficient of restitution of the WB-II model:

$$e_n = \sqrt{\frac{1}{1 + s\, v_{rn,imp} \sqrt{m_{eff} / k_l}}} \tag{2.99}$$

This equation indicates that the coefficient restitution is a decreasing function of impact velocity. With the same mathematics, the following equation can be obtained for the contact time:

$$t_{col} = \frac{\pi}{2} \frac{\sqrt{m_{eff}}\left(\sqrt{k_l} + \sqrt{k_l + s\sqrt{k_l m_{eff}}\, v_{rn,imp}}\right)}{k_l \sqrt{1 + s\sqrt{m_{eff} / k_l}\, v_{rn,imp}}} \tag{2.100}$$

It should be noted that s is an adjustable parameter that is obtained by fitting experimental results to the equation of either normal coefficient of restitution or contact time.

Thornton [73] proposed a nonlinear elastic perfectly plastic force-displacement model by assuming two phases for the compression stage: elastic and plastic deformations. In the elastic phase, it is assumed that the normal pressure distribution in the contact area follows the same distribution as the Hertz's theory (Equation 2.54) with the maximum pressure at the center of contact area. However, the elastic interaction at the contact area becomes plastic when the normal pressure at the center of contact area reaches the limiting contact pressure P_Y, which is a function of yield stress of the material. This is shown in Figure 2.16 in which the plastic deformation starts from the center of contact area and spreads toward its perimeter. In the plastic region ($0 \le r_{ca} \le r_p$), the pressure distribution is constant and equal to P_Y while outside the plastic region ($r_p \le r_{ca} \le a$) it still follows the Hertzian pressure distribution. Thus, we have the followings for the loading phase of collision. During the elastic compression, the normal force is given by:

$$\vec{f}_{ij}^{\,n} = \left(-\frac{4}{3} E_{eff} \sqrt{R_{eff}}\, \delta_n^{3/2}\right) \vec{n}_{ij} \quad \text{for} \quad \delta_n \le \delta_Y \tag{2.101}$$

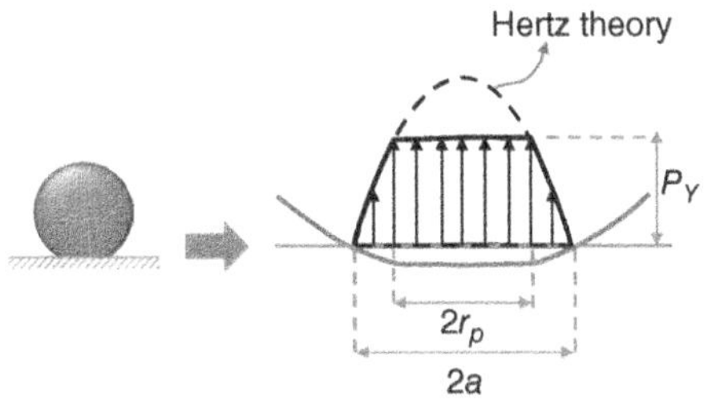

Figure 2.16 The normal pressure distribution in the contact area of an elastic perfectly plastic sphere based on the model of Thornton [73]

where δ_Y is the normal overlap at which plastic deformation starts and is calculated from:

$$\delta_Y = \left(\frac{\pi P_Y}{2E_{eff}}\right)^2 R_{eff} \tag{2.102}$$

When the pressure at the center of contact area reaches P_Y, plastic deformation occurs and the normal force is obtained by:

$$\vec{f}_{ij}^{\,n} = -\left(f_Y + \pi P_Y R_{eff}\left(\delta_n - \delta_Y\right)\right)\vec{n}_{ij} \quad \text{for } \delta_n > \delta_Y \tag{2.103}$$

In Equation 2.103, f_Y is the magnitude of normal force at which plastic deformation starts. Also, the contact radius is obtained from:

$$a = \left(\frac{3R_{eff} f_n^{Hertz}}{4E_{eff}}\right)^{1/3} \tag{2.104}$$

where f_n^{Hertz} is the equivalent elastic force that produces the same contact area.

Just before the start of restitution (unloading phase) normal overlap and normal force are recorded as the maximum normal overlap, $\delta_{n,max}$, and maximum normal force, $f_{n,max}$, to be used in the calculation of normal collision force in the unloading phase. The normal collision force for the whole unloading phase is expressed as:

$$\vec{f}_{ij}^{\,n} = \left(-\frac{4}{3}E_{eff}\sqrt{R_p}\left(\delta_n - \delta_p\right)^{3/2}\right)\vec{n}_{ij} \tag{2.105}$$

where R_p is calculated by:

$$R_p = \frac{4}{3}\frac{E_{eff}}{f_{n,\,max}}\left(\frac{2f_{n,\,max} + f_Y}{2\pi P_Y}\right)^{3/2} \tag{2.106}$$

In Equation 2.105, δ_p is the relative overlap at which normal force becomes zero and is given by:

$$\delta_p = \delta_{n,\,max} - \left(\frac{3f_{n,\max}}{4E_{eff}\sqrt{R_p}}\right)^{2/3} \tag{2.107}$$

The contact radius during unloading phase is calculated by:

$$a = \left(\frac{3R_{eff}\left|\vec{f}_{ij}^{\,n}\right|}{4E_{eff}}\right)^{1/3} \tag{2.108}$$

The following equations have been proposed for evaluating the normal restitution coefficient of perfectly elastic-plastic collisions [73, 75]:

$$e_n = \left(\frac{6\sqrt{3}}{5}\right)^{1/2}\left(1 - \frac{1}{6}\left(\frac{\bar{V}_Y}{v_{rn,imp}}\right)^2\right)^{1/2}\left(\frac{\bar{V}_Y / v_{rn,imp}}{\bar{V}_Y / v_{rn,imp} + 2\sqrt{\frac{6}{5} - \frac{1}{5}\left(\bar{V}_Y / v_{rn,imp}\right)^2}}\right)^{1/4} \tag{2.109}$$

where $\bar{V}_Y$ is the normal yield velocity and is expressed as:

$$\bar{V}_Y = \left(\frac{\pi}{2E_{eff}}\right)^2 \left(\frac{8\pi R_{eff}^3 P_Y^5}{15m_{eff}}\right)^{1/2} \quad (2.110)$$

When the normal impact velocity is lower than the yield velocity, no plastic deformation occurs and the collision undergoes elastic deformation while at velocities beyond the yield velocity, plastic deformation occurs.

For the tangential direction, Thornton *et al.* [61] used a tangential force-displacement model, which was initially derived from the MD theory for elastic impacts [76]. However, the tangential stiffness is larger for plastic collision since the contact area of the elastic perfectly plastic collision is larger than that of elastic collision at a given normal force. They accounted this effect by incorporating true contact radius for each collisional phase in the normal direction as described previously (Equations 2.57, 2.104, and 2.108). For calculating tangential force, an incremental approach is used where the new tangential force is obtained from the old tangential force $\vec{f}^{\,t}_{ij,old}$, tangential increment $\Delta\delta_t$, and tangential stiffness k_t which is a complex function of normal and tangential forces as well as material properties.

$$\vec{f}^{\,t}_{ij,new} = \vec{f}^{\,t}_{ij,old} + \left(k_t \Delta\delta_t\right)\vec{t}_{ij} \quad (2.111)$$

There are maximum three phases recognized in the tangential force–displacement curve. During the loading phase, where $\Delta\delta > 0$, the tangential stiffness is obtained from:

$$k_t = 8G_{eff} a\phi_1 + \mu\left(1-\phi_1\right)\frac{\Delta f_n}{\Delta\delta_t} \quad (2.112)$$

$$\phi_1 = \left(1 - \frac{f_t + \mu\Delta f_n}{\mu f_n}\right)^{1/3} \quad (2.113)$$

where $f_n = \left|\vec{f}^{\,n}_{ij}\right|$ and $f_t = \vec{f}^{\,t}_{ij}.\vec{t}_{ij}$. During the unloading phase ($\Delta\delta_t < 0$), the tangential force is calculated as follows:

$$k_t = 8G_{eff} a\phi_2 - \mu\left(1-\phi_2\right)\frac{\Delta f_n}{\Delta\delta_t} \quad (2.114)$$

$$\phi_2 = \left(1 - \frac{f_t^* - f_t + 2\mu\Delta f_n}{2\mu f_n}\right)^{1/3} \quad (2.115)$$

where f_t^* is the tangential force of loading reversal point (from loading to unloading) and should be continuously updated during the unloading phase by:

$$f_t^* = f_t^* + \mu\Delta f_n \quad (2.116)$$

After the unloading phase, there may be a reloading phase with $\Delta\delta_t > 0$ for which the tangential stiffness is calculated from:

$$k_t = 8G_{eff} a\phi_3 + \mu\left(1-\phi_3\right)\frac{\Delta f_n}{\Delta\delta_t} \quad (2.117)$$

$$\phi_3 = \left(1 - \frac{f_t - f_t^{**} + 2\mu\Delta f_n}{2\mu f_n}\right)^{1/3} \quad (2.118)$$

where f_t^{**} is the tangential force of unloading reversal point (from unloading to reloading) and needs to be continuously updated in the reloading phase by:

$$f_t^{**} = f_t^{**} - \mu\Delta f_n \quad (2.119)$$

The Coulomb's friction law is also used for the case of gross sliding (Equation 2.49). As stated by Thornton *et al.* [61], during the loading phase in the normal direction ($\Delta f_n > 0$), a problem occurs if $|\Delta f_t| < \mu\Delta f_n$ and the new tangential force does not lay on the force–displacement curve. To remedy this problem, it is suggested to use the following equation for the tangential direction:

$$k_t = 8G_{eff}a \quad (2.120)$$

2.4 Torque Expressions

The torque acting on contacting particles is comprised of two contributions: rotational torque caused by interparticle contacts in the tangential direction and rolling resistant torque that arises from the unevenness of the contact pressure distribution in the contact area. The former causes particles to rotate while the latter resists against the rotations of particles. The rotational torque is defined as:

$$\vec{M}_{ij}^t = R_i \vec{n}_{ij} \times \vec{f}_{ij}^c \quad (2.121)$$

Since the contact force is transformed from the contact point to the mass center of the particle, the outcome is the rotational torque on the particle.

The rolling resistance arises from several sources. The most important ones are plastic deformation and viscous hysteresis [24, 77, 78]. Consider a particle that is rolling on a horizontal flat surface. The rolling resistance torque caused by the rotation of particle on the flat surface slows down the particle rotation speed and finally makes it to stop. If such rolling resistance was not considered in the model, the particle rolling would continue for a long time [59]. There are some models proposed in literature for evaluating the rolling resistance torque [59, 79, 80]. Recently, Ai *et al.* reviewed different rolling resistance torque models and evaluated their success in simulating different granular flows [81].

2.4.1 Model A: Constant Torque Model

The rolling resistance torque in Model A formulation is defined as [59]:

$$\vec{M}_{ij}^r = -\mu_r R_{eff} \left|\vec{f}_{ij}^n\right| \hat{\omega}_{ij} \quad (2.122)$$

where μ_r is dimensionless coefficient of rolling resistance and $\hat{\omega}_{ij}$ is the unit vector of relative angular velocity of particles defined as:

$$\hat{\omega}_{ij} = \frac{\vec{\omega}_i - \vec{\omega}_j}{\left|\vec{\omega}_i - \vec{\omega}_j\right|} \quad (2.123)$$

In this model, a constant resistant torque (when $\left|\vec{f}_{ij}^{n}\right|$ does not change) is applied on colliding particles. The minus sign is used to emphasize that this torque opposes the relative rotation of the particle.

2.4.2 *Model B: Viscous Model*

The rolling resistance in Model B formulation is related to angular velocity of particles [78]:

$$\vec{M}_{ij}^{r} = -\mu_r' R_{eff} \left|\vec{f}_{ij}^{n}\right| \left|\vec{V}_{\omega}\right| \hat{\omega}_{ij} \tag{2.124}$$

where $\vec{V}_{\omega}$ is the relative rotational velocity at contact point between particles defined as:

$$\vec{V}_{\omega} = \left(\vec{\omega}_i \times R_i \vec{n}_{ij} - \vec{\omega}_j \times R_j \vec{n}_{ji}\right) \tag{2.125}$$

The advantage of this model over Model A is its proportionality to $\left|\vec{V}_{\omega}\right|$, which leads to zero resistant torque when particles do not roll on each other. This behavior is physically more realistic than Model A that applies a constant resistant torque on particles.

2.4.3 *Model C: Spring-Dashpot Model*

In Model C formulation, the rolling resistance torque is comprised of two contributions: spring torque and viscous damping torque.

$$\vec{M}_{ij}^{r} = \vec{M}_{el}^{r} + \vec{M}_{diss}^{r} \tag{2.126}$$

The spring torque $\vec{M}_{el}^{r}$ is defined like the hysteretic models described before. Thus, its implementation into the DEM simulation is possible. The spring torque is calculated based on the incremental procedure similar to that described in hysteretic models. In each time step, the spring torque is calculated as follows:

$$\vec{M}_{el,new}^{r} = \vec{M}_{el,old}^{r} + \Delta M_{el}^{r} \hat{\omega}_{ij} \tag{2.127}$$

where $\Delta\vec{M}_{el}^{r}$ is the change in the spring torque during each time step and is defined as:

$$\Delta M_{el}^{r} = -k_r \Delta\theta_r \tag{2.128}$$

in which $\Delta\theta_r$ is the incremental relative rotation between two particles and k_r is the rolling stiffness. There is a constraint on the spring torque. When $\left|\vec{M}_{el,\,new}^{r}\right| > M_r^m$, the full mobilization occurs and the elastic torque is calculated from:

$$\left|\vec{M}_{el,\,new}^{r}\right| = M_r^m \tag{2.129}$$

in which M_r^m is the mobilization torque and is defined as:

$$M_r^m = \mu_r'' R_{eff} \left|\vec{f}_{ij}^{n}\right| \tag{2.130}$$

The viscous damping torque depends on the rate of change in the relative rotation between two particles in contact, $d\theta_r/dt$, and the viscous damping constant, C_r:

$$\vec{M}^r_{diss} = \begin{cases} -\left(C_r \dfrac{d\theta_r}{dt}\right)\hat{\omega}_{ij} & \text{for } \left|\vec{M}^r_{el,new}\right| < M^m_r \\ -\left(\zeta C_r \dfrac{d\theta_r}{dt}\right)\hat{\omega}_{ij} & \text{otherwise} \end{cases} \tag{2.131}$$

where ζ is a constant that describes how the energy is dissipated due to damping term during full mobilization of particle. If $\zeta = 0$, no energy is dissipated in this condition. When $\zeta = 1$, the energy is dissipated during the whole period of rolling. The damping constant is defined as:

$$C_r = \eta_r \left(2\sqrt{I_r k_r}\right) \tag{2.132}$$

where η_r is the viscous damping ratio of rolling torque and I_r is calculated by:

$$I_r = \left(\frac{1}{I_i + m_i R_i^2} + \frac{1}{I_j + m_j R_j^2}\right)^{-1} \tag{2.133}$$

Model C formulation provides a more realistic view of the rolling resistant torque than Models A and B and it is generally applicable to quasi-static and dynamic granular flows [81].

2.5 Boundary and Initial Conditions

2.5.1 Boundary Conditions

Walls play a very important role in the dynamic behavior and static status of granular and multiphase flows. Solid bodies are confined in the system boundaries via the existence of walls. There are many practical applications like paddle mixers, granular conveyors, rotating drums, helical mixers, and V-blenders in which walls move during operation of the unit. In these cases, the convective motion of the whole system is caused by the motion of walls. Therefore, the interaction of wall with solid bodies plays an important role in the output of the discrete element simulation. Walls are stationary in most multiphase flows and the motion of solid bodies is governed by other forces like drag or buoyant forces. However, the interaction of walls with solid bodies still affects the output of the simulation.

In the previous section, we showed that dynamic friction coefficient μ and the ratio of tangential to normal spring stiffnesses κ are two important parameters that govern the mechanism of oblique collisions. Thus, choosing proper values for these parameters for the interaction between a solid body and a wall is vital. In many practical cases, we face a rough wall whose collision mechanism with a deformable solid is poorly understood. Usually, the friction coefficient and other collisional properties between solids and wall are measured experimentally and are directly incorporated into the model. However, this does not seem to remedy the problem of collision mechanism between a deformable body and a rough wall. In general, there are two methods for representing a wall in a discrete element simulation: granular wall and real wall.

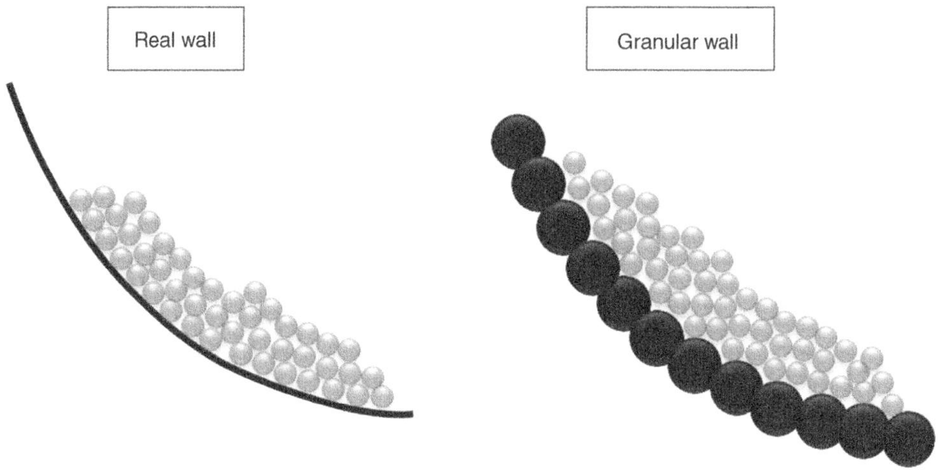

Figure 2.17 Representation of real and granular walls in the discrete-element methods

In defining the granular wall, the wall is built up from particles called *wall particles* in this text. In many situations, the motion of wall is not affected by the motion of solid bodies inside the system boundaries. Thus, the motion of wall particles cannot be calculated by Equation 2.24. Instead, their motion is controlled externally according to the translational and rotational velocities of the wall. In this definition, each wall particle may have its own size and mechanical properties. The interaction forces between a wall particle and a solid body are calculated based on the force-displacement models described before. A very important point that should be considered for the granular wall implementation is the relative size of wall particles to solid bodies (see Figure 2.17). Solid bodies may be locked between adjacent wall particles, artificially giving rise to the friction between wall and solid bodies. Thus, the size of wall particles and their mechanical properties influence the dynamic interaction between the wall and solid bodies. We can also take the advantage of this fact and produce a rough wall with an adjustable friction coefficient. The implementation of particulate wall into discrete element simulation is relatively straightforward. In addition, there is no limitation on the shape of the wall (flat, arc, circular, elliptic, etc.) that can be defined in the simulation.

In the real wall definition, the real shape or geometry of the wall is given by an analytical equation or plane sub-elements (see Figure 2.17). In this way, the interlock of solid bodies is prevented. All bodies near the wall are checked for possible collisions. To do this, the coordinate of the body is first transformed into the local wall coordinates and the distance is calculated. Then, if there is a positive overlap (collision), the contact point between wall and body is determined by proper methods. The same force-displacement models for particle–particle collisions are valid here. Wall is considered particle j with an infinite radius. Physical properties of the wall should be replaced by physical properties of particle j whenever necessary.

In some cases, the dimension of the system (simulation domain) is extended to a long distance in one or more directions and in the elongated dimension, the presence of boundary wall has the least effect on the overall dynamic behavior of granular or multiphase flow. In this situation, we can apply a periodic boundary condition instead of using a wall. Using the periodic boundary condition, we can mimic the dynamic behavior of the whole system with a smaller simulation domain and less number of particles. By applying this boundary condition in each dimension, each particle which leaves the system from one side is reinserted from the other side of the boundary. Consequently, interactions between particles at the opposite side should also be considered.

2.5.2 *Initial Condition*

To start the integration of equations of motion, it is necessary to define initial values for all variables involved in these equations. The initial conditions define values for position $\vec{x}_i$, velocity $\vec{v}_i$, rotational velocity $\vec{\omega}_i$, and angular position $\vec{\varphi}_j$ for all particles. The initial values assigned to each variable differ from case to case. However, the dynamic solution usually starts from a condition in which all particles are at rest and packed. There are various methods to obtain a packed assembly of solid particles. A very important question that we must answer is whether or not the initial condition (ordered or randomly packed particles) affects the simulation results. In general, when we are going to study the long-term dynamic characteristics of a granular (excluding static and quasi-static flows) or multiphase flow, the initial condition has minimal effects. However, the initial condition does affect the results when we are going to study the transient behavior of the system (from stationary/initial to long-term dynamic behavior) or when the simulation is static or quasi-static.

One option for initial positioning of particles in the system is to insert them one by one and side by side within the system boundaries. In this way, an ordered assembly of particles can be obtained. Maybe this would be the easiest and fastest way, but is not the best. Another option is to pack particles randomly within the system boundaries. The first method of random packing is to position all particles randomly in the simulation domain and allowing them to settle down by gravity. Coordinates of each new particle can be generated by a random number generator engine. For initial positioning of particles, we must ensure that any new inserted particle does not overlap with other particles. A proper criterion should be set for terminating the settling phase. For example, when the average kinetic energy of the system is less than a threshold, the program terminates the settling phase. In the second method, we can define a plane for particles through which particles are injected with a predefined rate and are allowed to fall under gravity and gradually fill the simulation domain.

Nomenclature

Symbol	Unit	Description
a	m	Radius of contact surface
CL_i	—	Contact list of particle i
C_n	—	Modification factor of viscous force in the ZZYn model
C_r	kg·m²/s	Viscous damping constant for rolling torque
C_t	—	Modification factor of viscous force in the ZZYt model
d	m	Surface separation distance of two particles
d_0	m	Minimum surface separation distance that particle interaction vanishes
D_{sw}	J	Minimum potential energy in the square-well model
e_n	—	Coefficients of restitution in the normal direction
e_t	—	Coefficients of restitution in the tangential direction

E_i	Pa	Young's modulus of particle i
E_{eff}	Pa	Effective Young's modulus
$\vec{f}^{n}_{diss}$	N	Normal dissipative/viscous force
$\vec{f}^{t}_{diss}$	N	Tangential dissipative/viscous force
$\vec{f}^{n}_{el}$	N	Normal elastic force
$\vec{f}^{t}_{el}$	N	Tangential elastic force
f_{Hertz}	N	Hertzian contact force
$\vec{f}^{c}_{ij}$	N	Contact force between particles i and j
$\vec{f}^{n}_{ij}$	N	Normal contact force between particles i and j
$\vec{f}^{t}_{ij}$	N	Tangential contact force between particles i and j
$\vec{f}^{t}_{ij,old}$	N	Tangential contact force at previous time step
$\vec{f}^{t}_{ij,new}$	N	Tangential contact force at current time step
f_{max}	N	Maximum normal force before unloading phase
f_n	N	Magnitude of normal contact force
$f_{n,max}$	N	Normal contact force before restitution
f^{*}_{n}	—	Ratio of normal collision force to maximum elastic normal force
f_t	N	Magnitude of tangential force
f^{*}_{t}	N	Tangential force of loading reversal point
f^{**}_{t}	N	Tangential force of unloading reversal point
f_{vdW}	N	Magnitude of van der Waals force
f_Y	N	Magnitude of normal force at which plastic deformation starts
$\vec{f}_i$	N	Sum of different forces on particle i
$\vec{f}^{ext}_{i}$	N	External forces acting on particle i
$\vec{f}^{f-p}_{i}$	N	Total fluid–particle interaction force on particle i
$\vec{f}^{p-p}_{ij}$	N	Interaction force between particles i and j
F_l	N	Constant normal force applied on particle
$\vec{g}$	m/s^2	Gravitational acceleration vector
G_{eff}	Pa	Effective shear modulus
G_i	Pa	Shear modulus of particle i
h_0, h_1	m	Initial height and maximum height after rebound
I	kg·m^2	Moment inertial of a particle
I_i	kg·m^2	Moment of inertia of particle i
I_r	kg·m^2	Equivalent moment of inertia in Equation 2.132
$\vec{J}$	kg·m/s	Impulse force vector
J_n	kg·m/s	Normal component of the impulse force
J_t	kg·m/s	Tangential component of the impulse force
$J_{n,coh}$	kg·m/s	Cohesive impulse force in the square-well method
$J_{n,reb}$	kg·m/s	Rebound impulse force in the square-well method
$J_{n,adh}$	kg·m/s	Adhesive impulse force in the square-well method
J_ω	kg·m^2/s	Angular impulse
$\tilde{k}_{Hertz}$	kg/(m$^{0.5}$s^2)	Hertzian-spring stiffness
k_l	N/m	Loading spring stiffness
$\tilde{k}_t$	N/m	Nonlinear spring stiffness in the tangential direction

k_n	N/m	Normal spring stiffness of linear spring
k_t	N/m	Tangential spring stiffness of linear spring
k_r	N·m	Rolling resistance spring stiffness
k_{ul}	N/m	Unloading spring stiffness
$\tilde{k}_{Hertz}$	kg/(m$^{0.5}$s^2)	Hertzian-spring stiffness defined in Equation 2.65
m, m_i	kg	Mass of particle, mass of particle i
m_{eff}	kg	Effective mass
$\vec{M}^r_{diss}$	N·m	Viscous rolling torque
$\vec{M}^r_{el}$	N·m	Elastic rolling torque
$\vec{M}^r_{el,new}$	N·m	Elastic rolling torque at the current time step
$\vec{M}^r_{el,old}$	N·m	Elastic rolling torque at the previous time step
$\vec{M}_i$	N·m	Sum of different torques that act on particle i
$\vec{M}^r_{ij}$	N·m	Rolling resistance torque on particle i due to contact with particle j
$\vec{M}^t_{ij}$	N·m	Tangential torque on particle i due to contact with particle j
M^m_r	N·m	Mobilization torque
$\vec{n}_{ij}$	—	Unit vector pointing from particle i to particle j
N	—	Total number of particles
P, P_{max}	Pa	Local and maximum normal pressure in contact surface
P_Y	Pa	Normal yield pressure at which plastic deformation starts
r_{ca}	m	Radial coordinate of contact area
r_p	m	Radius of circular plastic region
R_i	m	Radius of particle i
R_{eff}	m	Effective radius
s	1/m	Adjustable parameter in WB-II model
sgn	—	Sign function
t	s	Time
t_{col}	s	Collision time
$\vec{t}_{ij}$	—	Tangential vector at contact point
$\vec{v}_i$	m/s	Translational velocity of particle i
$\vec{v}_{ij}$	m/s	Relative velocity between particles i and j at contact point
$\vec{v}^0_{ij}$	m/s	Relative velocity of particles before collision
$\vec{v}^1_{ij}$	m/s	Relative velocity of particles after collision
$\vec{v}^c_{ij}$	m/s	Relative velocity of particles before the cohesive collision
$\vec{v}^n_{ij}$	m/s	Relative normal velocity between particles i and j at contact point
$\vec{v}^t_{ij}$	m/s	Relative tangential velocity between particles i and j at contact point
v_{rn}	m/s	Relative velocity at contact point in the normal direction
v_{rt}	m/s	Relative velocity at contact point in the tangential direction
$v_{rn,imp}$	m/s	Impact normal relative velocity
$v_{rn,reb}$	m/s	Rebound normal relative velocity
$V_{n,imp}$	m/s	Normal impact velocity to a flat wall
$V_{t,imp}$	m/s	Tangential impact velocity to a flat wall
$V_{n,reb}$	m/s	Normal rebound velocity from a flat wall
$V_{t,reb}$	m/s	Tangential rebound velocity from a flat wall

$V_{nc,reb}$	m/s	Rebound normal velocity at the contact point
$V_{tc,reb}$	m/s	Rebound tangential velocity at the contact point
$\bar{V}_Y$	m/s	Normal yield velocity
$\vec{V}_\omega$	m/s	Relative rotational velocity at the contact point
$\vec{x}_i$	m	Center position vector of particle *i*
Greek symbols		
α	—	Empirical parameter in Equation 2.63
β_0	—	Tangential coefficient of restitution at the contact point
δ_n	m	Normal overlap
$\delta_{n,0}$	m	Residual normal overlap after completing unloading phase
$\delta_{n,max}$	m	Normal overlap before restitution
δ_n^*	—	Ratio of normal overlap to maximum normal overlap of an elastic collision
δ_p	m	Normal overlap at which normal force becomes zero
δ_t	m	Tangential overlap
$\delta_{t,0}$	m	Tangential overlap in the previous time step
$\delta_{t,max}$	m	Tangential displacement at which sliding starts
δ_Y	m	Normal overlap at which plastic deformation starts
Δf_n	N	Incremental change in the normal force
$\Delta\vec{M}_{el}^r$	N·m	Incremental change in the elastic rolling torque
Δt_p	s	Time step for integrating the equation of motion of particle
$\Delta\delta_t$	m	Incremental change in tangential displacement
$\Delta\theta_r$	rad	Incremental change in the relative rotation between two particles
ζ	—	A constant in Equation 2.131b
η_1, η_2	Pa·s	Coefficients of shear and bulk deformation viscosities
η_n	kg/s	Normal damping coefficient in the LSD model
$\bar{\eta}_n$	kg/(s·m$^{0.5}$)	Normal damping coefficient for the KKn model in Equation 2.61
$\tilde{\eta}_n$	kg/(s·m$^{0.25}$)	Normal damping coefficient of the TTIn model in Equation 2.62
$\hat{\eta}_n$	kg/(s·m)	Normal damping coefficient of the ZZYn model in Equation 2.67
η_r	—	Viscous damping ratio of rolling torque
η_t	kg/s	Tangential damping coefficient in the LSD model
$\tilde{\eta}_t$	—	Tangential damping coefficient in Equation 2.74
θ_{imp}	°	Impact angle between particle path line and wall
κ	—	Ratio of tangential spring stiffness to normal spring stiffness
κ_0	1/s	Frequency of un-damped harmonic oscillator
μ	—	Coefficient of dynamic friction
μ_r	—	Coefficient of rolling resistance in model A
μ_r'	s/m	Coefficient of rolling resistance in model B
μ''	—	Coefficient of rolling resistance in model C
ρ_i	kg/m^3	Density of particle *i*
υ, υ_i	—	Poisson's ratio of particle *i*
$\vec{\varphi}_i$	rad	Angular position of particle *i*
ψ	1/s	Reduced damping coefficient
$\psi_{c,reb}$	—	Dimensionless rebound parameter in Equation 2.92
ψ_{imp}	—	Dimensionless impact parameter in Equations 2.53 and 2.91
$\vec{\omega}_i$	rad/s	Rotational velocity of particle *i*

$\vec{\omega}_i^0$	rad/s	Rotational velocity of particle *i* before collision
$\vec{\omega}_i^1$	rad/s	Rotational velocity of particle *i* after collision
ω_{imp}	rad/s	Impulse rotational velocity
ω_{reb}	rad/s	Rebound rotational velocity
$\hat{\omega}_{ij}$	—	Unit vector of relative angular velocity of particles *i* and *j*

References

[1] Campbell, C.S. and Brennen, C.E. (1985) Computer simulations of granular shear flows. *Journal of Fluid Mechanics*, **151**, 167–188.

[2] Gui, N. and Fan, J. (2009) Numerical simulation of motion of rigid spherical particles in a rotating tumbler with an inner wavelike surface. *Powder Technology*, **192**(2), 234–241.

[3] Richardson, D.C., Walsh, K.J., Murdoch, N., and Michel, P. (2011) Numerical simulations of granular dynamics: I. Hard-sphere discrete element method and tests. *Icarus*, **212**(1), 427–437.

[4] Hoomans, B.P.B., Kuipers, J.A.M., Briels, W.J., and van Swaaij, W.P.M. (1996) Discrete particle simulation of bubble and slug formation in a two-dimensional gas-fluidised bed: a hard-sphere approach. *Chemical Engineering Science*, **51**(1), 99–118.

[5] Hoomans, B.P.B., Kuipers, J.A.M., and van Swaaij, W.P.M. (2000) Granular dynamics simulation of segregation phenomena in bubbling gas-fluidised beds. *Powder Technology*, **109**, 41–48.

[6] Dahl, S.R. and Hrenya, C.M. (2004) Size segregation in rapid, granular flows with continuous size distributions. *Physics of Fluids*, **16**, 1–13.

[7] Dahl, S.R. and Hrenya, C.M. (2005) Size segregation in gas–solid fluidized beds with continuous size distributions. *Chemical Engineering Science*, **60**, 6658–6673.

[8] Li, J. and Kuipers, J.A.M. (2002) Effect of pressure on gas–solid flow behavior in dense gas-fluidized beds: a discrete particle simulation study. *Powder Technology*, **127**, 173–184.

[9] Li, J. and Kuipers, J.A.M. (2005) On the origin of heterogeneous structure in dense gas-solid flows. *Chemical Engineering Science*, **60**, 1251–1265.

[10] Link, J.M., Cuypers, L.A., Deen, N.G., and Kuipers, J.A.M. (2005) Flow regimes in a spout-fluid bed: a combined experimental and simulation study. *Chemical Engineering Science*, **60**, 3425–3442.

[11] Link, J.M., Godlieb, W., Deen, N.G., and Kuipers, J.A.M. (2007) Discrete element study of granulation in a spout-fluidized bed. *Chemical Engineering Science*, **62**, 195–207.

[12] Wu, C.L., Berrouk, A.S., and Nandakumar, K. (2010) An efficient chained-hash-table strategy for collision handling in hard-sphere discrete particle modeling. *Powder Technology*, **197**(1-2), 58–67.

[13] Poschel, T. and Schwager, T. (2005) *Computational Granular Dynamics: Models and Algorithms*, Springer-Verlag, Berlin/Heidelberg.

[14] Deen, N.G., Van Sint Annaland, M., Van der Hoef, M.A., and Kuipers, J.A.M. (2007) Review of discrete particle modeling of fluidized beds. *Chemical Engineering Science*, **62**, 28–44.

[15] Hamaker, H.C. (1937) The London van der Waals attraction between spherical particles. *Physica IV*, **10**, 1058–1080.

[16] Weber, M.W., Hoffman, D.K., and Hrenya, C.M. (2004) Discrete-particle simulations of cohesive granular flow using a square-well potential. *Granular Matter*, **6**, 239–254.

[17] Weber, M.W. and Hrenya, C.M. (2006) Square-well model for cohesion in fluidized beds. *Chemical Engineering Science*, **61**, 4511–4527.

[18] Cangialosi, F., Liberti, L., Notarnicola, M., and Stencel, J. (2006) Monte Carlo simulation of pneumatic tribo-charging in two-phase flow for high-inertia particles. *Powder Technology*, **165**(1), 39–51.

[19] Bi, X.T. (2011) Electrostatic phenomena in fluidized systems: present status of understanding and research needs. International Conference on Circulating Fluidized Beds and Fluidization Technology – CFB-10, Oregon.

[20] Matsusaka, S., Maruyama, H., Matsuyama, T., and Ghadiri, M. (2010) Triboelectric charging of powders: a review. *Chemical Engineering Science*, **65**, 5781–5807.

[21] Kosinski, P. and Hoffmann, A.C. (2010) An extension of the hard-sphere particle–particle collision model to study agglomeration. *Chemical Engineering Science*, **65**(10), 3231–3239.

[22] Seville, J.P.K., Willett, C.D., and Knight, P.C. (2000) Interparticle forces in fluidisation: a review. *Powder Technology*, **113**, 261–268.
[23] Mansourpour, Z., Mostoufi, N., and Sotudeh-Gharebagh, R. (2013) A mechanistic study of agglomeration in fluidised beds at elevated pressures. *The Canadian Journal of Chemical Engineering*, **91**(3), 560–569.
[24] Johnson, K.L. (1985) *Contact Mechanics*, Cambridge University Press, Cambridge.
[25] Aoki, K.M. and Akiyama, T. (1995) Simulation studies of pressure and density wave propagations in vertically vibrated beds of granules. *Physical Review E*, **52**(3), 3288–3291.
[26] Kruggel-Emden, H., Simsek, E., Rickelt, S., Wirtz, S., and Scherer, V. (2007) Review and extension of normal force models for the discrete element method. *Powder Technology*, **171**, 157–173.
[27] Kruggel-Emden, H., Wirtz, S., and Scherer, V. (2008) A study on tangential force laws applicable to the discrete element method (DEM) for materials with viscoelastic or plastic behavior. *Chemical Engineering Science*, **63**, 1523–1541.
[28] Di Renzo, A. and Di Maio, F.P. (2004) Comparison of contact-force models for the simulation of collisions in DEM-based granular flow codes. *Chemical Engineering Science*, **59**, 525–541.
[29] Di Renzo, A. and Di Maio, F.P. (2005) An improved integral non-linear model for the contact of particles in distinct element simulations. *Chemical Engineering Science*, **60**, 1303–1312.
[30] Stevens, A.B. and Hrenya, C.M. (2005) Comparison of soft-sphere models to measurements of collision properties during normal impacts. *Powder Technology*, **154**, 99–109.
[31] Thornton, C., Cummins, S.J., and Cleary, P.W. (2011) An investigation of the comparative behaviour of alternative contact force models during elastic collisions. *Powder Technology*, **210**, 189–197.
[32] Zhang, X. and Vu-Quoc, L. (2002) Modeling the dependence of the coefficient of restitution on the impact velocity in elasto-plastic collisions. *International Journal of Impact Engineering*, **27**, 317–341.
[33] Tomas, J. (2003) Mechanics of nanoparticle adhesion – a continuum approach. In *Particles on Surfaces 8: Detection Adhesion and Removal* (ed. K.L. Mittal), VSP.
[34] Cundall, P.A. and Strack, O.D.L. (1979) A discrete numerical model for granular assemblies. *Geotechnique*, **29**, 47–65.
[35] Mikami, T., Kamiya, H., and Horio, M. (1998) Numerical simulation of cohesive powder behavior in a fluidized bed. *Chemical Engineering Science*, **53**, 1927–1940.
[36] Takeuchi, S., Wang, S., and Rhodes, M. (2004) Discrete element simulation of a flat-bottomed spouted bed in the 3D cylindrical coordinate system. *Chemical Engineering Science*, **59**(17), 3495–3504.
[37] Mansourpour, Z., Karimi, S., Zarghami, R., Mostoufi, N., and Sotudeh-Gharebagh, R. (2010) Insights in hydrodynamics of bubbling fluidized beds at elevated pressure by DEM–CFD approach. *Particuology*, **8**, 407–414.
[38] Tagami, N., Mujumdar, A., and Horio, M. (2009) DEM simulation of polydisperse systems of particles in a fluidized bed. *Particuology*, **7**, 9–18.
[39] Gibson, L.M., Gopalan, B., Pisupati, S.V., and Shadle, L.J. (2013) Image analysis measurements of particle coefficient of restitution for coal gasification applications. *Powder Technology*, **247**, 30–43.
[40] Kharaz, A.H., Gorham, D.A., and Salman, A.D. (2001) An experimental study of the elastic rebound of spheres. *Powder Technology*, **120**, 281–291.
[41] van Beek, M.C., Rindt, C.C.M., Wijers, J.G., and van Steenhoven, A.A. (2006) Rebound characteristics for 50-μm particles impacting a powdery deposit. *Powder Technology*, **165**(2), 53–64.
[42] Dong, H. and Moys, M.H. (2006) Experimental study of oblique impacts with initial spin. *Powder Technology*, **161**(1), 22–31.
[43] Lorenz, A., Tuozzolo, C., and Louge, M.Y. (1997) Measurements of impact properties of small, nearly spherical particles. *Experimental Mechanics*, **37**(3), 292–298.
[44] Luding, S., Rosato, A.D., and Dave, R.N. (2000) Measurement of collisional properties of spheres using high-speed video analysis. *Physical Review E*, **56**, 5717–5725.
[45] Maw, N., Barber, J.R., and Fawcett, J.N. (1976) The oblique impact of elastic spheres. *Wear*, **38**, 101–114.
[46] Tsuji, Y., Tanaka, T., and Ishida, T. (1992) Lagrangian numerical simulation of plug flow of cohesionless particles in a horizontal pipe. *Powder Technology*, **71**, 239–250.
[47] Di Maio, F.P. and Di Renzo, A. (2004) Analytical solution for the problem of frictional-elastic collisions of spherical particles using the linear model. *Chemical Engineering Science*, **59**, 3461–3475.
[48] Zheng, Q.J., Zhu, H.P., and Yu, A.B. (2012) Finite element analysis of the contact forces between a viscoelastic sphere and rigid plane. *Powder Technology*, **226**, 130–142.

[49] Norouzi, H.R., Mostoufi, N., and Sotudeh-Gharebagh, R. (2012) Effect of fines on segregation of binary mixtures in gas–solid fluidized beds. *Powder Technology*, **225**, 7–20.

[50] Schutyser, M.A., Briels, W.J., Rinzema, A., and Boom, R.M. (2003) Numerical simulation and PEPT measurements of a 3D conical helical-blade mixer: a high potential solids mixer for solid-state fermentation. *Biotechnology and Bioengineering*, **84**(1), 29–39.

[51] Zhao, X.L., Li, S.Q., Liu, G.Q., Song, Q., and Yao, Q. (2008) Flow patterns of solids in a two-dimensional spouted bed with draft plates: PIV measurement and DEM simulations. *Powder Technology*, **183**(1), 79–87.

[52] Zhang, Y., Jin, B., Zhong, W., Ren, B., and Xiao, R. (2010) DEM simulation of particle mixing in flat-bottom spout-fluid bed. *Chemical Engineering Research and Design*, **88**(5-6), 757–771.

[53] Wellmann, C., Lillie, C., and Wriggers, P. (2008) Comparison of the macroscopic behavior of granular materials modeled by different constitutive equations on the microscale. *Finite Elements in Analysis and Design*, **44**(5), 259–271.

[54] Hertz, H. (1882) Ueber die Berührung fester elastischer Körper (On contact between elastic bodies). *Journal für die reine und angewandte Mathematik*, **92**, 156–171.

[55] Mindlin, R.D. and Deresiewicz, H. (1953) Elastic spheres in contact under varying oblique forces. *Journal of Applied Mechanics*, **20**, 327–344.

[56] Vu-Quoc, L. and Zhang, X. (1999) An accurate and efficient tangential force-displacement model for elastic frictional contact in particle-flow simulations. *Mechanics of Materials*, **31**, 235–269.

[57] Thornton, C. and Yin, K.K. (1991) Impact of elastic spheres with and without adhesion. *Powder Technology*, **65**, 153–166.

[58] Langston, P.A., Tuzun, U., and Heyes, D.M. (1994) Continuous potential discrete particle simulations of stress and velocity-fields in hoppers-transition from fluid to granular flow. *Chemical Engineering Science*, **49**, 1259–1275.

[59] Zhou, Y.C., Wright, B.D., Yang, R.Y., Xu, B.H., and Yu, A.B. (1999) Rolling friction in the dynamic simulation of sandpile formation. *Physica A*, **269**, 536–553.

[60] Zhu, H.P. and Yu, A.B. (2003) The effects of wall and rolling resistance on the couple stress of granular materials in vertical flow. *Physica A*, **325**(3–4), 347–360.

[61] Thornton, C., Cummins, S.J., and Cleary, P.W. (2013) An investigation of the comparative behaviour of alternative contact force models during inelastic collisions. *Powder Technology*, **233**, 30–46.

[62] Vu-Quoc, L. and Zhang, X. (1999) An elastoplastic contact force-displacement model in the normal force–displacement model in the normal direction: displacement-driven version. *Proceedings of the Mathematical Physical & Engineering Science*, **455**, 4013–4044.

[63] Timoshenko, S.P. and Goodier, J.N. (1970) *Theory of Elasticity*, McGraw-Hill, New York.

[64] Hunt, K.H. and Grossley, F.R.E. (1975) Coefficient of restitution interpreted as damping in vibro impact. *ASME Journal of Applied Mechanics*, **7**, 440–445.

[65] Kuwabara, G. and Kono, K. (1987) Restitution coefficient in collision between two spheres. *Japanese Journal of Applied Physics*, **26**, 1230–1233.

[66] Brilliantov, N.V., Spahn, F., Hertzsch, J.-M., and Pöschel, T. (1996) Model for collisions in granular gases. *Physical Review E*, **53**(5), 5382–5392.

[67] Malone, K.F. and Xu, B.H. (2008) Determination of contact parameters for discrete element method simulation of granular systems. *Particuology*, **6**, 521–528.

[68] Langston, P.A., Tüzün, U., and Heyes, D.M. (1995) Discrete element simulation of granular flow in 2D and 3D hoppers: dependence of discharge rate and wall stress on particle interactions. *Chemical Engineering Science*, **50**, 967–987.

[69] Feng, Y.Q., Xu, B.H., Zhang, S.J., Yu, A.B., and Zulli, P. (2004) Discrete particle simulation of gas fluidization of particle mixtures. *AIChE Journal*, **50**(8), 1713–1728.

[70] Dintwa, E., Van Zeebroeck, M., Tijskens, E., and Ramon, H. (2005) Determination of parameters of a tangential contact force model for viscoelastic spheroids (fruits) using a rheometer device. *Biosystems Engineering*, **91**(3), 321–327.

[71] Davies, R.M. (1949) The determination of static and dynamic yield stresses using a steel ball. *Proceedings of the Royal Society of London A: Mathematical, Physical and Engineering Sciences*, **197**(1050), 416–432.

[72] Sadd, M.H., Tai, Q.M., and Shukla, A. (1993) A contact law effects on wave-propagation in particulate materials using distinct element modeling. *International Journal of Non-Linear Mechanics*, **28**, 251–265.

[73] Thornton, C. (1997) Coefficient of restitution for collinear collisions of elastic perfectly plastic spheres. *Journal of Applied Mechanics*, **64**, 383–386.

[74] Walton, O.R. and Braun, R.L. (1986) Viscosity, granular temperature and stress calculations for shearing assemblies of inelastic frictional disks. *Journal of Rheology*, **30**, 949–980.

[75] Wu, C.Y., Thornton, C., and Li, L.Y. (2003) Coefficients of restitution for elastoplastic oblique impacts. *Advanced Powder Technology*, **14**(4), 435–448.

[76] Thornton, C. and Randall, C.W. (1988) Applications of theoretical contact mechanics to solid particle system simulation. In *Micromechanics of Granular Materials* (eds M. Satake and J.T. Jenkins), Elsevier, Amsterdam.

[77] Brilliantov, N.V. and Poeschel, T. (1999) Rolling as a “continuing collision”. *European Physics Journal B*, **12**, 299–301.

[78] Brilliantov, N.V. and Poschel, T. (1998) Rolling friction of a viscous sphere on a hard plane. *Europhysics Letters*, **42**(5), 511–516.

[79] Iwashita, K. and Oda, M. (1998) Rolling resistance at contacts in simulation of shear band development by DEM. *ASCE Journal of Engineering Mechanics*, **124**, 285-292.

[80] Jiang, M.J., Yu, H.S., and Harris, D. (2005) A novel discrete model for granular material incorporating rolling resistance. *Computers and Geotechnics*, **32**, 340–357.

[81] Ai, J., Chen, J., Rotter, J.M., and Ooi, J.Y. (2011) Assessment of rolling resistance models in discrete element simulations. *Powder Technology*, **206**, 269–282.

3

DEM Implementation

Developing a general purpose discrete element method (DEM) code is the ultimate goal of scientists and engineers dealing with granular flow simulations in which thousands to millions of distinct particles interact with each other through physical and non-physical contacts. Such a code should include robust solvers to handle complex geometries of walls, internals, and particles, as well as any type of linear and non-linear interaction. At the same times, it should efficiently utilize hardware resources to make simulating these systems with a large number of particles possible for building comprehensive knowledge and skills. Therefore, numerical implementation of DEM formulations is crucial for successful generation of new tools to achieve this goal.

In this chapter, the computational view of DEM is first addressed following with the structure of the accompanied numerical code. The chapter then brings in practical implementation in which important details of the code with more focus on contact search algorithms and integration methods as well as wall implantation are discussed. Furthermore, we have intentionally compared various search algorithms and integration methods with the aim of choosing efficient techniques in future simulations. In order to help the reader to discover the implementation of algorithms, some code lines are given in this chapter to ease understanding of the techniques presented. The chapter ends with emphasis on serial to parallel calculations applied to DEM.

3.1 Computational View

Computational efficiency of a DEM code is crucial. In the soft-sphere simulation, dynamic characteristics of the system are numerically estimated by an iterative time-integration of equations of motion for each individual particle. Typical simulations involve thousands to

Coupled CFD-DEM Modeling: Formulation, Implementation and Application to Multiphase Flows, First Edition.
Hamid Reza Norouzi, Reza Zarghami, Rahmat Sotudeh-Gharebagh and Navid Mostoufi.

Companion website: www.wiley.com/go/norouzi/CFD-DEM

millions of distinct particles that interact with each other through physical and non-physical contact. The number of these interactions is of the order of number of particles in the system. An efficient contact search algorithm should be implemented to find pairs of particles that interact with each other. We discussed different models for evaluating contact force interactions in Chapter 2. It is also necessary to save the contact history data[1] of all contacting pairs in memory. This history is then used to calculate normal and tangential contact forces. Other interaction forces between particles are also calculated. The state of the system in each time step is determined through calculating linear and rotational velocities as well as position of particles. To this end, accurate methods should be used to integrate equations of motion (Newton's and the Euler's second laws of motion) for each particle. Thus, it is clear that a large amount of computations should be performed in each time step of the simulation.

Developing a general purpose DEM code for granular flow requires a framework for the explicit time integration of equations of motion. This framework requires us to choose a small enough time step for which a disturbance wave propagates a distance not farther than the distance between centers of contacting particles. The critical time step based on the Rayleigh analysis for particles with density ρ, shear modulus G, Poisson's ratio ν, and radius R_p is obtained from [1]:

$$\Delta t_{crit} = \frac{\pi R_p}{\chi}\sqrt{\frac{\rho}{G}} \tag{3.1}$$

in which χ is approximated to be:

$$\chi = 0.1631\nu + 0.8766 \tag{3.2}$$

It is obvious that a fraction of the critical time step should be used as the time step in order to have a stable time integration.

The same analysis can be performed for a single degree-of-freedom elastic mass-spring system with mass m and spring stiffness k. The undamped natural frequency of this mass-spring system is [2]:

$$f_{nat} = \frac{1}{2\pi}\sqrt{\frac{k}{m}} \tag{3.3}$$

It is common to use angular velocity, ω_{nat}, instead of the natural frequency due to its simplicity, when solving equations analytically. This velocity is related to the natural frequency by $\omega_{nat} = 2\pi f_{nat}$. The time step in an explicit integration should be a fraction of the natural period (inverse of the natural frequency) of the system to ensure that each contact is processed in several time steps and the disturbance wave is not propagated to a distance farther than the particles in contact. The critical time step, therefore, becomes proportional to the natural period of the system:

$$\Delta t_{crit} \propto 2\pi\sqrt{\frac{m}{k}} \tag{3.4}$$

[1] Tangential overlap is the least necessary contact parameter that should be saved for each contact pair. However, some contact force models need more parameters to be saved as the contact history (see Chapter 2).

In discrete element simulations, a fraction of natural period of the mass-spring system is chosen as the time step for the numerical integration. So far, various time step fractions have been suggested in the literature, ranging from 0.1 to $1/\pi$ [2–5]. In a more precise analysis, the minimum integration time step also depends on the integration accuracy and packing conditions [2].

Based on these Equations 3.1–3.4 and material properties of the granular material, the time step for numerical integration generally falls in the range of 10^{-7}–10^{-5} s. Now consider that we want to simulate a case for 100 s real time. With this short time step, the program needs to repeat these explained computations millions of times. Thus, it is vital for a DEM code to be computationally efficient and this requires special attention. In fact, calculations should be performed as fast as possible with a high accuracy for feasible simulations. Ideally, the computation time of each time step should be linearly proportional to the number of particles. Enhancing the computational efficiency of a DEM code, therefore, becomes more pronounced when it is supposed to simulate granular and multiphase flows containing a large number of particles. Although it may be argued that the growing progress in the computational resources can overcome this problem, a large scale simulation still needs a lot of computational load and its execution still takes several hours per 1 s of real time. Moreover, having an efficient code, we can simulate granular flows with normal computers and at lower computational costs.

A number of methods have been utilized to reduce the computational load of a DEM code and to make this method viable for simulation of large scale systems. We cover the following methods in this chapter:

- *Efficient contact search algorithms*: a sequence of operations should be carried out to detect all interacting body pairs, either physical or non-physical. The contact search is performed frequently in the dynamic discrete element simulations, either every time step or each several time steps. The goal here is to implement a contact search algorithm in which the CPU time varies linearly with the number of bodies. CPU[2] time and RAM[3] requirements should be insensitive to the packing condition and size of the system. We present different contact search algorithms in Section 3.3.
- *High order integration methods*: utilizing high order and stable integrating methods permits us to use longer time steps and to prevent instability in a dynamic simulation. Importantly, the high order numerical integration also reduces artificial dissipation or generation of kinetic energy due to numerical errors. We cover different numerical integration methods in Section 3.4.
- *Selecting small values for spring stiffness*: we can use a smaller value for the spring stiffness than its real value in some occasions. This allows us to select a larger time step and to accelerate the simulation. However, this is not always a valid choice as to be discussed in Section 3.5.
- *Parallel programing*: parallelization means to breakdown the whole problem into a set of smaller parts and solving them simultaneously. Depending on the architecture of the computational unit(s), different types of parallelizing can be utilized to speed up the calculations. We give some general guidelines on common parallelization methods and how to convert a serial code to a parallel code in Section 3.7.

---

[2] Central processing unit.

[3] Random access memory.

3.2 Program Structure[4]

In this part we review the main steps that should be performed for each iteration in a discrete element code. Figure 3.1 shows these steps for a process containing non-cohesive spherical particles without heat/mass transfer and fluid inertial effects. The first step is the simulation setup in which particle data, geometry, and properties of particles and walls are read into the program. Then, all necessary components for a DEM simulation are initialized. For particle data, it is required to define all initial position vectors, initial velocity vectors, diameter, and type/property type of particles. The type of particle is an integer number that determines the physical property associated with that particle. By using the property type, we can simulate a mixture of particles with size distribution and different physical properties.

The shape of container and moving parts should be defined in the program. Any complex shape should be decomposed into simple triangles or plane quadrangles (called wall elements) and read into the program. For example, to define a cube, we need to decompose the cube into six square planes as its faces. In some cases, the whole container is moving (translational or rotational) or some of its parts are moving (like blades in a mixer). To simulate these, it is possible to define a translational velocity vector and a rotational velocity around an arbitrary axis for those moving parts. The geometry/wall definition and how it is implemented will be discussed in Section 3.6.

Physical property of particles and walls should be supplied to the program. These properties are then used for calculating contact forces between colliding particles and between particle and wall. We must specify physical properties of particles and walls, like density, Young's modulus, shear modulus, and Poisson's ratio (density is not required for the wall). In addition, interaction properties for particle-particle and particle-wall contacts should be provided. They are dynamic friction factor, rolling friction factor, normal coefficient of restitution, and tangential coefficient of restitution.

After importing/initializing all necessary data into the program, program initializes all the necessary components for performing the discrete element simulation. These components are declared in type `DEMSystem` in file *`g_DEMSystem.f90`*.[5]

- *Particle data*: program allocates enough memory to store the ID,[6] property type, position and diameter, translational velocity, rotational velocity, total contact force on particle, total torque on particle, linear, rotational accelerations, and so on, for all particles. These vectors only store the corresponding data of the current time and all data for previous time step are overwritten in each time step.
- *Geometry object*: stores data of all wall elements (triangle or plane quadrangle). If moving wall components exist, their translational or rotational velocities are set here. In addition, it provides some methods to check if a particle is in contact with wall. Type `Geometry` and its associated methods are defined in file *`g_Geometry.f90`* and will be discussed in detail in Section 3.6.
- *Property object*: stores all physical properties associated with particles and walls as well as their interaction parameters. Type `PhysicalProperty` and its associated methods are defined in file *`g_Prtcl_Property.f90`*.

[4] The related files of code are located in: http://www.wiley.com/go/norouzi/CFD-DEM

[5] We only bring those lines that are essential and help to understand algorithms better.

[6] In this book and the code, ID stands for the identification of particle that starts from 1 for the first particle and ends at N for the last particle.

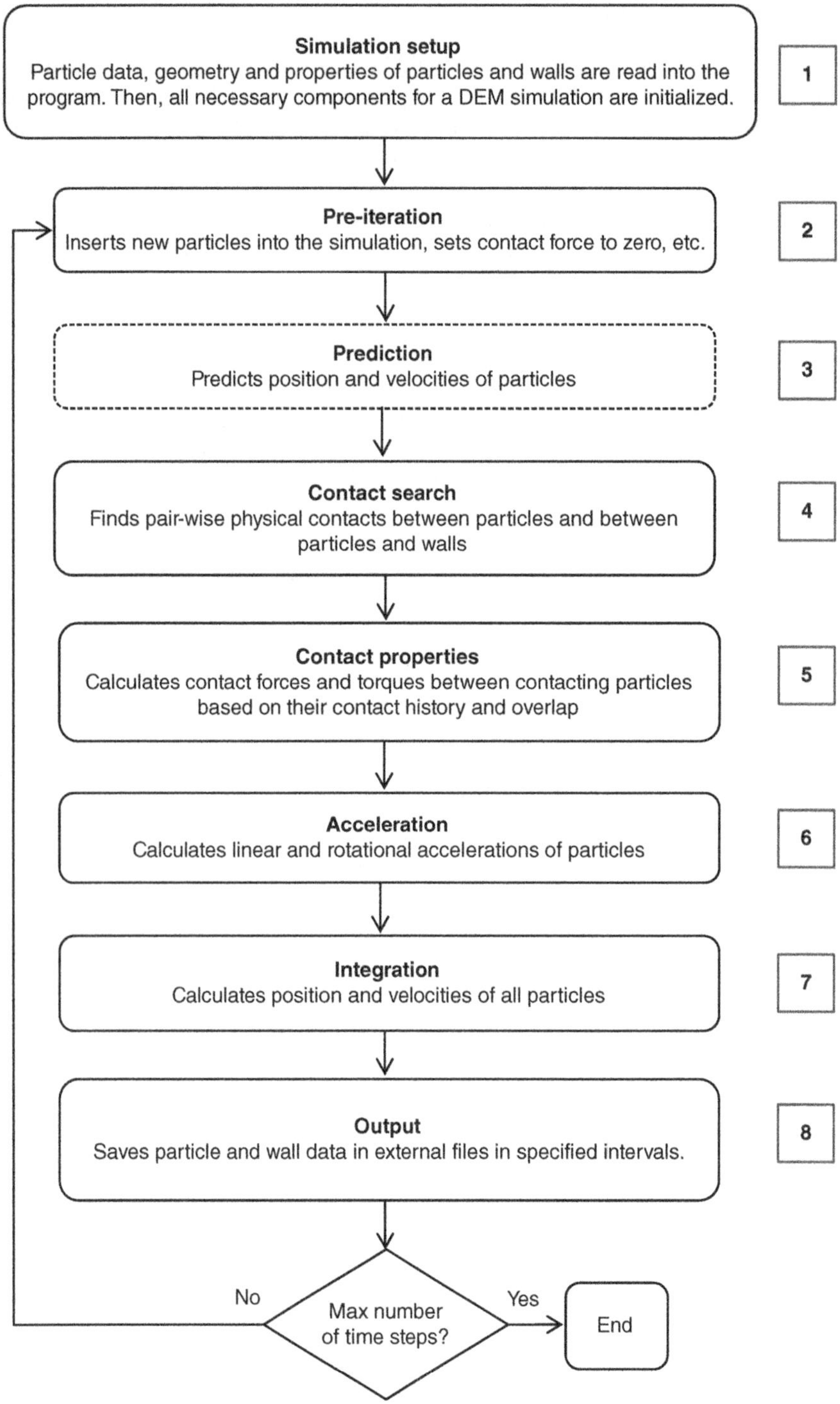

Figure 3.1 Algorithm of the DEM program for the flow of non-cohesive spherical particles without heat or mass transfer and fluid inertial effects

- *Contact search objects*: particle–particle or particle–wall contact search objects are responsible for managing the required memory (allocating and de-allocating the memory) and carrying out operations for the contact search between particles, and between particles and walls, respectively. A number of broad contact search algorithms are implemented in the code and will be discussed in detail in Section 3.3. The definition of type `ContactSearch` is in file *`g_ContactSearch.f90`* and is provided for performing particle-particle contact search. Definition of type `ContactSearchPW` is in file *`g_ContactSearchPW.f90`*, which is provided for performing particle-wall contact search.
- *Contact list objects*: all contact pairs (particles or particle and wall in contact) are found after the contact search step. Each pair and its associated data should be saved in a list (container) to be used for calculating the contact force and torque between colliding objects. Some methods are provided to add new pairs to the list, to find an existing pair and to delete a pair. As we showed in Chapter 2, the contact history (or possibly other necessary parameters) should be saved in the memory for each contact pair. These data are then used to calculate the contact force. For example, for the linear spring-dashpot (LSD) or the non-linear visco-elastic models, the tangential displacement between colliding pairs should be saved in each time step. The type `ContactList` is declared in file *`g_Prtcl_ContactList.f90`* to maintain and manage a list of contact pairs and their associated contact history data.
- *Contact force object*: this object calculates the contact force and torques for those contact pairs in the particle–particle and particle–wall contact lists. Two types are declared in the program to calculate contact force and torques. The first type is `LSD_ContactForce` declared in the file *`g_Prtcl_LSD_Model.f90`* and calculates the contact force based on the LSD model with limited or non-limited tangential displacement. The second type is `NonLin_ContactForce` declared in the file *`g_Prtcl_NonLin_Model.f90`* and calculates the contact force based on the non-linear viscoelastic force with limited or non-limited tangential displacement. Other contact force models can also be implemented in the code by extending the existing types and defining necessary methods to calculate contact force.
- *Integration objects*: type `prtcl_Integration` is declared in the file *`g_Prtcl_Integration.f90`* for integrating translational equation of motion for each particle. A number of integration schemes are implemented in the program and one of them should be selected by the user. We will discuss various integration schemes and their properties in Section 3.4. This object allocates additional memory, if the selected integration scheme requires additional data for performing integration. The type `prtcl_rot_Integration` is derived from type `prtcl_Integration` for the rotational equation of motion (spherical particle) and functions similarly.

After initializing all components of the discrete element simulation, the program lunches the particle iteration loop (steps 2–8 in Figure 3.1). Program Listing 3.1 illustrates the code for the main iteration loop. The iteration starts with pre-iteration operations. These operations may include inserting new particles into the simulation (if any particle should be inserted), setting contact forces, and torques to zero, updating the list of neighboring particles and list of walls, and so on (lines 12–14). The next step is the prediction step in which position and velocities (translational and rotation) of particles at the next time step are predicted using the history of position, velocity, and acceleration of particles (lines 17–20). The program performs this step if the integration scheme is one of the predictor-corrector schemes, otherwise, the program bypasses this step and the rest of operations are carried out based on non-predicted position and velocities.

The program performs particle-particle contact search according to one of the algorithms implemented in the code in lines 23–25 and finds particle–wall contacts in lines 28–30. After execution of these lines, two contact lists are obtained: particle–particle and particle–wall. Having these lists, the program calculates contact properties (contact force and torques) of particle–particle and particle–wall pairs using their contact history and overlap (lines 33–40). Linear and rotational accelerations of all particles are calculated in lines 43–45 according to total force and torque acting on particles. Using the calculated accelerations and based on the selected integration scheme, position, and velocities of all particles at the next time step are calculated using a proper integration formula (lines 48–51). The last step is to save particle and wall data at specified time intervals in external files as the output. Two sets of data files can be produced here, one set for wall data and another for particle data. The program can produce output files in two different formats that can be imported into visualizing packages for post processing. The first format is the vtk format that can be visualized in ParaView. This is a free software package under BSD license, which is available in UNIX and Windows operating systems[7] [6]. The second format is the plt format that can be visualized in Tecplot. This is a commercial software package which is available in UNIX and Windows operating systems.[8]

The program executes the iteration loop for a certain number of times specified by the user. Depending on number of particles and number of specified iterations, the execution of the program may take between several minutes to several days. The program also creates a log file for the simulation in which execution events, warnings, errors, and execution time of different parts of the program are recorded. It can be seen from Program Listing 3.1, every main step of the loop is enclosed by two subroutine calls used to record execution time of these steps.

In the program, we always deal with scalar and vector variables. For example, when describing the position or velocity of a particle, we need a vector of float type and for describing its id or property type, scalar of integer type. Most of the operations in the program are performed on these vector and scalar variables. It is therefore important to declare them and define necessary mathematical operations for them. Definitions of basic data types and valid operations applicable to them are found in file *g_TypeDef.f90*. For scalar variables, `real(RK)` and `integer(IK)` should be used for float and integer variable types, respectively. Since they are intrinsic data types in FORTRAN standard, all mathematical operations and type castings are available by default.

New types are declared for vector variables as illustrated in Program Listing 3.2. Type `real3` is used for representing variables like velocity, force, and any triple vector of type float; type `integer3` for a triple of integer variables as a vector like the cell index of particles and type `real4` for a set of four float variables as a vector (position and diameter). Various necessary operators are defined for these types, like (+) for summation, (−) for subtraction, (*) for multiplication, (/) for division, (=) for assignment and type casting, (`.dot.`) for dot product of two vectors, (`.cross.`) for cross product of two vectors, and so on. These operators are frequently used in the whole program.

When setting up a simulation case, you need to select among different available methods and algorithms in the program, that is, you need to select the contact search algorithm or integration scheme for solving translational and rotational equations of motion. The constants used for selecting among various algorithms are defined in file *g_Prtcl_DefaultValues.f90*.

---

[7] Visit www.paraview.org (accessed November 2015).

[8] Visit www.tecplot.com (accessed November 2015).

Program Listing 3.1 Definition of the iteration loop subroutine in the file g_DEMSystem.f90.

```
subroutine DEMS_iterate(this, numiter)
    implicit none
    class(DEMSystem) this
    integer(IK),intent(in):: numiter ! number of time steps

    . . . some code . . .

    do i=1,numIter

        call this%m_total_timer%start()

        ! pre-iteration adjustments
        call this%m_pre_iter_timer%start()
          call this%preIteration()
        call this%m_pre_iter_timer%finish()

        ! predicts position and velocity based on the previous time steps
        call this%m_prediction_timer%start()
          call this%Lin_Integration%predict()
          call this%rot_Integration%predict()
        call this%m_prediction_timer%finish()

        ! finds contacts between particels
        call this%m_ContSearchPP_timer%start()
          call this%Cont_Search%FindContacts()
        call this%m_ContSearchPP_timer%finish()

        ! finds contacts between particles and walls
        call this%m_ContSearchPW_timer%start()
          call this%Cont_Search_PW%FindContacts(this%numPrtcl ,
this%prtcl_dpos , this%prtcl_ids)
        call this%m_ContSearchPW_timer%finish()

        ! calculates contact force and torques between particles
        call this%m_ForcePP_timer%start()
          call this%m_cont_force%AllContactForce_PP()
        call this%m_ForcePP_timer%finish()

        ! calculates contact force and torques between particles and walls
        call this%m_ForcePW_timer%start()
          call this%m_cont_force%AllContactForce_PW()
        call this%m_ForcePW_timer%finish()

        ! calculates linear and angular accelerations
        call this%m_Acceleration_timer%start()
          call this%clc_Acceleration()
        call this%m_Acceleration_timer%finish()

        ! corrects position and velocities
        call this%m_integration_timer%start()
```

```
            call this%Lin_Integration%correct()
            call this%rot_Integration%correct()
        call this%m_integration_timer%finish()

        ! moves walls if a wall is moving
        call this%m_Geometry%move_walls(this%prtcl_dt)

        this%iterNumber = this%iterNumber + 1

        ! writes result to the output file (code is not presented here)
        . . . some code . . .

        call this%m_total_timer%finish()

        ! output to log file and terminal/command window (code is not presented here)
        . . . some code . . .

    end do

end subroutine
```

Program Listing 3.2 Declaration of basic vector variables in file g_TypeDef.f90.

```
. . . some code . . .
type real3
    real(RK)    x
    real(RK)    y
    real(RK)    z
end type

type integer3
    integer(IK) x
    integer(IK) y
    integer(IK) z
end type

type real4
    real(RK)    x
    real(RK)    y
    real(RK)    z
    real(RK)    w
end type
. . . some code . . .
```

3.3 Contact Search Algorithms

DEM simulations involve a large number of particles in physical and/or non-physical contact. A physical contact between two particles occurs when their surfaces touch and they interact with each other through their collision area. A non-physical contact happens when particles interact with each other through long rang forces and their separation distance is less than a threshold. We refer to these two types of interaction as contact in this chapter. Finding all contacts is necessary for successful simulations of granular materials. It is not possible to calculate interaction force and torques in each time step without ample information on pair-wise contact interactions.

Consider a system containing N discrete particles. A simple scheme to find contact between particles, called all particle pairs, is to test all possible pair-wise distances between particles. If their distance is less than a threshold, they are in contact (either physical or non-physical). Thus, $N(N-1)/2$ tests should be performed to obtain a final contact list that contains a list of pairs of contacting particles. Implementation of this scheme is very easy. A loop should be performed over all particles and each particle would be tested against the remaining particles through another loop. Based on this method, number of contact searches for a system comprising 100 particles would be 4950. This number increases to 499,500 for a system of 1000 particles. It is obvious that the computational cost (CPU time) to perform such a contact test dramatically increases with the number of particles in the system.

We can apply the "all-particle-pairs" method in the small to medium scale, static, or quasi-static DEM simulations in which the relative positions of bodies are not or rarely changed. In this way, the program constructs the contact list only once or few times during the whole simulation. Thus, the computation time of constructing the contact list becomes a small portion of the total time of the simulation. In a dynamic DEM simulation, bodies move significantly (as in most granular and multiphase flows) and the contact list should be updated frequently, even in every time step in many cases. Therefore, this method is not feasible for dynamic simulations, even for small scale simulations, and we should implement a more efficient method for contact detection.

Various contact detection algorithms have been developed to reduce the contact detection time of dynamic discrete element simulations. A contact detection algorithm is a collection of operations and methods that should be done step-by-step to find all contacting particles and to prevent redundant contact tests between particles which are far from each other. A contact detection algorithm must be robust, CPU and RAM efficient, and easy to implement [7]. A robust algorithm should detect all existing contacts. CPU efficiency corresponds to the total computation time for detecting all contacts that must be as short as possible. The computation time of a contact detection algorithm is expressed in terms of N since the number of operations required for completing the contact detection phase is related to the number of particles. In the worst situation, an algorithm may take $O(N^2)$ operations (quadratic), like the all particle pairs method. In the best case, it may take $O(N)$ operations (linear), which is the best known performance today. Alternatively, it may take $O(N\log(N))$ operations (logarithmic), which is between these two limits. Obviously, the linear variation of contact detection time with the number of particles is a goal.

The RAM efficiency is related to the total memory required for allocating variables used in all steps of the contact detection algorithm. This memory requirement must be as small as possible. The RAM requirement should be proportional to the number of particles in an ideal algorithm, neither to the size of simulation domain nor to the packing density. In large scale simulations comprised of hundreds of thousands or millions of particles, the memory allocated for the contact detection algorithm may be a limiting factor for two possible reasons. First, the size of physical memory available for calculations may not be large enough for allocating sufficient space for all variables in the contact detection algorithm. Second, the time

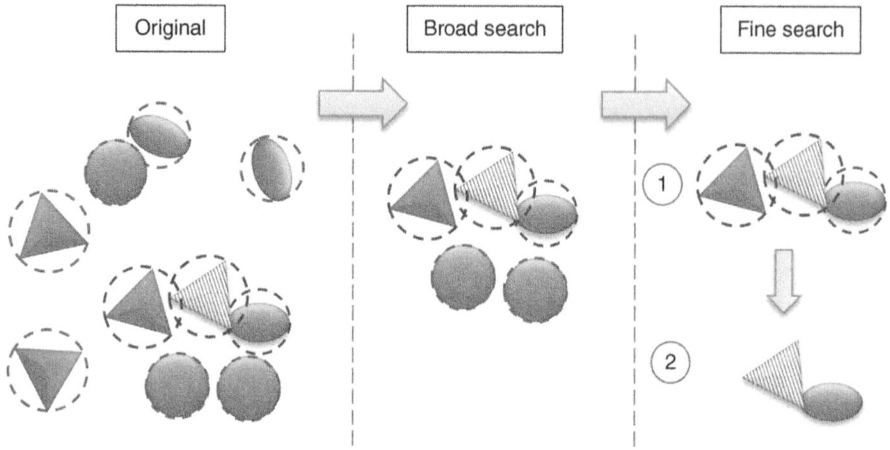

Figure 3.2 Different steps in a contact detection algorithm for finding contacts between a target particle (cross-hatched body) and other particles in the system. Left-hand section shows all particles with bounding spheres. Middle section shows the process of broad search that identifies all particles near the target particle. Right-hand section shows the fine search phase. First, the particles whose bonding spheres touch each other are determined. Then, geometric resolution is performed to find the exact contact point

required to access various pieces of a large allocated memory in different addresses becomes long due to limited band width of the memory and frequent cache misses. This degrades the overall performance of the algorithm, although the number of operations is not changed. Therefore, a memory management should be considered to reduce the allocated memory and to maintain coalesced memory accesses.

A contact detection algorithm can be divided into two separate phases, broad (neighbor) search and fine (exact) search. In the broad search, a list of particles in the neighborhood of the target particle (those that may be in contact with that particle, called contact candidates) is established (the neighbor list). In the fine search, all contact candidates in the neighbor list are tested and pairs in contact are added to the final contact list. We use the concept of bounding sphere to generalize the broad search to all geometric shapes.

Figure 3.2 illustrates the concept of bounding sphere. Each particle, regardless of its shape, can be completely surrounded by a sphere. The bounding sphere is insensitive to the shape and orientation of the particle and is represented by a center point and radius. In this situation, the problem of broad search is to find the potential contacts between spheres with known coordinates and radius. The bounding sphere is constructed only once during the simulation for each particle. This ensures us that there is no computational overhead associated with the use of bounding spheres. Using bounding sphere instead of the real shape of particle in the broad search enables us to develop scalable and general codes which can handle particles with any arbitrary shape.

The fine search phase starts with testing the contact between bounding spheres. If the bounding spheres touch each other, particles with their exact shape would be then checked for contact (see Figure 3.2). The method of finding contact point and contact area (overlap) varies depending on the shape of each particle. Although real particles are usually non-spherical, the simplest and most commonly used shape for a particle is spherical. However, in some practical applications, which involve non-spherical bodies, assuming spherical shape for particles is not promising. The shape of a particle can be represented by an analytical equation like ellipsoids [8] and superellipsoids [9] (see Chapter 4). In the case of irregular shapes (mostly with sharp edges), the surface of particle can be covered by triangular or quadrilateral elements and the fine search would be performed between these elements. The computational load of fine search depends on the way we represent the shape of bodies. In general, complex shapes require more calculations to determine the overlap, contact area, and normal vector of the contact plane.

We only cover the broad search in this chapter, since the fine search for spherical particles is very straightforward. Broad search algorithms presented here are categorized into three groups: cell-based, sort-based, and tree-based algorithms. Many broad search algorithms have been developed, each suitable for a specific condition. For example, some are sensitive to the size variation, some suit both large and small domains and some work well for dense packing conditions. We tried to describe algorithms that are simple and applicable to different conditions. These algorithms are robust enough to cover all intended applications of this book and can provide a basic knowledge to understand complex contact search algorithms. In general, the broad search imposes the largest computation time to DEM simulations (between 35 and 80% of the total computation time[9]). It can therefore be a computational bottleneck. The majority of this section is dedicated to broad search algorithms and the fine search is discussed at the end.

3.3.1 *Definition of Problem*

The aim of the broad search is to find pairs of spheres with a known radius and position that may have an overlap. Spheres are bounding boxes around each body with identical sizes or a size distribution (see Figure 3.3). Some algorithms suit mono-sized spheres, but we consider spheres with a size distribution to generalize our implementation. All spheres are contained in a rectangular box called simulation domain. Each sphere in the system has a unique id number[10] that is used to identify spheres in the code. Obviously, the id number of the first body is 1, that of the second body is 2, and so on. Two points $(x_{min}, y_{min}, z_{min})$ and $(x_{max}, y_{max}, z_{max})$ determine lower and upper corners of the simulation domain. All entities of the simulation, like particles and walls, should be inside this domain. Thus, the following constraints should be always held during the simulation:

$$\begin{cases} x_{min} < x_i < x_{max} \\ y_{min} < y_i < y_{max} \\ z_{min} < z_i < z_{max} \end{cases} \tag{3.5}$$

where x_i, y_i, and z_i represent the Cartesian coordinates of the center of sphere i.

[9] This range has been obtained based on our own experience when simulating granular flows at various operating conditions.

[10] Identification number.

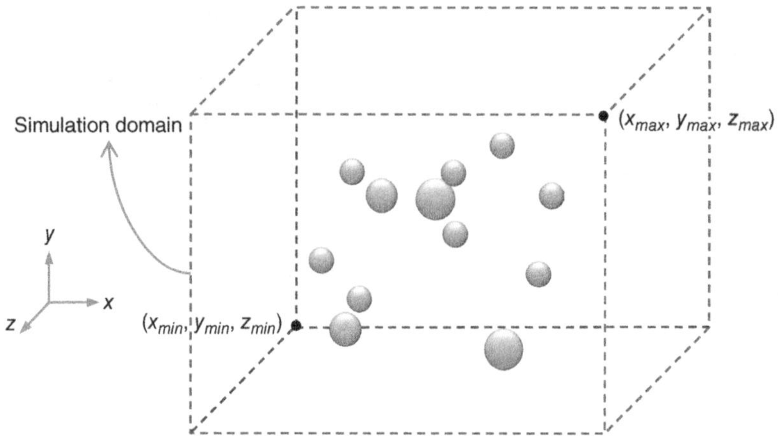

Figure 3.3 Bounding spheres in 3D Cartesian coordinates enclosed by simulation domain

The broad search phase generates an initial list of sphere pairs that may have contact (the neighbor list). This list must contain all contacting pairs. The final contact list is then obtained by performing fine search for all pairs in the initial list. Accuracy is defined as the ratio of number of pairs in the final contact list to that in the initial list. The end goal of a broad search algorithm is to use minimum computational resources for obtaining an initial list whose number of pairs is very close to the number of pairs in the final contact list (high accuracy).

As mentioned earlier, the order of the computation cost of the broad search varies between N and N^2. However, computational cost of the fine search is not necessarily proportional to N and mostly depends on the packing condition of the system and the geometric complexity of particles. In the case of dense packing, in which particles are close to each other, the number of candidate contacts can be of the order of N while for a loose packing this can be reduced to a fraction of N. Moreover, complex shapes need more operations to find their contact point and overlap. A broad search algorithm with high accuracy avoids many redundant geometric resolution tests between non-contacting bodies. The importance of choosing an accurate algorithm is more highlighted when dealing with particles with complex shapes.

3.3.2 Cell-Based Algorithms

In the cell-based algorithms, the simulation domain is discretized into cube cells (square cells in a 2D space) with the same size of dx. Real coordinates of spheres are converted into integer coordinates (ix, iy, iz) by the following equations[11]:

$$ix = \text{int}\left(\frac{x_i - x_{min}}{dx}\right) + 1 \qquad (3.6)$$

[11] Array index in FORTRAN starts from 1 (although this is modifiable) and array index in standard C and C++ starts from zero. We add 1 to the particle integer coordinate to be consistent with the FOTRAN convention.

$$iy = \text{int}\left(\frac{y_i - y_{min}}{dx}\right) + 1 \tag{3.7}$$

$$iz = \text{int}\left(\frac{z_i - z_{min}}{dx}\right) + 1 \tag{3.8}$$

After integerization, spheres are mapped onto the cells according to their integer coordinates. If we choose the size of the cube (dx) as large as the diameter of the largest sphere in the system, we can ensure that each sphere can have contact with spheres in the same cell and adjacent cells[12] (neighboring cells). In this way, we localize the contact detection test of the target sphere to the spheres in these cells. This reduces the number of fine contact searches and increases the accuracy of the contact search algorithm. Integerization of sphere centroids is done in all cell-based algorithms presented in this chapter. The differences between these algorithms are in the way that spheres are mapped onto the cells and the way a particle is accessed in a cell.

Program Listing 3.3 illustrates the code for integerizing the coordinates of spheres. Program gets input data (coordinates and diameter) of sphere i in line 13 and calculates its integer coordinates in lines 14–16. The condition in line 19 checks whether this sphere is in the simulation domain. The program returns integer coordinates of sphere if it is in the simulation domain (line 27) and deletes it if it is out of the simulation domain (line 23). Deletion of sphere/particle is performed by setting the flag of particle to `Pflg_deleted`.

Program Listing 3.3 Code for integerizing the coordinates of spheres (bounding boxes) defined in file g_Prtcl_CellBased.f90.

```
01  subroutine Cell_boxIndex(this)
02      implicit none
03      class(CellBased):: this
04
05      . . . some code . . .
06
07      ! loops over all particles
08      do n = 1, nPrtcl
09
10          ! performs the calculations for particles in simulation domain only
11          if( this%bndg_flag( this%getMemIndx(n) ) >= Pflg_inDomain ) then
12
13              BndgBox = this%getBndgBox_i(n)
14              ind%x = (BndgBox%x - minDomain%x) / this%dx + 1
15              ind%y = (BndgBox%y - minDomain%y) / this%dx + 1
16              ind%z = (BndgBox%z - minDomain%z) / this%dx + 1
17
18              ! checks if the particle is within simulation domain boundaries
19              if( ind%x <= 0_IK .or. ind%x >= this%nx+1 .or. &
20                  ind%y <= 0_IK .or. ind%y >= this%ny+1 .or. &
```

[12] Some cell-based algorithms allow using a cell size smaller than the diameter of spheres. For example, refer to the CGRID algorithm [10] or cell-based bounding box algorithm [11].

```
                ind%z <= 0_IK .or. ind%z >= this%nz+1 ) then
                ! if not, deletes it
                this%bndg_flag( this%getMemIndx(n) ) = Pflg_deleted

            else
                ! if so, returns index
                this%box_index(n) = ind

            end if

        end if

    end do

end subroutine
```

3.3.2.1 NBS Contact Search Algorithm

The name no binary search (NBS) has been given to this algorithm because it detects all contacts only once. This algorithm consists of two main steps: mapping spheres onto the cells and the broad contact search. In the first step, all spheres are mapped onto the cube cells according to their integer coordinates. For this purpose, a three-dimensional array of integers, called the *Head array*, can be used whose elements represent cells with indices (ix, iy, iz). Obviously, the size of the Head array in each dimension should be (nx, ny, nz) to provide enough space in the memory for all cells. There may be many empty cells which can be represented by a -1 in the corresponding element of the Head array. Otherwise, the value of the Head array element shows the id of the sphere that it contains. There are occasions in which more than one sphere resides in a cell. In this case, the Head array cannot store ids of all spheres in a cell and an additional Next array is required to store the ids of the rest of spheres in that cell. The Next array is a one-dimensional array of integers of size N. Head and Next arrays together create a linked list that can be used to map spheres onto the cube cells.

Consider the particulate system shown in Figure 3.4 in 2D space. The simulation domain is discretized into cells with equal size ($id = 1$). Each cell may be empty or contain one or more than one sphere. To this point, the integer coordinates of all spheres have been calculated. The procedure to construct the linked list is illustrated in Program Listing 3.4. Line 3 nullifies the Head array by setting each element of `HeadList` to -1. Then, a loop over all spheres (those in the simulation domain) starts and the linked list is constructed in lines 12 and 13. The corresponding linked list is illustrated in Figure 3.5. Since it is hard to show the particulate system and the constructed linked list in 3D space, we used 2D space for these illustrations.

In the second step, the initial contact list is constructed. The program performs contact tests between spheres in the target cell and those in the adjacent cells (neighboring cells). The target cell has eight neighbors in 2D space and 26 neighbors in 3D space. If the program performs the contact test for all neighboring cells, each contact would be detected twice.

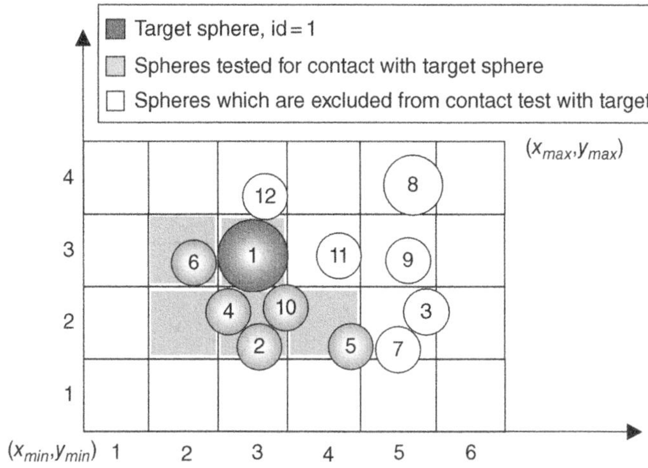

Figure 3.4 Discretized simulation domain and bounding sphere in 2D space

Program Listing 3.4 The loop for creating the linked list in the NBS algorithm in file g_Prtcl_NBS.f90.

```
. . . some code . . .

this%HeadList = -1

do n = 1, nPrtcl

    ! for particles which are in the simulation domain
    if( this%bndg_flag( this%getMemIndx(n) ) >= Pflg_inDomain ) then
        ! gets integer coordinates of sphere
        ind = this%getIndex(n)
        ! pushes the box number into the head of the list
        this%NextList(n) = this%HeadList(ind%x,ind%y,ind%z)
        this%HeadList(ind%x,ind%y,ind%z) = n
    end if

end do
```

This is due to the fact that all cells are considered as the target cells in the broad search. To avoid identifying a contact twice, the NBS mask is applied on the neighboring cells. Contact tests should be performed between the target cell and the neighboring cells determined by the NBS mask. The NBS masks for 2D and 3D spaces are illustrated in Figure 3.6. According to the NSB, contact tests should be done between the target cell and 4 neighboring cells in 2D space and 13 neighboring cells in 3D space.

$$\text{Head} = \begin{bmatrix} -1 & -1 & -1 & -1 & -1 & -1 \\ -1 & -1 & 10 & 5 & 7 & -1 \\ -1 & 6 & 1 & 11 & 9 & -1 \\ -1 & -1 & 12 & -1 & 8 & -1 \end{bmatrix}$$

Array index	1	2	3	4	5	6	7	8	9	10	11	12
Next =	−1	−1	−1	2	−1	−1	3	−1	−1	4	−1	−1

Figure 3.5 The linked list constructed based on the NBS method for the particulate system shown in Figure 3.4

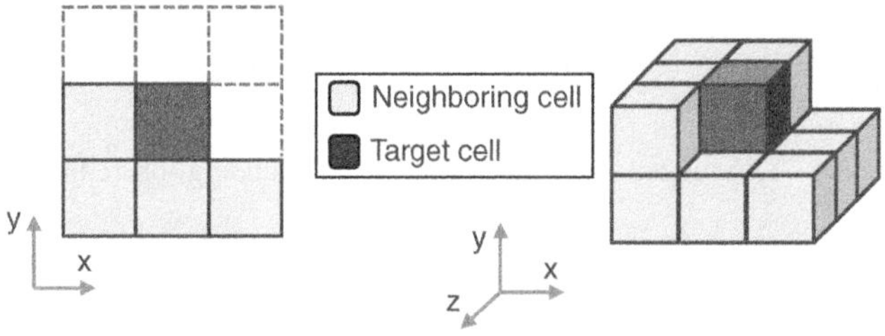

Figure 3.6 The NBS masks in 2D space (a) and 3D space (b)

If we consider $(0, 0)$ as coordinates of the target cell in 2D space, the NBS mask includes following cells:

$$(-1,0),(-1,-1),(0,-1),(1,-1) \tag{3.9}$$

Similarly, if we consider $(0, 0, 0)$ as the coordinates of the target cell in 3D space, the NBS mask includes following cells:

$$\begin{aligned}&(0,0,-1),(-1,0,\text{-}1),(-1,0,0),(-1,0,1),(0,-1,-1),(0,-1,0),(0,-1,1)\\&(-1,-1,-1),(-1,-1,0),(-1,-1,1),(1,-1,-1),(1,-1,0),(1,-1,1)\end{aligned} \tag{3.10}$$

We illustrate the algorithm for performing the broad search in Program Listing 3.5. Since the actual code lines are very lengthy, we do not present them here. Instead, we present the algorithm in a form very similar to the programming language. This method includes a loop over all spheres. Line 3 checks if the containing cell of particle i is newly considered as the target cell. This condition guarantees that each non-empty cell is considered as the target cell only once. In lines 4–11, each sphere in the target cell is checked for a possible contact with the remaining spheres in the target cell and with all neighboring cells determined by the NBS mask.

It is relatively easy to show that the computation time of this algorithm is proportional to N. To find the integer coordinates of spheres, a loop is performed over all spheres. It involves some read and write (R&W) operations (i.e., three elements x, y, and z and flag

Program Listing 3.5 Loop for performing broad search in the NBS algorithm in file g_Prtcl_NBS.f90.

```
loop over all spheres i = 1, N
    Get integer coordinates of sphere i
    if (Is this the first time that this cell is considered as target)
        loop over all spheres in target cell (sphere m)
            loop over remaining spheres in target cell ( sphere n)
                perform fine search for sphere pair m and n
            end loop
            loop over spheres in neighbor cells determined by NBS mask (n)
                perform fine search for sphere pair m and n
            end loop
        end loop
    end if
end loop
```

of particle), seven logical comparisons, six summations, and three divisions per sphere. Therefore, we have:

$$T_1 \propto (6\,summations + 3\,divisions + x\,R\,\&\,W + 7\,logics)N \tag{3.11}$$

In the next step, to construct the linked list, a loop is performed on all spheres which includes one read from memory to get integer coordinates (three elements *ix*, *iy*, and *iz*) and two substitutions (read and write to the memory) are performed per sphere. Thus, we have:

$$T_2 \propto (x\,reads + 2\,subs)N \tag{3.12}$$

The final step in the broad search phase is application of NBS mask to find potential contacts. In an ideal case, it is assumed that non-empty cells contain only one sphere in average. A loop is performed on each sphere and this sphere is checked against 13 neighboring cells (in 3D) for possible contacts. If we assume a similar computational load for each of these 13 contact checks, the required computation time is linearly related to N:

$$T_3 \propto 13N \tag{3.13}$$

Equations 3.11–3.13 show that the total CPU time of broad search is proportional to the number of spheres in the system.

As we discussed before, the program should allocate a 3D integer array with size of (nx, ny, nz) and a 1D integer array with the size of N. Therefore, the RAM requirement for mapping particles onto cells in this algorithm is:

$$RAM \propto (nx \times ny \times nz) + N \tag{3.14}$$

It shows that the RAM requirement is a cubic function of simulation domain size and a linear function of number of particles in the system. This cubic term becomes very large when the ratio of simulation domain length to the cell size increases. For a small system with dense packing, where the number of particles is comparable to the number of cells, this RAM requirement is not problematic. However, for a large system with loose packing, where the number of particles is small compared to the number of cells, the RAM allocation can be very high and much of the RAM space remains useless (−1 elements in Head and Next arrays in Figure 3.5). Memory latencies associated with scattered accesses to a large chunk of memory, as well as frequent cache misses, increases the computation time. Munjiza and Andrew [12] and Munjiza [7] addressed this problem by introducing singly connected linked lists along each axis. In the next section, we describe this method and refer to it as NBS–Munjiza.

3.3.2.2 NBS-Munjiza Contact Search Algorithm

The Munjiza algorithm is similar to the original NBS algorithm except the way particles are mapped onto cells. Thereby, the traversal of the linked lists and accessing spheres in cells are different. Instead of using a 3D Head array and a 1D Next array to construct a linked list, separate linked lists are constructed along each axis. Spheres are accessed in a particular cell through these linked lists. This algorithm has two main steps: mapping spheres onto cells and broad search.

Spheres are first mapped onto rows of cells (y-axis). This operation forms linked lists, called Y-lists. It contains ny linked lists and each list belongs to a row. The Y-lists are constructed by two one-dimensional integer arrays, called `HeadY` and `NextY` of sizes ny and N, respectively. We describe the main framework of this algorithm for 2D space in illustrations and describe its implementation in both 2D and 3D. Consider the particulate system shown in Figure 3.4 in 2D space. The constructed Y-lists for this system are shown in Figure 3.7. These are constructed by looping over all spheres. Each sphere is pushed into a row of Y-list according to its integer y-coordinate. Let's consider the third row of cells ($iy = 3$). Sphere 1 is first placed into this row, followed by spheres 6, 9, and 11. `HeadY` contains the last sphere placed into each row. For example, sphere 11 is the last sphere placed into the third row, thus, `HeadY(3) = 11`. For each sphere, `NextY` contains the id of the next sphere in the linked list. The number −1 in both arrays indicates termination of the linked list. For example, `HeadY(1) = -1` means that there is no sphere in the first row of cells (see Figure 3.4) and `NextY(1) = -1` means that sphere 1 is the last sphere in the linked list of row 3 (see Figure 3.7). Traversing the linked list of each row is simple. For example, consider the third row with `HeadY(3) = 11`. This means that the first sphere in this list is sphere 11 and id of the next sphere is stored in `NextY(11)`, which is 9. The id of the next sphere is stored in `NextY(9)`, which is 6. This procedure is continued until −1 is reached, which means that there is no extra sphere in the list (here, `NextY(1) = -1`).

Program Listing 3.6 shows the code for constructing Y-lists. First line nullifies `HeadY`. Lines 3–12 loop over all spheres to construct Y-Lists based on the integer y-coordinate of spheres. The condition in line 5 ensures that the sphere is in the simulation domain.

To this point, spheres are mapped onto rows. To access all spheres in the cell (ix, iy), it is necessary to construct linked lists of all spheres in the row iy. These lists are called X-lists and are constructed according to integer x-coordinate of spheres (ix) in a similar way to what was described for Y-lists. By constructing X-lists of row iy, we can access all spheres that

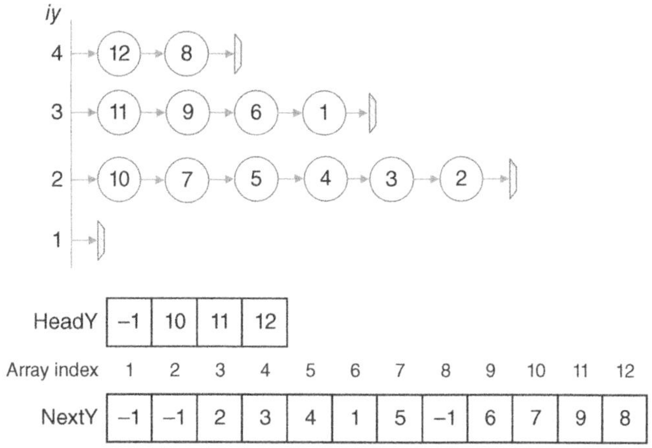

Figure 3.7 Constructed Y-lists of the particulate system of Figure 3.4

Program Listing 3.6 Loop for constructing Y-Lists in file g_Prtcl_NBS_Munjiza.f90.

```
. . . some code . . .
this%HeadY = -1

do n = 1, nPrtcl

    ! if sphere is in the simulation domain
    if( this%bndg_flag( this%getMemIndx(n) ) >= Pflg_inDomain ) then
        box = this%getIndex(n)   ! gets the integer coordinates of sphere
        iy  = box%y              ! integer y-coordinate
        this%NextY( n ) = this%HeadY(iy)
        this%HeadY( iy ) = n
    end if

end do
```

reside in cells of this row. X-lists comprise of one-dimensional arrays HeadX and NextX with sizes nx and N, respectively. Figure 3.8 shows X-lists of the second row of the particulate system shown in Figure 3.4.

To perform a broad search for a sphere in the target cell (ix, iy), it is necessary to have X-Lists of rows iy and $iy-1$ (see Figure 3.6). Therefore, there should be two X-Lists, one for the row iy and the other for the row $iy-1$. The algorithm of broad search is shown in Program Listing 3.7.

In these lines, `XLists0` belongs to X-lists of row $iy-1$ and `XLists` to X-lists of row iy. The program constructs Y-lists in line 1 and nullifies `XLists0` in line 2. A loop over all rows of the simulation domain is performed to start contact search. Obviously, the contact check should be carried out for non-empty rows, which is checked at line 6. If a row is non-empty, the program loops on each column of row iy and sets the cell (ix, iy) as the target cell. In lines 10–20, the NBS mask is used to check for contact between particles in both target and neighbor cells.

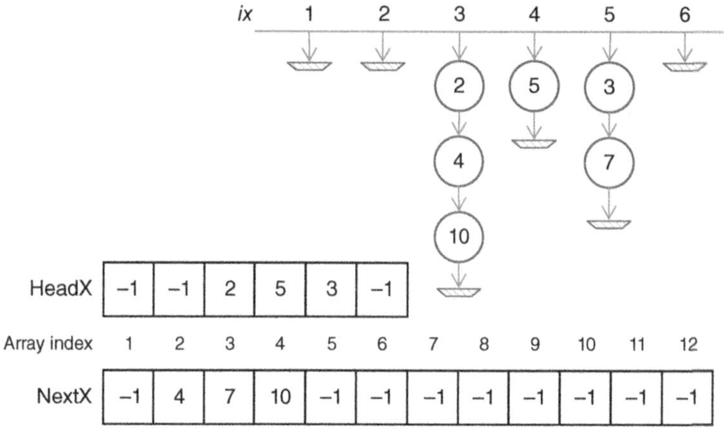

Figure 3.8 X-lists of the second row of the particulate system of Figure 3.4

Program Listing 3.7 Broad search loop in 2D space based on the NBS-Munjiza algorithm.

```
construct Ylists
nullify XLists0
loop over all rows iy = 1, ny
  construct the XLists of row iy

  if (row iy contains any sphere)

    loop over all columns ix = 1, nx

      loop over all spheres in target cell (ix,iy)

        loop over remaining spheres in target cell
          perform fine search for spheres pair
        end loop

        loop over all spheres in the adjacent cells determined by NBS mask
        perform fine search for spheres pair
        end loop

      end loop

    end loop

  end if

  XLists0 <= XLists
end loop
```

In the original implementation proposed by Munjiza and Andrews [12], a loop is performed over all spheres. The reason is to prevent considering an empty cell as the target cell and to reduce the calculation time. In our implementation, however, the loop is performed over all rows. The condition at line 6 assures that non-empty cells are not met. In this way, the program performs fewer loops (ny compared to N) and at the same time, non-empty cells are not met. This is beneficial from the programming point of view. The broad search starts from the first row and ends at the last row. For each row, the program only needs to construct the X-lists of row iy (the target row) and the X-lists of row $iy-1$ are obtained by executing the assignment statement in line 26.

In a 3D simulation, we need another set of linked lists, named Z-lists, which maps spheres on the z-axis. The Z-lists comprise of two one-dimensional arrays of integers HeadZ and NextZ with sizes nz and N, respectively. For locating a sphere in the cell (ix, iy, iz), first the Y-lists should be constructed based on the integer y-coordinate (iy) of spheres. Then, X-lists of layer iy should be constructed based on the integer x-coordinate (ix) of spheres. Finally, the Z-lists of column ix should be constructed from the constructed X-lists that contain all spheres in the ix^{th} column of layer iy. Now, we have access to all spheres in the cell (ix, iy, iz) through iz^{th} element of HeadZ. The NBS mask for 3D space is shown in Figure 3.6. To locate all spheres in the target cell and cells around the target cell, five sets of Z-lists should be constructed at the same time. Three of them belong to columns $ix-1$, ix, and $ix+1$ of the layer $iy-1$ and two of them to columns $ix-1$ and ix of the layer iy. Obviously, there should be two sets of X-lists, one for layer iy and one for layer $iy-1$. Moreover, one set of Y-lists is required for all spheres in the system. The implementation of the broad search step in 3D is somehow similar to that for 2D space, with the difference that it further maps spheres on the z-axis. The algorithm of the NBS–Munjiza contact search in 3D space is presented in Program Listing 3.8.

Program Listing 3.8 Broad search loop in 3D space based on the NBS-Munjiza algorithm.

```
construct Ylists
nullify XLists0

loop over all layers iy = 1, ny
  construct the XLists of layer iy

  if (layer iy contains any sphere)
    nullify Zlists0 and Zlists
    construct Zlists0 of column ix = 1

    loop over all columns ix = 1, nx
      construct Zlists of column ix
      construct Zlists0 of column ix+1

      if(column ix contains any sphere)

        loop over all spheres in target cell (ix,iy,iz)

          loop over remaining spheres in target cell
            perform fine search for spheres pair
          end loop
```

```

            loop over all spheres in the adjacent cells determined by NBS mask
            perform fine search for spheres pair
             end loop

           end loop

           Zlists of column ix-1  <= Zlists of column ix
           Zlists0 of column ix-1 <= Zlists0 of column ix
           Zlists0 of column ix   <= Zlists0 of column ix+1

         end if

       end loop

     end if

     XLists0 <= XLists
   end loop
```

Two vectors of length N (NextY and NextX), one vector with size of ny and two vectors with size of nx should be allocated for this algorithm in 2D implementation. Therefore, the RAM requirement of this algorithm is given by:

$$RAM_{2D} \propto 2N + ny + 2nx \tag{3.15}$$

Similarly, in a 3D space we have:

$$RAM_{3D} \propto 3N + ny + 2nx + 5nz \tag{3.16}$$

It can be seen in Equations 3.15 and 3.16 that the RAM requirement is a linear function of the number of particles and the simulation domain length. The CPU time is also a linear function of the number of bodies in the system. In fact, for constructing all linked lists, loops are performed over all spheres in the system. Also, for performing the broad search, all non-empty cells (hence, all spheres) are considered as the target sphere. Therefore, we can conclude that the CPU time is proportional to the number of bodies in the system.

The cell size in NBS and NBS–Munjiza algorithms should be at least as large as the largest sphere. Ideally, the average number of spheres in non-empty cells is close to one for mono-sized systems and a target sphere in the target cell is checked for possible contact against 13 spheres in the adjacent cells. Now, consider a system containing spheres of varying size. If we keep the size of the cell as large as the largest sphere, we can use these algorithms for contact search. However, the efficiency degrades significantly as the size distribution of spheres broadens. The average number of spheres in a non-empty cell increases when the size distribution of spheres becomes wider and each sphere in the target cell is checked for contact against far more than 13 spheres in adjacent cells. Most of these checks are between pairs which are not in contact, hence, we have many redundant contact checks. Consequently, the accuracy is decreased and the CPU time increases noticeably. To overcome this problem, Peters *et al.* [13] proposed a method that can work in combination with basic cell-based

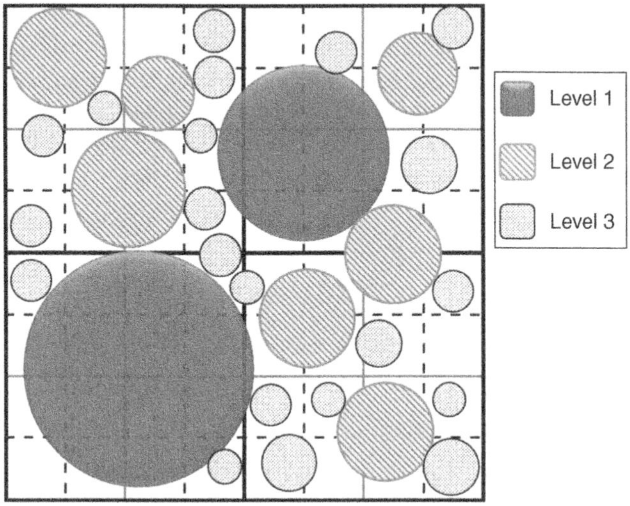

Figure 3.9 A hierarchical contact search with variation of sphere size. Three levels with different cell sizes are considered and spheres are assigned to each level based on their size

contact search algorithms. We describe this method and the way it can be combined with the basic contact search methods, like NBS and NBS–Munjiza, in the following section.

3.3.2.3 Hierarchical Contact Search Algorithms

The basic cell-based algorithms are robust, scalable to large simulations and, at the same time, easy to implement. The drawback of these algorithms is that thy use cells with a fixed size. This is a disadvantage when the system contains spheres with a wide size distribution. The cell size should be large enough to keep the largest sphere inside. However, at the same time more small spheres can be found in each cell and the average number of spheres in each non-empty cell increases. Peters *et al.* [13] proposed a hierarchical search algorithm for these systems. This algorithm can be used in conjunction with NBS or NBS–Munjiza algorithms to generalize them for multi-sized systems. Given a hierarchy containing L grid levels, each has cells with fixed size of l_i^{cell}. Grids are arranged in the decreasing cell size order, with the largest grid at level 1 and smallest at level L. Spheres are assigned to each level based on their size. Large spheres are assigned to grid level with large cell size and small spheres to grid level with small cell size. In this way, the problem of the contact search algorithm would be divided into several smaller contact search problems, in which the size of spheres does not change very much. The use of multiple grid levels requires different broad search steps.

Consider a system containing spheres with different sizes as shown in Figure 3.9. A particle may have many neighboring small particles (potential contacts), but few number of neighboring particles of the same size or larger. The hierarchal search algorithm takes the advantage of this fact to reduce redundant contact checks and to increase the accuracy of the basic contact search algorithm. The simulation domain is discretized into L levels of grid cells. The size of the first grid level, l_1^{cell}, is as large as the largest sphere. The next level is created using a cell size that is half of l_1^{cell}. Thus, each cell in level 1 contains four sub-cells of level 2 (eight sub-cells in a 3D space). Again, cells of level 2 can be further divided into four sub-cells and the third grid-level is constructed. Accordingly, each cell of level 1 contains 16 sub-cells of level 3. The number of constructed grid levels is optional and can be continued up to a level whose size is not smaller than the smallest sphere. For every grid level, a size range is defined and used for assigning spheres to each level. The size range for every level is:

$$\begin{cases} \left(l_2^{cell}, l_1^{cell}\right] & level1 \\ \left(l_3^{cell}, l_2^{cell}\right] & level2 \\ \left(l_4^{cell}, l_3^{cell}\right] & level3 \end{cases} \tag{3.17}$$

with

$$l_i^{cell} = \left(\frac{1}{2}\right)^{i-1} d_{max} \tag{3.18}$$

where d_{max} is the size of the largest sphere. We assign spheres to each grid level according to their size. The size of the spheres should fall within the range of that level.

Once L grid levels are constructed and spheres are assigned to them, we obtain L smaller contact search problems. These sub-problems are independent of each other and each has its identical set of spheres and cell sizes. Spheres of each level are mapped onto cells in the similar fashion described for basic contact search algorithms like NBS and NBS-Munjiza. The broad search of target spheres in level i involves two steps. In the first step, a broad search is performed for all spheres in level i. As mentioned in previous sections, the contact check is done between spheres in the target cell and all neighboring cells determined by the NBS mask. In the second step, cross-level contact searches are performed. Each sphere of the current level i is considered as the target sphere and the contact is checked between the target sphere and all surrounding larger spheres in lower levels (lower levels contain larger spheres). For example, to perform the cross-level search for level 3, contact tests are done between every sphere of level 3 and its neighboring spheres from levels 2 and 1. Figure 3.10 shows a system with two grid levels. The target sphere belongs to level 2 and we are supposed to find its larger neighboring spheres in level 1. It is almost straightforward to determine larger neighboring spheres of the target sphere in level i. We first find the cell of grid level 1 on which the target sphere is located and consider it as the cross-level target cell. Now, all spheres (larger spheres) in this cell and neighboring cells (shaded cells) are checked for contact with the target sphere. It should be noted that we do not apply the NBS mask here since contact checks are performed between small spheres and larger spheres in lower levels (levels with larger cells). This hinders repeated contact checks. The process of broad search can be further accelerated by excluding levels with no sphere from contact search.

Other researchers have used the concept of multi-level grids and have developed contact search algorithms for systems with size distribution [14–16]. Ogarko and Luding [16] combined multi-level grids with basic cell based algorithms. In their implementation, sizes of levels are independent of each other. They optimized the number of required levels and cell sizes for spheres with a wide size distribution. Lu *et al.* [15] combined classical octree contact

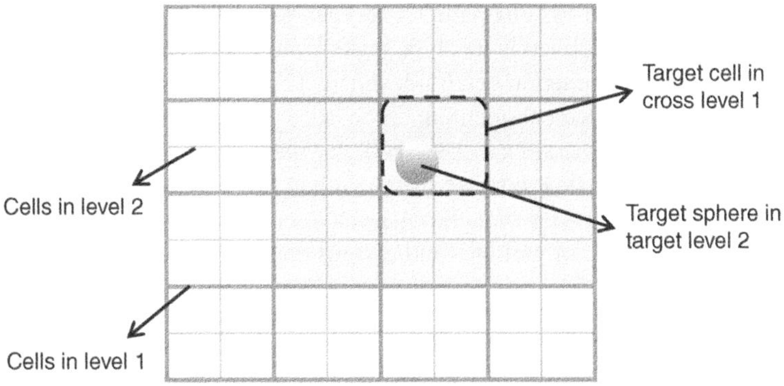

Figure 3.10 Method of performing cross-search between a target sphere of target level (smaller grid) and spheres of cross level (larger grid). The target sphere is mapped onto the cross-level grid and target cell in the cross level is determined. All larger spheres in this cell and its surrounding cells (nine shaded cells in total) are checked for possible contacts

search with multi-level classification of spheres based on their size. In their algorithm, called multi-octree, spheres are assigned to each level of octree based on their size. Their algorithm is faster than the original octree algorithm due to increased accuracy.

3.3.2.4 Code Description and Efficiency of Algorithms

The previously mentioned cell-based algorithms have been implemented in the program of this book. When setting up a simulation using this code, you can select the contact search algorithm from the following:

- *NBS algorithm*: Type `NBS`, defined in file *`g_Prtcl_NBS.f90`*, provides the methods for performing broad contact search based on the NBS algorithm. It manages allocation and de-allocation of the memory (Head and Next arrays) required for the contact search. For finding particle–particle contacts, the method `ContactSearch()` should be invoked. This algorithm can be selected using `CSM_NBS` keyword when setting up the simulation.
- *NBS-Munjiza algorithm*: Type `NBS_Munjiza` provides the functionality for performing contact search based on the NBS-Munjiza algorithm. It is defined in file *`g_Prtcl_NBS_Munjiza.f90`* and allocates/de-allocates the memory required for constructing one set of Y-lists, two sets of X-lists, and five sets of Z-lists. By invoking method `ContactSearch()`, the program finds particle–particle contacts. This method can be activated in the program using `CSM_NBS_Munjiza` keyword when setting up the simulation.
- *Hierarchical method combined with NBS*: This algorithm is an extension to the NBS algorithm. It decomposes the contact search problem into smaller sub-problems by creating multiple grids and assigning particles to these grids according to their size. Then, these sub-problems use the NBS algorithm for performing the contact search. Type `NBS_Hrchl` is declared in file *`g_Prtcl_Hrchl_NBS.f90`* and provides the functionality for performing the hierarchical contact search. When it is selected as the contact search algorithm in the program (using keyword `CSM_NBS_Hrchl`), it creates enough number of grid levels according to the

size range of particles and assigns particles to each gird based on their size. The method `ContactSearch()` is defined in this type to perform particle–particle contact search.

- *Hierarchical method combined with NBS-Munjiza*: This algorithm is an extension to NBS-Munjiza using the hierarchical method. Type `NBS_Munjiza_Hrchl` is declared in file *`g_Prtcl_Hrchl_Munjiza.f90`* and the method `ContactSearch()` in this type performs particle–particle contact search. When it is activated in the program (using the keyword `CSM_NBS_Munjiza_Hrchl`), it creates enough number of gird levels according to the size range of particles and assigns them to each grid based on their size.

It is very important to determine the efficiency of these algorithms. For this purpose, we carried out a set of simulations with 4-mm particles (mono-sized). Number of particles in these simulations was varied between 1000 and 50,000. In each simulation, 30,000 iterations were performed and the computation time of contact search stage and the memory requirement were recorded. For this set, we compared the performance of the NBS and NBS–Munjiza algorithms. Variations of computation time of contact search as a function of number of particles for both algorithms are illustrated in Figure 3.11a. It can be seen in this figure that the computation time of contact search increases linearly with the number of particles in both algorithms, which confirms that these algorithms consist of $O(N)$ operations. Under these simulation conditions, the computation time of the NBS–Munjiza method is around 15% lower than that of NBS method in all cases. As we discussed earlier, this is related to the overhead of scattered memory accesses and frequent cache misses in the NSB algorithm.

Figure 3.11b compares the memory requirement of both algorithms. This figure demonstrates that the NBS algorithm needs more memory than NBS–Munjiza. This increases the probability of cache misses since the CPU is trying to accesses a larger chunk of memory. It is worth noting that the memory requirement of both algorithms varies linearly with the number of particles, as previously shown in Equations 3.14 and 3.16.

To test the efficiency of the hierarchical algorithm, we performed a set of simulations with multi-sized particles. The size ratio (ratio of largest to smallest particle) was varied between

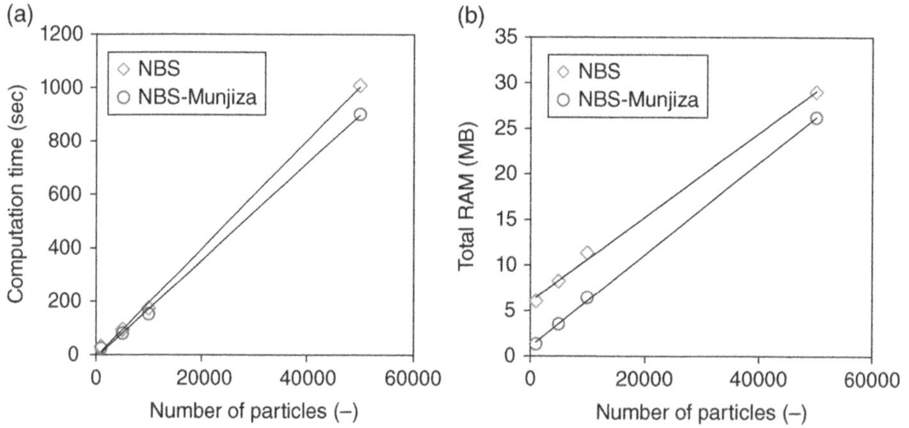

Figure 3.11 Variation of (a) computation time of contact search stage and (b) total RAM requirement for simulation versus number of particles for two different cell-based algorithms. Total number of iterations was 30,000

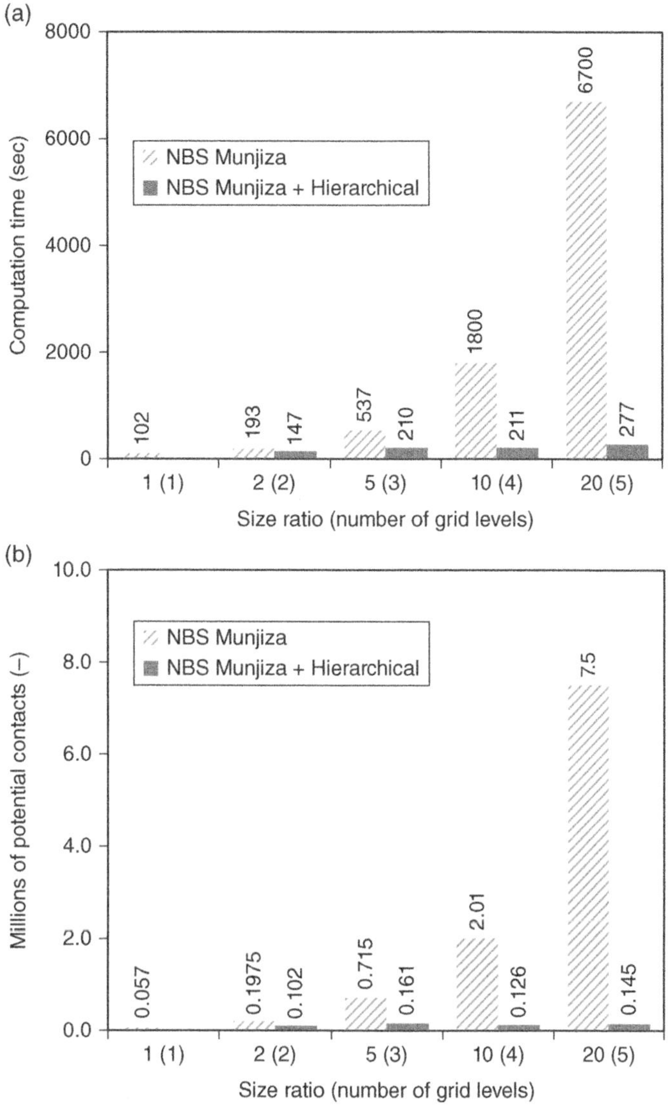

Figure 3.12 Variation of (a) computation time of contact search and (b) average potential contacts for 30,000 iterations of simulations containing 5000 particles versus the size ratio. Numbers in the parentheses show the number of grid levels used in each simulation

1 and 20 and the total number of particles was 5000. The size of smallest particle was 2 mm in all cases. In each simulation, 30,000 iterations were performed and the total computation time of contact search and number of potential contacts (initial contact list) were recorded. Figure 3.12a shows the computation time of NBS–Munjiza and the hierarchical algorithms versus the size ratio of particles. The computation time of NBS–Munjiza algorithm increases exponentially with the size ratio while the computation time of the hierarchical algorithm remains almost constant.

The numbers of potential contacts in both algorithms are compared in Figure 3.12b. This figure illustrates that the hierarchical algorithm maintains the number of potential contacts almost constant when the size ratio increases. However, in the NBS-Munjiza algorithm, only one grid level is used and the average number of particles per cell increases by increasing the size ratio, which results in a significant increase in the number of potential contacts. In fact, most of the computation time is consumed for redundant contact checks. The memory requirement for allocating additional grid levels is a small portion (around 10%) of the total memory requirement of the simulation. It can be concluded, therefore, that the hierarchical method, combined with basic contact search algorithms, can maintain the efficiency of the contact detection algorithm in a reasonable range when the size ratio of particles changes. It should be noted that the performance of this algorithm is affected by changing the shape of the size distribution. We did not study this effect here and leave it to the interested reader to check.

3.3.3 Sort-Based Algorithms

In cell-based algorithms described previously, the linked list concept was used to map spheres onto cells and to access them whenever needed. For example, in the NBS algorithm, Head and Next arrays are used for mapping particles onto cells and storing the identifiers of spheres in each cell. Mapping of particles can be performed by sorting arrays instead of linked lists [7]. For each coordinate axis, one integer array is used (X and Y for a 2D space and adding Z for a 3D space). These arrays are used to store the projected integer coordinates of center of spheres on each axis. The integer coordinates of spheres are calculated by Equations 3.6–3.8. A one-dimensional array, named the ID array, is also used to store identifiers of spheres. In total, 3N (4N in a 3D space) integer numbers should be allocated for this algorithm. Mapping is done through a two-stage sorting. In the first stage, X, Y, and ID arrays are sorted based on the integer x-coordinate of spheres. In the second stage, sorting is performed based on the integer y-coordinate of spheres on the segments of arrays with the same integer x-coordinate. In a 3D space, a further sorting stage should be performed based on z-coordinate of spheres on array segments with the same integer x- and y-coordinates. After the sorting phase, spatially close spheres can be found by applying a simultaneous search on X and Y arrays (X, Y, and Z arrays in 3D). In the broad search step, a loop over all spheres is performed. The neighboring spheres of a target sphere can be determined by a search operation (e.g., binary search algorithm) on the ordered data in the coordinate arrays.

Various sorting algorithms can be used here such as the quick sort algorithm which requires $O(N \log N)$ operations. The memory requirement of this contact search algorithm is low and does not depend on the system size (number of cells). The CPU time of this algorithm is proportional to the performance of sort and search algorithms. Therefore, the CPU time is proportional to $O(N \log N)$ and is insensitive to packing condition of the system. The performance of this algorithm degrades when there is a variation in size of spheres (typically the size of the ratio is larger than 2, as shown in Figure 3.12). This is due to the use of cells with fixed size, as it is the case for basic cell-based algorithms.

The problem of using grid cells can be remedied by eliminating cells and representing a sphere by its bounds in the space (four extremities in 2D space and six in 3D space). Double ended spatial sorting (DESS) algorithm proposed by Perkins and Williams [17] uses this technique. It is an efficient algorithm that is not sensitive to size of spheres. In this algorithm, sphere bounds are projected onto Cartesian axes. A bidirectional sorted list of start and end bounds is maintained for every coordinate axis. Contact search between a target sphere and its surrounding particles is performed using these sorted lists. This algorithm consists of two main steps:

1. Establishment of bidirectional sorted lists of spheres' bounds
2. Broad search over all spheres.

We describe these steps for 2D space and leave it to the reader to extend them to 3D space.

3.3.3.1 Bidirectional Mapping

Consider the particulate system with a size distribution as shown in Figure 3.13. Lower and upper bounds of each sphere are first projected on a given axis. For example, lower and upper bounds of sphere A are projected on the x-axis by:

$$LB_A = x_A - \frac{1}{2}d_A \tag{3.19a}$$

$$UB_A = x_A + \frac{1}{2}d_A \tag{3.19b}$$

Instead of allocating a unique id to each sphere (center of sphere), we assign a unique bound id to the bound of each sphere. For example, bound ids of lower and upper bounds of sphere A are identified by Al and Au, respectively. Thus, for each sphere, two pairs (Al, LB_A) and (Au, UB_A) should be calculated. Each pair contains the id of the sphere, whether it is a lower or upper bound of the sphere and its projected bound. The same procedure is followed for all spheres and a list of 2N extremities is obtained for each axis. The rank/order of each bound is obtained by sorting this list by projected bound values. For example, as shown in Figure 3.13, rank of

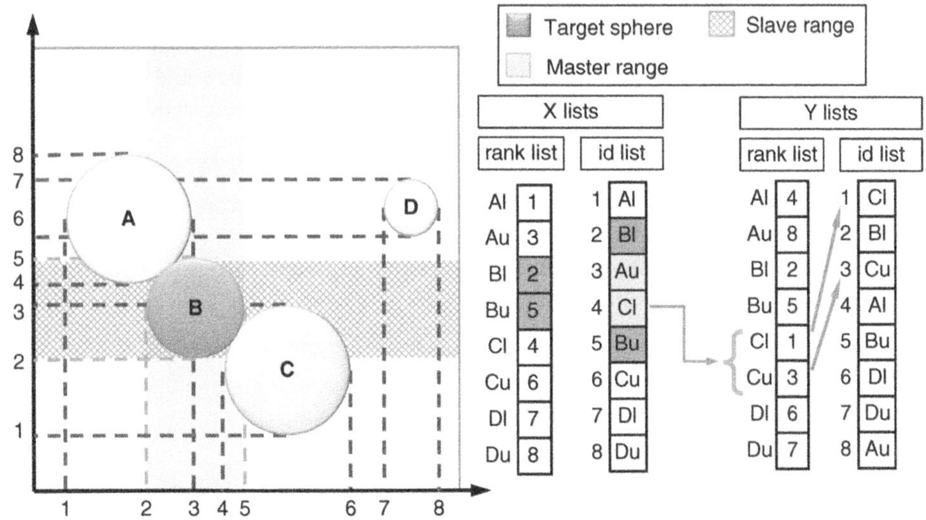

Figure 3.13 Schematics of the DESS algorithm in 2D space

the lower bound of sphere A is 1 on x-axis, while the rank of its upper bound is 3. To establish the bidirectional mapping between each sphere and its bound positions, two lists should be constructed: id list and rank list. The id list is constructed by sequentially putting bound ids of sorted bounds in a 1D array of size $2N$.

Using the id list, we can determine the order of object bounds on each axis by traversing this list. The rank list maps bound id to the sorted rank. For example, lower and upper ranks of sphere A can be easily obtained by checking the first two elements of the rank list. For generating the rank list, no sorting operation is required. This list is generated by traversing the id list element-by-element and updating the corresponding element in the rank list. Once the id and rank lists are constructed for each axis, the broad search can be performed. Before proceeding to this step, we must choose one axis as the master axis and another as the slave axis. In the illustrated example, the x-axis is chosen as the master axis and the y-axis as the slave axis.

3.3.3.2 Broad Search

A loop over all spheres is performed and each sphere is considered once as the target sphere in the broad search step. Then, a set of candidate spheres, which has a potential contact with the target sphere, is determined. Here, we describe the procedure of broad search for the target sphere B. The rank list on the x-axis (master axis) gives the lower and upper ranks of sphere B. Having lower and upper ranks of sphere B, we can go through the id list and determine the master range of sphere B. For example, the master range of sphere B is $(2,5)$. Bounds Au and Cl are in the master range. Objects with a lower bound in the master range should be considered for checking their intersection with the target sphere on the slave axis. Objects with upper bound are excluded from this check to avoid repeated contact checks. Therefore, only Cl should be checked for possible intersection with the target sphere in the slave axis. Lower and upper ranks of sphere C in the rank list of slave axis are $(1,3)$. If at least one of the bounds of the target sphere B lies within this range, we can conclude that the sphere C has a potential contact with the target sphere B. The lower bound of sphere B exists in the slave range of sphere C (in the id list of the slave axis). Thus, spheres B and C should be added to the initial contact list as the potential contacts.

One of the axes should be selected as the master axis before starting the broad search step. This selection depends on how the bounds are distributed on each axis. The cumulative range width of objects in the rank list should be calculated for each axis. The axis with the lowest cumulative range width is selected as the master axis. This would result in a better CPU performance.

Two lists should be constructed for every axis. Each list is a 1D array of integers with the length $2N$. In addition, a 1D array of real variables with the length of $2N$ is required for storing the projections of object extremities. Thus, the memory requirements of this algorithm in 2D and 3D spaces are obtained from[13]:

$$RAM_{2D} \propto 2\times(4N+2N)=12N \tag{3.20a}$$

$$RAM_{3D} \propto 3\times\left(4N+2N\right)=18N \tag{3.20b}$$

[13] These equations are presented based on the assumption that real variables are single-precision and are represented by 32 bits. It should be noted that 64 bits are required for double-precision real variables and the above equation would become slightly different.

These expressions show that the memory is scaled linearly with number of spheres in the system and is not sensitive to the system size.

CPU time of the DESS algorithm strongly depends on the performance of the sorting algorithm used in construction of the id list. For constructing this list on each axis, $2N$ pairs should be sorted. The sorting is performed on real values. Most sorting algorithms require $O(N \log N)$ operations, which makes the contact search phase to be of the same order. In the best case, if the sorting algorithm scales with N, the whole contact search phase would also scale with N. Some sort algorithms exhibit $O(N \log N)$ performances on average conditions and $O(N)$ performances when data are nearly sorted. Sorting algorithms, like block sort, comb sort, and binary tree sort, follow this behavior. We can use this property to obtain linear performance in the contact search. Once spheres are sorted based on their bound, ids of spheres can be exchanged based on their spatial location. Then, the bound pairs are nearly sorted in the next time step [18]. This rearrangement of ids should be performed every several time steps due to the fact that the movement of spheres in successive time steps is not significant.[14]

Two-dimensional numerical tests show that the performance of the DESS algorithm is not affected by variations in the spheres size or system size [17]. When the size ratio of spheres becomes greater than 2, this algorithm shows a better performance for small to medium scale problems when compared with the NBS–Munjiza algorithm. However, in a large scale simulation ($N > 10^5$) with moderate size distribution, the NBS–Munjiza performs better.

The sort-based algorithms are very efficient for shared-memory parallelization since they offer a non-locking contact search. In addition, efficient parallel algorithms exist for sorting a large amount of data [20–22]. Locks or atomic operations should be used in parallel implementation of many basic contact search algorithms, like cell-based algorithms, to prevent racing condition (we will show this in Section 3.7). These operations degrade computational efficiency of the contact search algorithm and should be minimized in the parallel implementation. As we showed before, in a sort-based contact search one or more sorting steps are performed in a sort-based contact search algorithm. For sorting, a parallel algorithm can be used to leverage the maximum computational power of all cores in a machine with multiprocessor architecture.

3.3.4 Tree-Based Broad Search Algorithms

The concept of a tree can be used for spatial partitioning of the simulation domain and positioning particles into the tree nodes for carrying out the contact search. Tree data structure is a collection of nodes that are distributed at different levels. Each parent node has a number of children, known as *child nodes*. Each child node has its own children, and so on. We want to use the binary tree (each parent node has two children) for spatial partitioning of the simulation domain. This algorithm is described in the following for 2D space for the sake of simplicity and it can be easily extended to 3D space.

[14] The same technique can be considered for memory locations. Objects and their associated data are rearranged in memory locations based on their spatial locations [19]. In this way, memory accesses are coalesced and a cache-friendly accessing of memory is provided.

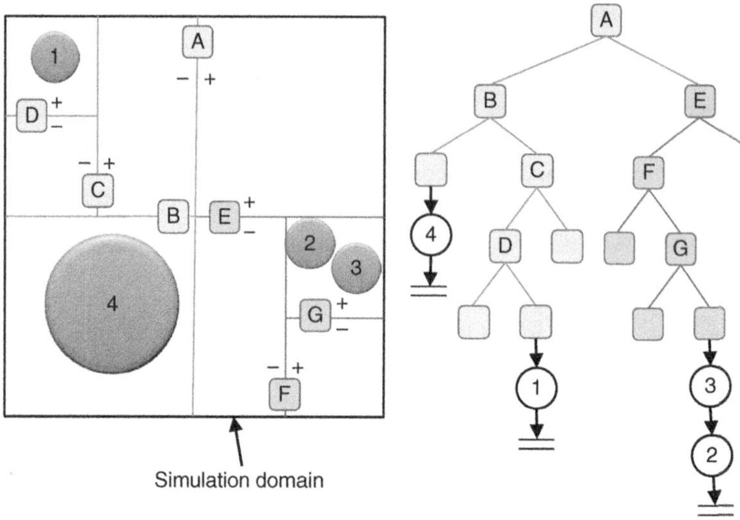

Figure 3.14 Partitioning of the simulation domain and positioning of particles into a binary tree

Figure 3.14 shows a particulate system with its constructed binary tree. Root node represents the smallest square that completely encloses all particles. This square is first partitioned into two subspaces by a partitioning line. These two partitions are two children of the node, called *negative* and *positive partitions*. The negative partition is assigned as the first child and the positive partition as the second child. Each child is recursively partitioned into two subspaces by the partitioning lines. This partitioning is continued up to a level at which the size of subspace becomes comparable with the size of particle. For example, let's see how to locate particle 1 in the tree. The tree has four levels. The root node is first divided by line *A* into two subspaces. The center of particle 1 is in the negative partition, so it belongs to the first child. This partition is further divided by line *B* and the particle is positioned in the positive partition (second child). The partitioning continues up to the fourth level at which the size of the particle becomes comparable to the size of the subspace. The corresponding node at the fourth level points to a linked list that holds id of particles in that node. Other particles are mapped onto the tree nodes in a similar manner. The number of required operations per particle for constructing a binary tree is proportional to $\log_2(N)$. Thus, for constructing the whole tree for all particles, $O(N\log_2(N))$ operations are required.

The binary tree can be created by pre-allocating a zero-based 1D array of nodes with the size of $2^n - 1$, where n is the number of levels. The root node is `bTree(0)` and its children are `bTree(1)` and `bTree(2)`. In general, for every parent node `bTree(i)`, there are two children `bTree(2*i+1)` and `bTree(2*i+2)`. If we create the tree in this way, we will have a complete tree (a tree with all nodes). Since there is a logical order between nodes (array indices), the particle can be directly inserted into the corresponding node once its node position is determined by the recursive partitioning. Therefore, it is not necessary to traverse the tree from the root node. Similarly, particles can be directly accessed by calculating the array index of node in `bTree` from the node position.

A linked list should be used to keep the ids of particles in each node. The linked list of particles at each node can be created by either dynamic linked lists or by a static linked list for all

	0	1	2	3	4		19	20	21	22	23	24	25	26	
bTree	−1	−1	−1	4	−1	....	−1	1	−1	−1	−1	−1	−1	3	...

	1	2	3	4
Next	−1	−1	2	−1

Figure 3.15 Binary tree and static linked list of the particulate system shown in Figure 3.14

particles. If a static linked list is used, the binary tree with n levels can be constructed by allocating a 1D array of integers (`bTree`) with the size $2^n - 1$ for the tree and a 1D array of integers (`Next`) with the size N for holding the linked list. All elements of `bTree` are initially −1, indicating that the tree contains no particle. Node value in the array `bTree` indicates the head of linked list and elements of the `Next` array represent the id of the next particle in the linked list. The list is terminated if a −1 is encountered. Binary tree and static linked list of the particulate system shown in Figure 3.14 are illustrated in Figure 3.15.

If we wish to use dynamic linked lists for holding particles in each node, the binary tree and the linked list should be created differently. For this purpose, the type `Object` should be defined as in Program Listing 3.9:

Program Listing 3.9 Definition of type Object for creating the dynamic linked list.

```
type Object
    integer(IK) id   ! id of the particle
    type(Object),pointer:: NextO  ! pointer to the next particle in the list
end type
```

The tree should be created by allocating a 1D array of `Object` with the size $2^n - 1$. In each node, `id` holds the identifier of the first particle in that node (−1 if the node is empty) and `NextO` points to the next particle, if more than one particle exists in the node.

Implementation of a tree as a complete tree has two problems:

- A complete binary tree with n levels requires $2^n - 1$ nodes. Therefore, a complete binary tree with, say 20 levels, requires 1 million nodes. This restricts the number of levels, that is, the ratio of the length of simulation domain to the length of the smallest subspace. For example, in a 3D simulation with a tree consisted of 25 levels, the length ratio becomes 256 and 33.5 million nodes are required. Therefore, this algorithm requires a large amount of memory for a large system with small particles.
- In many situations, some parts of the simulation domain are empty and the corresponding nodes in those parts are never used. Therefore, they are not needed to be created in the tree.

Using pointers instead of pre-allocated arrays enables us to create a dynamic tree. A dynamic binary tree can be updated dynamically, nodes can be added as needed or deleted when not required and number of levels can be changed during the run time. Using a dynamic tree for a dilute system is very beneficial, since a large amount of memory can be saved by not allocating memory for unused nodes. The type defined in program listing 3.10 should be used to create a dynamic tree:

Program Listing 3.10 Definition of type Object for creating dynamic linked list.

```
type Node
    type(Node),  pointer:: parent ! parent node (for bi-directional tree)
    type(Node),  pointer:: child1 ! first child node
    type(Node),  pointer:: child2 ! second child node
    type(Object),pointer:: Prtcls ! pointer to particles in the node
end type
```

Pointers `child1` and `child2` point to the children of the node and `Prtcls` to the first particle in this node. Pointer `parent` points to the parent node and is required for bi-directional traversing of the tree (top-down and down-top traversing).

After constructing the tree, particles are mapped on a hierarchy of nodes/subspaces with different sizes. Large particles reside on nodes in top levels and small particles on the nodes in bottom levels. This is very similar to the cell-based hierarchical contact search method after mapping the particles. The contact search is done by performing a loop over all particles. Node position of each particle is previously computed by recursive partitioning. At the same level, particles in the node are check against particles in the adjacent nodes. Thereafter, each particle is checked against particles in the adjacent nodes in the upper levels (larger particles). Each non-empty node should be processed once during the contact search.

For performing the contact search, the binary tree should be constructed according to the new position of particles at each time step. We can construct the tree first by deleting it and building it from the scratch, then inserting all particles in it. In this way, the tree should be traversed N times, which is not a very efficient method. An alternative approach is to update the node position of particle whose position is changed significantly. For this purpose, the node position of particles at the time of their insertion into the tree should be saved. In the subsequent iterations, if the new node position of the particle differs from the saved one, the particle should be deleted and re-inserted based on its new node position. In this way, most of the particles which have not moved considerably are excluded from being inserted into the tree.

The quadtree can be used instead of binary tree broad search in 2D space (octree in 3D space). In quadtree, each parent node has four children; thus, the space is divided into four equal subspaces in each partitioning stage (see Figure 3.16). Likewise, each subspace is partitioned into four subspaces. Quadtree of the particulate system of Figure 3.14 is illustrated in Figure 3.16. For a complete tree with n levels, the quadtree needs $(4^n - 1)/(4-1)$ nodes and the octree needs $(8^n - 1)/(8-1)$ nodes. Consider that the ratio of simulation domain length to smallest subspace is 256 in a 2D simulation. In this system, if using a complete binary tree, we need a binary tree with 17 levels that requires around 131,000 nodes; while for a complete quadtree, we need a tree with 9 levels that requires around 87,000 nodes. This shows that using quadtree instead of binary tree leads to a huge reduction in the memory requirement. In a 3D case with the same length ratio, even more memory would be saved. In the case of dynamic tree implementation, using quadtree instead of binary tree is still favorable since a dynamic quadtree still requires less memory than a dynamic binary tree.

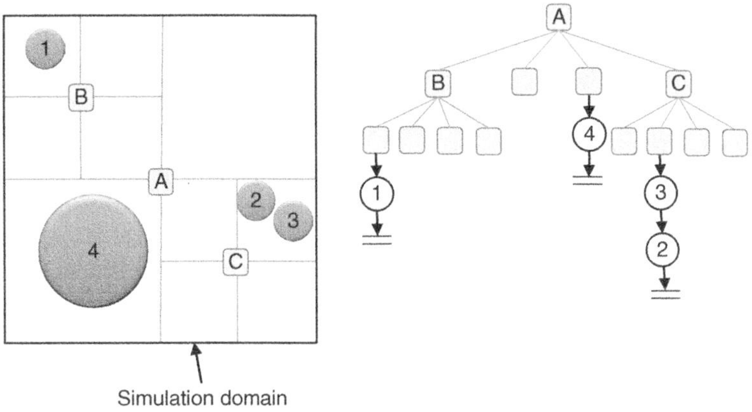

Figure 3.16 Quadtree of the particulate system shown in Figure 3.14

3.3.5 *Fine Search for Spherical Particles*

After the broad search phase, all pair-wise contact candidates (potential contact pairs) are obtained. A contact candidate is a pair of particles that may have an overlap between bounding spheres. The overlap of bounding spheres i and j is simply calculated as:

$$ovrlp = 0.5\left(d_i + d_j\right) - \left|\vec{x}_i - \vec{x}_j\right| \tag{3.21}$$

Bounding spheres with positive overlap are in contact. For spherical particles, the bounding sphere exactly fits the particle surface. Therefore, no further calculation is required for finding the exact overlap between spherical particles.

3.4 Integration Methods

In the framework of the DEM, a large set of coupled ordinary differential equations should be solved over time to obtain the dynamic behavior of the granular flow. The following equations describe translational and rotational motions for a spherical particle (motion with six degrees of freedom in three dimensions):

$$m_i \frac{d\vec{v}_i}{dt} = m_i \frac{d^2\vec{x}_i}{dt^2} = \sum_{j \in CL_i} \vec{f}_{ij}^{\,p-p} + \vec{f}_i^{\,f-p} + \vec{f}_i^{\,ext} \tag{3.22a}$$

$$I_i \frac{d\vec{\omega}_i}{dt} = I_i \frac{d^2\vec{\varphi}_i}{dt^2} = \sum_{j \in CL_i} \left(\vec{M}_{ij}^t + \vec{M}_{ij}^r\right) \tag{3.22b}$$

For a system containing N discrete particles in 3D space, $6N$ coupled ordinary differential equations should be solved. It is evident that a numerical solution is the only choice here. Various integration methods can be found in the literature and can be classified into implicit and explicit solvers. The choice between implicit and explicit solvers highly depends on the type and size of the problem.

In general, the implicit scheme involves construction of global stiffness matrix and solving a set of linear or non-linear equations (depending on the contact force model). Therefore, the RAM requirement and CPU time for performing such calculations in large scale simulations become critical. In addition, when the contact force model is changed in the simulation, the previously implemented numerical method may fail in solving the new set of non-linear equations [23]. In addition, inclusion of long-range forces, fluid-particle interaction forces is far from possible in this scheme. Therefore, application of an implicit integration scheme is limited to small-scale problems within the context of discontinuous deformation analysis.

In contrast, the explicit scheme is better suited for dynamic simulation of granular flows. In this scheme, due to small time step of integration, discontinuities associated with changes between contact and rebound states are less challenging. The contact force on a particle is obtained by evaluating the force existing between the particle and its surrounding colliding particles and the stiffness matrix is not required. In addition, for different contact force models and interaction forces, the same integration method can be utilized. Therefore, the explicit methods are the best choice for a general-purpose large-scale discrete element simulation.

In this section, we present a wide range of explicit time integration schemes. They are categorized into three different groups, namely, single-step, multi-step, and predictor-corrector methods. In general, higher order methods are more complex and computationally more expensive than lower order methods. There are two important motivations to select higher order methods over lower order ones. The first motivation is the accuracy of the integration method. Due to non-linearity of equations and sharp changes in the contact force (when particles come into contact and when they rebound), some integration methods cannot accurately compute the position and velocity, even though the time step is very small. This leads to dissipation or amplification of kinetic energy of particles due to numerical errors. The second motivation is related to the computation time of simulation. It has been shown that all explicit integration schemes are conditionally stable [24]. This means that the time step should be small enough to avoid instability in the dynamic solution of system.

In general, by increasing the order of integration method to a certain level (we will later see that increasing the order does not lead to selecting larger time steps, after a certain order), a larger time step can be selected. With adapting a larger time step, fewer numbers of iterations are required for 1 s of real time simulation. However, higher order methods are computationally more expensive than lower order ones since they use extra memory for saving velocity and acceleration history and sometimes require multiple evaluations of forces acting on particles. Therefore, the speed-up gain may be hindered by additional computational load associated with the integration method. We will discuss both accuracy and computation time of integration at the end of this section.

Table 3.1 Various explicit numerical integration methods used in discrete element simulations

Integration method	Abbreviations	Force evaluations per time step	Extra variables[a]	Accuracy order Position	Accuracy order Velocity
Single-step					
Forward Euler	FE	1	0	1	1
Modified Euler	ME	1	0	2	1
Taylor second order	TY2	1	0	2	1
Taylor third order	TY3	1	1	3	2
Taylor fourth order	TY4	1	2	4	3
Central difference	CD	1	0	2	2
Position Verlet	PV	1	0	2	2
Runge–Kutta fourth order	RK4	4	8	4	4
Multi-step					
Velocity Verlet	VE	1	1	3	2
Adams–Bashforth second order	AB2	1	2	2	2
Adams–Bashforth third order	AB3	1	4	3	3
Adams–Bashforth fourth order	AB4	1	6	4	4
Adams–Bashforth fifth order	AB5	1	8	5	5
Predictor-corrector					
Adams–Moulton third order	AB2AM3	1	5	3	3
Adams–Moulton fourth order	AB3AM4	1	7	4	4
Adams–Moulton fifth order	AB4AM5	1	9	5	5
Gear third order	Gear3	1	4	3	3
Gear fourth order	Gear4	1	5	4	4
Gear fifth order	Gear5	1	6	5	5

[a] Number of vector variables that should be saved in the memory in addition to position, velocity, and acceleration in the current time step.

We rearrange equations of motion (3.22) to the following form for integration:

$$\vec{a} = \frac{\mathrm{d}\vec{v}}{\mathrm{d}t} = \frac{\mathrm{d}^2\vec{x}}{\mathrm{d}t^2} = \frac{1}{m_i}\left(\sum_{j \in CL_i} \vec{f}_{ij}^{\,p-p} + \vec{f}_i^{\,f-p} + \vec{f}_i^{\,ext}\right) \tag{3.23a}$$

$$\frac{\mathrm{d}\vec{\omega}}{\mathrm{d}t} = \frac{\mathrm{d}^2\vec{\varphi}}{\mathrm{d}t^2} = \frac{1}{I_i}\left(\sum_{j \in C.L.} \left(\vec{M}_{ij}^{t} + \vec{M}_{ij}^{r}\right)\right) \tag{3.23b}$$

Note that we have dropped particle indices from these equations for legibility of integration expressions presented next. We also add the notation n to these variables as the counter of the time step. Accordingly, $\vec{v}_n$ represents the velocity vector of particle i at current time step and $\vec{v}_{n+1}$ at the next time step. Abbreviations, accuracy order, required number of force evaluations per time step, and number of extra vector variables that should be saved in the memory (in addition to position, velocity, and acceleration vectors) are summarized in Table 3.1 for the methods considered in this chapter.

3.4.1 Single-Step Methods

The first method introduced here is the *forward Euler* method (FE). In this method, position and velocity of particles in the next time step $(t+\Delta t_p)$ are computed from position and velocity in the current time step from:

$$\vec{v}_{n+1} = \vec{v}_n + \vec{a}_n \Delta t_p \tag{3.24a}$$

$$\vec{x}_{n+1} = \vec{x}_n + \vec{v}_n \Delta t_p \tag{3.24b}$$

where $\vec{a}_n$ is the acceleration of particle in the current time step and evaluated by positions and velocities of particle at the current time step (Equation 3.32a,b). Here, Δt_p is the time step of numerical integration for translational and rotational equations of motion of particles. Only the first two terms in the Taylor series expansion are used in this method. Therefore, the integration has the first order accuracy for both position and velocity.[15] This is a very common integration method [5, 25–30] due to its simplicity and no need to save additional variables in the memory.

Implementation of the FE method is illustrated in Program Listing 3.11. Vector variables `pos`, `vel`, and `acc` contain position, velocity, and acceleration of particle at the current time step before the subroutine is invoked. New position and velocity of particle are calculated and returned in lines 8 and 9.

It is possible to use the newly calculated velocity for calculating position in the next time step [31]:

$$\vec{v}_{n+1} = \vec{v}_n + \vec{a}_n \Delta t_p \tag{3.25a}$$

$$\vec{x}_{n+1} = \vec{x}_n + \vec{v}_{n+1} \Delta t_p \tag{3.25b}$$

Program Listing 3.11 Implementation of the forward Euler method in file g_Prtcl_OneStepIntegration.f90.

```
01  subroutine FE_Integrate(this , acc , dt , pos, vel )
02      implicit none
03      class(ForwardEuler) this
04      type(real3),intent(in)    :: acc    ! acceleration
05      real(RK),    intent(in)    :: dt     ! time step
06      type(real3),intent(inout):: pos, vel
07
08      pos = pos + vel * dt ! position equation
09      vel = vel + acc * dt ! velocity equation
10
11  end subroutine
```

[15] An integration method with accuracy of n has a local truncation error of $O(\Delta t_t^{n+1})$ and a global truncation error of $O(\Delta t_t^{n})$.

Program Listing 3.12 Implementation of the modified Euler method in file g_Prtcl_OneStepIntegration.f90.

```
01  subroutine BE_Integrate(this , acc , dt , pos, vel )
02      implicit none
03      class(ModifiedEuler) this
04      type(real3),intent(in)    :: acc   ! acceleration
05      real(RK),    intent(in)   :: dt    ! time step
06      type(real3),intent(inout):: pos, vel
07
08      vel = vel + acc * dt ! velocity equation
09      pos = pos + vel * dt ! position equation
10
11  end subroutine
```

Program Listing 3.13 Implementation of the TY2 method in file g_Prtcl_OneStepIntegration.f90.

```
01  subroutine T2_Integrate(this , acc , dt , pos, vel )
02      !  . . . as before
03
04      pos = pos + vel * dt + 0.5_RK * acc * dt * dt
05      vel = vel + acc * dt
06
07  end subroutine
```

This is called the *modified Euler* integration scheme (ME). This improves the accuracy of integration for calculating the position vector. Implementation of this method is illustrated in Program Listing 3.12. The velocity of particle at the next time step is first calculated in line 8. Then, in line 9, the position of particle at the next time step is calculated based on the new velocity of particle.

We can keep more terms of the Taylor series expansion for calculating position and velocity. In this way, a higher order of accuracy can be obtained. In the TY2 integration method (TY stands for Taylor and 2 for keeping the first two terms in the expansion), velocity is calculated using the first two terms in the Taylor series while three terms of the Taylor series are kept for the position [32]:

$$\vec{v}_{n+1} = \vec{v}_n + \vec{a}_n \Delta t_p \tag{3.26a}$$

$$\vec{x}_{n+1} = \vec{x}_n + \vec{v}_n \Delta t_p + \tfrac{1}{2} \vec{a}_n \Delta t_p^2 \tag{3.26b}$$

In this way, a second order accuracy is obtained in the position expression while the accuracy of the velocity expression remains of the first order. Application of this method is also straightforward and does not need additional variables to be saved in the memory. Program listing 3.13 illustrates the implementation of the TY2 method. In lines 4 and 5, the new position and velocity of particle are calculated based on Equations 3.26b and 3.26a, respectively.

Program Listing 3.14 Implementation of the TY3 method in file g_Prtcl_OneStepIntegration.f90.

```
01   subroutine T3_Integrate(this , acc , dt , pos, vel )
02       !  . . . as before
03       !// locals
04       type(real3) b
05
06       b = (acc-this%a0) !b = (acc-this%a0)/dt
07       pos = pos + vel * dt + 0.5_RK * acc* dt*dt + 0.1666667_RK*b*dt*dt
08       vel = vel + acc * dt + 0.5_RK * b * dt
09       this%a0 = acc
10
11   end subroutine
```

Higher order accuracies can be obtained by keeping more terms in the Taylor series expansion. In the TY3 method (TY for Taylor and 3 for keeping the first three terms in the expansion), the first derivative of the acceleration is used and position and velocity formulas are of the third and second order accuracy, respectively [33]:

$$\vec{v}_{n+1} = \vec{v}_n + \vec{a}_n \Delta t_p + \tfrac{1}{2} \vec{b}_n \Delta t_p^2 \tag{3.27a}$$

$$\vec{x}_{n+1} = \vec{x}_n + \vec{v}_n \Delta t_p + \tfrac{1}{2} \vec{a}_n \Delta t_p^2 + \tfrac{1}{6} \vec{b}_n \Delta t_p^3 \tag{3.27b}$$

$$\vec{b}_n = \mathrm{d}\vec{a}_n / \mathrm{d}t \tag{3.27c}$$

in which $\vec{b}_n$ is the first time derivative of acceleration. To calculate this value, the acceleration of particle at the previous time step should be saved in the memory. The code for this method is illustrated in Program Listing 3.14. At line 6, the first derivative of acceleration (backward) is calculated from current and previous time step accelerations. Then, position and velocity of particle are calculated in lines 7 and 8. Value of acceleration at the current time step is then saved in the memory at line 9 for being used in the next time step.

In TY4 method (TY for Taylor and 4 for keeping the first four terms in the expansion), the second time derivative of the acceleration in integration formulas for the position and velocity are used. This results in the fourth and third order accuracy for the position and velocity, respectively. These formulas are:

$$\vec{v}_{n+1} = \vec{v}_n + \vec{a}_n \Delta t_p + \tfrac{1}{2} \vec{b}_n \Delta t_p^2 + \tfrac{1}{6} \vec{c}_n \Delta t_p^3 \tag{3.28a}$$

$$x_{n+1} = x_n + v_n \Delta t_p + \tfrac{1}{2} a_n \Delta t_p^2 + \tfrac{1}{6} b_n \Delta t_p^3 + \tfrac{1}{24} c_n \Delta t_p^4 \tag{3.28b}$$

where $\vec{c}_n = \mathrm{d}\vec{b}_n / \mathrm{d}t$ is the second order derivative of the acceleration. In the TY4 method, $\vec{a}_{n-1}$ and $\vec{b}_{n-1}$, which are the acceleration and the first derivative of the acceleration, should be saved in the memory for calculating $\vec{b}_n$ and $\vec{c}_n$. The code for this method is illustrated in Program Listing 3.15. In lines 7 and 8, the first and second time derivatives of acceleration

Program Listing 3.15 Implementation of the TY4 method in file g_Prtcl_OneStepIntegration.f90.

```
01   subroutine T4_Integrate(this , acc , dt , pos, vel )
02       ! . . . as before
03
04       !// locals
05       type(real3) b, c
06
07       b = (acc-this%a0) !b = (acc-this%a0)/dt
08       c = (b - this%b0) !c = (b - this%b0)/dt
09       pos = pos + vel * dt + 0.5_RK * acc* dt*dt + 0.1666667_RK*b*dt*dt +
     0.04166666667_RK*c*dt**3.0_RK
10       vel = vel + acc * dt + 0.5_RK * b * dt + 0.1666667_RK*c*dt*dt
11       this%a0 = acc
12       this%b0 = b
13
14   end subroutine
```

are calculated and in line 9 and 10, the position and velocity of particle at the next time step are obtained. In lines 11 and 12, the acceleration and the first time derivative of acceleration are saved for being used in the next time step.

Another method, first used by Cundall and Strack [4], is the *central difference* method (CD). This method has been also applied by other researchers in their DEM simulations [2, 34, 35]. In the CD method, velocity is calculated at the middle of time step $(t+\frac{1}{2}\Delta t_p)$ while the new position is calculated at $t+\Delta t_p$. The forces acting on a particle depend on its position and velocity and must be calculated based on the values at the previous time step of velocity $(t-\frac{1}{2}\Delta t_p)$ and the current time step of position (t). The integration expressions read as:

$$\vec{v}_{n+1/2} = \vec{v}_{n-1/2} + \vec{a}\left(\vec{x}_n, \vec{v}_{n-1/2}\right)\Delta t_p \tag{3.29a}$$

$$\vec{x}_{n+1} = \vec{x}_n + \vec{v}_{n+1/2}\Delta t_p \tag{3.29b}$$

Another method, which is very similar to the central different method, is called the *position Verlet* integration method (PV) [24]. The integration expressions of this second order method are:

$$\vec{v}_{n+1} = \vec{v}_n + \vec{a}\left(x_{n-1/2}, v_n\right)\Delta t_p \tag{3.30a}$$

$$\vec{x}_{n+1/2} = \vec{x}_{n-1/2} + \vec{v}_n\Delta t_p \tag{3.30b}$$

It can be seen in these equations that the acceleration is calculated by the position of the particle at $t-\frac{1}{2}\Delta t_p$ and its current velocity. Then, the new position vector at $t-\frac{1}{2}\Delta t_p$ is calculated using its current velocity.

The last method among single-step algorithms is the *fourth order Runge–Kutta* (RK4) [36]. This method requires four force evaluations in each time step and saves eight additional variables in the memory. Although the RK4 method is computationally expensive, it has been selected by some researcher [37, 38] due to its high order of accuracy. Expressions for position and velocity in this method are:

$$\text{first step}\begin{cases}\vec{a}_n(\vec{x}_n,\vec{v}_n)\\ \vec{x}_1=\vec{x}_n+\frac{1}{2}\vec{v}_n\Delta t_p\\ \vec{v}_1=\vec{v}_n+\frac{1}{2}\vec{a}_n\Delta t_p\end{cases}\quad \text{second step}\begin{cases}\vec{a}_1(\vec{x}_1,\vec{v}_1)\\ \vec{x}_2=\vec{x}_n+\frac{1}{2}\vec{v}_1\Delta t_p\\ \vec{v}_2=\vec{v}_n+\frac{1}{2}\vec{a}_1\Delta t_p\end{cases}$$
$$\text{third step}\begin{cases}\vec{a}_2(\vec{x}_2,\vec{v}_2)\\ \vec{x}_3=\vec{x}_n+\vec{v}_2\Delta t_p\\ \vec{v}_3=\vec{v}_n+\vec{a}_2\Delta t_p\end{cases}\quad \text{fourth step}\begin{cases}\vec{a}_3(\vec{x}_3,\vec{v}_3)\\ \vec{x}_{n+1}=\vec{x}_n+\frac{1}{6}(\vec{v}_n+2\vec{v}_1+2\vec{v}_2+\vec{v}_3)\Delta t_p\\ \vec{v}_{n+1}=\vec{v}_n+\frac{1}{6}(\vec{a}_n+2\vec{a}_1+2\vec{a}_2+\vec{a}_3)\Delta t_p\end{cases} \tag{3.31}$$

The first step starts with the evaluation of the acceleration at the current time, then obtaining new position and velocity vectors, $\vec{x}_1$ and $\vec{v}_1$, at $t+\frac{1}{2}\Delta t_p$. In the second step, the acceleration is evaluated using $\vec{x}_1$ and $\vec{v}_1$. Then, $\vec{x}_2$ and $\vec{v}_2$ are calculated with new acceleration at time $t+\frac{1}{2}\Delta t_p$. In the fourth step, the position and velocity vectors at the next time stem are evaluated.

3.4.2 Multi-Step Algorithms

Integration methods that use velocities and accelerations of previous time step(s) to estimate velocity and position vectors at $t+\Delta t_p$ are called multi-step methods. The first multi-step integration method presented here (VE) is derived by *Verlet* [39] and used by other researchers in their DEM simulations [40]. In this method, the position vector at $t-\Delta t_p$ is used for calculating the new position vector. This expression has an accuracy of the third order for position. For calculating the velocity, central time derivative of position (second order accuracy) is used.

$$\vec{x}_{n+1}=2\vec{x}_n-\vec{x}_{n-1}+\vec{a}_n\Delta t_p^2 \tag{3.32a}$$

$$\vec{v}_{n+1}=\frac{\vec{x}_{n+1}-\vec{x}_{n-1}}{2\Delta t_p} \tag{3.32b}$$

Adams–Bashforth integration methods with various accuracy orders is a family of multi-step algorithms [41, 42]. The general expressions for position and velocity in this method are as follows:

$$\vec{v}_{n+1}=\vec{v}_n+(\lambda_1\vec{a}_n-\lambda_2\vec{a}_{n-1}+\lambda_3\vec{a}_{n-2}-\lambda_4\vec{a}_{n-3}+\lambda_5\vec{a}_{n-4})\Delta t_p \tag{3.33a}$$

$$\vec{x}_{n+1}=\vec{x}_n+(\lambda_1\vec{v}_n-\lambda_2\vec{v}_{n-1}+\lambda_3\vec{v}_{n-2}-\lambda_4\vec{v}_{n-3}+\lambda_5\vec{v}_{n-4})\Delta t_p \tag{3.33b}$$

where $\lambda_1-\lambda_5$ are Adams–Bashforth integration coefficients, listed in Table 3.2 for various orders. The abbreviation used for this family is AB (which stands for Adams–Bashforth)

Table 3.2 Adams–Bashforth integration coefficients for various accuracy orders

Method	λ_1	λ_2	λ_3	λ_4	λ_5
AB2	$\frac{3}{2}$	$\frac{1}{2}$	—	—	—
AB3	$\frac{23}{12}$	$\frac{16}{12}$	$\frac{5}{12}$	—	—
AB4	$\frac{55}{24}$	$\frac{59}{24}$	$\frac{37}{24}$	$\frac{9}{24}$	—
AB5	$\frac{1901}{720}$	$\frac{2774}{720}$	$\frac{2616}{720}$	$\frac{1274}{720}$	$\frac{251}{720}$

Program Listing 3.16 Implementation of the AB4 method in file g_Prtcl_MultiPointIntegration.f90.

```
01  subroutine AB4_Integrate(this , acc , dt , pos, vel )
02      ! . . . as before
03
04      vn = vel
05
06      pos = pos + ( AB3_coeff(1) * vn  - AB3_coeff(2) * this%v1 + AB3_
    coeff(3) * this%v2 ) * dt
07      vel = vel + ( AB3_coeff(1) * acc - AB3_coeff(2) * this%a1 + AB3_
    coeff(3) * this%a2 ) * dt
08
09      ! shifts vectors one time step back
10      this%v2 = this%v1
11      this%v1 = vn  ! as velocity at time step n-1 used in the next time step
12
13      this%a2 = this%a1
14      this%a1 = acc ! as velocity at time step n-1 used in the next time step
15
16  end subroutine
```

followed by an integer which is the order of accuracy. It can be seen in this table that the history of position and velocity vectors at previous time steps should be saved for being used in these expressions. For example, four vector variables, two position vectors and two velocity vectors at previous time steps, should be saved in the memory when using the AB3 method (see also Table 3.1). Similar to the single-step algorithms, calculation of the force acting on a particle should be done once and based on position and velocity vectors in the current time step.

The code for AB4 integration method is shown in Program Listing 3.16 as an example for this family. In lines 6 and 7, position and velocity of particle are calculated according to Equations 3.33a and 3.33b. Variables `v1`, `v2`, and `v3` represent velocity of particle at time steps $n-1$, $n-2$, and $n-3$, respectively, and `a1`, `a2`, and `a3` are acceleration of the particle at these time steps. In lines 10–14, the saved velocity and acceleration vectors are shifted one time step back in time and velocity and acceleration at the current time step are saved into the memory.

3.4.3 Predictor-Corrector Methods

Another class of integration methods is called the predictor-corrector. These methods basically work in two steps. In the first step, called prediction, position and velocity vectors at $t+\Delta t_p$ are predicted based on a single- or multi-step explicit method. These variables are saved in the memory. Then, predicted position and velocity vectors are used to evaluate forces acting on the particle and the predicted acceleration $\vec{a}^p_{n+1}$ is calculated from Equation 3.23a,b. In the second step, the corrector expression, which is usually implicit, is used to calculate the corrected (final) velocity and position vectors.

The explicit *Adams–Bashforth* method can be used in conjunction with the implicit *Adams–Moulton* method to perform prediction and correction steps, respectively. To construct a predictor-corrector integration method, the accuracy of Adams–Moulton method should be an order higher than that of the Adams–Bashforth method. We introduce AB2AM3, AB3AM4, and AB4AM5 methods here. The abbreviation used for this family is as follows: AB (which stands for Adams–Bashforth), followed by its order, plus AM (which stands for Adams–Moulton), followed by its order. In this method, Equations 3.33a and 3.33b are used to obtain predicted values of position and velocity. Then, after evaluation of forces acting on the particle, the predicted acceleration is calculated from Equation 3.23a,b. The final value of position and velocity vectors are obtained from the following corrector equations:

$$\vec{v}_{n+1} = \vec{v}_n + \left(\delta_1 \vec{a}^p_{n+1} + \delta_2 \vec{a}_n - \delta_3 \vec{a}_{n-1} + \delta_4 \vec{a}_{n-2} - \delta_5 \vec{a}_{n-3}\right)\Delta t_p \tag{3.34a}$$

$$\vec{x}_{n+1} = \vec{x}_n + \left(\delta_1 \vec{v}^p_{n+1} + \delta_2 \vec{v}_n - \delta_3 \vec{v}_{n-1} + \delta_4 \vec{v}_{n-2} - \delta_5 \vec{v}_{n-3}\right)\Delta t_p \tag{3.34b}$$

where $\delta_1 - \delta_5$ are Adams–Moulton integration coefficients, listed in Table 3.3 for various orders of integration.

The code of correction step of AB3AM4 is shown in Program Listing 3.17. Note that the code for the prediction step is very similar to the code in Program Listing 3.16 and we do not explain it here again. To this point, the prediction step is finished and the acceleration has been obtained based on the predicted position and velocity. Therefore, when this subroutine is invoked, `acc`, `pos`, and `vel` contain the predicted values. In lines 7 and 8, position and velocity of the particle are corrected. Corrected values are then saved in the memory for being used in the next time step in lines 11–19. At the end, the corrected values of position and velocity are returned.

Another family of predictor-corrector integration methods is the *Gear's algorithm* that has been widely applied for molecular dynamics and discrete element simulations [43–46].

Table 3.3 Adams–Moulton coefficients for various accuracy orders of integration

Method	δ_1	δ_2	δ_3	δ_4	δ_5
AM3	$\frac{5}{12}$	$\frac{8}{12}$	$\frac{1}{12}$	—	—
AM4	$\frac{9}{24}$	$\frac{19}{24}$	$\frac{5}{24}$	$\frac{1}{24}$	—
AM5	$\frac{251}{720}$	$\frac{646}{720}$	$\frac{264}{720}$	$\frac{106}{720}$	$\frac{19}{720}$

Program Listing 3.17 Implementation of the correction step of AB3AM4 method in file g_Prtcl_TwoStepIntegration.f90.

```
subroutine AB3AM4_correct(this, acc , dt, pos ,vel)
    ! . . . as before

    ! pos, vel and acc contain predicted values of position, velocity and
acceleration
    ! corrects position and velocity
    c_pos  = this%x + ( AM4_coeff(1)*vel + AM4_coeff(2)*this%v - AM4_coeff(3)*this%v1 + AM4_coeff(4)*this%v2 ) * dt
    c_vel  = this%v + ( AM4_coeff(1)*acc + AM4_coeff(2)*this%a - AM4_coeff(3)*this%a1 + AM4_coeff(4)*this%a2 ) * dt

    ! shifts variables one time step back to be used in next iteration
    this%x = c_pos

    this%v2 = this%v1
    this%v1 = this%v
    this%v = c_vel

    this%a2 = this%a1
    this%a1 = this%a
    this%a = acc

    ! returns corrected values
    pos = c_pos
    vel = c_vel

end subroutine
```

Different orders of Gear algorithm are available. In the following, we illustrate formulation for up to the fifth order. First, the predictor calculates the position, velocity, and higher order derivatives of position of the particle at time $t+\Delta t_p$ using the Taylor series expansion. The prediction expressions are as follows:

$$\vec{x}^p_{n+1} = \vec{x}_n + \vec{v}_n \Delta t_p + \tfrac{1}{2}\vec{a}_n \Delta t_p^2 + \tfrac{1}{6}\vec{b}_n \Delta t_p^3 + \tfrac{1}{24}\vec{c}_n \Delta t_p^4 + \tfrac{1}{120}\vec{d}_n \Delta t_p^5 \tag{3.35a}$$

$$\vec{v}^p_{n+1} = \vec{v}_n + \vec{a}_n \Delta t_p + \tfrac{1}{2}\vec{b}_n \Delta t_p^2 + \tfrac{1}{6}\vec{c}_n \Delta t_p^3 + \tfrac{1}{24}\vec{d}_n \Delta t_p^4 \tag{3.35b}$$

$$\vec{a}^p_{n+1} = \vec{a}_n + \vec{b}_n \Delta t_p + \tfrac{1}{2}\vec{c}_n \Delta t_p^2 + \tfrac{1}{6}\vec{d}_n \Delta t_p^3 \tag{3.35c}$$

$$\vec{b}^p_{n+1} = \vec{b}_n + \vec{c}_n \Delta t_p + \tfrac{1}{2}\vec{d}_n \Delta t_p^2 \tag{3.35d}$$

$$\vec{c}^p_{n+1} = \vec{c}_n + \vec{d}_n \Delta t_p \tag{3.35e}$$

$$\vec{d}^p_{n+1} = \vec{d}_n \tag{3.35f}$$

with

$$\vec{b}_n = d\vec{a}_n/dt, \vec{c}_n = d\vec{b}_n/dt, \vec{d}_n = d\vec{c}_n/dt \tag{3.36}$$

By setting $\vec{d}_n$ to zero, this method reduces to the fourth order and by further setting $\vec{c}_n$ to zero, it reduces to the third order. The predicted position and velocity vectors are used for evaluating all forces acting on the particle. Then, using Equation 3.23a,b, the corrected acceleration is obtained. If the prediction of velocity and position by Equations 3.35a and 3.35b are precise enough, the predicted and corrected accelerations would be very close. The difference between these values describes deviation of the predicted position and velocity and other quantities from their correct values. Therefore, we have:

$$\Delta\vec{a}_{n+1} = \vec{a}_{n+1}^{corr} - \vec{a}_{n+1}^{p} \tag{3.37}$$

In the correction step, the corrector expressions improve the predicted values according to the following equations:

$$\begin{bmatrix} \vec{x}_{n+1} \\ \vec{v}_{n+1} \\ \vec{a}_{n+1} \\ \vec{b}_{n+1} \\ \vec{c}_{n+1} \\ \vec{d}_{n+1} \end{bmatrix} = \begin{bmatrix} \vec{x}_{n+1}^{p} \\ \vec{v}_{n+1}^{p} \\ \vec{a}_{n+1}^{p} \\ \vec{b}_{n+1}^{p} \\ \vec{c}_{n+1}^{p} \\ \vec{d}_{n+1}^{p} \end{bmatrix} + \begin{bmatrix} \eta_1 \Delta t_p^2 \\ \eta_2 \Delta t_p \\ \eta_3 \\ \eta_4 \Delta t_p^{-1} \\ \eta_5 \Delta t_p^{-2} \\ \eta_6 \Delta t_p^{-3} \end{bmatrix} \Delta\vec{a}_{n+1} \tag{3.38}$$

here, $\eta_1 - \eta_6$ are Gear's method integration coefficients listed in Table 3.4 for various orders of integrations.

Program Listing 3.18 illustrates the code for prediction and correction steps of the Gear3 method. In the prediction step, predicted value of position, velocity, and acceleration are computed based on Equations 3.35a–3.35c. In the correction step, program calculates $\Delta\vec{a}_{n+1}$ in line 21 and corrects the position, velocity, acceleration, and the first derivative of the acceleration.

3.4.4 Evaluation of Integration Methods

A robust numerical integration method should ideally produce accurate results with the minimum usage of the computational resources. Some methods are relatively simple and use minimal computational resources while having a poor accuracy. Some of them are very

Table 3.4 Coefficients of the Gear's method for various order of integration

Method	η_1	η_2	η_3	η_4	η_5	η_6
Gear3	$\frac{1}{12}$	$\frac{5}{12}$	1	1	—	—
Gear4	$\frac{19}{240}$	$\frac{3}{8}$	1	$\frac{3}{2}$	1	—
Gear5	$\frac{3}{32}$	$\frac{251}{720}$	1	$\frac{11}{6}$	2	1

Program Listing 3.18 Implementation of the prediction and correction steps of Gear3 method in file g_Prtcl_TwoStepIntegration.f90.

```
subroutine G3_predict(this, dt, pos, vel )
    ! . . . as before

    ! predicts the position and velocity and acceleration and keeps them
    this%x = this%x + this%v * dt +  0.5_RK * dt2 * this%a +
0.1666666666666667_RK * dt2*dt * this%b
    this%v = this%v + this%a * dt +  0.5_RK * dt2 * this%b
    this%a = this%a + this%b * dt

    ! returns predicted position and velocity
    pos = this%x
    vel = this%v

end subroutine

!**********************************************************************
  **
! Gear3 correction method
!**********************************************************************
  **
subroutine G3_correct(this, acc , dt, pos ,vel )

    ! evaluates the deviations from corrected value
    delta_a = acc - this%a

    ! corrects position, velocity, acceleration and etc.
    this%x = this%x + G3_coeff(1) * dt*dt * delta_a
    this%v = this%v + G3_coeff(2) * dt    * delta_a
    this%a = this%a + G3_coeff(3)         * delta_a
    this%b = this%b + G3_coeff(4) / dt    * delta_a

    ! returns corrected values
    pos = this%x
    vel = this%v

end subroutine
```

accurate while being computationally expensive. Many comparisons have been made on accuracy and efficiency of integration methods for a single particle impact [2, 24, 33, 47]. However, we cannot make a general statement on the best and the most robust integration method that suits all types of discrete element simulations. The best integration method is case-specific. Accordingly, we discuss accuracy and computational efficiency separately and give some general guidelines for choosing the suitable integration method.

3.4.4.1 Accuracy

High accuracy is a must for an integration method. With an inaccurate method, some of the kinetic energy of colliding particles may be dissipated or amplified due to numerical error. This may lead to erroneous results and sometimes abnormal behavior of the granular flow.

In some cases, repeated amplification of kinetic energy during the simulation leads to the instability of the solution and artificially high local velocity of particles. Therefore, the integration method should be accurate enough to maintain the numerical error less than a certain value (e.g., less than 1%). We define the time step resolution as follows:

$$\alpha = \frac{t_{col}}{\Delta t_p} \tag{3.39}$$

where t_{col} is the collision time.

Different values of resolution have been used by researchers. In general, results become more accurate when the resolution increases. Tuley *et al.* [47] reported that all types of integration methods give accurate results for $\alpha > 300$. Kruggel-Emden *et al.* [33] tested the accuracy of different integration methods for an oblique impact at various contact angels and compared the results with exact analytical solution in terms of normal and tangential coefficients of restitution. Results of their comparison are reported in Figure 3.17 for normal coefficient of restitution. The maximum allowable error was considered to be 1% average absolute relative deviation. This figure shows that FE and TY2 methods do not go below the maximum acceptable error when $0 < \alpha \leq 100$, while the CD method produces acceptable results when $\alpha > 75$. Among single-step methods, solutions with less than the maximum allowable error can be achieved by RK4, TY3, and TY4 when $\alpha > 10$. Multi-step methods, like AB4 and AB5, as well as the whole family of Gear method, are stable and all provide acceptable solution when $\alpha > 8$. It is worth mentioning that the average relative error of tangential coefficient of restitution was always less than the error of normal coefficient of restitution in their study and we do not further discuss it here.

Using a fixed time step for integration introduces an additional error since detecting the start of a collision as well as its real rebound instance would not be realized correctly. Figure 3.18 shows the difference between continuous motion of a particle in the real world and discontinuous motion of a particle (due to stepwise integration) in the numerical simulation. The collision actually starts between t_1 and t_2 in the real world while it is first detected at t_2 in the simulation (due to using fixed time step). At this point, some kinetic energy should have been converted to potential energy (in the spring) and dissipated (due to viscous loses). It is also possible that the moment of rebound would be detected incorrectly. This also introduces additional errors because some of the potential energy incorrectly converts back into the kinetic energy.

By considering the fact that the exact start of a collision may not be detected as a result of using a fixed time step, a larger time step resolution should be employed to obtain an acceptable error [33]. Tuley *et al.* [47] also studied the accuracy of some integration methods by considering this issue. Among the methods tested, they showed that the multi-step methods (like AB2 and AB4) and predictor-corrector methods (like AB3AM4) are more stable than others and the accuracy of the results does not change noticeably when the moment of collision is not detected precisely in the simulation. This is due to the fact that the *history* of transition from free motion to collision is used in the integration equations of the Adams–Bashforth and Adams–Bashforth–Multon families. They also showed that single point methods (even RK4) are only accurate when the moment of collision is precisely detected. This is due to the fact that single-step methods only use the information at the current time step.

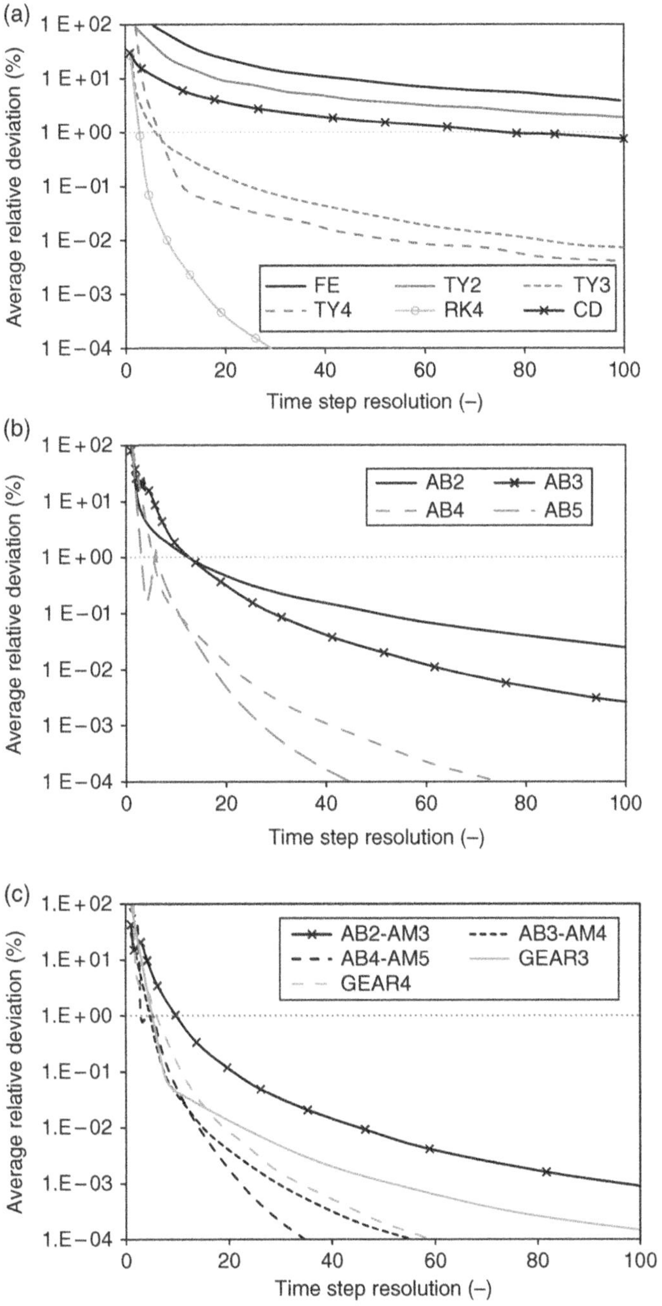

Figure 3.17 The average relative deviation of normal coefficient of restitution for oblique impact between exact solution and results of various numerical integrators (a) single-step, (b) multi-step, and (c) predictor-corrector methods. Integration was started at the incipient of collision. Data were adapted from Kruggel-Emden *et al.* [33]

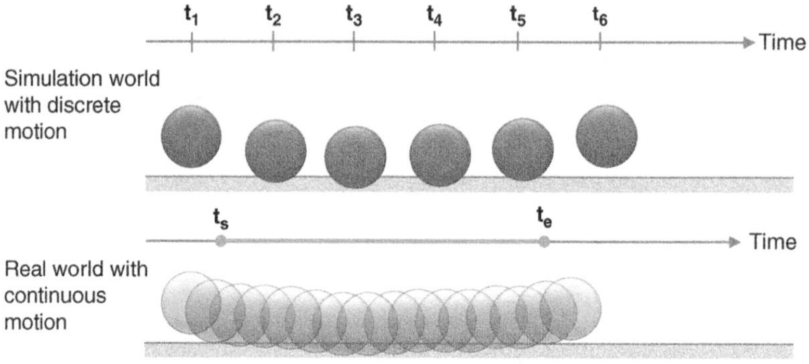

Figure 3.18 Effect of using a fixed time step in DEM simulations on detection of correct moments of collision and rebound

3.4.4.2 Computational Efficiency

Complex and accurate integration methods require ample RAM and CPU operations. In general, computational overhead associated with a complex integration method are grouped into:

1. Computations associated with recalculation of forces acting on particles. Recalculation of forces includes steps like contact force detection and calculating contact forces between all contacting pairs.
2. Computations associated with performing a predictor step in case of predictor-corrector methods.
3. Computations associated with calculating parameters that appear in the integration expression, like $\vec{b}_n$ in TY3.

Computation time of integration methods (those implemented in the numerical code that accompanies this book) in a real discrete element simulation is not known to us. Using the numerical code, we performed a set of simulation runs for settling of 8000 particles with size of 7 mm in a cylindrical container to examine this issue. The simulation conditions in all simulation were the same with the exception of the integration method. All simulations were continued for 30,000 iterations. Figure 3.19 shows the computation time of integration step for single-step, multi-step, and predictor-corrector (correction and prediction) methods. This figure demonstrates that FE and ME methods have the lowest computation time among all methods, while the predictor-corrector methods have the highest computation time.

In the single-step methods, the computation time of integration step increase with increasing the accuracy of the method and it almost doubles in TY4 method with respect to FE method. Increase in the computation time is associated with using a more complex integration formula (which requires more mathematical operations) and accessing extra variables stored in the memory. The integration time of the Adams–Bashforth family is in the range of single step methods and it increases with accuracy of the method. All predictor-corrector methods have longer computation times than single-step and multi-step methods. This is due to the additional computations for the prediction step of these methods. The computation times of

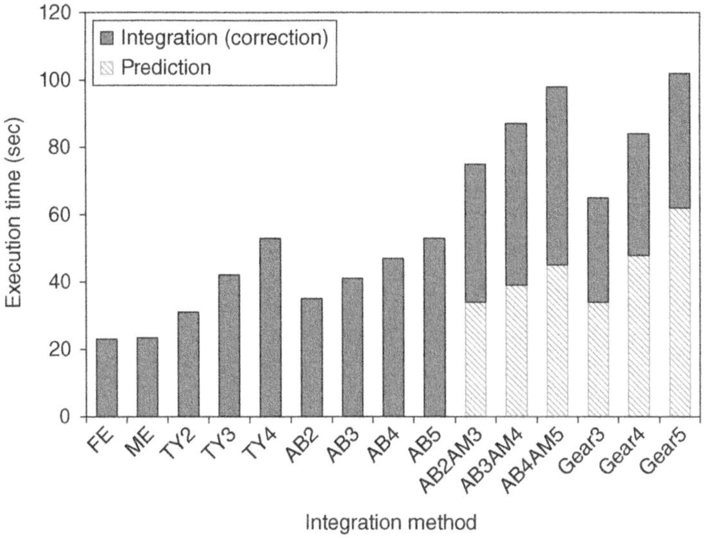

Figure 3.19 Execution time of integration step for various methods in a DEM simulation with 8000 spherical particles and 30,000 iterations

Gear5 and AB4AM5 methods are the highest. They are five times greater than the FE method and 2.5 times greater than the TY4 and AB4 methods.

Now, the question is whether it is feasible to use high order methods, like AB4, or any of predictor corrector methods in a DEM simulation. To answer this question, we first should know what fraction of total computation time is dedicated to the integration time in a simulation. This time is in the range of 10–17% of the total time in single-step and multi-step methods and 21–26% in predictor-corrector methods (in the simulations performed in this section). These values reveal that even with a sharp increase in the computation time of the integration step, it has a low effect on the total computation time of simulation and using high order methods or predictor corrector methods is feasible.

To better compare the performance of the integration methods, we compare computation time of integration step, memory requirement, and minimum required time step resolution of each method in Figure 3.20 after normalization. The single step methods usually require high time step resolutions and low memory and computation time. In contrast, multi-step methods require more memory space with lower time step resolution and slightly higher computation time when comparing them to single step methods. Finally, the predictor-corrector methods require the most memory and computation time with lowest time step resolution.

3.5 Spring Stiffness

In a DEM simulation, a fraction of collision time (inverse of the collision time) is chosen for integration time step. Accuracy of results depends on the type of integration method and time step resolution. By increasing the collision time and keeping the time step resolution constant,

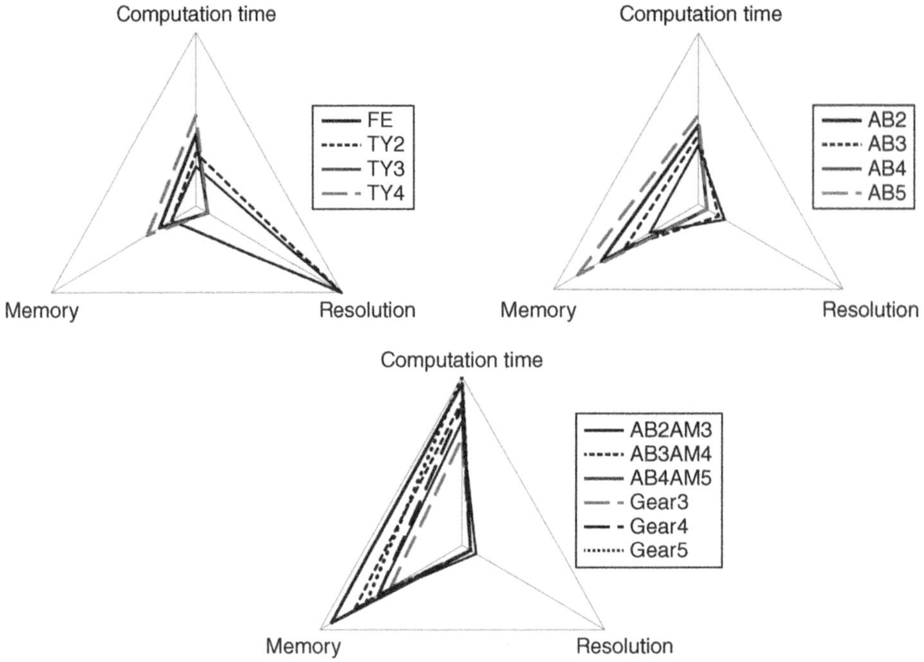

Figure 3.20 Comparison of memory requirement, computation time of integration step, and required time step resolution for different integration methods. All parameters are normalized with respect to the corresponding maximum value

the accuracy remains constant. This leads to a larger time step, based on Equation 3.39, and decreasing the computational load of simulation per 1 s of simulation. The collision time of an elastic contact in the LSD force model is proportional to the square root of the ratio of the mass to the spring stiffness of the particle. There are two main approaches for increasing the collision time: density-scaling and selecting low values of spring stiffness.

In the density scaling (also known as mass scaling), the density of particles is artificially increased to an extremely high value. This approach is mainly applicable to a quasi-static problem in which the kinetic energy of the system is very low compared to the total energy of system and the inertial force plays a minor role in the physical system. In many granular and multiphase flows, the kinetic energy of particles is of the same order of magnitude of the total energy of the system. Also, other forces, like gravity and buoyancy, play an important role in the flow. Therefore, the mass scaling cannot be applied to such systems.

The other technique for increasing the contact time is to reduce the spring stiffness (Young's modulus for non-linear contact force models model). The spring stiffness is typically between 10^6 and 10^8 N/m. However, many studies can be found in which very small values of spring stiffness (like 800 N/m) were used. In these studies, the DEM simulation can satisfactorily predict the characteristic behavior of the flow in macro-scale, like spout-fluid bed hydrodynamics [48], bubble diameter in gas-solid fluidized beds [44, 49], segregation and mixing [50–52], and flow pattern in hoppers [53, 54]. However, we know

Table 3.5 Relationships for evaluating collision time, maximum normal overlap, and maximum normal force in an elastic contact between two spherical particles for linear and non-linear contact models

Parameter	Linear model		Non-linear model	
t_{col}	$\pi\sqrt{\frac{m_{eff}}{k_n}}$	(3.40a)	$2.86\left(\frac{m_{eff}^2}{R_{eff}E_{eff}^2 v_{rn,imp}}\right)^{1/5}$	(3.41a)
$\delta_{n,max}$	$v_{rn,imp}\sqrt{\frac{m_{eff}}{k_n}}$	(3.40b)	$\left(\frac{225}{256}\frac{m_{eff}^2 v_{rn,imp}^4}{E_{eff}^2 R_{eff}}\right)^{1/5}$	(3.41b)
$f_{n,max}$	$v_{rn,imp}\sqrt{m_{eff}k_n}$	(3.40c)	$\left(\frac{500}{144}m_{eff}^3 E_{eff}^2 R_{eff} v_{rn,imp}^6\right)^{1/5}$	(3.41c)

that the contact properties are sensitive to the spring stiffness and we need to answer two questions here: how the reduction of stiffness affects microscopic and macroscopic characteristics of the impact? And, in which circumstance can we select a smaller value for the spring stiffness?

The dynamics of normal and oblique impacts of a particle with a flat wall depends on the coefficient of restitution, the dynamic friction coefficient, and the ratio of tangential to normal stiffnesses, κ (see Chapter 2). Hence, the rebound angle, the rebound velocity, and the tangential restitution coefficient are not affected by the value of normal spring stiffness. However, the collision time, the maximum overlap, and the maximum contact force depend on the spring stiffness. Table 3.5 lists expressions for evaluating the collision time, the maximum overlap, and the contact force in a normal elastic contact of two spherical particles for both linear and non-linear contact models. These parameters are strong functions of spring stiffness or Young's modulus. For example, when the stiffness is reduced 100 times (a very regular reduction in DEM simulations), collision time and maximum overlap increase 10 times, while the contact force decreases 10 times.

Table 3.6 lists variation of these parameters for different values of spring stiffness and impact velocities. Values in this table belong to the normal contact of two elastic spheres with the diameter of 2 mm and the density of 2500 kg/m^3. For the lowest stiffness, the maximum overlap is around 8% of the diameter at the impact velocity of 1 m/s, while it increases to 54% at the impact velocity of 10 m/s. The percentage of overlap decreases to less than 1% for $k_n \approx 10^4$ N/m at 1 m/s and for $k_n \approx 10^6$ N/m at 10 m/s. Similar changes can be observed for collision time and maximum contact force. Concerning the first question (effect of spring stiffness on the characteristics of the impact), microscopic characteristics of granular flow, like collision time, contact force, and maximum overlap, are considerably affected when reducing the spring stiffness while rebound velocity and contact angle are not. For the second question (circumstances at which we can reduce the spring stiffness), we grouped different flow problems into two main groups: flows in which the maximum overlap affects results and those in which collision time and maximum force affect the results. These are discussed in the following.

Table 3.6 Variation of contact time, maximum overlap, and maximum contact force as a function of material stiffness and impact velocity in elastic collisions of two spherical particles with density of 2500 kg/m³ and diameter of 2 mm

Impact velocity (m/s)	$E_{eff}(Pa)$	$k_n(N/m)$	Linear $\hat{\delta}_{n,max}(\%)^a$	Linear $t_{col}(s)$	Linear $f_{n,max}(N)$	Non-linear $\hat{\delta}_{n,max}(\%)^a$	Non-linear $t_{col}(s)$	Non-linear $f_{n,max}(N)$
1	5.0×10^5	1.8×10^2	8.46	5.3×10^{-4}	0.03	9.04	5.3×10^{-4}	0.04
	5.0×10^6	1.2×10^3	3.37	2.1×10^{-4}	0.08	3.60	2.1×10^{-4}	0.09
	5.0×10^7	7.3×10^3	1.34	8.4×10^{-5}	0.20	1.43	8.4×10^{-5}	0.23
	5.0×10^8	4.6×10^4	0.53	3.4×10^{-5}	0.49	0.57	3.4×10^{-5}	0.57
	5.0×10^9	2.9×10^5	0.21	1.3×10^{-5}	1.23	0.23	1.3×10^{-5}	1.44
	5.0×10^{10}	1.8×10^6	0.08	5.3×10^{-6}	3.10	0.09	5.3×10^{-6}	3.62
	5.0×10^{11}	1.2×10^7	0.03	2.1×10^{-6}	7.78	0.04	2.1×10^{-6}	9.10
10	5.0×10^5	4.6×10^2	53.35	3.4×10^{-4}	0.49	57.01	3.4×10^{-4}	0.57
	5.0×10^6	2.9×10^3	21.24	1.3×10^{-4}	1.23	22.70	1.3×10^{-4}	1.44
	5.0×10^7	1.8×10^4	8.46	5.3×10^{-5}	3.10	9.04	5.3×10^{-5}	3.62
	5.0×10^8	1.2×10^5	3.37	2.1×10^{-5}	7.78	3.60	2.1×10^{-5}	9.10
	5.0×10^9	7.3×10^5	1.34	8.4×10^{-6}	19.53	1.43	8.4×10^{-6}	22.85
	5.0×10^{10}	4.6×10^6	0.53	3.4×10^{-6}	49.07	0.57	3.4×10^{-6}	57.40
	5.0×10^{11}	2.9×10^7	0.21	1.3×10^{-6}	123.26	0.23	1.3×10^{-6}	144.18

[a] The normal overlap divided by the particle diameter and multiplied by 100.

3.5.1 *Maximum Overlap*

Controlling the maximum overlap of colliding particles is essential in simulation of granular and multiphase flows. Li *et al.* [55] studied packing of spherical particles with a spring stiffness between 400 and 8000 N/m. They showed that at very small spring stiffness, the packed bed voidage decreases considerably (from 0.55 to 0.35) along the depth of the bed due to excessive overlaps between particles in lower layers. However, such a trend was not observed in simulations with spring stiffness of 8000 N/m for which the voidage was almost constant in the bed. This shows that choosing a very small value for stiffness leads to a remarkable change in the porosity distribution, hence, affecting the coordinate number, heat transfer characteristics, and permeability of the bed.

Excessive overlaps between particles should also be avoided in simulation of multiphase flows. There is a mutual interaction between solid and fluid phases via porosity and interphase momentum transfer. When the level of overlap between particles is high, the local porosity would be underestimated. This affects the fluid velocity, pressure fields, and interphase momentum transfer. The artificial change in the local porosity due to excessive overlap of particles results in large deviations in the output of simulation. As a rule of thumb, the maximum overlap should not exceed about 2% of the particle diameter. This value is arbitrary and may vary from an application to another. Thus, the value of spring stiffness should be chosen based on the maximum acceptable interparticle overlap [56].

In order to select the spring stiffness, first, the maximum relative velocity (particle-particle or particle-wall) in the simulation should be estimated. Then, the value of spring stiffness (Young's modulus) can be estimated by considering a maximum level of interparticle overlap and the maximum relative velocity according to Equation 3.40b or 3.41b. For example, consider particles with diameter of 2 mm, density of 2500 kg/m^3 and maximum relative velocity of 1 m/s. These conditions are typical in flow of particles in most bubbling fluidized beds, hoppers, and tumbling blenders. By considering that the maximum interparticle overlap should be 2% of the particle diameter, the spring stiffness should be of the order of 10^3 N/m in the simulation. That is why DEM simulations of regular granular and multiphase flows with small values of stiffness can satisfactorily reproduce the macroscopic behavior of the actual flow, even though the microscopic contact parameters are altered significantly.

3.5.2 Collision Time and Maximum Contact Force

Although macroscopic characteristics (such as velocity profile) of regular granular and multiphase flows are not affected by reducing the spring stiffness, other phenomena that are linked to micro-scale parameters are influenced by this reduction. For example, heat transfer in the particle-scale in granular media [57, 58] and sintering of hot granules [59, 60] are affected by micro-scale parameters such as contact time and maximum contact force. Particle-particle heat transfer depends on material properties, contact radius, and contact time of particles. Therefore, estimated particle-particle heat transfer rate can be altered by orders of magnitude if the value of spring stiffness is changed. Zhou *et al.* [61] proposed a method to mitigate the effect of using a small Young's modulus in their thermal discrete element simulation in fluidized beds. By considering a small Young's modulus (as it is used in discrete element simulations), contact area and collision time in the simulation become greater than those in the real world. They corrected these parameters in the simulation by scaling them to the actual values (using the actual value of Young's modulus) and used these corrected values in the heat transfer equations. They compared the result of their corrected heat transfer equation with experimental results on particle-particle heat transfer coefficient and confirmed that this correction works well to predict the heat transfer coefficient when using small values of Young's modulus in the simulation.

In an agglomerating fluidized bed in which hot granules (e.g., polyethylene or polypropylene) exists, particles form agglomerates due to formation of permanent solid bridge. According to the proposed mechanism (sintering), growth rate of the solid neck is a function of material properties, temperature, contact time, and contact radius [60]. A comparison should be made between the strength of cohesive solid bridge and maximum contact force to decide whether or not a permanent bridge would be formed between the colliding particles. Therefore, the rate of agglomerate formation is determined by the competition between the formation rate of solid neck and the maximum contact force. A change in the spring stiffness alters both formation rate of solid neck and maximum contact force, hence, changes the rate of agglomeration in the simulation. These effects should be considered in the formulation of agglomeration phenomena.

3.6 Wall Implementation

A very important part of a discrete element simulation is definition of walls. Particles are confined within the simulation domain by walls. In addition, moving walls transfer kinetic energy to particles and cause their motion (mostly convective). An algorithm should be developed for handling interactions between particles and walls. Particle-wall interactions include normal and tangential contact forces, cohesion, heat exchange, and momentum exchange. For quantifying these interactions, we should calculate distance of particles from wall, contact point, level of overlap, and normal vector at the contact point as well as velocity of the wall at the contact point.

A wall can be defined by a set of analytical equations and its surface can be either closed or open with an active side. This type of geometry representation restricts the program to standard geometries which can be defined by analytical equations. Examples of such surfaces are sphere, cone, cube, ellipsoid, cylinder, torus, and so on (e.g., Katok and Climenhaga [62]). Analytical equation of the surface is an expression which maps 3D points to real numbers, $F(x,y,z):\mathbb{R}^3 \to \mathbb{R}$. Points for which $F(x,y,z)<0$ reside in the interior side of the surface, $F(x,y,z)=0$ determines the points on the surface and points for which $F(x,y,z)>0$ fall outside the wall. The process of contact detection is to find the intersection of the body/particle with the wall based on the analytical equation of the wall. One drawback of this method is that each set of analytical equations needs a different numerical procedure to find the intersection point and in some cases the numerical procedure fails.

Spherical solid particle may be used to construct the wall. A granular wall is built of wall particles with a specific size. We described wall particles and their potential use in discrete element simulations in Chapter 2. The position of wall particles is determined throughout the simulation by the wall's translational and rotational velocities. In this case, the contact between particle and wall is treated in the same fashion as for a particle-particle contact and no additional numerical procedure is required. The process of finding a contact between wall and particles does not depend on the shape of the wall and almost every arbitrary shape can be considered in this case. The use of a granular wall is not very common and has some difficulties, which include multiple contacts between a particle and wall, interlocks of particles between wall particles, physically unrealistic particle-wall friction, and large computational overheads associated with the use of wall particles that sometimes doubles the computation time.

It is possible to represent a wall explicitly by plane triangular or quadrilateral elements, called the *wall element*. These wall elements are represented by their vertex points and a normal vector. Since walls with any geometric shape can be covered by these elements (surface meshing or tessellating), we would not be restricted to standard shapes. This method also does not have the problem of granular wall concerning particle interlocks and artificial particle-wall friction. The process of contact detection is to find the contact point between particles and wall elements Available methods for such contact detection are robust, simple, and computationally efficient [63]. Figure 3.21 shows different possible shapes of wall produced by meshing the surface of the wall. Surface meshes of tote-blender and V-blender are produced by functions available in the program (methods defined in type `Geometry` in file *`g_Geometry.f90`*) and those of bladed mixer and screw blade meshes are imported into the program from external files with STL format[16] (method `add_stl_file` defined in type `Geometry` in file *`g_Geometry.f90`*).

The method of detecting particle-wall contact varies depending on the way we represent walls and particles in the program. Application of DEM to real-scale and complex industrial flows is increasing and requires complex wall geometries, while at the same time needs implementing a general contact detection method. Among the previously mentioned methods, the

[16]A file with STL (standard tessellation language) format contains representation of a surface in 3D space. The surface is tessellated into oriented triangles, each of which is described by a unit normal vector and three vertex points.

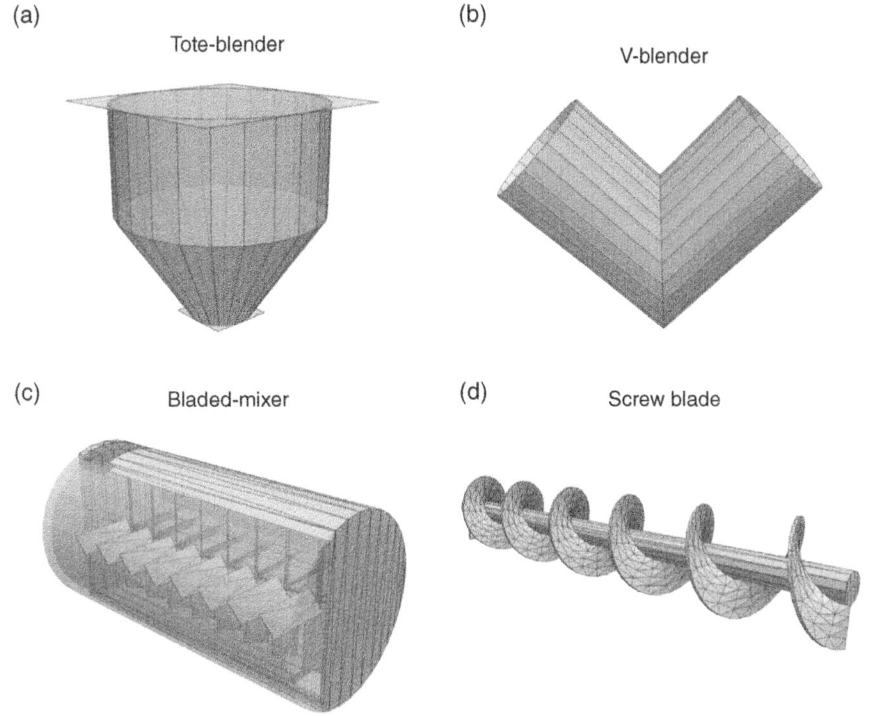

Figure 3.21 Various wall geometries that can be created directly in the program (a,b) or can be imported from external files with STL format (c,d)

first two methods (analytical equation and granular wall) have some drawbacks that limit their applications to simple geometries and granular flows. In contrast, representing the wall with plane triangles or quadrilaterals (explicit representation) is a promising method. Therefore, we only focus on the explicit representation method and develop methodologies required for defining walls, performing contact detection between a sphere and a wall and defining moving walls.

3.6.1 Definition of Wall Elements

A plane triangular wall element is defined by three non-collinear points P_1, P_2, and P_3 in 3D space. The unit normal vector of this plane is defined by:

$$\vec{n}_w = \frac{(P_2 - P_1)\times(P_3 - P_1)}{\left|(P_2 - P_1)\times(P_3 - P_1)\right|} \tag{3.42}$$

Equation of the plane can be defined either by a normal unit vector and a point on the plane or by an implicit equation as follows:

$$n_{w,x}x + n_{w,y}y + n_{w,z}z + d = 0 \tag{3.43}$$

where $n_{w,x}$, $n_{w,y}$, and $n_{w,z}$ are components of unit normal vector of the plane, $\vec{n}_w$, and d is given by:

$$d = -\left(\vec{n}_w \cdot P_1\right) \tag{3.44}$$

The front (active) side of the wall is determined by the right-hand rule when starting from the first point, P_1, and traversing next points of the plane in order (P_2 and P_3).

A plane quadrilateral is defined by non-collinear points P_1, P_2, P_3, and P_4 in 3D space. The unit normal vector of this plane is given by Equation 3.42 and its implicit form by Equation 3.43. Since, by definition, all points of the wall element should lay on a plane, the fourth point should satisfy the plane equation. If this point does not satisfy the plane equation, the supplied points do not form a plane quadrilateral.

Program Listing 3.19 shows a part of `PlaneWall` definition that is used to represent a stationary wall element in the program. A short description of each data member is also provided. Data members `L1`, `L2`, `L3`, and `L4` are parametric 3D lines that form the four edges of the plane. These lines are used for edge contact detection. Data member `bothSide` can be used to make both sides of the plane active. In this way, contact detection is performed for particles that are on both sides of the plane.

Program Listing 3.20 illustrates the definition of the member function `PW_CreateWall` that creates a plane wall element in the program. Lines 15–23 save necessary data of the wall in the memory. Lines 26–28 calculates the normal vector of the plane based on Equation 3.42 and the parameter d in the implicit equation of plane is calculated in line 31. In lines 34–39, the program checks if the fourth point lies on the plane and in lines 41–44, four parametric lines of plane edges are created.

Program Listing 3.19 Definition of type PlaneWall in file g_PlaneWall.f90.

```
type PlaneWall

        integer(IK) user_id  ! user supplied wall id
        integer(IK) wall_id  ! program generated wall id
        integer(IK) wall_prop_Type  ! property type of wall material
        logical bothSide     ! if true, both sides are active
        real(RK)    d        ! d in the implicit equation of plane
        type(real3) n        ! normal vector
        type(real3) P1       ! first point
        type(real3) P2       ! second point
        type(real3) P3       ! third point
        type(real3) P4       ! fourth point
        type(p_line) L1      ! line connects P1 to P2
        type(p_line) L2      ! line connects P2 to P3
        type(p_line) L3      ! line connects P3 to P4
        type(p_line) L4      ! line connects P4 to P1

    contains
. . .
end type
```

Program Listing 3.20 Definition of member function PW_CreateWall in file g_PlaneWall.f90.

```
logical function PW_CreateWall(this, p1, p2, p3, p4 , user_id, wall_
id, prop_type, both)
    implicit none
    class(PlaneWall)           this
    type(real3),intent(in) :: p1, p2, p3, p4
    integer(IK),intent(in) :: user_id, wall_id, prop_type
    logical    ,intent(in) :: both

    !// locals
    real(RK) val
    type(real3) ln

    !// body

    ! assignments
    this%P1 = p1
    this%P2 = p2
    this%P3 = p3
    this%P4 = p4

    this%wall_id = wall_id
    this%user_id = user_id
    this%wall_prop_Type = prop_type
    this%bothSide = both

    !normal vector of plane
    ln = (p2-p1) .cross. (p3-p1)
    val = norm( ln ) ! norm of the vector
    this%n = ln/val  ! unit normal vector

    ! d in the implicit equstion: ax+by+cz+d = 0
    this%d = - (this%n .dot. p1)

    !checks whether the fourth point lies on the plane!
    if( abs( (this%n .dot. p4) + this%d ) < 0.00001_RK ) then
        PW_CreateWall = .true.
    else
        PW_CreateWall = .false.
        return
    end if
    ! parametric lines
    call this%L1%setLine(p1,p2) ! p1 --> p2
    call this%L2%setLine(p2,p3) ! p2 --> p3
    call this%L3%setLine(p3,p4) ! p3 --> p4
    call this%L4%setLine(p4,p1) ! p4 --> p1

end function
```

3.6.2 *Contact Detection*

Particle–wall contact detection comprises of two main phases. In the first phase, a list of particles/spheres[17] that are in certain proximity of a wall element is established. These particles may have physical contact with the wall element (very similar to broad search phase in particle–particle contact search). Such a particle list is established for each wall element. In this way, a large number of particles would be excluded from the exact contact test and a lot of computational effort would be saved. In the second phase, all particles in the particle list of each wall element are tested against that wall to find the contact point and overlap (similar to fine search in the particle–particle contact search).

3.6.2.1 Generating the List of Neighbor Particle

The following steps should be performed to generate the neighboring particle list of a wall element:

- *Step 1*: The smallest axis-aligned hexahedron which encloses all points of the wall element is found (see Figure 3.22). A hexahedron can be defined by two of its diagonal corner points.
- *Step 2*: This hexahedron is dilated by R_{min} (minimum proximity distance) and an enlarged hexahedron is created. Now, we can confidently state that the minimum distance between points on the wall element and enlarged cube faces is larger than or equal to R_{min}.
- *Step 3*: Particles whose centers are inside the cube are considered as potential contacts and are added to the particle list of the wall element.

The valid range for R_{min} is $[r_{\max}, \infty)$, where r_{max} is radius of the largest particle in the system. With $R_{min} = r_{max}$, the list should be updated at every time step and with $R_{min} = \infty$ all particles in the system are included in the particle list and the list should be updated only once during the entire simulation. If we choose a value greater than r_{max} (e.g., three or four times r_{max}), we can use a simple property of the DEM dynamics to reduce the computation time; particles move a very short distance in the space from one time step to another and the list of neighboring particles of a wall changes slowly (i.e., in several time steps). Therefore, we only need to update the neighboring particle list of the wall every several time steps.

A simple procedure is implemented in the code for calculating the moment of the next update of the neighbor particle lists. Assuming that the maximum absolute velocity of particles and walls are $V_{p,max}$ and $V_{w,max}$, respectively, the maximum absolute relative velocity in the whole system becomes:

$$V_{max} = V_{p,max} + V_{w,max} \tag{3.45}$$

[17] If a particle is spherical, the bounding sphere would become the particle itself. We explain particle–wall contact search for spherical particles. This can be extended to use for non-spherical particles using the bounding box concept.

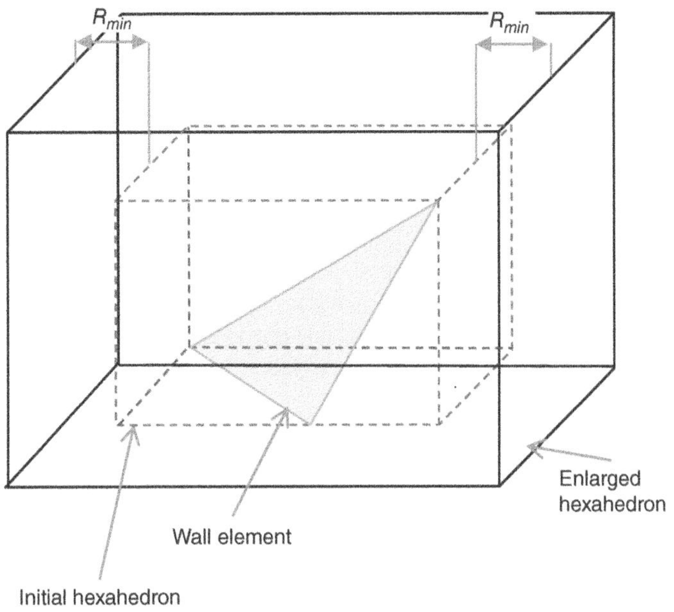

Figure 3.22 The initial hexahedron around the wall element and the enlarged hexahedron for finding the particle list of the wall element

Considering the maximum acceleration of particles to be a_{max}, the maximum relative displacement (free motion without contact) in the whole system can be estimated from:

$$\Delta x_{max} = \frac{1}{2} a_{max} t^2 + V_{max} t \tag{3.46}$$

By setting $\Delta x_{max} = R_{min}$ in Equation 3.46 and solving it, two roots would be obtained. The positive root is taken as the time that a particle can escape from or enter to the neighborhood of a wall element. The neighbor particle lists should be updated after passing this time from the previous update.

3.6.2.2 Fine Search for Particle-Wall Contact

In the fine contact search, the program performs a two-step test to determine whether or not a spherical particle, with center C and radius R, is in contact with a wall element. In the first step, the closest distance of particle center to the wall element (assuming that the wall element is an infinite plane wall) should be calculated. Given a plane with a unit normal vector $\vec{n}_w$ and a point P_1, the closest distance to the plane can be calculated from:

$$dist = \vec{n}_w \cdot (C - P_1) \tag{3.47}$$

When the plane is represented by Equation 3.43, the closest distance to the plane is given by:

$$dist = \vec{n}_w \cdot C + d \tag{3.48}$$

If *dist* is positive, the particle center is on the front/active side of the plane, otherwise, it is on the back side of the plane and should be excluded from the contact search. If the distance is shorter than the sphere radius, there would be a possible contact between the particle and the wall element. We should then go to the second step of the test.

In the second step, the exact contact point of sphere with the wall element should be found. For a spherical particle against a plane triangle, seven situations can be expected: contact point within the boundaries, contact point on one of the edges (three possibilities), and contact point at a corner point (three possibilities). Similarly, there are nine possibilities for a contact between a spherical particle and a plane quadrilateral element. The most probable contact type is the one that the contact point is within the boundaries of the wall element. The next possible situation is contact on edges and the least probable one is the contact on a corner point. In the following, we present methods for detecting the contact point with one of these components in the order of their possibilities.

Contact point of a sphere with a plane is the orthogonal projection of the sphere center onto the plane. For a given plane with a unit normal vector $\vec{n}_w$ and point P_1, the contact point P is obtained from:

$$P = C - \tau \vec{n}_w \tag{3.49}$$

with

$$\tau = \vec{n}_w \cdot (C - P_1) \tag{3.50}$$

If we consider the implicit equation of the plane, it is given by:

$$P = C - \left(\vec{n}_w \cdot C + d\right)\vec{n}_w \tag{3.51}$$

Once the contact point is found, we should determine if the contact point is within the boundaries of the wall.

Test a Point against a Triangle

Consider a triangle with three vertex points on a plane. Any point P within this plane can be expressed by a parametric equation in barycentric coordinates of the point. If we define the two independent vectors of this triangle as follows:

$$\vec{v}_0 = P_2 - P_1 \tag{3.52a}$$

$$\vec{v}_1 = P_3 - P_1 \tag{3.52b}$$

we can specify any point on the plane of triangle by:

$$P = P_1 + \upsilon\, \vec{v}_0 + \omega\, \vec{v}_1 \tag{3.53}$$

By defining $\vec{v}_2 = P - P_1$ we have:

$$\vec{v}_2 = \upsilon\, \vec{v}_0 + \omega\, \vec{v}_1 \tag{3.54}$$

where υ and ω are barycentric coordinates of the point.[18] A system of two linear algebraic equations is obtained by taking dot products of $\vec{v}_0$ and $\vec{v}_1$ with Equation 3.54:

$$\upsilon\left(\vec{v}_0.\vec{v}_0\right)+\omega\left(\vec{v}_0.\vec{v}_1\right)=\vec{v}_0.\vec{v}_2 \tag{3.55a}$$

$$\upsilon\left(\vec{v}_1.\vec{v}_0\right)+\omega\left(\vec{v}_1.\vec{v}_1\right)=\vec{v}_1.\vec{v}_2 \tag{3.55b}$$

Solving this system of equations by the Cramer's rule gives the barycentric coordinates of the contact point. It can be seen in Figure 3.23 that the edges of triangle divide the triangle plane into seven regions that include one interior, three vertex regions, and three edge regions. Depending on the value of barycentric coordinates, we can find the location of point P on the triangle plane (see Figure 3.23). We seek to determine if the point lies inside the triangle (region 1), which holds the following conditions:

$$\upsilon \geq 0 \text{ and } \omega \geq 0 \text{ and } \upsilon+\omega \leq 1$$

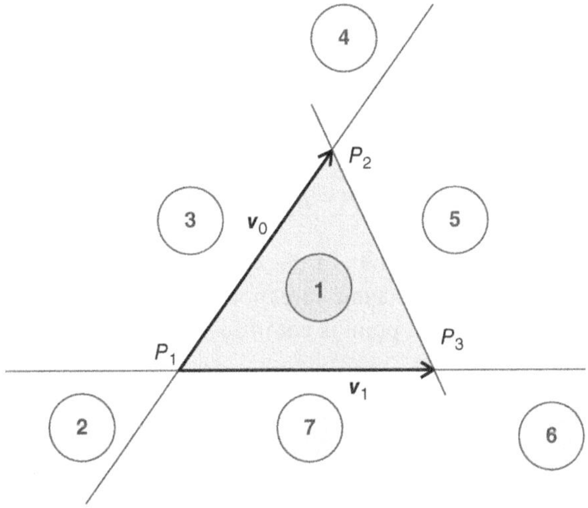

Region	Conditions
1	$\upsilon \geq 0$ and $\omega \geq 0$ and $\upsilon+\omega \leq 1$
2	$\upsilon \leq 0$ and $\omega \leq 0$ and $\upsilon+\omega < 0$
3	$\upsilon > 0$ and $\omega < 0$ and $\upsilon+\omega < 1$
4	$\upsilon \geq 1$ and $\omega \leq 0$ and $\upsilon+\omega > 1$
5	$\upsilon \geq 0$ and $\omega \geq 0$ and $\upsilon+\omega > 1$
6	$\upsilon \leq 0$ and $\omega \geq 1$ and $\upsilon+\omega > 1$
7	$\upsilon < 0$ and $\omega > 0$ and $\upsilon+\omega < 1$

Figure 3.23 The parametric representation of a triangle with barycentric coordinates that divides the plane into seven regions

[18] The standard form of the parametric equation of a triangle in barycentric coordinates is $P = uP_1 + \upsilon P_2 + \omega P_3$ with $u+\upsilon+\omega = 1$ and barycentric coordinates (u, υ, ω).

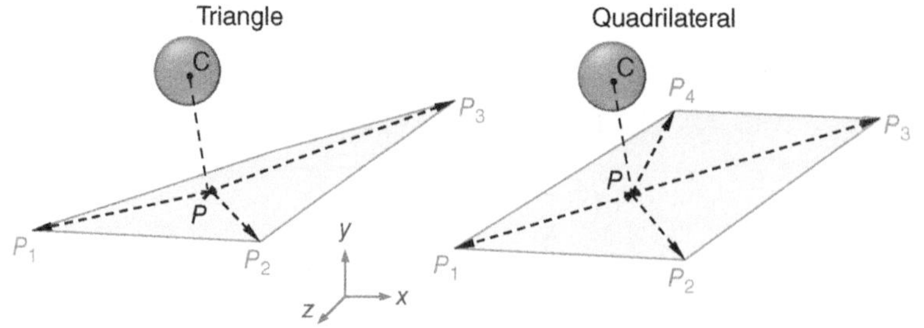

Figure 3.24 The process of contact detection between a spherical particle and triangular and quadrilateral wall elements

There is another method, shown in Figure 3.24, to test if the point P is inside a triangle. By considering P as the origin, we get three sub-triangles (P, P_1, P_2), (P, P_2, P_3), and (P, P_3, P_1). If the normal vectors of these sub-triangles, point to the same direction in the space, the point is inside the triangle. The following calculation should be done to test this condition:

$$\vec{u} = (P_1 - P) \times (P_2 - P) \quad (3.56a)$$

$$\vec{v} = (P_2 - P) \times (P_3 - P) \quad (3.56b)$$

$$\vec{w} = (P_3 - P) \times (P_1 - P) \quad (3.56c)$$

where $\vec{u}$, $\vec{v}$, and $\vec{w}$ are the normal vectors of the sub-triangles. When the dot product of two vectors is positive, they point to the same direction in the space. Therefore, if each of the following conditions is satisfied, the point is not inside the triangle:

$$\vec{u} \cdot \vec{v} < 0, \text{ or } \vec{u} \cdot \vec{w} < 0 \quad (3.57)$$

These conditions also cover the points located on the boundaries. In this case, the dot product of two vectors is zero because either of these vectors is zero.

Test a Point against a Quadrilateral

We extend the previously described methodology to test if the contact point P is within the boundaries of the quadrilateral with four non-collinear vertex points (see Figure 3.24). Point P is located inside the quadrilateral if the normal vectors of four sub-triangles, constructed by (P, P_1, P_2), (P, P_2, P_3), (P, P_3, P_4), and (P, P_4, P_1), point to the same direction in the space. To check this condition, the normal vectors of sub-triangles are calculated:

$$\vec{u} = (P_1 - P) \times (P_2 - P) \quad (3.58a)$$

$$\vec{v} = (P_2 - P) \times (P_3 - P) \quad (3.58b)$$

$$\vec{w} = (P_3 - P) \times (P_4 - P) \quad (3.58c)$$

$$\vec{x} = (P_4 - P) \times (P_1 - P) \quad (3.58d)$$

where $\vec{u}$, $\vec{v}$, $\vec{w}$, and $\vec{x}$ are normal vectors of sub-triangles. Similar to what we saw for triangle, if each of the following conditions is satisfied, point P is not inside the quadrilateral boundaries:

$$\vec{u}\cdot\vec{v}<0,\text{ or }\vec{u}\cdot\vec{w}<0,\text{ or }\vec{u}\cdot\vec{x}<0 \tag{3.59}$$

The use of cross product in this test imposes a high computational load to the program. Instead, we can use Lagrange's identity to convert the cross product into dot products. For example, for the first condition we have:

$$\vec{u}\cdot\vec{v}=\left[(P_1-P)\cdot(P_2-P)\right]\left[(P_2-P)\cdot(P_3-P)\right]-\left[(P_2-P)\cdot(P_2-P)\right]\left[(P_1-P)\cdot(P_3-P)\right] \tag{3.60}$$

Program Listing 3.21 shows implementation of this test into the program. This code tests if the projected point p is located within the boundaries of a planar quadrilateral (the code for testing a point against triangle is similar to this code). The program calculates the cross products according to the Lagrange's identity and checks the first condition of Equation 3.59 in lines 18–22. If this condition is satisfied, the program returns false value, meaning that the point is outside the wall element. The second condition in Equation 3.59 is tested in lines 26–29 and the third condition in lines 33 and 34. If none of the conditions is satisfied, the program returns a true value, meaning that the point is within the boundaries of the wall element.

Program Listing 3.21 Code for determining whether or not a point is in the plane boundaries defined in file g_PlaneWall.f90.

```
01  logical function PW_IsInPlane( this, p )
02      implicit none
03      class(PlaneWall)          this
04      type(real3),intent(in):: p
05
06      !// locals
07      real(RK) p1p2, p2p3, p2p2, p1p3, p2p4, p1p4, p1p1
08      type(real3) p1p, p2p, p3p, p4p
09
10      !// body
11      PW_IsInPlane = .false.
12      p1p = this%P1-p;
13      p2p = this%P2-p;
14      p3p = this%P3-p;
15
16      ! first condition u.v < 0
17      ! u.v= [(p1-p)x(p2-p)].[(p2-p)x(p3-p)]= (p1p.p2p)(p2p.p3p) - (p2p.p2p)(p1p.p3p)
18      p1p2 = p1p .dot. p2p
19      p2p3 = p2p .dot. p3p
20      p2p2 = p2p .dot. p2p
21      p1p3 = p1p .dot. p3p
22      if( p1p2*p2p3 - p2p2*p1p3 < 0.0_RK ) return
23
24     ! second condition u.w < 0
```

```
        ! u.w= [(p1-p)x(p2-p)].[(p3-p)x(p4-p)]= (p1p.p3p)(p2p.p4p) - (p2p.p3p)(p1p.p4p)
        p4p = this%P4-p;
        p2p4 = p2p .dot. p4p
        p1p4 = p1p .dot. p4p
        if( p1p3*p2p4 - p2p3*p1p4 < 0.0_RK ) return

        ! third condition u.x < 0
        ! u.x= [(p1-p)x(p2-p)].[(p4-p)x(p1-p)]= (p1p.p4p)(p2p.p1p) - (p2p.p4p)(p1p.p1p)
        p1p1 = p1p .dot. p1p
        if( p1p4*p1p2 - p2p4*p1p1 < 0.0_RK ) return

        PW_IsInPlane = .true.
        return

    end function
```

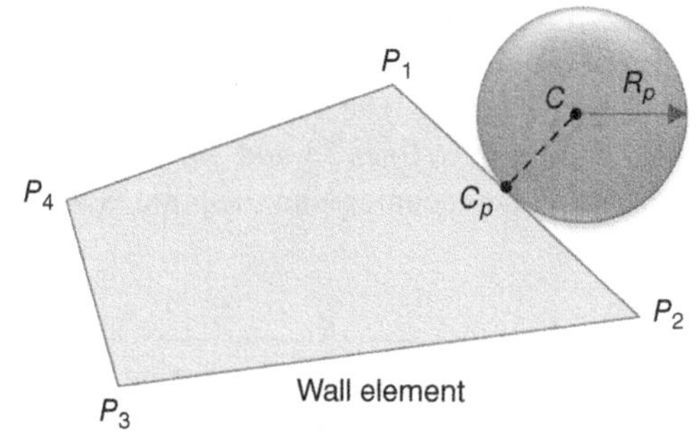

Figure 3.25 Testing a sphere against one edge of the wall element

Test a Particle against Edges and Corners

Each edge of the wall element can be described by a parametric line in 3D space as follows:

$$L(r) = (P_2 - P_1)r + P_1 \tag{3.61}$$

where P_1 and P_2 are two corner points of the edge and r is a real number. This equation gives points on the line, between P_1 and P_2 if $0 \le r \le 1$, before P_1 if $r < 0$ and after P_2 if $r > 1$. As shown in Figure 3.25, the center of particle is first projected on the line to check the location of a spherical particle with radius of R_p against this edge:

$$r = \frac{(C - P_1).(P_2 - P_1)}{(P_2 - P_1).(P_2 - P_1)} \tag{3.62}$$

for $0 \le r \le 1$, the projected point is located between points P_1 and P_2, and the contact point is obtained by substituting r into Equation 3.61. If the distance between center point and projected point is less than the radius of particle, the particle is in contact with the edge. It is also possible that the particle is in contact with one of the corner points. If either of the following conditions is held, then the contact of particle with corner points should be checked.

$$\begin{cases} \dfrac{-R_P}{|P_2 - P_1|} \le r < 0 & \text{contact check with } P_1 \quad (3.63a) \\ 1 < r \le 1 + \dfrac{R_P}{|P_2 - P_1|} & \text{contact check with } P_2 \quad (3.63b) \end{cases}$$

Program Listing 3.22 illustrates the code for determining if a sphere has a contact point with a wall element. First, the distance and the overlap between particle and wall are calculated in lines 17–23. If the overlap is positive, particle may be in contact with the wall element. The program first checks if the particle has contact with the surface (within the boundaries of the wall). For this purpose, the orthogonal projection of the center of sphere onto the wall is first found by invoking method `NearestPointOnPlane` and is passed to method `IsInPlane` in line 27. `IsInPlane` (described in Program Listing 3.21) returns true if the projected point is located within the boundaries. Otherwise, the program checks the particle against edges and corners in line 32.

Program Listing 3.22 Code for determining whether a sphere is in contact with a wall element in file g_PlaneWall.f90.

```
01  logical function PW_isInContact(this, box )
02      implicit none
03      class(PlaneWall)          this
04      type(real4),intent(in):: box   ! sphere
05
06      !// loclas
07      real(RK)     rad, dist, ovrlp
08      type(real3) p
09
10      !// body
11
12      rad = 0.5_RK * box%w
13      p = box ! conversion from real4 to real3
14      PW_isInContact = .false.
15
16      ! first check if there is an overlap between sphere and wall
17      dist = this%PointFromPlane(p)
18      if( this%bothside) then
19          ovrlp = rad - abs(dist)
20      else
21          if( dist < 0.0_RK ) return ! in-active side of wall
22          ovrlp = rad - dist
23      end if
24
25      if( ovrlp .gt. 0.0_RK) then
26          ! now checks if the contact point is located within the plane boundaries
27          if( this%IsInPlane( this%NearestPointOnPlane(p) ) ) then
28              PW_isInContact = .true.
```

```
                return
            end if
            !checks if the particle has contact with edges or corner points
            if( this%IsOnLines(box) ) then
                PW_isInContact = .true.
                return
            end if

        end if

    end function
```

In many situations, a coplanar wall comprises of a number of wall elements. The previously mentioned method for detecting the contact point may lead to multiple contacts (contact of a particle with multiple wall elements of a coplanar wall) when the particle is located between common edges or common corners of two adjacent wall elements. This issue has been addressed elsewhere [63, 64] and a solution to avoid this situation is proposed. The code does not prevent this case, although the probability of occurrence of such condition and its effect on final results is not clear to us.

3.6.3 Moving Wall

According to the explicit representation of wall, a wall is comprised of one or more plane triangle or quadrilateral elements (wall elements). A wall, and hence its constituent wall elements, can be stationary or moving. Locations of wall elements are fixed during a simulation with stationary walls while they change in a simulation involving moving walls. In general, motion of the wall in a discrete element simulation can be described by (i) a predefined velocity for the wall, (ii) specifying external forces or moments acting on the wall, or (iii) in response to the force experienced by particles (two-way momentum coupling between wall and particles). Predefined velocity for the wall is used when it moves as a result of an external actuator like shaft (e.g., rotating drums and mixers). In this case, the influence of particle load on the wall is negligible and can be ignored. In some circumstances, particle load affects the movement of wall. In this case, the rigid body dynamics should be applied to obtain the movement of wall as a result of particle load as well as predefined external force and moments on the wall, if any.

In this book (and the program), we only consider the moving wall with a predefined velocity. To represent a moving wall element in the program, we need to add rotational and translational velocities to the stationary wall element (`PlaneWall`). Therefore, type `mvng_PlaneWall` is derived from type `PlaneWall` and new functionalities are added to it to handle wall motion in the program (see Program Listing 3.23). Data member `trans_vel` is used for defining translational velocity for the wall and data members `rot_vel` and `rot_line` are used to define the rotational velocity around an arbitrary axis, respectively.

As shown in Program Listing 3.23, motion of the wall is described by three translational velocities in the x-, y-, and z-directions and a rotational velocity around an arbitrary axis in 3D space. A combination of these velocities can be used to obtain the desired movement pattern of the wall. Translational velocities should be defined with respect to a space-fixed frame (global frame of reference) and the rotational velocity with respect to the axis of rotation. Direction of the rotation obeys the right-hand rule in the positive direction of the axis of rotation. In the

Program Listing 3.23 Definition of type mvng_PlaneWall in file g_PlaneWall.f90 to represent moving wall elements.

```
type, extends(PlaneWall):: mvng_PlaneWall
    logical       :: moving    = .false. ! if it linearly moves
    logical       :: rotating  = .false. ! if it rotates
    type(real3)   :: trans_vel = zero_r3 ! translational velocity, initially zero
    real(RK)      :: rot_vel   = 0.0_RK  ! rotational velocity, initially zero
    type(p_line):: rot_line              ! rotation axis

    contains ! associated methods

    . . .

end type !// moving plane wall
```

program, wall elements only move from one time step to another and they are assumed to be fixed during each time step. In spite of this assumption, we still need the velocity of the wall at the contact point to calculate the particle-wall contact force. The program should calculate the wall velocity in the space-fixed frame based on specified velocities for the wall.

The program moves a wall according to its translational and rotational motions in each time step. If translational velocity of the wall element is $\vec{V}_{wall}$, the displacement of an arbitrary point P on it during a time step becomes:

$$P^{n+1} = P^{n} + \vec{V}_{wall}\Delta t_{p} \tag{3.64}$$

Also, wall points are rotated by $d\theta$ around axis of rotation in each time step:

$$d\theta = \omega_{wall}\Delta t_{p} \tag{3.65}$$

where ω_{wall} is rotational velocity of the wall. Axis of rotation in the program is a parametric line defined by two points O_1 and O_2. The normalized direction vector of this line is:

$$\vec{n} = \frac{O_2 - O_1}{\left|O_2 - O_1\right|} \tag{3.66}$$

The following equation is used to calculate new coordinates of point $O(x, y, z)$ on the wall around axis of rotation by the angle of $d\theta$:

$$\begin{bmatrix} x' \\ y' \\ z' \end{bmatrix} = \begin{bmatrix} \left(a\left(n_y^2 + n_z^2\right) - n_x\left(bn_y + cn_z - n_x x - n_y y - n_z z\right)\right)\left(1 - \cos d\theta\right) + x\cos d\theta \\ +\left(-cn_y + bn_z - n_z y + n_y z\right)\sin d\theta \\ \left(b\left(n_x^2 + n_z^2\right) - n_y\left(an_x + cn_z - n_x x - n_y y - n_z z\right)\right)\left(1 - \cos d\theta\right) + y\cos d\theta \\ +\left(cn_x - an_z + n_z x - n_x z\right)\sin d\theta \\ \left(c\left(n_x^2 + n_y^2\right) - n_z\left(an_x + bn_y - n_x x - n_y y - n_z z\right)\right)\left(1 - \cos d\theta\right) + z\cos d\theta \\ +\left(-bn_x + an_y - n_y x + n_x y\right)\sin d\theta \end{bmatrix} \tag{3.67}$$

in which (a, b, c) is the coordinates of point O_1 and (x', y', z') is the new coordinates of point O as a result of rotation.

To move a wall element, program moves its corner points based on its translational and rotational velocities and updates all related variables of the wall element, like its normal vector and parametric equation of its edges. The corner points first undergo a translational motion according to Equation 3.64, then a rotational motion according to Equation 3.67.

3.7 Parallelization

Parallelization falls into two categories: data parallelization and task parallelization.[19] Data parallelization means doing the same task on different sets of data, which is shared between processing elements (processors or threads). This can be done on modern processors, which are capable of pipelining or graphical processing units (GPUs). Task parallelization is doing different tasks on the same or different data on each processing element simultaneously. These processing elements communicate with one another by special functions through exchanging data and signals. It should be noted that the parallelization of a real problem (DEM in our case) is not resolved by purely utilizing either of these models. Many parallel models take advantage of both to achieve the maximum speed up. The most common parallelization models on CPUs are distributed-memory message passing interface (MPI) and shared-memory multiprocessing, like OpenMP®.

In this section, we give an overview on these two parallelization models. We do not get into the details of parallel programming here since it is beyond the scope of this book. The interested reader is referred to [66–69] for MPI and OpenMP® programming models. This section also provides some useful tips on how to parallelize a serial code and to decide which programming model is suitable for the application. Some basics of two parallelization models and their strengths and weak points are also discussed.

3.7.1 Distributed Memory Parallelization

Distributed memory computer architecture comprises of connected individual computational nodes, each having its own memory and processing unit. The computational node itself may be a shared-memory computer. In the MPI model, the program is copied to all processors and they are executed concurrently. The execution flows of program in the processors are controlled by synchronization messages. Processors can communicate with each other and exchange data by calling send/receive routines. Although the MPI is designed for distributed memory architecture, it also can be executed on shared-memory computers, which exploit low communication latency between processors. If the program is well designed, the parallel code is scalable, meaning that it can be executed on one processor to more than thousands processors. A scalable code is also portable, which means that it can be implemented on any parallel computer in which processors can independently execute codes and

[19] The classification presented here is arbitrary. Other types of classification for the level of parallelization can be found in [65].

communicate with each other via message passing. With the aid of parallelization, a system containing several millions of particles can be simulated within a reasonable computation time [70, 71]. This enables us to broaden the application of DEM to larger and more complex systems.

The most challenging part of a distributed memory parallel programming is the load distribution among processors. In molecular dynamics simulations, the computational load can be distributed among processors by utilizing atom decomposition, force decomposition, and space/domain decomposition techniques [72]. Domain decomposition [73–75], mirror domain (data replication) [43], and particle subset [76] methods are the most common approaches [77, 78] used for discrete element simulations. Among them, domain decomposition is the most efficient and common approach for distributing computational load between processors.

In the domain decomposition method, the simulation domain is subdivided into distinct regions. Particles are assigned to these regions based on their spatial positions. A processor is assigned to each domain and particles' data are kept in the dedicated memory of the processor. At boundaries between two regions, particles interact with each other via contact or non-contact (long-range) forces. These regions should have overlap to share particle data between processors. Particles in the overlapping regions are called ghost particles. The extent of this overlap depends on the type of interaction forces between particles. Obviously, the length of overlapping regions is more for long-range forces, in which the interaction forces act in longer distances (i.e., several times of particle diameter), in comparison with contact force in which interactions exist at the particle surface. As particles move from one region to another, all particle data, including position, velocity, radius, contact history, and so on, should be transferred to the corresponding processor. Therefore, each processor performs typical calculation steps of a DEM problem for a given set of particles (which is a fraction of the total number of particles) within a certain domain (local simulation domain) and parallelization can be done based on previously developed serial code with addition of special functionalities for data transfer among processors, processors synchronization, and load balancing as shown in Figure 3.26.

Granularity of the problem is very important for efficient scaling of massage passing model. Granularity is the ratio of computational load to communications. Based on this definition, a coarse-grained problem involves a large amount of computations in each processor with a few communications among them. In contrast, a fine-grained problem involves a few computations with many communications. Low communication bandwidth between nodes/processors and overheads associated with it make the data transfer between processors a time consuming operation. There are some factors that decrease the granularity of a DEM simulation:

- *Substantial movement of particles*: It is required to perform data transfer between processors when particles move from one region to another. The more significant the movement of particles, the more data transfer is required.
- *Increase in the extent of overlap of regions*: The extent of overlap of regions depends on the interaction distance between particles. The number of ghost particles increases by enlarging the overlapping regions. Hence, the size of data to be transferred for sharing particle data between processors increases. The extent of overlap of regions depends on the size of largest particle in problems without long-range forces. In such situations, wide size distribution of

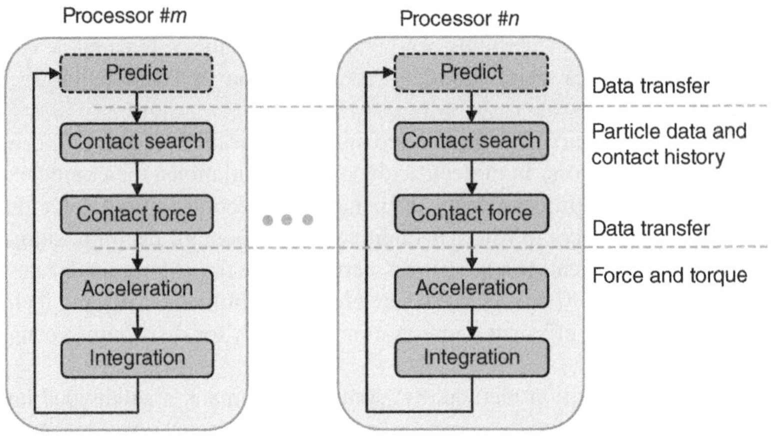

Figure 3.26 Message passing parallelization of a typical DEM code based on the space decomposition approach with overlapping regions. Each processor has its own set of particles and simulation domain. Each processor performs typical steps of a serial code independently to find the trajectory and velocities of its dedicated particles. Dashed lines show the points that processors exchange data of particles in the overlapping region. The output module is not shown here. Writing to the file is logically serial and data from all processors are sent to a target processor and then the target processor writes data into a file

particles also increases the data exchange load between processors since more small particles can reside in overlapping regions.

- *Increase in the number of processors for a certain problem*: Efficiency of parallelization decreases as the number of processors is increased (relative to the total number of particles). This is mainly due to the fact that a larger fraction of particles should be considered as ghost particles when larger number of processors is used.

The scaling behavior and efficiency of parallelization improves with increasing the granularity of the problem. Efficiency of the parallelization is the ratio of speedup to the number of processors, defined as:

$$\varepsilon_{Np} = \frac{t_{serial}}{N_P \, t_{parallel}} \times 100 \tag{3.68}$$

where t_{serial} and $t_{parallel}$ are run times of serial and parallel versions of the code, respectively, and N_p is the number of processors. Fleissner and Eberhard [79] showed that with increasing the problem size (number of particles), a better scalability is obtained in dynamic discrete element simulations. Shojaaee *et al.* [80] obtained the same results in quasi-static simulations where the efficiency remained around 100% for 1 million particles on 256 processors.

Load balancing and minimizing the overheads associated with it are the most challenging aspects of the distributed-memory programming. Computational load of each processor is directly connected to how the entire simulation domain is divided into distinct regions. A simple way for doing this is to divide the simulation domain into N_p regions with equal volumes. This can be a good decomposition for problems in which the entire simulation domain is homogeneously filled with particles during the simulation. However, heterogeneity in the spatial distribution of particles dramatically unbalances the computational load of processors and leads to idle time of processors with lower loads. Fluidization of particles, discharge of vessels, particle flow in a double cone blender, and settling of particles are all examples in which the spatial distribution of particles dynamically changes during the simulation.

Some methods for load balancing can be found in literature [79, 80]. Generally, the space decomposition and particle distribution among processors should be done in such a way that an almost equal number of particles are assigned to each processor while at the same time, the number of contact points between particles of two adjacent regions should be as few as possible [76]. Redistribution of particles can be done after completion of each time step. However, since the movement of particles is very limited in each time step, this should be done every several time steps. Figure 3.27 shows a typical domain decomposition and redistribution of particles among processors. The redistribution is done by moving the boundaries of each region. When the standard deviation of computation time of processors exceeds a predefined value, this redistribution can be performed.

3.7.2 Shared-Memory Parallelization

The other parallelization model is accomplished on the shared-memory multiprocessor architecture. OpenMP® is a programming interface that supports shared-memory parallelization in compilers like C++ and FORTRAN. It is a collection of compiler directives, runtime routines, environmental variables, and synchronization and work-sharing constructs. This interface is a thread-based programing in which the threads are executed concurrently. All threads have access to variables in the shared-memory (with application life time). In addition, a thread can have its own private data which is accessible within that thread (with thread life time). OpenMP® programming model allows the user to use the same source code lines with both OpenMP-enabled and normal compilers. This is accomplished by using special compiler directives that a normal compiler misses them. To convert a serial code into a parallel one, we should identify parallelizable parts of the code and start from the most time-consuming part. At the end, the final code is composed of serial and parallel parts/regions, which are executed one after another (see Figure 3.28).

Before we give the idea of parallelization of a DEM code, we illustrate how a serial code is modified to perform tasks in parallel. Suppose that integer coordinates of all spheres should be calculated. The serial code looks like this:

Program Listing 3.24 A simple serial loop for calculating integer coordinates of particles.

```
do i= 1, N
     ix = (pos(i)%x - xmin)/dx + 1
     iy = (pos(i)%y - ymin)/dx + 1
     iz = (pos(i)%z - zmin)/dx + 1
end do
```

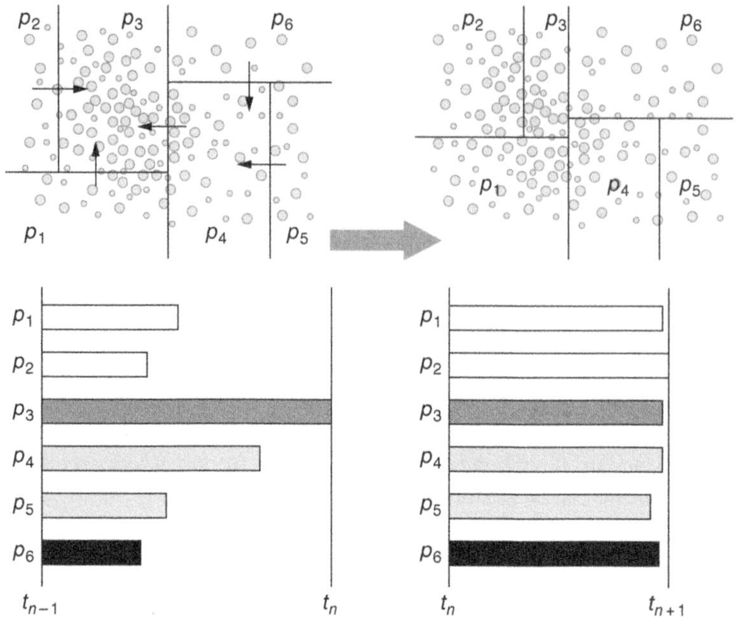

Figure 3.27 Typical orthogonal recursive bisection domain decomposition for six processors (left). The computational load of processors is diverse that leads to idle time of some processors. The boundaries are shifted to redistribute particles among processors (right). The computational load of processors is balanced now. (Source: Fleissner, F. and Eberhard, P. (2007) Load Balanced Parallel Simulation of Particle-Fluid DEM-SPH Systems with Moving Boundaries. In *Parallel Computing: Architectures, Algorithms and Applications, Vol. 38* (ed. C. Bischof, M. Bucker, P. Gibbon, G.R. Joubert, T. Lippert, B. Mohr, and F. Peters), John von Neumann Institute for Computing, Julich. [79] with permission.)

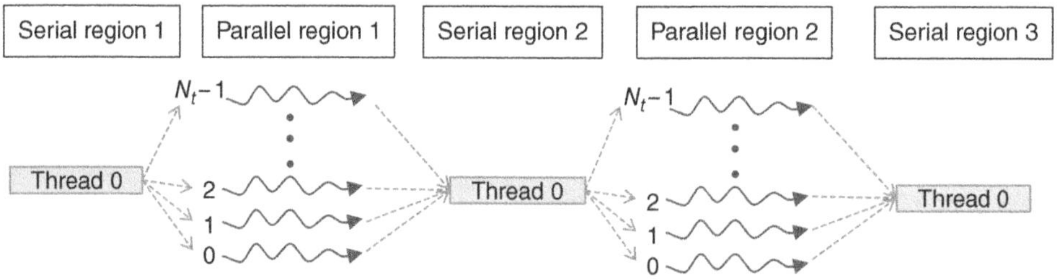

Figure 3.28 The execution paths of serial and parallel regions of the code. Execution proceeds with thread 0 in serial region. Once this thread reaches the parallel region, it creates a team of N_t threads that are executed concurrently and thread 0 becomes the master thread in this team. At the end of parallel region, all private variables are deleted and all threads except the master thread are destroyed. Thread 0 continues the execution of the next serial region. At the end of the parallel region, master thread waits for other threads to finish their execution (implicit synchronization)

In this code, the integer coordinates of sphere are calculated one after another. The parallel code is shown in Program Listing 3.25:

Program Listing 3.25 Parallelized loop shown in Program Listing 3.24.

```
01  !$OMP PARALLEL DO DEFAULT (SHARED) PRIVATE(i, ix, iy, iz)
02  do i= 1, N
03      ix = (pos(i)%x - xmin)/dx + 1
04      iy = (pos(i)%y - ymin)/dx + 1
05      iz = (pos(i)%z - zmin)/dx + 1
06  end do
07  !$OMP END PARALLEL DO
```

Clauses at lines 1 and 7 are a compiler directive-pair that indicate the parallel region. All lines that start with `!$OMP` are omitted by a normal compiler while they are read by an OpenMP-capable compiler. The directive-pair `!$OMP PARALLEL DO`/`!$OMP END PARALLEL DO` makes the following do loop to be executed in parallel. In this example, the work is distributed over different threads and each thread computes a portion of iterations. For example, when the total number of spheres is 1000 ($N = 1000$) and four threads are executing in the parallel region, each thread executes 250 iterations of the do loop. This simple solution only applies to situations in which threads perform independent tasks and there is no true dependency.[20] As we will discuss in the following paragraph, many do-loops in a DEM code have this property and they can be parallelized in a similar way. The clauses `DEFAULT (SHARED)` and `PRIVATE` are used to define the shared and private variables in the parallel region, respectively. These clauses tell the compiler that all variables used in this parallel region are shared by default except variables `i`, `ix`, `iy`, and `iz`, which are private variables.

We showed different modules of a DEM code in Figure 3.1. Among them, modules 3–7 possess more than 95% of the computational load. Fortunately, we can parallelize them directly or by making changes to the algorithm. Module 8 involves write operations to the file. These operations are serial in nature and we cannot parallelize them. Modules 3, 6, and 7 involve loops over all spheres. Since there is no data dependency between iterations of the loops, these loops can be parallelized in a way similar to what we explained in the previous paragraph. Modules 4 and 5 involve steps that may lead to data racing conflicts. Data racing conflict occurs when two different threads try to concurrently perform read and write operations to the same memory location (variable). This causes incorrect results. In module 5, the contact properties, including contact force and torque of contacting pairs, are calculated. Calculated values should be added to the total force acting on the particle. This summation involves read and write operations. If two threads want to update the total contact force of one particle at the same time, data racing conflict occurs. This can be resolve using atomic or reduction operations. In Program Listing 3.26, we show that how the atomic add can be

[20] Data dependency arises in different situations. For example, it exists when one thread needs a variable that is calculated in another thread or another statement. There are three types of data dependencies including true, anti, and output dependencies. All dependencies can be parallelized with additional consideration except true dependency. In some cases, true dependency can also be removed by reduction operations and the remaining cases with true dependency, change in the algorithm is required for parallelization.

Program Listing 3.26 Atomic operation for adding the contact force, fc, to the total contact force of particle *i*, force(i).

```
. . . // previous lines //
!$OMP ATOMIC
force(i)%x = force(i)%x + fc%x
!$OMP ATOMIC
force(i)%y = force(i)%y + fc%y
!$OMP ATOMIC
force(i)%z = force(i)%z + fc%z
. . . //next lines//
```

exploited to overcome the possible data racing among threads when adding the contact force, fc, to the total contact force of particle *i*, force (i).

All pairwise contacts should be found and added to the contact list in module 4. This step possesses the largest computational load and is the most challenging step in parallelization. Cell-based algorithms described in Section 3.3.2 cannot be used in a parallel code in their present form. Data racing conflict may occur when two different threads try to concurrently insert two particles in the same cell. This does not happen often, but we must avoid it since the constructed linked list may miss some particles in it. One can solve this problem using locks. The thread that inserts a particle in the linked list of a cell sets a lock for that cell (using unique flags for each cell) and prevents other threads to insert particles in the same cell until the operation is complete. Using locks imposes some overhead on the execution of program and slows down the program. In addition, it is possible to face deadlocks that cause program failure. Sort-based algorithms described in Section 3.3.3 can be used instead. The mapping step involves sorting data pairs in these algorithms. Sorting data is the computational bottleneck in these algorithms. We can use high performance parallel sort algorithms, like radix sort and quick sort, to accelerate the contact detection phase. Other sort-based algorithms, which are developed for the shared-memory multiprocessor architecture, can be found in literature [20, 21, 81, 82].

In each module of a DEM code, loop level load distribution is performed in the OpenMP® model. Therefore, it does not have load balancing difficulties of the MPI model. OpenMP® shows scalability comparable to the MPI model. Although many parts of the DEM can be directly parallelized, some parts need some changes in the algorithm. We should take care of data dependencies and data racing conflicts and remove them if possible. The MPI can be executed on both shared- and distributed-memory architecture, while the OpenMP® model is restricted to shared-memory architecture. This may impose some limitations on available memory resources.

A question that arises here is how much speedup we can gain by converting a serial code into a parallel one. For example, can we expect a 64× speedup when executing the parallel code with 64 processors? Usually, not all computations associated with a program are parallelizable and a fraction of the code remains serial. Therefore, the maximum expected speedup should be lower than 64×. Amdahl's law gives the maximum speed up of a parallel program for the same serial program that is executing on N_t threads as follows [83]:

$$speedup(N_t) = \frac{1}{f + \frac{1}{N_t}(1-f)} \tag{3.69}$$

where f is the fraction of computations that is strictly serial. Therefore, considering that 95% of computations are performed in parallel, we must expect 15.4 × speed up when we execute it on 64 processors. The real performance is still lower than this value due to limited memory/communication band width and idle time of processors due to unbalanced computational loads, locks, and synchronization among processors.

Nomenclature

Symbol	Unit	Description
$\vec{a}$	m/s^2	Linear acceleration vector of particle
a_{max}	m/s^2	Maximum particle acceleration in the system
$\vec{a}_n$	m/s^2	Acceleration vector of particle at current time step (time step n)
$\vec{a}_{n+1}^{corr}$	m/s^2	Corrected acceleration vector at the next time step
$\vec{a}_{n+1}^{p}$	m/s^2	Predicted acceleration vector at the next time step
Al	—	Identifier of the lower bound of bounding sphere A
Au	—	Identifier of the upper bound of bounding sphere A
$\vec{b}_n$	m/s^3	First derivatives of acceleration vector
$\vec{b}_{n+1}^{p}$	m/s^3	Predicted value of the first derivatives of acceleration vector at the next time step
$\vec{c}_n$	m/s^4	Second derivatives of acceleration vector
$\vec{c}_{n+1}^{p}$	m/s^4	Predicted value of the second derivatives of acceleration vector at the next time step
C	m	Coordinates of particle center point
CL_i	—	Contact list of particle i
d	m	Nearest distance of plane with origin of coordinates system
d_i	m	Diameter of bounding sphere i
$dist$	m	Closest distance between a point and a plane
d_{max}	m	Diameter of the largest bounding sphere in the system
$\vec{d}_n$	m/s^5	Third derivatives of acceleration vector
$\vec{d}_{n+1}^{p}$	m/s^5	Predicted value of the third derivatives of acceleration vector at next time step
dx	m	Size of cells in cell-based algorithms
$d\theta$	rad	Incremental change in rotation of a wall around rotation axis during one time step
E_{eff}	Pa	Effective Young's modulus
f	—	Fraction of computations that is strictly serial
f_{nat}	Hz	Un-damped natural frequency
$f_{n,max}$	N	Maximum normal contact force
$\vec{f}_i^{ext}$	N	External forces acting on particle i
$\vec{f}_i^{f-p}$	N	Total fluid-particle interaction force on particle i
$\vec{f}_{ij}^{p-p}$	N	Interaction force between particles i and j
G	Pa	Shear modulus
ix	—	Integerized x-coordinate of bounding sphere
iy	—	Integerized y-coordinate of bounding sphere
iz	—	Integerized z-coordinate of bounding sphere
I_i	kg·m^2	Moment of inertia of particle i

k, k_n	N/m	Spring stiffness
l_i^{cell}	m	Cell size in grid level i
L	—	Number of grid levels
$L(r)$	m	Function of a parametric line
LB_A	m	Lower bound of bounding sphere A
m	kg	Mass of particle
m_i	kg	Mass of particle i
m_{eff}	kg	Effective mass
$\vec{M}_{ij}^r$	N·m	Rolling resistance torque on particle i due to contact with particle j
$\vec{M}_{ij}^t$	N·m	Tangential torque on particle i due to contact with particle j
n	—	Number of levels in a tree
$\vec{n}$	—	Normalized direction vector of parametric line
$\vec{n}_w$	—	Normal vector of plane
$n_{w,x}, n_{w,y}, n_{w,z}$	—	Components of unit normal vector $\vec{n}_w$
nx	—	Number of cells in the x-direction
ny	—	Number of cells in the y-direction
nz	—	Number of cells in the z-direction
N	—	Number of particles in the system
N_P	—	Number of processors
N_t	—	Number of threads
$ovrlp$	m	Overlap of bounding spheres i and j
O_1, O_2	m	Coordinates of the start and end points of the axis of rotation of a wall
P	m	Contact point
P_1, P_2, P_3, P_4	m	Corner points of a wall element
P^n	m	Coordinates of an arbitrary point on wall at time step n
r	—	Parameter variable in the parametric equation of line
r_{max}	m	Radius of largest sphere in system
R_p	m	Particle radius
RAM	B	RAM requirement
RAM_{2D}	B	RAM requirement for 2D simulation
RAM_{3D}	B	RAM requirement for 3D simulation
R_{eff}	m	Effective radius
R_{min}	m	Minimum proximity distance
t	s	Time
t_{col}	s	Collision time
$t_{parallel}$	s	Computational time of the parallel code
t_{serial}	s	Computational time of the serial code
T_1	s	Computational time related to indexing of particles
T_2	s	Computational time related to mapping of particles
T_3	s	Computational time related to broad search
UB_A	m	Upper bound of bounding sphere A
(u, υ, ω)	—	Barycentric coordinates
$\vec{u}, \vec{v}, \vec{w}, \vec{x}$	m	Normal vectors in sub-triangles in testing a point against wall algorithm
$\vec{v}_0, \vec{v}_1$	m	Vectors defining two edges of a triangular plane
$\vec{v}_n$	m/s	Velocity of particle at the current time step (time step n)
$\vec{v}_{n+1}^P$	m/s	Predicted velocity at the next time step
$v_{rn,imp}$	m/s	Relative normal velocity at impact

V_{max}	m/s	Maximum absolute relative velocity
$V_{w,max}$	m/s	Maximum velocity magnitude of wall points in the system
$V_{p,max}$	m/s	Maximum velocity magnitude of particles in the system
$\vec{x}_i$	m	Vector center position of particle or bounding sphere i
x_i	m	x-Coordinate of bounding sphere i
$\vec{x}_n$	m	Coordinates of particle at current time step (time step n)
x_{max}	m	Maximum x-coordinate of simulation domain
x_{min}	m	Minimum x-coordinate of simulation domain
y_i	m	y-Coordinate of bounding sphere i
y_{max}	m	Maximum y-coordinate of simulation domain
y_{min}	m	Minimum y-coordinate of simulation domain
z_i	m	z-Coordinate of bounding sphere i
z_{max}	m	Maximum z-coordinate of simulation domain
z_{min}	m	Minimum z-coordinate of simulation domain

Greek symbols

α	—	Time-step resolution
$\delta_1 - \delta_5$	—	Adams–Moulton integration coefficients
$\Delta\vec{a}_{n+1}$	m/s^2	Deviation of acceleration
Δx_{max}	m	Maximum relative displacement
$\delta_{n,max}$	m	Normal relative overlap
$\hat{\delta}_{n,max}$	—	Dimensionless normal overlap
Δt_{crit}	s	Critical integration time step
Δt_p	s	Integration time step for particle
ε_{Np}	—	Efficiency of parallelization on N_p processors
$\eta_1 - \eta_6$	—	Gear's method integration coefficients
κ	—	Ratio of tangential spring stiffness to normal spring stiffness
$\lambda_1 - \lambda_5$	—	Adams–Bashforth integration coefficients
ν	—	Poisson's ratio
ρ	kg/m^3	Density of particle
τ	m	Parameter in Equation 3.49
$\vec{\varphi}_i, \vec{\varphi}$	rad	Angular position of particle i
χ	—	Parameter in Equation 3.1
ω_{wall}	rad/s	Rotational speed of wall around the axis of rotation
$\vec{\omega}_i, \vec{\omega}$	rad/s	Rotational velocity of particle i
ω_{nat}	rad/s	Un-damped natural angular velocity

References

[1] Li, Y., Xu, Y., and Thornton, C. (2005) A comparison of discrete element simulations and experiments for "sandpiles" composed of spherical particles. *Powder Technology*, **160**(3), 219–228.

[2] O'Sullivan, C. and Bray, J.D. (2004) Selecting a suitable time step for discrete element simulations that use the central difference time integration scheme. *Engineering Computations*, **21**(2–4), 278–303.

[3] Langston, P.A., Tuzun, U., and Heyes, D.M. (1994) Continuous potential discrete particle simulations of stress and velocity-fields in hoppers-transition from fluid to granular flow. *Chemical Engineering Science*, **49**, 1259–1275.

[4] Cundall, P.A. and Strack, O.D.L. (1979) A discrete numerical model for granular assemblies. *Geotechnique*, **29**, 47–65.

[5] Tsuji, Y., Kawaguchi, T., and Tanaka, T. (1993) Discrete particle simulation of two-dimensional fluidized bed. *Powder Technology*, **77**(1), 79–87.

[6] Ahrens, J., Geveci, B., and Law, C. (2005) ParaView: an end-user tool for large data visualization. In *Visualization Handbook* (eds C.D. Hansen and C.R. Johnson), Academic Press/Elsevier, Oxford.

[7] Munjiza, A. (2004) *The Combined Finite-Discrete Element Method*, John Wiley & Sons, Ltd., Chichester.

[8] Zhou, Z.Y., Pinson, D., Zou, R.P., and Yu, A.B. (2011) Discrete particle simulation of gas fluidization of ellipsoidal particles. *Chemical Engineering Science*, **66**, 6128–6145.

[9] Hilton, J.E., Mason, L.R., and P.W. Cleary (2010) Dynamics of gas–solid fluidised beds with non-spherical particle geometry. *Chemical Engineering Science*, **65**, 1584–1596.

[10] Williams, J.R., Perkins, E., and Cook, B. (2004) A contact algorithm for partitioning arbitrary sized objects. *Engineering Computations*, **21**(2–4), 235–248.

[11] Walizer, L.E. and Peters, J.F. (2011) A bounding box search algorithm for DEM simulation. *Computer Physics Communications*, **182**(2), 281–288.

[12] Munjiza, A. and Andrews, K.R.F. (1998) NBS contact detection algorithm for bodies of similar size. *International Journal for Numerical Methods in Engineering*, **43**, 131–149.

[13] Peters, J.F., Kala, R., and Maier, R.S. (2009) A hierarchical search algorithm for discrete element method of greatly differing particle sizes. *Engineering Computations*, **26**(6), 621–634.

[14] He, K., Dong, S., and Zhou, Z. (2007) Multigrid contact detection method. *Physical Review E*, **75**(3), 036710.

[15] Lu, L., Gu, Z., Lei, K., Wang, S., and Kase, K. (2010) An efficient algorithm for detecting particle contact in non-uniform size particulate system. *Particuology*, **8**(2), 127–132.

[16] Ogarko, V. and Luding, S. (2012) A fast multilevel algorithm for contact detection of arbitrarily polydisperse objects. *Computer Physics Communications*, **183**(4), 931–936.

[17] Perkins, E. and Williams, J.R. (2001) A fast contact detection algorithm insensitive to object sizes. *Engineering Computations*, **18**, 48–61.

[18] Li, C.F., Feng, Y.T., and Owen, D.R.J. (2006) SMB: collision detection based on temporal coherence. *Computer Methods in Applied Mechanics and Engineering*, **195**(19–22), 2252–2269.

[19] Yao, Z., Wang, J., Liu, G., and Cheng, M. (2004) Improved neighbor list algorithm in molecular simulations using cell decomposition and data sorting method. *Computer Physics Communications*, **161**, 27–35.

[20] Mazhar, H., Heyn, T., and Negrut, D. (2011) A scalable parallel method for large collision detection problems. *Multibody System Dynamics*, **26**(1), 37–55.

[21] Pabst, S., Koch, A., and Straßer, W. (2010) Fast and scalable CPU/GPU collision detection for rigid and deformable surfaces. *Computer Graphics Forum*, **29**(5), 1605–1612.

[22] Green, S., (2012) Particle simulation using CUDA, in *CUDA Documentation*, Nvidia Corporation.

[23] Belytschko, T., Liu, W.K., Moran, B., and Elkhodary, K. (2013) *Nonlinear Finite Elements for Continua and Structures*, 2nd edn, John Wiley & Sons, Ltd., Chichester.

[24] Rougier, E., Munjiza, A., and John, N.W.M. (2004) Numerical comparison of some explicit time integration schemes used in DEM, FEM/DEM and molecular dynamics. *International Journal for Numerical Methods in Engineering*, **61**, 856–879.

[25] Di Renzo, A. and Di Maio, F.P. (2004) Comparison of contact-force models for the simulation of collisions in DEM-based granular flow codes. *Chemical Engineering Science*, **59**, 525–541.

[26] Hoomans, B.P.B., Kuipers, J.A.M., and van Swaaij, W.P.M. (2000) Granular dynamics simulation of segregation phenomena in bubbling gas-fluidised beds. *Powder Technology*, **109**, 41–48.

[27] Mansourpour, Z., Karimi, S., Zarghami, R., Mostoufi, N., and Sotudeh-Gharebagh, R. (2010) Insights in hydrodynamics of bubbling fluidized beds at elevated pressure by DEM–CFD approach. *Particuology*, **8**, 407–414.

[28] Mansourpour, Z., Mostoufi, N., and Sotudeh-Gharebagh, R. (2013) Mechanistic study of agglomerating fluidized bed by numerical simulation. *Canadian Journal of Chemical Engineering*, **91**, 560–569.

[29] Norouzi, H.R., Mostoufi, N., Mansourpour, Z., Sotudeh-Gharebagh, R., and Chaouki, J. (2011) Characterization of solids mixing patterns in bubbling fluidized beds. *Chemical Engineering Research and Design*, **89**, 817–826.

[30] Tsuji, Y., Tanaka, T., and Ishida, T. (1992) Lagrangian numerical simulation of plug flow of cohesionless particles in a horizontal pipe. *Powder Technology*, **71**, 239–250.

[31] Cromer, A. (1981) Stable solutions using the Euler approximation. *American Journal of Physics*, **49**(5), 455–459.

[32] Taguchi, Y.-H. (1992) Powder turbulence: direct onset of turbulent flow. *Journal de Physique II France*, **2**, 2103–2114.

[33] Kruggel-Emden, H., Sturm, M., Wirtz, S., and Scherer, V. (2008) Selection of an appropriate time integration scheme for the discrete element method (DEM). *Computers and Chemical Engineering*, **32**, 2263–2279.

[34] Kwapinska, M., Saage, G., and Sotsas, E. (2006) Mixing of particles in rotary drums: a comparison of discrete element simulations with experimental results and penetration models for thermal processes. *Powder Technology*, **161**(1), 69–78.

[35] Dang, H.K. and Meguid, M.A. (2013) An efficient finite-discrete element method for quasi-static nonlinear soil-structure interaction problems. *International Journal for Numerical and Analytical Methods in Geomechanics*, **37**(2), 130–149.

[36] Ferziger, J.H. and Peric, M. (2002) *Computational Methods for Fluid Dynamics*, 3rd edn, Springer-Verlag, Berlin/Heidelberg.

[37] Melheim, J.A. (2005) Cluster integration method in Lagrangian particle dynamics. *Computer Physics Communications*, **171**(3), 155–161.

[38] Shida, K., Suzuki, R., and Kawai, T. (1997) Numerical error of total energy. dependence on timestep. *Computer Physics Communications*, **102**(1–3), 59–67.

[39] Verlet, L. (1967) Computer experiments on classical fluids. I. Thermodynamical properties of Lennard–Jones molecules. *Physical Review*, **159**(1), 98–103.

[40] Pandey, P., Song, Y., Kayihan, F., and Turton, R. (2006) Simulation of particle movement in a pan coating device using discrete element modeling and its comparison with video-imaging experiments. *Powder Technology*, **161**(2), 79–88.

[41] Takeuchi, S., Wang, S., and Rhodes, M. (2004) Discrete element simulation of a flat-bottomed spouted bed in the 3D cylindrical coordinate system. *Chemical Engineering Science*, **59**(17), 3495–3504.

[42] Marshall, J.S. (2009) Discrete-element modeling of particulate aerosol flows. *Journal of Computational Physics*, **228**(5), 1541–1561.

[43] Poschel, T. and Schwager, T. (2005) *Computational Granular Dynamics: Models and Algorithms*, Springer-Verlag, Berlin/Heidelberg.

[44] Hassani, M.A., Zarghami, R., Norouzi, H.R., and Mostoufi, N. (2013) Numerical investigation of effect of electrostatic forces on the hydrodynamics of gas-solid fluidized beds. *Powder Technology*, **246**, 16–25.

[45] Peters, B., Dziugys, A., Hunsinger, H., and Krebs, L. (2005) An approach to qualify the intensity of mixing on a forward acting grate. *Chemical Engineering Science*, **60**(6), 1649–1659.

[46] Balevičius, R., Džiugys, A., Kačianauskas, R., Maknickas, A., and Vislavičius, K. (2006) Investigation of performance of programming approaches and languages used for numerical simulation of granular material by the discrete element method. *Computer Physics Communications*, **175**(6), 404–415.

[47] Tuley, R., Danby, M., Shrimpton, J., and Palmer, M. (2010) On the optimal numerical time integration for Lagrangian DEM within implicit flow solvers. *Computers and Chemical Engineering*, **34**, 886–899.

[48] Zhang, Y., Jin, B., Zhong, W., Ren, B., and Xiao, R. (2010) DEM simulation of particle mixing in flat-bottom spout-fluid bed. *Chemical Engineering Research and Design*, **88**(5–6), 757–771.

[49] Kaneko, Y., Shiojima, T., and Horio, M. (1999) DEM simulation of fluidized beds for gas-phase olefin polymerization. *Chemical Engineering Science*, **54**, 5809–5821.

[50] Norouzi, H.R., Mostoufi, N., and Sotudeh-Gharebagh, R. (2012) Effect of fines on segregation of binary mixtures in gas–solid fluidized beds. *Powder Technology*, **225**, 7–20.

[51] Kuo, H.P., Knight, P.C., Parker, D.J., Tsuji, Y., Adams, M.J., and Seville, J.P.K. (2002) The influence of DEM simulation parameters on the particle behaviour in a V-mixer. *Chemical Engineering Science*, **57**(17), 3621–3638.

[52] Di Renzo, A., Di Maio, F.P., Girimonte, R., and Formisani, B. (2008) DEM simulation of the mixing equilibrium in fluidized beds of two solids differing in density. *Powder Technology*, **184**(2), 214–223.

[53] Li, J., Langston, P.A., Webb, C., and Dyakowski, T. (2004) Flow of sphero-disc particles in rectangular hoppers—a DEM and experimental comparison in 3D. *Chemical Engineering Science*, **59**(24), 5917–5929.

[54] Ketterhagen, W.R., Curtis, J.S., Wassgren, C.R., Kong, A., Narayan, P.J., and Hancock, B.C. (2007) Granular segregation in discharging cylindrical hoppers: a discrete element and experimental study. *Chemical Engineering Science*, **62**(22), 6423–6439.

[55] Li, Q., Rudolph, V., Wang, F., Kajikawa, S.-I., and Horio, M. (2001) Axial porosity distribution in a packed bed of deformable particles: a numerical study based on DEM. In *Handbook of Conveying and Handling of Particulate Solids* (eds A. Levy and H. Kalman), Elsevier Science B.V., Amsterdam.
[56] Malone, K.F. and Xu, B.H. (2008) Determination of contact parameters for discrete element method simulations of granular systems. *Particuology*, **6**, 521–528.
[57] Zhou, Z.Y., Yu, A.B., and Zulli, P. (2009) Particle scale study of heat transfer in packed and bubbling fluidized beds. *AIChE Journal*, **55**, 868–884.
[58] Malone, K.F. and Xu, B.H. (2008) Particle-scale simulation of heat transfer in liquid-fluidised beds. *Powder Technology*, **184**, 189–204.
[59] Mansourpour, Z., Mostoufi, N., and Sotudeh-Gharebagh, R. (2013) A mechanistic study of agglomeration in fluidised beds at elevated pressures. *The Canadian Journal of Chemical Engineering*, **91**(3), 560–569.
[60] Seville, J.P.K., Willett, C.D., and Knight, P.C. (2000) Interparticle forces in fluidisation: a review. *Powder Technology*, **113**, 261–268.
[61] Zhou, Z.Y., Yu, A.B., and Zulli, P. (2010) A new computational method for studying heat transfer in fluid bed reactors. *Powder Technology*, **197**, 102–110.
[62] Katok, A. and Climenhaga, V. (2008) *Lectures on Surfaces: (Almost) Everything You Wanted to Know About Them*, American Mathematical Society, Pennsylvania.
[63] Hu, L., Hu, G.M., Fang, Z.Q., and Zhang, Y. (2013) A new algorithm for contact detection between spherical particle and triangulated mesh boundary in discrete element method simulations. *International Journal for Numerical Methods in Engineering*, **94**(8), 787–804.
[64] Kremmer, M. and Favier, J.F. (2001) A method for representing boundaries in discrete element modelling – part I: geometry and contact detection. *International Journal for Numerical Methods in Engineering*, **51**(12), 1407–1421.
[65] Rauber, T. and Runger, G. (2013) *Parallel Programming for Multicore and Cluster Systems*, 2nd edn, Springer, Dordrecht/Heidelberg/London/New York.
[66] Chandra, R., Menon, R., Dagum, L., Kohr, D., Maydan, D., and McDonald, J. (2000) *Parallel Programming in OpenMP*, Morgan Kaufmann (Academic Press), Burlington, MA.
[67] Pacheco, P. (2011) *An Introduction to Parallel Programming*, Morgan Kaufmann (Elsevier), Burlington, MA.
[68] Quinn, M.J. (2003) *Parallel Programming in C with MPI and OpenMP*, McGraw-Hill, New York.
[69] Hatcher, P.J. and Quinn, M.J. (1991) *Data-Parallel Programming on MIMD Computers*, The MIT Press, Massachusetts.
[70] Radeke, C.A., Glasser, B.J., and Khinast, J.G. (2010) Large-scale powder mixer simulations using massively parallel GPU architectures. *Chemical Engineering Science*, **65**, 6435–6442.
[71] Capecelatro, J. and Desjardins, O. (2013) An Euler–Lagrange strategy for simulating particle-laden flows. *Journal of Computational Physics*, **238**, 1–31.
[72] Plimpton, S. (1995) Fast parallel algorithms for short range molecular dynamics. *Journal of Computational Physics*, **117**, 1–19.
[73] Tsuji, T., Yabumoto, K., and Tanaka, T. (2008) Spontaneous structures in three-dimensional bubbling gas-fluidized bed by parallel DEM–CFD coupling simulation. *Powder Technology*, **184**, 132–140.
[74] Munjiza, A., Walther, J.H., and Sbalzarini, I.F. (2009) Large-scale parallel discrete element simulations of granular flow. *Engineering Computations*, **26**(6), 688–697.
[75] Sbalzarini, I.F., Walther, J.H., Bergdorf, M., Hieber, S.E., Kotsalis, E.M., and Koumoutsakos, P. (2006) PPM – A highly efficient parallel particle–mesh library for the simulation of continuum systems. *Journal of Computational Physics*, **215**(2), 566–588.
[76] Kafui, D.K., Johnson, S., Thornton, C., and Seville, J.P.K. (2011) Parallelization of a Lagrangian–Eulerian DEM/CFD code for application to fluidized beds. *Powder Technology*, **207**, 270–278.
[77] Schafer, B.C., Quigley, S.F., and Chan, A.H.C. (2004) Acceleration of the Discrete Element Method (DEM) on a reconfigurable co-processor. *Computers and Structures*, **82**, 1707–1718.
[78] Smith, B., Bjorstad, P., and W., Gropp (1996) *Domain Decomposition: Parallel Multilevel Methods for Elliptic Partial Differential Equations*, Cambridge University Press, Cambridge.
[79] Fleissner, F. and Eberhard, P. (2007) Load balanced parallel simulation of particle-fluid DEM-SPH systems with moving boundaries. in *Parallel Computing: Architectures, Algorithms and Applications*, Vol. **38** (eds C. Bischof,

M. Bucker, P. Gibbon, G.R. Joubert, T. Lippert, B. Mohr, and F. Peters), John von Neumann Institute for Computing, Julich.

[80] Shojaaee, Z., Shaebani, M.R., Brendel, L., Torok, J., and Wolf, D.E. (2012) An adaptive hierarchical domain decomposition method for parallel contact dynamics simulations of granular materials. *Journal of Computational Physics*, **231**(2), 612–628.

[81] Chen, J. and Matuttis, H.-G. (2013) Optimization and OpenMP parallelization of a discrete element code for convex polyhedra on multi-core machines. *International Journal of Modern Physics C*, **24**(2), 1350001.

[82] Grand, S.L. (2007) Broad-phase collision detection with CUDA. in *GPU Gems 3* (ed. H. Nguyen), Addison-Wesley Professional, Indiana.

[83] Amdahl, G.M. (1967) Validity of the single processor approach to achieving large scale computing capabilities. AFIPS Conference Proceedings.

4

Non-Spherical Particles

In Chapters 2 and 3 we addressed the basic formulations and implementation of discrete-element method (DEM) modeling for spherical particles. The particles were assumed spherical with smooth surface, as their geometries are represented by their center points and radii. Computational costs associated with representing their geometries, tracking them, and finding the contact plane/point are low. Therefore, simulation with a large number of spherical particles is possible with today's computational facilities, letting us simulate granular flows in pilot and industrial scales.

Many practical granular flows involve non-spherical particles that may be near round (sphericity close to unity) or with irregular shapes. Estimating the shape of near round particles with spheres in these flows is a reasonable assumption for reproducing the characteristics of such granular flows. However, this assumption is far from reality for particles with irregular shape, like ovals, cylinders, ellipsoids or particles with sharp edges, flat surfaces, and so on. Using non-spherical particles in the DEM, as compared with spherical particles, requires additional computational costs due to implementing more complex algorithms for contact detection, tracking particles, and contact force calculations. This is compulsory for performing accurate DEM simulation, even though the costs are high.

In this chapter, first we discuss the effect of particle shape and various approaches for representing particle shape in the DEM. Then, the basic formulas of kinematics and dynamics of a non-spherical rigid body in 3D space are presented. Euler angles and quaternions are discussed as two main tools for representing particle orientation and transformations between different frames of references. Thereafter, we will have a look at superellipsoid and multi-sphere methods as two different methods for representing non-spherical particles in a simulation. Algorithms of finding contact plane, evaluating contact forces in each method are also addressed.

Coupled CFD-DEM Modeling: Formulation, Implementation and Application to Multiphase Flows, First Edition.
Hamid Reza Norouzi, Reza Zarghami, Rahmat Sotudeh-Gharebagh and Navid Mostoufi.

Companion website: www.wiley.com/go/norouzi/CFD-DEM

4.1 Shape Representation

Many experimental observations and simulations have proved the significant effect of particle shape on powder behavior. For example, Höhner *et al.* [1] performed DEM simulations to study the granular flow of spherical and non-spherical particles in a discharging hopper. They used multi-sphere and polyhedral methods to represent shape. Their results showed that increasing the particle angularity (from a particle with smooth surface like a sphere, to a particle with edges like a cube) reduces the mass flow rate out of the hopper orifice and increases the chance of formation of arch-like void structure above the orifice. It was shown that non-spherical particles almost have higher rolling resistance than spherical particle and their flowability is less than spherical particles [2, 3]. The heap formed by non-spherical particles has larger static angle of repose than that formed by spherical particle. In addition, non-spherical particles in rotating drums show larger dynamic angle of repose [4]. The packing behavior (packing density and coordination number) and fluidization behavior of non-spherical particles are also different from spherical particles [5]. These examples clearly show the significance of particle shape on the behavior of powders.

The differences between dynamics of spherical and non-spherical particles stem from some sources. The only mechanism that acts against the rotation of a spherical particle (in the absence of fluid viscous effects) is rolling friction while the rotation of a non-spherical particle is hindered by interlocks with other particles. As we get far from the spherical shape, more interlocks occur between particles, which hinder the rolling tendency. The symmetry of the sphere imposes no preferred orientation in the assembly of particles, while some non-spherical particles (those with large aspect ratios) have a preferred orientation. Accordingly, the coordination number and the interaction forces of a non-spherical particle are different than those of a sphere. Due to the symmetry of the contacting spheres, the normal force (which acts along the line connecting the two centers) does not contribute to the torque acting on a sphere while it contributes to the torque acting on a non-spherical particle. These are the main sources of deviation of the dynamics of spherical particles from that of non-spherical particles [6, 7]. As an example, we illustrated the rolling of a spherical and a non-spherical particle over a slope in Figure 4.1. Both particles are initially at rest and are allowed to roll down on the surface due to gravity. It can be seen that the sphere travels along a straight line, while the non-spherical particle travels on a zigzag path and its orientation continuously changes. This figure clearly shows that how a simple rolling could be different for the spherical and non-spherical particle.

In the implementation of DEM for non-spherical particles, four issues should be addressed: the kinematics of the particle, the method of presenting particle shape to DEM, calculation of the contact force, and finding the contact plane between particles. Equation of rotational motion of a non-spherical particle is different from that of a sphere. A method should be therefore considered for defining particle orientation in 3D space and solving Euler's equations. A sphere can be represented by its center point and radius in the computer, while shape representation of a non-spherical particle (exact or approximate) requires additional data. Shape representation is still one of the key challenges in the DEM modeling of granular flows. According to the way that particle shape is represented, the method of finding a contact plane between particles and force calculation should be changed.

A variety of methods for representing particle shapes can be found in the literature. They are categorized into two main groups: single-element and multi-element particle approaches.

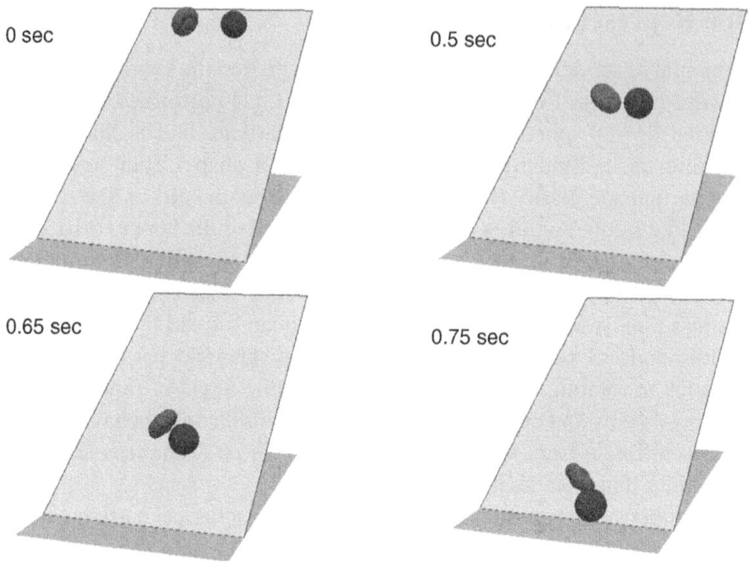

Figure 4.1 Rolling of a spherical and a non-spherical particle over a slope. Both particles are initially at rest and they are allowed to roll down the surface due to gravity

In the single-element approach, the surface of the particle is defined by a set of continuous analytical equations. In this way, a particle with smooth surface and no bump is obtained. Using a *superellipsoid* (or in a more general term, superquadrics) is the most important method in this group. A superellipsoid is a smooth surface defined by an implicit equation, in form of $F(x, y, z) = 0$ in 3D space. By changing the parameters in this equation, different shapes with a closed surface can be obtained. Superellipsoids have been used in various DEM simulation of granular flows like hopper discharge [3, 8], flow pattern of spherical and non-spherical particles in screw feeders [9] and gas fluidization of ellipsoidal particles [5, 10]. Although this method can provide an accurate estimation of the particle surface, not all shapes can be represented by this method. Especially, a particle with sharp edges cannot be represented by this family of equations.

Perhaps the most challenging issue associated with the use of the family of superellipsoids is detection of contact between two particles. This requires solving a set of non-linear equations numerically by methods like the Newton's method or the secant method [11]. Usually, initial guess for starting the iterative solution affects number of iterations required to find the contact plane. Sometimes, the iterative solution may even diverge. Therefore, a suitable algorithm for determining the initial guess should be used. The problem with the initial guess and larger number of iterations become serious when particles first come into contact. In the next time steps, since particles do not move significantly relative to each other, the initial guess would be the contact plane in the previous time step and a fewer number of iterations would be required. In some regions of surface where the slope is gentle, the iterative method needs extra iterations to converge into the solution and in regions with a steep slope, the solution may diverge and relaxation may be required. A number of contact detection algorithms for superellipsoidal/ellipsoidal particles can be found in the literature [11–14].

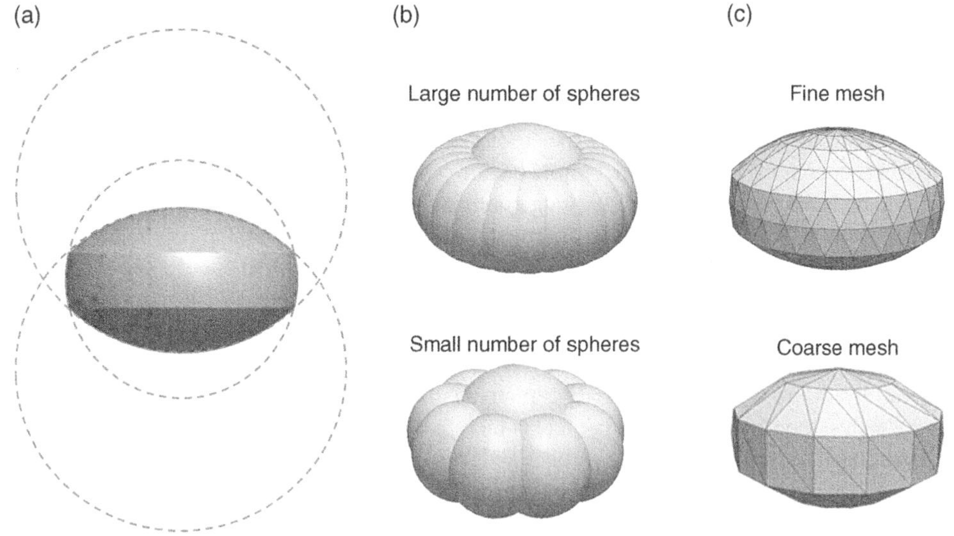

Figure 4.2 Three different methods to representing the shape of a bi-convex tablet (a) intersecting surfaces: the surface of tablet is formed by intersecting surfaces of three spheres, (b) multi-sphere: the surface of tablet is approximated by gluing small spheres, and (c) polyhedral: the surface of tablet is covered by triangles

Smoothed surface of particles can be resembled by intersecting surfaces of spheres or other simple geometric shapes. For example, Song *et al.* [15] represented the surface of a standard round tablet with three spherical surfaces (see Figure 4.2a). They developed a contact search algorithm that contains nine criteria for different possible contact points (three for particle-wall and six for particle–particle). The advantage of the method of intersecting shapes is obtaining a smooth surface with edges, which accurately represents the real surface. However, from one shape to another, the method for representing the shape and contact detection algorithm should be changed.

In the multi-element particle approach, complex surface of the particle is constructed from smaller sub-elements (with simple shapes) that are connected to each other. Among different methods in this group, the multi-sphere method, also called glued spheres or clustered spheres, is one of the most common one. The complex surface of a particle can be formed by overlapped spheres. By increasing the number of spheres used in the construction of the complex particle, the real surface would be represented more accurately. However, regardless of the number of spheres, the resultant surface would be bumpy and without a sharp edge or vertex (see Figure 4.2b). Multi-sphere DEM simulations have been used for studying the dynamics of tablets with various shapes in pan coaters [16–18], agglomerating fluidized beds [19], and mixing of corn-shaped particles in spouted beds [20, 21] as well as filling/discharge of rice grains to/from a hopper and their pilling [1, 22].

A DEM code for spherical particle can be easily extended to handle multi-sphere method. The contact search and contact force calculations are similar, although calculation of force needs some modifications, which will be discussed later in this chapter. The multi-sphere method is not limited to a particular shape or a family of geometric shapes and theoretically can produce any complex surface, except surfaces with sharp edges. By increasing the number of spheres in a particle, the particle surface is represented more accurately, while at the same time it increases the computational costs and errors due to multiple contact points. Thus, there is a trade-off between accuracy and computational cost associated with the multi-sphere method. Similar to multi-sphere particles, other rigid particles with a simple geometry can be glued to each other to form more complex shapes. For example, spherocylinders can be glued side by side to produce long rods. The packing behavior of long rods is of great interest in adsorbing catalytic beds, metal casting, and studying pore structure of polymeric fibers [23–25].

Particles with irregular shapes and sharp edges can be generated by the polyhedral method. Surface of a polyhedral particle consists of triangular sub-elements. Other polygons also can be used instead of triangle. These triangles can be rigidly connected to each other to create a rigid body (see Figure 4.2c). Alternatively, they can be connected to each other by elastic, deformable, dissipative bounds to create deformable shapes. The shape of a deformable particle can be changed by external forces and bending moments acting on it. By increasing the number of sub-elements, the surface of a real particle is more accurately approximated, but it needs higher computational resources for storing vertices and face data in the memory and performing contact detection tests between triangles that belong to different particles. The polyhedral method is very similar to the multi-sphere method. Interaction between two particles is calculated based on the interaction of triangles that constitute these particles while in the multi-sphere method this is done through the constituent spheres. The differences between polyhedral and multi-sphere methods are contact detection methods and calculation of contact force between sub-elements. A number of algorithms have been suggested for contact detection and contact force calculations in the literature [26–28]. The reader is referred to [29, 30] for more details on various aspects of simulating the flow of particles with irregular shapes using the polyhedral method.

4.2 Kinematics and Dynamics of a Rigid Body

As it was discussed earlier, a non-spherical particle may be a single body whose surface is defined by analytical equations or a body composed of sub-elements that are rigidly connected to each other. Therefore, the basic principles of the kinematics and dynamics of the non-spherical rigid body apply to both groups. However, each group may require additional considerations that will be specifically discussed.

The position of a non-spherical rigid body in the space is defined by the location of its center of mass and its orientation. In Cartesian coordinates, two different coordinates systems are required. First, an orthogonal body-fixed frame (local coordinates) located on the center of mass and moves and rotates with the rigid body.[1] Second, an orthogonal space-fixed frame (global coordinates) that is inertial and it does not move or rotate in the space. In a DEM simulation, all variable such as position, orientation, and velocities of the particle, as well as location of walls, are defined in this frame. The transformation of vector variables between the body-fixed and space-fixed frames are frequently performed to compute contact force and torque as well as to solve equations of motion. The location and orientation of the body-fixed frame (hence the body itself) is defined relative to the space-fixed frame. Therefore, a convention should be used to transform variables from the space-fixed frame to the body-fixed frame and vice versa. This can be done using the transformation matrix, which will be presented in the following section.

[1] The body-fixed frame should also coincide with principal axes of the rigid body.

4.2.1 Euler Angles and Transformation Matrix

In 2D space, three scalar variables (three degrees of freedom) are required to define the rigid body motion: two for coordinates of the center of mass and one for the rotation angle θ. The rotation by the angle θ around the center of mass is simply done using the following rotation matrix:

$$A_\theta = \begin{bmatrix} \cos\theta & -\sin\theta \\ \sin\theta & \cos\theta \end{bmatrix} \tag{4.1}$$

In 3D space, six scalar variables (six degrees of freedom) are required for defining the rigid body motion: three for the coordinates of the center of mass and three angles for specifying the orientation of the body in the space. Defining the orientation of body in the space and transformation between the space- and body-fixed frames in 3D space is not as simple as 2D space. There is a variety of methods to define the orientation of a body in 3D space. The most well-known is by using Euler angles.

Defining the orientation of the body-fixed frame and transformation of vector variables from the space-fixed frame to the body-fixed frame and vice versa are possible using Euler angles. According to the classical Euler angles, the orientation of the body-fixed frame is defined by three Euler angles ϕ, θ, and ψ. Consider the spaced-fixed frame XYZ. It is desirable to perform a set of rotations to transform this frame to the body-fixed frame xyz. For this purpose, the following steps should be carried out in sequence:

- First, a rotation is done by ϕ around Z-axis that results in a new coordinate system $x'y'z'$ as shown in Figure 4.3a.
- Then, in the $x'y'z'$ coordinates system, the second rotation is done by θ around x'-axis, which gives a new coordinate system $x''y''z''$ as shown in Figure 4.3b.
- Finally, in the $x''y''z''$ coordinates system, the third rotation is done by ψ around z''-axis and gives the body-fixed frame coordinates xyz as shown in Figure 4.3c.

The rotation matrices for these steps are as follows:

$$A_\phi = \begin{pmatrix} \cos\phi & -\sin\phi & 0 \\ \sin\phi & \cos\phi & 0 \\ 0 & 0 & 1 \end{pmatrix} \qquad 0 \le \phi < 2\pi \tag{4.2a}$$

$$A_\theta = \begin{pmatrix} 1 & 0 & 0 \\ 0 & \cos\theta & \sin\theta \\ 0 & -\sin\theta & \cos\theta \end{pmatrix} \qquad 0 \le \theta \le \pi \tag{4.2b}$$

$$A_\psi = \begin{pmatrix} \cos\psi & -\sin\psi & 0 \\ \sin\psi & \cos\psi & 0 \\ 0 & 0 & 1 \end{pmatrix} \qquad 0 \le \psi < 2\pi \tag{4.2c}$$

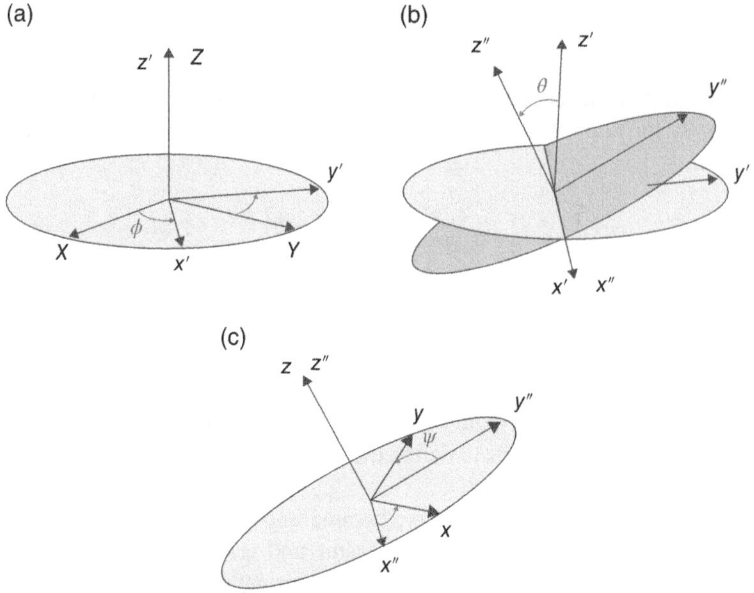

Figure 4.3 Three subsequent rotations (zxz convention) based on Euler angles. (a) rotation by ϕ around the z-axis, (b) rotation by θ around the x′-axis, and (c) rotation by ψ around teh z″-axis

In general, one can combine these three rotations into a single rotation by the following matrix:

$$
\begin{aligned}
A &= A_\psi A_\theta A_\phi \\
&= \begin{bmatrix} \cos\phi\cos\psi - \sin\phi\cos\theta\sin\psi & \sin\phi\cos\psi + \cos\phi\cos\theta\sin\psi & \sin\theta\sin\psi \\ -\cos\phi\sin\psi - \sin\phi\cos\theta\cos\psi & -\sin\phi\sin\psi + \cos\phi\cos\theta\cos\psi & \sin\theta\cos\psi \\ \sin\phi\sin\theta & -\cos\phi\sin\theta & \cos\theta \end{bmatrix}
\end{aligned} \tag{4.3}
$$

where A is the transformation matrix that transforms an arbitrary vector $\vec{x}^s$ in the space-fixed frame to $\vec{x}^b$ in the body-fixed frame by the following equation:

$$\vec{x}^b = A\vec{x}^s \tag{4.4}$$

The reverse of this rotation can be done to transform an arbitrary vector $\vec{x}^b$ from the body-fixed frame to the space-fixed frame. This is done by multiplying two sides of Equation 4.4 by A^{-1}:

$$\vec{x}^s = A^{-1}\vec{x}^b \tag{4.5}$$

Since the rotation matrix is orthogonal, $A^{-1} = A^T$. Thus, we can use the transpose of A instead of its inverse, which is less computationally intensive.

Each element of the transformation matrix A is a direct cosine of the angle between an axis of the body-fixed frame and the space-fixed frame. For example, A_{xZ} specifies the cosine of the angle between the x-axis of the body-fixed frame and Z-axis of the space-fixed frame. Therefore, the transformation matrix also describes the relative orientation of the two-coordinate systems with nine parameters. For instance, the first column of A, transforms the unit vector *i* of the space-fixed frame to the body-fixed frame. Similarly, the second column transforms the unit vector *j* and the third column, the unit vector *k* [31].

We are also interested in knowing the relation between the angular velocity of body in the body-fixed frame and incremental change (time derivatives) of Euler angles. The absolute angular velocity in the body-fixed frame is the vector sum of time derivatives change of Euler angles:

$$\omega^b = \dot{\phi} + \dot{\theta} + \dot{\psi} \tag{4.6}$$

It is relatively easy to show that [31]:

$$\omega_x^b = \dot{\phi}\sin\theta\sin\psi + \dot{\theta}\cos\psi \tag{4.7a}$$

$$\omega_y^b = \dot{\phi}\sin\theta\cos\psi - \dot{\theta}\sin\psi \tag{4.7b}$$

$$\omega_z^b = \dot{\psi} + \dot{\phi}\cos\theta \tag{4.7c}$$

If we solve these three equations for $\dot{\phi}$, $\dot{\theta}$, and $\dot{\psi}$, we get time derivatives of Euler angles in terms of fixed-body angular velocity:

$$\dot{\phi} = \frac{\omega_x^b \sin\psi + \omega_y^b \cos\psi}{\sin\theta} \tag{4.8a}$$

$$\dot{\theta} = \omega_x^b \cos\psi - \omega_y^b \sin\psi \tag{4.8b}$$

$$\dot{\psi} = \omega_z^b - \dot{\phi}\cos\theta \tag{4.8c}$$

Equations 4.8a–4.8c are important relations, since the angular equation of motion is solved in the orthogonal body-fixed frame and $\vec{\omega}^b$ is obtained, while non-orthogonal Euler angles are used to represent the orientation. Thus, these equations are used to obtain new orientation of the body in each time-step of simulation by integrating time derivatives of Euler angles.

4.2.2 Equations of Motion

The motion of a body in 3D space is defined by translational motion of the center of mass and rotational motion about the center of mass. The general form of the translational equation of motion of the center mass is:

$$m_i \vec{a}_i = m_i \frac{d\vec{v}_i}{dt} = \vec{f}_i \tag{4.9}$$

where $\vec{f}_i$ is sum of forces (gravity, contact force, fluid-particle force, etc.) acting on the center of mass of the body in the space-fixed frame and $\vec{v}_i$ is the translational velocity of the center of mass in the space-fixed frame. If we compare this equation with Equation 2.24a, we find out that they both are similar and we can solve the translational motion of the center of mass of a non-spherical body similar to a spherical particle.

The rotational motion of a rigid body in 3D space is given by:

$$\dot{\vec{L}} = \vec{M} \tag{4.10}$$

where $\dot{\vec{L}}$ is the rate of change of angular momentum and $\vec{M}$ is sum of torques acting on a particle. The reference point to evaluate torques and angular momentum should be similar and it can be any arbitrary point that moves with the center of mass of the body. It is more common to choose the center of mass as the reference point. Then, the angular momentum reads as:

$$\vec{L}^b = I\,\vec{\omega}^b \tag{4.11}$$

where I is the inertia tensor of the body at its center of mass and is defined as:

$$I = \begin{bmatrix} I_{xx} & -I_{xy} & -I_{xz} \\ -I_{yx} & I_{yy} & -I_{yx} \\ -I_{zx} & -I_{zy} & I_{zz} \end{bmatrix} \tag{4.12}$$

Diagonal elements of this matrix are called centroidal mass moment of inertia and off-diagonal elements are called centroidal mass product of inertia. The derivative of angular momentum of the rigid body is obtained by [32]:

$$\dot{\vec{L}}^b = I\dot{\vec{\omega}}^b + \vec{\omega}^b \times I\vec{\omega}^b \tag{4.13}$$

Therefore, the equation of rotational motion around its center of mass in the body-fixed frame becomes:

$$\vec{M}^b = I\dot{\vec{\omega}}^b + \vec{\omega}^b \times I\vec{\omega}^b \tag{4.14}$$

and in matrix form:

$$\begin{bmatrix} M_x^b \\ M_y^b \\ M_z^b \end{bmatrix} = I \begin{bmatrix} \dot{\omega}_x^b \\ \dot{\omega}_y^b \\ \dot{\omega}_z^b \end{bmatrix} + \begin{bmatrix} 0 & -\omega_z^b & \omega_y^b \\ \omega_z^b & 0 & -\omega_x^b \\ -\omega_y^b & \omega_x^b & 0 \end{bmatrix} I \begin{bmatrix} \omega_x^b \\ \omega_y^b \\ \omega_z^b \end{bmatrix} \tag{4.15}$$

This set of equations should be solved to obtain the rotational velocity of body in the fixed-body frame. If the reference point is kept on the center of mass and the body-fixed coordinates is aligned to coincide with principal axes of the body, then the off-diagonal elements of the inertia tensor vanish and Equation 4.15 is simplified to the well-known Euler's equations for rotational motion:

$$M_x^b = \hat{I}_1\dot{\omega}_x^b + \left(\hat{I}_3 - \hat{I}_2\right)\omega_y^b\omega_z^b \tag{4.16a}$$

$$M_y^b = \hat{I}_2\dot{\omega}_y^b + \left(\hat{I}_1 - \hat{I}_3\right)\omega_x^b\omega_z^b \tag{4.16b}$$

$$M_z^b = \hat{I}_3\dot{\omega}_z^b + \left(\hat{I}_2 - \hat{I}_1\right)\omega_x^b\omega_y^b \tag{4.16c}$$

where $\hat{I}_1$, $\hat{I}_2$, and $\hat{I}_3$ are principal central moments of inertia about the center mass of body. Since Euler's equations are simpler and involve less non-linear terms, we choose the orientation of the body-fixed frame in a way that it coincides with the principal axes of the body. As it is seen in Equations 4.16a–4.16c, all vectors are defined in the body-fixed frame, while torques may be calculated in the space-fixed frame by the following equation:

$$\vec{M}^s = \sum_j \vec{r}_{CM-j} \times \vec{f}_j^s \tag{4.17}$$

where $\vec{r}_{CM-j}$ is the vector pointing from the center of mass of the of body to its contact point with body j in the space-fixed frame and $\vec{f}_j^s$ is the contact force between bodies in space-fixed frame. To obtain the torque acting on the body in the body-fixed frame, the transformation matrix can be used:

$$\vec{M}^b = A\vec{M}^s \tag{4.18}$$

In a special case, such as a spherical body, $\hat{I}_1 = \hat{I}_2 = \hat{I}_3 = \frac{2}{5}mR^2$ and due to the symmetric geometry of sphere, the space-fixed frame can always be the principal axes of the sphere, hence A becomes a unit matrix. Therefore, all vectors in the body-fixed and space-fixed frames become equal and Equations 4.16a–4.16c are reduced to the rotational equation of motion of sphere as shown in Equation 2.24b.

Now consider a non-spherical body which interacts with other bodies in the space-fixed (global) frame. We want to develop a step-wise procedure similar to what has been proposed for spherical bodies to track position and orientation of all bodies in each time step. Based on the explicit scheme, knowing all variables at time t and assuming a constant value for them from t to $t+\Delta t$, the new state of body can be calculated according to steps shown in Figure 4.4. These steps should be repeated for all bodies in the system to obtain new state of the system at $t+\Delta t$.

According to this procedure, using Euler angles is relatively easy and straightforward. However, a big problem may occur when Euler angles are used to define the orientation of the body-fixed frame. Equations 4.8a–4.8c are used for computing the time derivatives of Euler angles. For $\theta = 0$, $\dot{\phi}$, and $\dot{\psi}$ become singular. This means that numerical integration cannot be performed to obtain new orientation of the body and then re-compute the new transformation matrix. A simple solution that may come to mind is that we limit values of $\dot{\phi}$ and $\dot{\psi}$ (e.g., a high value instead of infinity) and perform the numerical integration. In this way, the problem is solved numerically, but the conservation of angular momentum is violated at the same time. Therefore, this is not a good solution. Other solutions are proposed to remedy this problem. We discuss two methods here: using the transformation matrix to define the orientation instead of Euler angles and using quaternions to describe orientation.

As we discussed earlier, elements of the transformation matrix are direct cosine of angles between axes of the body-fixed frame and axes of the space-fixed frame. Thus, the transformation matrix describes the orientation of the body-fixed frame relative to the space-fixed frame with nine parameters. If the time derivative of the transformation matrix (dA/dt) is calculated, a new transformation matrix (new orientation of body) at time $t+\Delta t$ is obtained by numerical integration. In contrast to the Euler angles method, in which the time derivative of Euler angles is used to obtain new orientation, time derivative of the transformation matrix is used in this method. This solution remedies the problem of singularity associated with the derivative of Euler angles for $\theta = 0$. The transformation matrix uses nine parameters for describing the orientation

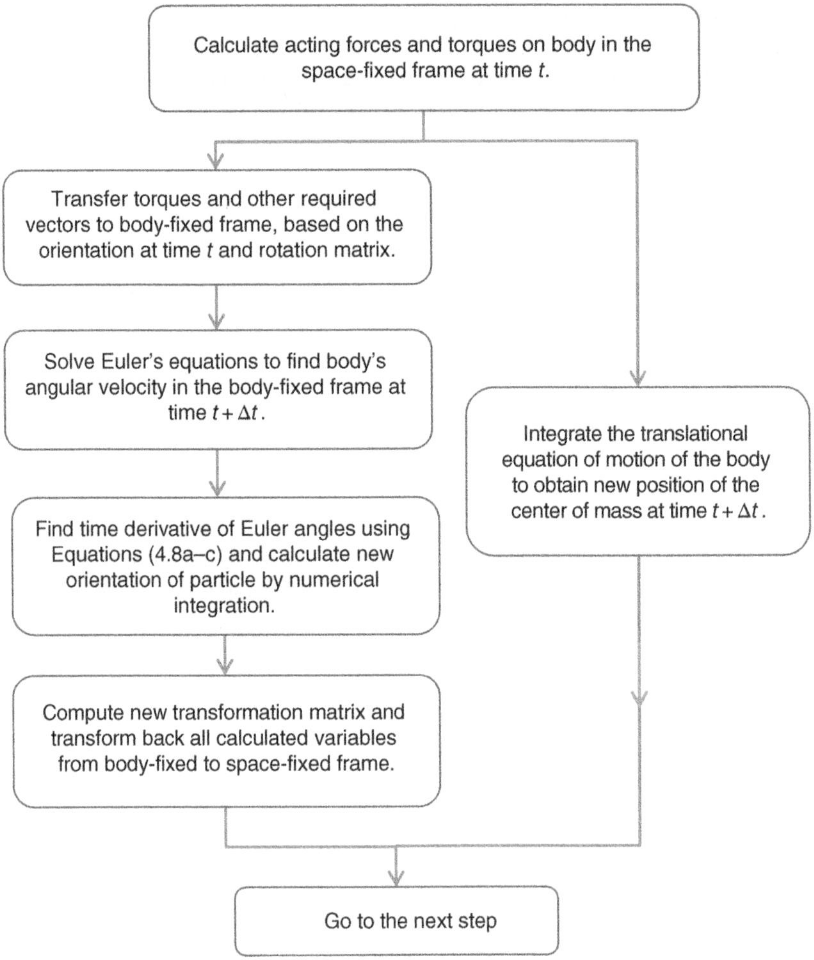

Figure 4.4 Steps to integrate/solve equations of motion of a non-spherical rigid particle in 3D space. This flow chart corresponds to steps 5–7 in Figure 3.1

of body in 3D space. This means that it requires six constraints (independent equations) to make the system determined. These constraints come from the orthogonal property of the transformation matrix. In an orthogonal matrix, the scalar product of non-similar columns should be zero and the scalar product of each column with itself should be one. If dA/dt is integrated to obtain the new orientation of the body in each time step, very small numerical errors are introduced in the elements of A. These small numerical errors are accumulated over time and matrix A loses its orthogonality and hence it would be no longer a transformation matrix. The problem of numerical errors can be mitigated by applying high-order integration methods. Munjiza *et al.* [33] used transformation matrix to define the orientation of bodies in their DEM simulations. They reformulated time evolution of the transformation matrix to be used in fourth order Runge–Kutta numerical integration. Their step-wise procedure for numerical integration is complex and can be implemented rather hard into a regular DEM code. In addition, it involves too many matrix operations, which degrades the computational efficiency. Another solution to the problem of Euler angles is to use quaternions to define the orientation of the body-fixed frame. In the following section, we describe this method in detail.

4.2.3 Quaternions for Rigid Body Dynamics

Quaternion is a vector in 4D space that extends a complex number into a higher dimension. A quaternion is defined using a scalar value s and a 3D vector $\vec{v}$ as follows:

$$q = (s, \vec{v}) \quad \text{or} \quad q = s\tilde{e}_0 + v_1\tilde{e}_1 + v_2\tilde{e}_2 + v_3\tilde{e}_3 \tag{4.19}$$

where $(\tilde{e}_0, \tilde{e}_1, \tilde{e}_2, \tilde{e}_3)$ are elements of the quaternion space that should hold the following relation:

$$\tilde{e}_0^2 + \tilde{e}_1^2 + \tilde{e}_2^2 + \tilde{e}_3^2 = 1 \tag{4.20}$$

We briefly present some basic properties of quaternions that are used for our application here. The interested reader who wants to know more about quaternions and their properties is referred to [34]. The product of two quaternions is defined as follows:

$$q_1.q_2 = (s_1, \vec{v}_1).(s_2, \vec{v}_2) = (s_1 s_2 - \vec{v}_1.\vec{v}_2,\ s_1\vec{v}_2 + s_2\vec{v}_1 + \vec{v}_1 \times \vec{v}_2) \tag{4.21}$$

Quaternion product is not commutative, meaning $q_1.q_2 \neq q_2 q_1$. We can also define the quaternion in terms of a rotation angle α around vector $\vec{u}$:

$$q_\alpha = \left(\cos\frac{\alpha}{2},\ \vec{u}\sin\frac{\alpha}{2}\right) \tag{4.22}$$

The following expression is used to rotate and arbitrary vector $\vec{r}$ by angle α around vector $\vec{u}$ [34]:

$$\vec{r}' = q_\alpha.(0, \vec{r}).q_\alpha^* \tag{4.23}$$

where q_α^* is the conjugate of q_α and is defined as:

$$q_\alpha^* = \left(\cos\frac{\alpha}{2},\ -\vec{u}\sin\frac{\alpha}{2}\right) \tag{4.24}$$

We can also define the components of the unit quaternion in terms of Euler angles:

$$\tilde{e}_0 = \cos\frac{\theta}{2}\cos\frac{\phi+\psi}{2} \tag{4.25a}$$

$$\tilde{e}_1 = \sin\frac{\theta}{2}\cos\frac{\phi-\psi}{2} \tag{4.25b}$$

$$\tilde{e}_2 = \sin\frac{\theta}{2}\sin\frac{\phi-\psi}{2} \tag{4.25c}$$

$$\tilde{e}_3 = \cos\frac{\theta}{2}\sin\frac{\phi+\psi}{2} \tag{4.25d}$$

In this way, the 3D orientation of the body-fixed frame is mapped into the components of the unit quaternion. Therefore, quaternions are capable of representing the orientation of a body in 3D space by four parameters (contrary to the transformation matrix with nine parameters). The rotation matrix A can be expressed in terms of components of the unit quaternion:

$$A=\begin{bmatrix} \tilde{e}_0^2+\tilde{e}_1^2-\tilde{e}_2^2-\tilde{e}_3^2 & 2(\tilde{e}_1\tilde{e}_2+\tilde{e}_0\tilde{e}_3) & 2(\tilde{e}_1\tilde{e}_3-\tilde{e}_0\tilde{e}_2) \\ 2(\tilde{e}_1\tilde{e}_2-\tilde{e}_0\tilde{e}_3) & \tilde{e}_0^2-\tilde{e}_1^2+\tilde{e}_2^2-\tilde{e}_3^2 & 2(\tilde{e}_2\tilde{e}_3+\tilde{e}_0\tilde{e}_1) \\ 2(\tilde{e}_1\tilde{e}_3+\tilde{e}_0\tilde{e}_2) & 2(\tilde{e}_2\tilde{e}_3-\tilde{e}_0\tilde{e}_1) & \tilde{e}_0^2-\tilde{e}_1^2-\tilde{e}_2^2+\tilde{e}_3^2 \end{bmatrix} \tag{4.26}$$

To transform an arbitrary vector, $\vec{r}$, between the space-fixed frame and the body-fixed frame, one can use the transformation matrix obtained from Equation 4.26 or use the quaternion and its conjugate:

$$\vec{r}^{\,b}=A\vec{r}^{\,s} \tag{4.27a}$$

$$\left(0,\vec{r}^{\,b}\right)=q^*.\left(0,\vec{r}^{\,s}\right).q \tag{4.27b}$$

and from the body-fixed frame to the space-fixed frame:

$$\vec{r}^{\,s}=A^T\vec{r}^{\,b} \tag{4.28a}$$

$$\left(0,\vec{r}^{\,s}\right)=q.\left(0,\vec{r}^{\,b}\right).q^* \tag{4.28b}$$

In addition to the transformation relations, we also need a time derivative of the quaternion to update the instantaneous orientation of the body-fixed frame. The time derivative of quaternion is given by:

$$\dot{q}=\frac{1}{2}q.\left(0,\vec{\omega}^b\right)=\frac{1}{2}\begin{bmatrix} -\tilde{e}_1\omega_x^b-\tilde{e}_2\omega_z^b-\tilde{e}_3\omega_z^b \\ \tilde{e}_0\omega_x^b+\tilde{e}_2\omega_z^b-\tilde{e}_3\omega_y^b \\ \tilde{e}_0\omega_y^b-\tilde{e}_1\omega_z^b+\tilde{e}_3\omega_x^b \\ \tilde{e}_0\omega_z^b+\tilde{e}_1\omega_y^b-\tilde{e}_2\omega_x^b \end{bmatrix} \tag{4.29}$$

Having the time derivative of quaternion as a function of angular velocity of the body-fixed frame, the orientation of the fixed frame can be calculated by numerical integration at each time step. Quaternion does not have the problem of singularity in Euler angles and uses four parameters to specify orientation (contrary to transformation matrix with nine parameters). Although numerical errors also accumulate in the components of quaternion during the integration, it can be frequently normalized during simulation to hold Equation 4.20.

4.3 Superellipsoids

Using analytical equation of superellipsoids is one of the methods for representing particle shape in the DEM model. In this section, we present the general form of mathematical equation of superellipsoids and their geometric properties. We also discuss the method of evaluating contact forces and torques between superellipsoidal particles based on the developed models for spherical particles. At the end, some algorithms for finding the contact plane between superellipsoidal particles are presented.

The edge of a particle in 2D space and the surface of a particle in 3D space can be defined by a set of analytical equations. In the DEM simulation of granular flows, superellipse (2D) and superellipsoid (3D) are mostly used to represent the particle analytically. A superellipse can be defined by a closed curve with the following equation:

$$\left(\frac{x^b}{a}\right)^m + \left(\frac{y^b}{b}\right)^m = 1 \tag{4.30}$$

where a and b are positive real numbers that show the size of superellipse on major and minor axes, and m is a positive number that shows the shape of curve and is calculated by:

$$m = \frac{p}{q} > 0, \text{ where} \begin{cases} p \text{ is an even positive integer} \\ q \text{ is an odd positive integer} \end{cases} \tag{4.31}$$

Figure 4.5 shows different superellipses produced by Equation 4.30 for $a = b = 1$. For $m = 2$, equation of a circle is obtained. By increasing the value of m, the shape becomes more rectangular and as $m \to \infty$, a rectangle is obtained. As can be seen in this figure, the superellipse equation produces interesting shapes in 2D, but our main interest is to produce shapes in 3D. Therefore, we dedicate the remaining of this section to superellipsoids.

Superellipsoids are a type of geometric shapes that belong to a more general family of geometric shapes named superquadrics. In addition to superellipsoids, superhyperboloids

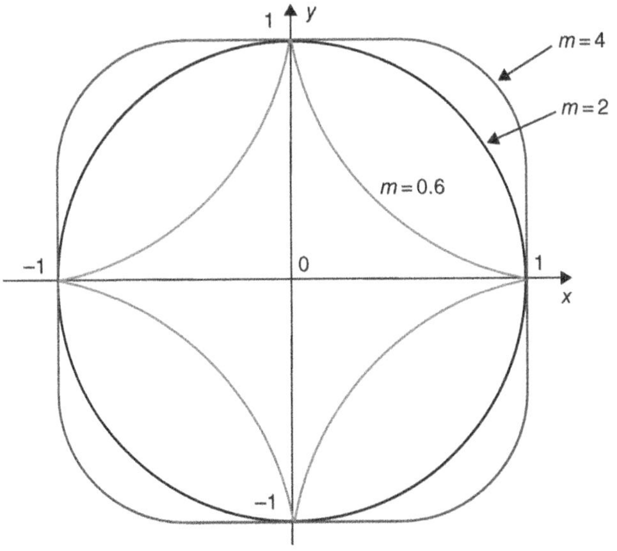

Figure 4.5 A superellipse can change gradually from a star-shape to a circle and to a square in the limit by changing the value of m from near zero to infinity in Equation 4.30 with $a = b = 1$

of one and two pieces and supertoroids are also categorized in this family. We define an equation of a superellipsoid in the body-fixed frame that coincides with its principal axes. In this way, all off-diagonal elements of the inertia tensor become zero. The parametric equation of the surface of a superellipsoid in the body-fixed frame is expressed as [35]:

$$\begin{bmatrix} x^b \\ y^b \\ z^b \end{bmatrix} = \begin{bmatrix} a\cos^{\varepsilon_1}\gamma \ \cos^{\varepsilon_2}\lambda \\ b\cos^{\varepsilon_1}\gamma \ \sin^{\varepsilon_2}\lambda \\ c\sin^{\varepsilon_1}\gamma \end{bmatrix}, \quad \begin{matrix} -\pi/2 \le \gamma \le \pi/2 \\ -\pi \le \lambda < \pi \end{matrix} \tag{4.32}$$

It should be noted that the terms with exponents ε_1 and ε_2 are singed power functions, meaning that $\sin^{\varepsilon_2}\lambda = \text{sign}(\sin\lambda)|\sin\lambda|^{\varepsilon_2}$. In this equation, parameters a, b, and c are positive real numbers that show the size of body on the coordinate axes and exponents ε_1 and ε_2 are positive real numbers that determine the shape of the superellipsoid cross-section. ε_2 determines the shape of superellipsoid cross section from the top (a plane parallel to xy plane) and ε_1 determines the side-view cross section of a superellipsoid.

There are five parameters for defining the surface of a superellipsoid in the body-fixed frame. Figure 4.6 shows different shapes that can be produced by changing values of a, b, c, ε_1, and ε_2. The number of shape types that can be produced by changing these parameters is far greater than is illustrated in this figure. Generally, with ε_1 and ε_2 near one, a smooth shape without any edges or round edges is obtained. However, exponents with values much lower or much higher than one, produce shapes with sharp edges and vertices.

Alternatively, the equation of the superellipsoid can be expressed in implicit form in the body-fixed coordinates [35]:

$$F\left(x^b, y^b, z^b\right) = \left(\left(\frac{x^b}{a}\right)^{\frac{2}{\varepsilon_2}} + \left(\frac{y^b}{b}\right)^{\frac{2}{\varepsilon_2}}\right)^{\frac{\varepsilon_2}{\varepsilon_1}} + \left(\frac{z^b}{c}\right)^{\frac{2}{\varepsilon_1}} - 1 \tag{4.33}$$

This equation represents all points inside, outside, or on the surface of a superellipsoid. Thus, it can be use it to check if a given point (x^b, y^b, z^b) is on the surface of the superellipsoid or not. If $F = 0$, the point is on the surface of the shape, if $F < 0$ the point is inside the shape, and if $F > 0$, it is outside the shape.

4.3.1 Contact Forces

Consider two arbitrary superellipsoids that are in contact, as shown in Figure 4.7. All the vectors and variables shown on this figure are in the space-fixed frame. We are aimed to calculate contact forces and torques acting on these particles.

We have established well the methodologies of computing contact force and torques between two spherical particles in Chapter 2 and now extend these methodologies to superellipsoidal particles. Obviously, we only bring the equations that are different from those used for spherical particles. The contact force at contact point C between two particles is comprised of normal and

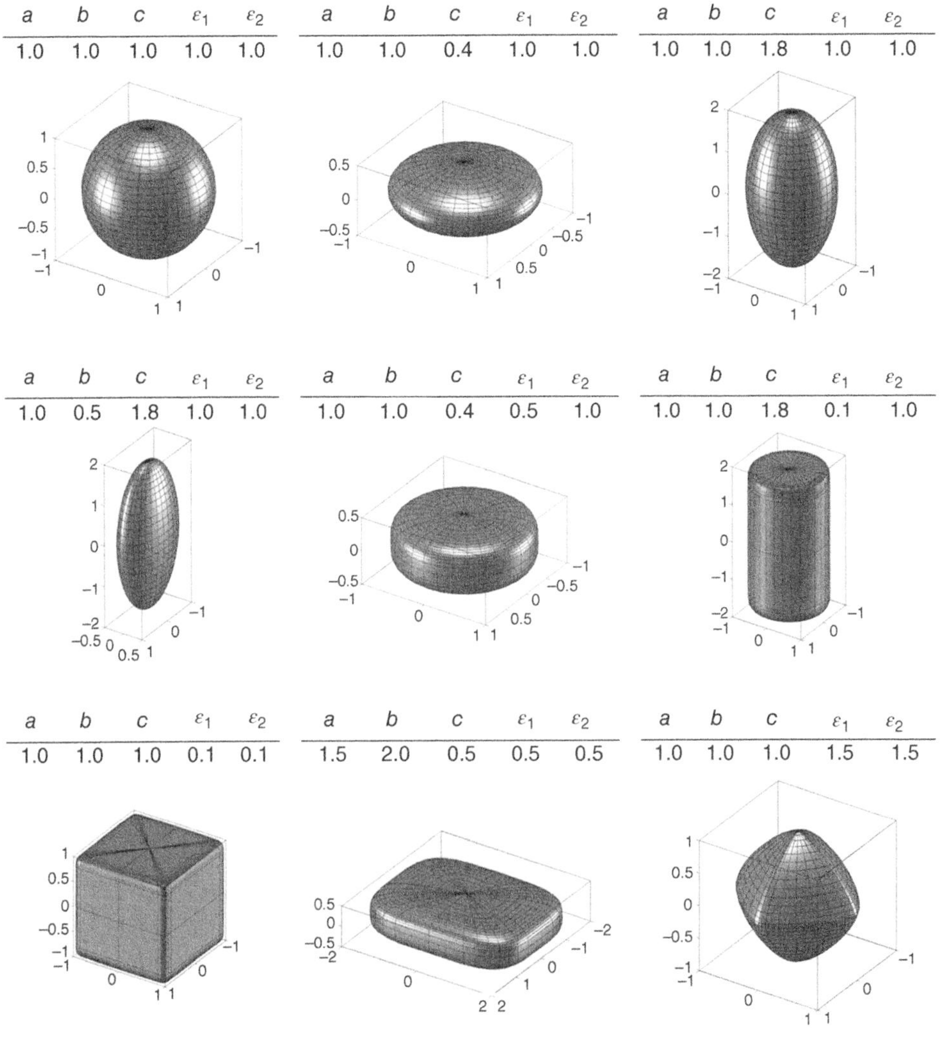

Figure 4.6 Some shapes that can be generated by superellipsoid equation by changing values of size parameters a, b, c, and exponents ε_1 and ε_2

tangential contact forces. The tangential overlap, normal, and tangential relative velocities are required to calculate contact force and torques. The velocity of particle at the contact point in the space-fixed frame is given by:

$$\vec{v}^s_{c,i} = \vec{v}^s_{CM,i} + \vec{\omega}^s_i \times \vec{R}^s_{C,ij} \tag{4.34}$$

or

$$\vec{v}^s_{c,i} = \vec{v}^s_{CM,i} + A^T\left(\vec{\omega}^b_i \times \vec{R}^b_{C,ij}\right) \tag{4.35}$$

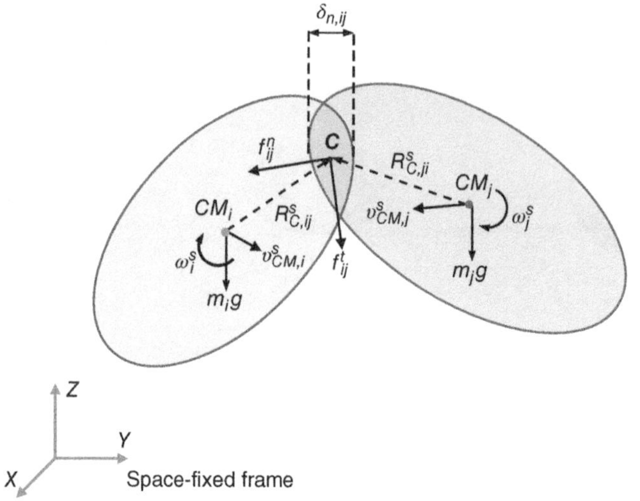

Figure 4.7 Schematic illustration of colliding superellipsoid i, j, and the forces acting on them in the space-fixed frame

where $\vec{R}^s_{C,ij}$ is the vector pointing from the center of mass of particle i to the contact point C, $\vec{v}^s_{CM,i}$ is the linear velocity of center of mass of particle in the space-fixed frame, and $\vec{\omega}^s_i$ is the rotational velocity of particle around its center of mass in the space-fixed frame. In Equation 4.35, A^T is the transformation matrix which transforms vector variables from the body-fixed frame to the space-fixed frame.[2] The relative velocity of contacting particles is then computed from:

$$\vec{v}^s_{ij} = \vec{v}^s_{c,i} - \vec{v}^s_{c,j} \tag{4.36}$$

Normal and tangential components of relative velocity at the contact point are calculated from Equations 2.29 and 2.30. Having the tangential velocity at the contact point, the tangential overlap can be evaluated by the integral in Equation 2.32. Thereafter, the contact force between two colliding particles can be calculated. The net contact force acting on the particle in the space-fixed frame, $\vec{f}^s_{cont,i}$, is the sum of the contact forces acting on the particle from all surrounding colliding particles:

$$\vec{f}^s_{cont,i} = \sum_{j \in CL_i} \vec{f}^s_{cont,ij} \tag{4.37}$$

where CL_i is the list of surrounding particles that are in contact with particle i and $\vec{f}^s_{cont,ij}$ is the contact force acting between particles i and j.

Different linear and non-linear viscoelastic models have been presented for spherical particles. Both can be used. In the case of linear models, we need normal and tangential overlaps as well as normal and tangential relative velocities. Assuming constant spring stiffnesses in

[2] If one uses a quaternion for defining orientation of the body-fixed frame, the quaternion products can be used for transformation.

the normal and tangential directions, we can use models developed for spherical particles. This assumption is inadequate, since the curvature of a non-spherical particle changes with the location of contact force and causes some deviations from the real behavior of contacting particles. Recently, Zheng *et al.* [36] studied contact forces between two viscoelastic ellipsoidal particles by finite-element method analysis. They found that linear model is accurate enough to predict contact properties at macroscopic level (like rebound velocity) while this model fails in predicting microscopic contact properties, like contact time or maximum contact force. This clearly suggests that the linear model can be used (with acceptable accuracy) in simulations in which microscopic contact properties are unimportant.

We also can extend non-linear force models to non-spherical particles [37, 38]. Based on the Hertz's theory, contact force of two colliding particles (assuming that the overlap is much smaller than the sizes of particles) depends on the geometry of contact area, not their shape. The contact area of two ellipsoids is an ellipse. The geometry of contact area is characterized by curvatures of the contact point. Having curvatures at the contact point, the effective radius can be calculated. With effective radius calculated, non-linear contact force models of spherical particles can be utilized to calculate contact force between two non-spherical particles. Accuracy of the calculated contact force can be further enhanced by including some correction factors in the Hertz's theory as described by Hale [38] for elastic contacts and by Zheng *et al.* [36] for viscoelastic contacts.

4.3.2 Effective Radius and Curvatures

The calculation of contact force needs computation of the effective radius of contacting particles. When two spherical particles are in contact, the effective radius, R^*, is calculated from the following equation that only needs radii of the two particles.

$$R^* = \frac{R_i R_j}{R_i + R_j} \tag{4.38}$$

where R_i are R_j radii of particles. However, for superellipsoids, the curvature of the particle in each dimension may be different. The following equation can be used to calculate the effective radius of two non-spherical particles:

$$R^* = \frac{1}{2\sqrt{A'B'}} \tag{4.39}$$

where A' and B' can be obtained by solving the following two equations simultaneously:

$$2(A'+B') = \frac{1}{R_i'} + \frac{1}{R_i''} + \frac{1}{R_j'} + \frac{1}{R_j''} \tag{4.40}$$

$$4(A'-B')^2 = \left(\frac{1}{R_i'} - \frac{1}{R_i''}\right)^2 + \left(\frac{1}{R_j'} - \frac{1}{R_j''}\right)^2 + 2\left(\frac{1}{R_i'} - \frac{1}{R_i''}\right)\left(\frac{1}{R_j'} - \frac{1}{R_j''}\right)\cos(2\beta) \tag{4.41}$$

where $\frac{1}{R_i'}$ and $\frac{1}{R_i''}$ are curvatures of the surface of ellipsoid and β is the inclined angle defined as the angle between the principal directions of two colliding particles at contact point in the space-fixed frame. We follow the method described by Harris [39] to calculate curvature at the contact point on the surface of an superellipsoid.

Suppose that the surface can be analytically defined by a parametric equation as[3]:

$$\vec{r} = \vec{r}(u,v) \tag{4.42}$$

where u and v are two parameters that vary in ranges that cover the whole surface. Alternatively, all points on the surface can be represented by the position vectors as:

$$\vec{r} = \left(x(u,v),\, y(u,v),\, z(u,v)\right) \tag{4.43}$$

The principal curvatures of surface can be determined from the first and the second fundamental forms of the surface. Consider point P on the surface for which the principal curvatures should be calculated. The first differential of the parametric curve $\vec{r}$ can be expressed as:

$$d\vec{r} = \frac{\partial \vec{r}}{\partial u} du + \frac{\partial \vec{r}}{\partial v} dv \tag{4.44}$$

The first fundamental form of the surface is defined as the dot product of the differential vector:

$$d\vec{r}.d\vec{r} = Edu^2 + 2Fdudv + Gdv^2 \tag{4.45}$$

where E, F, and G are coefficients of the first fundamental form. Since $\frac{\partial \vec{r}}{\partial u}$ and $\frac{\partial \vec{r}}{\partial v}$ are two tangent vectors along iso-parametric curves r_u (u-parameter curve) and r_v (v-parameter curve) on the surface. The normal vector at point P is obtained by cross product of these vectors:

$$\vec{n} = \frac{\frac{\partial \vec{r}}{\partial u} \times \frac{\partial \vec{r}}{\partial v}}{\left|\frac{\partial \vec{r}}{\partial u} \times \frac{\partial \vec{r}}{\partial v}\right|} \tag{4.46}$$

The second fundamental form of the surface is obtained by:

$$\frac{\partial^2 \vec{r}}{\partial u^2}.\vec{n}du^2 + 2\frac{\partial^2 \vec{r}}{\partial u \partial v}.\vec{n}dudv + \frac{\partial^2 \vec{r}}{\partial v^2}.\vec{n}dv^2 \tag{4.47}$$

or in the compressed form:

$$Ldu^2 + 2Mdudv + Ndv^2 \tag{4.48}$$

where L, M, and N are coefficients of the second fundamental form. It can be shown that the normal curvature at point P is the ratio of the first and the second fundamental forms of the surface. By defining the direction parameter λ as:

$$\lambda = \frac{dv}{du} \tag{4.49}$$

[3] As we mentioned before, equation of superellipsoids are defined in the body-fixed frame. In this subsection, we drop the superscript b from all symbols for the sake of clarity.

The curvature equation can be written as:

$$\kappa = \frac{L + 2M\lambda + N\lambda^2}{E + 2F\lambda + G\lambda^2} \quad (4.50)$$

The curvature depends on the direction parameter. The extrema of the curvature with regard to λ gives two principal curvatures along two principal directions. The following equation should be solved to find the extreme points of Equation 4.50:

$$(EM - FL) + (EN - GL)\lambda + (FN - GM)\lambda^2 = 0 \quad (4.51)$$

This equation is second order with respect to λ. It can be shown that this equation has either two real roots or a repeated root. In the case of a repeated root, the normal curvature is similar in all directions (umbilic point). In the two real roots case, the solution gives two direction parameters and two principal curvatures are obtained by substituting them in Equation 4.50.

Now let's see how the curvature of two superellipsoidal particles can be calculated at the contact point in the body-fixed frame. First, the implicit equation of superellipsoid should be rearranged to obtain the parametric equation of superellipsoid similar to Equation 4.43:

$$\vec{r} = \left(x, y, c\left(1 - \left(\left(\frac{x}{a}\right)^{\frac{2}{\varepsilon_2}} + \left(\frac{y}{b}\right)^{\frac{2}{\varepsilon_2}}\right)^{\frac{\varepsilon_2}{\varepsilon_1}}\right)^{\frac{\varepsilon_1}{2}} \right) \quad (4.52)$$

with $u = x$ and $v = y$. To obtain normal curvatures at the contact point the coefficients of the first and second fundamental forms should be calculated. Calculating these coefficients in the general form of the superellipsoid equation leads to a very long expression. We, therefore, present this procedure for a simpler equation, like an ellipsoid. By setting $\varepsilon_1 = \varepsilon_2 = 1$, Equation 4.52 is reduced to the parametric equation of an ellipsoid:

$$\vec{r} = \left(x, y, c\sqrt{1 - \left(\frac{x}{a}\right)^2 - \left(\frac{y}{b}\right)^2} \right) \quad (4.53)$$

The first and second derivatives of the ellipsoid equation are as follows:

$$\frac{d\vec{r}}{dx} = \left(1, 0, -\frac{c^2 x}{a^2 z}\right) \quad (4.54)$$

$$\frac{d^2\vec{r}}{dx^2} = \left(0, 0, -\frac{c^2}{a^2 z}\left(1 + \frac{c^2 x^2}{a^2 z^2}\right)\right) \quad (4.55)$$

$$\frac{d^2\vec{r}}{dxdy} = \left(0, 0, -\frac{c^4 xy}{a^2 b^2 z^3}\right) \quad (4.56)$$

$$\frac{d\vec{r}}{dy} = \left(0, 1, -\frac{c^2 y}{a^2 z}\right) \tag{4.57}$$

$$\frac{d^2\vec{r}}{dy^2} = \left(0, 0, -\frac{c^2}{b^2 z}\left(1 + \frac{c^2 y^2}{b^2 z^2}\right)\right) \tag{4.58}$$

$$\frac{d\vec{r}}{dx} \times \frac{d\vec{r}}{dy} = \left(\frac{c^2 x}{a^2 z}, \frac{c^2 y}{b^2 z}, 1\right) \tag{4.59}$$

By substituting these expressions into Equations 4.44 and 4.47 and rearranging, the following expressions are obtained for coefficients of the fundamental forms:

$$E = 1 + \frac{c^4 x^2}{a^4 z^2} \tag{4.60}$$

$$G = 1 + \frac{c^4 y^2}{b^4 z^2} \tag{4.61}$$

$$M = \frac{c^4 xy}{a^2 b^2 m z^3} \tag{4.62}$$

$$F = \frac{c^4 xy}{a^2 b^2 z^2} \tag{4.63}$$

$$L = \frac{c^4 \left(b^2 - y^2\right)}{a^2 b^2 m z^3} \tag{4.64}$$

$$N = \frac{c^4 \left(a^2 - x^2\right)}{a^2 b^2 m z^3} \tag{4.65}$$

with

$$m = \sqrt{1 + \frac{c^4 x^2}{a^4 z^2} + \frac{c^4 y^2}{b^4 z^2}} \tag{4.66}$$

We also need to calculate the principal direction vectors of colliding particles to obtain β in Equation 4.41. By definition, β is the angle between the principal directions of two colliding particles at the contact point in the space-fixed frame. So, we need to calculate the principal direction vector of two particles at the contact point in the space-fixed frame. Solving of Equation 4.51 gives two roots. Considering one of the roots, one of the principal direction vectors in the body-fixed frame can be obtained by:

$$\vec{C} = \cos(\theta)\frac{d\vec{r}}{dx} + \sin(\theta)\frac{d\vec{r}}{dy} \tag{4.67}$$

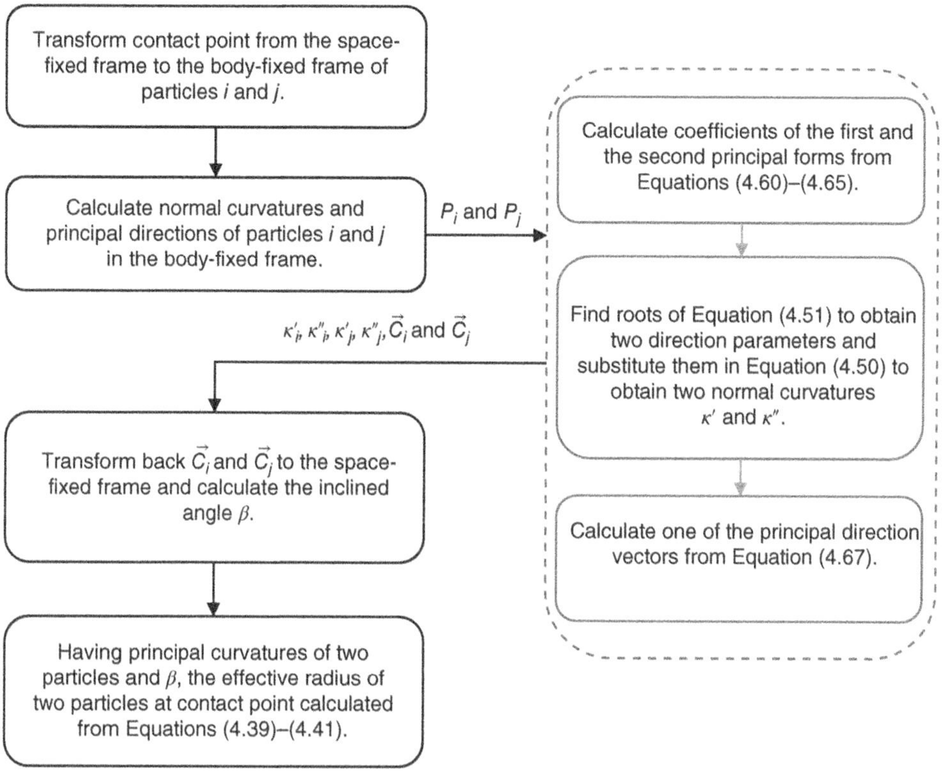

Figure 4.8 The procedure of calculating the curvatures of two ellipsoidal particles and the inclined angle at their contact point

where $\theta = \tan^{-1}(\lambda)$. After finding the principal direction vector of two colliding particles in their body-fixed frames, $\vec{C}_i$ and $\vec{C}_j$, they should be first transformed back to the space-fixed frame (gives $\vec{C}_i^s$ and $\vec{C}_j^s$) and the angle between them is then calculated from:

$$\cos\beta = \frac{\vec{C}_i^s . \vec{C}_j^s}{\left|\vec{C}_i^s\right|\left|\vec{C}_j^s\right|} \tag{4.68}$$

A stepwise procedure is developed to better understand the procedure of finding the curvature between two ellipsoidal particles. Figure 4.8 summarizes all steps required to calculate the curvatures of two colliding ellipsoidal particles and inclined angle at their contact point. Having the curvatures and inclined angle, the effective radius and the contact force between particles can be calculated.

4.3.3 *Torque Calculations*

Contact of non-spherical particles with each other and with wall can cause torque on them. Contact forces are calculated in the space-fixed frame. The resulting contact torque in the space-fixed frame is calculated from:

$$\vec{M}^s_{cont,i} = \sum_{j \in CL_i} \left(\vec{R}^s_{C,ij} \times \vec{f}^s_{cont,ij} \right) \tag{4.69}$$

It is also possible to include rolling torques, $\vec{M}^s_{r,i}$, using Equation 2.122 or 2.124. All variables used in the calculation of the rolling torque should be in the space-fixed frame. The total torque acting on the particle is sum of contact and rolling torques:

$$\vec{M}_i^s = \vec{M}^s_{cont,i} + \vec{M}^s_{r,i} \tag{4.70}$$

To solve the Euler's equations, the total torque on the particle should be transformed to its body-fixed frame using either the rotation matrix or the quaternion products.

It is also needed to calculate mass and principal moments of inertia of superellipsoids. The mass of a superellipsoid with density ρ_i is given by Jaklic *et al.* [35]:

$$m_i = 2\rho_i abc\varepsilon_1\varepsilon_2 B\left(\frac{\varepsilon_1}{2}+1,\varepsilon_1\right)B\left(\frac{\varepsilon_2}{2},\frac{\varepsilon_2}{2}\right) \tag{4.71}$$

where $B(x, y)$ is the beta function. The principal moments of inertia read as follows:

$$\hat{I}_1 = \frac{1}{2}\rho_i abc\varepsilon_1\varepsilon_2 \left[b^2 B\left(\frac{3}{2}\varepsilon_2,\frac{1}{2}\varepsilon_2\right)B\left(\frac{1}{2}\varepsilon_1,2\varepsilon_1+1\right)+4c^2 B\left(\frac{1}{2}\varepsilon_2,\frac{1}{2}\varepsilon_2+1\right)B\left(\frac{3}{2}\varepsilon_1,\varepsilon_1+1\right)\right] \tag{4.72}$$

$$\hat{I}_2 = \frac{1}{2}\rho_i abc\varepsilon_1\varepsilon_2 \left[a^2 B\left(\frac{3}{2}\varepsilon_2,\frac{1}{2}\varepsilon_2\right)B\left(\frac{1}{2}\varepsilon_1,2\varepsilon_1+1\right)+4c^2 B\left(\frac{1}{2}\varepsilon_2,\frac{1}{2}\varepsilon_2+1\right)B\left(\frac{3}{2}\varepsilon_1,\varepsilon_1+1\right)\right] \tag{4.73}$$

$$\hat{I}_3 = \frac{1}{2}\rho_i abc\varepsilon_1\varepsilon_2\left(a^2+b^2\right)B\left(\frac{3}{2}\varepsilon_2,\frac{1}{2}\varepsilon_2\right)B\left(\frac{1}{2}\varepsilon_1,2\varepsilon_1+1\right) \tag{4.74}$$

4.3.4 Contact Detection

At the end of the broad search phase, a list of potential contact pairs is generated. In the fine search/contact detection phase, particle pairs with overlapping bounding spheres are determined first (see Figure 3.2). If particles are spherical, the bounding sphere fits the particle surface. Therefore, the fine search ends at this stage for spherical particles. In case of non-spherical particles with touching bounding spheres, a further step, called geometric resolution, is required to determine if particles collide. In this section, we present three algorithms for geometric resolution of superellipsoidal particles.

The first algorithm, called geometrical potential algorithm, was proposed by Ting [40] and Ng [41]. This algorithm was first applied to two dimensional elliptic particles, then used for three-dimensional ellipsoidal particles [14]. Consider that the surface equation of particle i in the space-fixed frame (global frame) is given by $f_i(x,y,z)=0$ and the surface equation of particle j by $g_j(x,y,z)=0$. Equation $f_i(x,y,z)=f$ represents a group of geometric similar superellipsoids with varying size and the same origin. For example, $f<0$ produces a smaller superellipsoid than the original (inside the original surface) and $f>0$ a bigger one

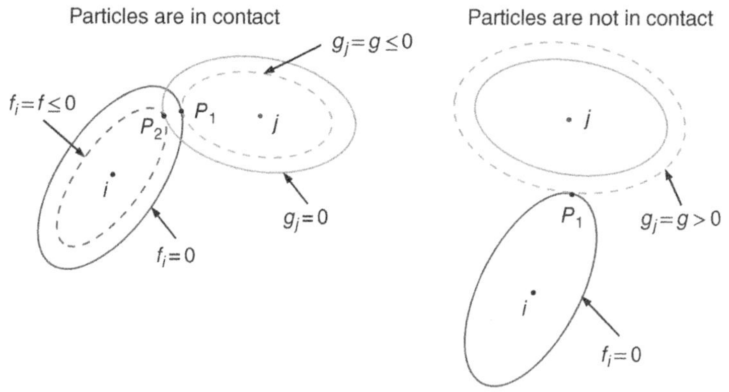

Figure 4.9 Contact detection of superellipsoids according to the geometrical potential concept

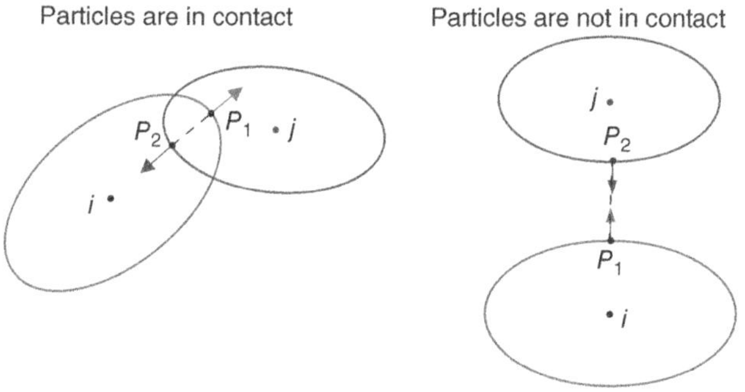

Figure 4.10 Contact detection of superellipsoids according to the common normal concept

(out of the original surface). It can be seen in Figure 4.9 that there is a point (P_1) on the surface of particle i that has the lowest potential in particle j (inner most surface in particle j, i.e., the dashed curve in particle j). If $g_i(x_1, y_1, z_1) \leq 0$, P_1 is inside or on the surface of particle j and two particles are in contact. In this case, the second point (P_2) should be found. This can be obtained by finding a point on particle j that produces the lowest potential in particle i (inner most surface in particle i). The line connecting these two points can be considered as the normal vector and the midpoint as the contact point. Mathematically speaking, this problem is the minimization of g_j subject to constraint $f_i = 0$ to obtain P_1 and the minimization of f_i subject to constraint $g_j = 0$ to obtain P_2. It is clear that if two particles are not in contact, Point 1 holds $g_j(x_1, y_1, z_1) > 0$ and is outside particle j (see Figure 4.9). Lin and Ng [14] formulated this algorithm for ellipsoids and obtained two sixth-degree polynomial equations, the solutions of which give P_1 and P_2.

Another algorithm is based on the common normal concept (see Figure 4.10). Two points on surfaces of particles i and j can be found in a way that normal vectors of the surface on these points are parallel to the line connecting these two points. The mid-point of this line is considered as the contact point. According to Lin and Ng [14], the following equations should be solved to find such points, P_1 and P_2, on particles:

$$f_i\left(x_1, y_1, z_1\right) = 0 \tag{4.75a}$$

$$g_j\left(x_2, y_2, z_2\right) = 0 \tag{4.75b}$$

$$\frac{\dfrac{\partial f_i\left(x_1, y_1, z_1\right)}{\partial x}}{\Delta_i} + \frac{\dfrac{\partial g_j\left(x_2, y_2, z_2\right)}{\partial x}}{\Delta_j} = 0 \tag{4.75c}$$

$$\frac{\dfrac{\partial f_i\left(x_1, y_1, z_1\right)}{\partial y}}{\Delta_i} + \frac{\dfrac{\partial g_j\left(x_2, y_2, z_2\right)}{\partial y}}{\Delta_j} = 0 \tag{4.75d}$$

$$\frac{\left|x_2 - x_1\right|}{\Delta} - \frac{\left|\dfrac{\partial f_i\left(x_1, y_1, z_1\right)}{\partial x}\right|}{\Delta_i} = 0 \tag{4.75e}$$

$$\frac{\left|y_2 - y_1\right|}{\Delta} - \frac{\left|\dfrac{\partial g_j\left(x_2, y_2, z_2\right)}{\partial y}\right|}{\Delta_j} = 0 \tag{4.75f}$$

where f_i and g_j are surface equation of particles i and j, respectively, in the space-fixed frame. Equations 4.75a and 4.75b ensure that points P_1 and P_2 are always on the surface of two particles, Equations 4.75c and 4.75d ensure that surface normal vectors on points P_1 and P_2 are anti-parallel and Equations 4.75e and 4.75f ensure that the line connecting these two points is in the same direction of normal vectors. In these equations:

$$\Delta = \left|P_1 - P_2\right| \tag{4.76}$$

$$\Delta_i = \left[\left(\frac{\partial f_i\left(x_1, y_1, z_1\right)}{\partial x}\right)^2 + \left(\frac{\partial f_i\left(x_1, y_1, z_1\right)}{\partial y}\right)^2 + \left(\frac{\partial f_i\left(x_1, y_1, z_1\right)}{\partial z}\right)^2\right]^{1/2} \tag{4.77}$$

$$\Delta_j = \left[\left(\frac{\partial g_j\left(x_2, y_2, z_2\right)}{\partial x}\right)^2 + \left(\frac{\partial g_j\left(x_2, y_2, z_2\right)}{\partial y}\right)^2 + \left(\frac{\partial g_j\left(x_2, y_2, z_2\right)}{\partial z}\right)^2\right]^{1/2} \tag{4.78}$$

if $f_i\left(x_2, y_2, z_2\right) < 0$ and $g_j\left(x_1, y_1, z_1\right) < 0$, particles are in contact, otherwise they are not.

The last algorithm is proposed for ellipsoidal particles by Hopkins [11]. The surface of an ellipsoid in the body-fixed frame is given by the following equation:

$$\left(\frac{x^b}{a}\right)^2 + \left(\frac{y^b}{b}\right)^2 + \left(\frac{z^b}{c}\right)^2 - 1 = 0 \tag{4.79}$$

where parameters a, b, and c reflect the extension of particle along each coordinate axis. In this algorithm, this surface is called the constraint surface of particle. Another surface is created by dilating the constraint surface by δ. This is the actual surface of the particle. The normal (shortest) distance between every point of dilated and constraint surfaces is δ. The equation of the dilated surface is given by:

$$\left(\frac{x^b}{a+\delta}\right)^2 + \left(\frac{y^b}{b+\delta}\right)^2 + \left(\frac{z^b}{c+\delta}\right)^2 - 1 = 0 \tag{4.80}$$

An arbitrary vector, which starts from a point on the constraint surface of particle i and ends at a point on the constraint surface of particle j, is defined as:

$$\vec{d} = C_j + \vec{P}_j - \left(C_i + \vec{P}_i\right) \tag{4.81}$$

where C_i and C_j are coordinates of particle centers in the space-fixed frame, and $\vec{P}_i$ and $\vec{P}_j$ are the vectors that connect center points of particles to the end points of vector $\vec{d}$ on the constraint surfaces (see Figure 4.11).

To find the possible contact point, the end points of this vector are allowed to slide on the constraint surface. Obviously these end points are restricted to be on the constraint surfaces. The shortest distance between these two particles is found by minimizing the length of vector $\vec{d}$. This means that, vector $\vec{d}$ is perpendicular to the surfaces of two particles at the minimum condition. If the length of this vector is less than 2δ, particles are in contact. In this condition, normal vector at contact point is parallel to vector $\vec{d}$ and the contact point is the midpoint of this vector.

Hopkins [11] proposed the elastic band algorithm to solve this constraint minimization problem. The end points of vector $\vec{d}$ should move in the direction of its tangential part with

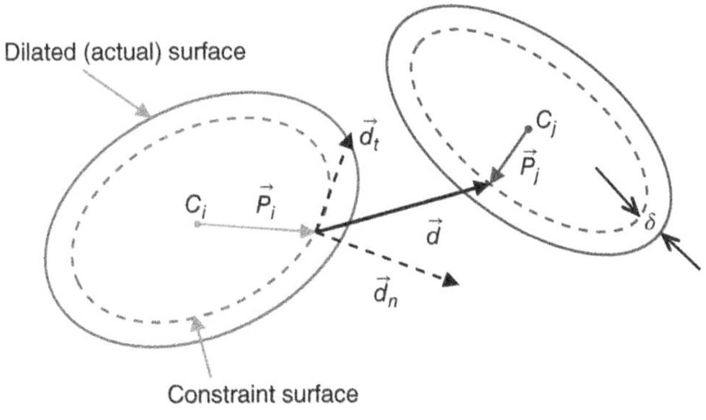

Figure 4.11 Schematic of dilated surface algorithm to find the possible contact point between two ellipsoidal particles

respect to the surface (see Figure 4.11). The tangential part of vector $\vec{d}$ with respect to the constraint surface is obtained by:

$$\vec{d}_t = \vec{d} - \left(\vec{d}.\vec{n}\right)\vec{n} \tag{4.82}$$

where $\vec{n}$ is the unit normal vector of the surface at point P in the space-fixed frame. Point P is called the sliding point and is located at the end point of either vectors $\vec{P}_i$ or $\vec{P}_j$. To obtain the normal vector in the space-fixed frame, it is first calculated in the body-fixed frame and then transformed to the space-fixed frame using either transformation matrix or quaternion product. Point P is relocated to its new location by moving a fraction of $\vec{d}_t$:

$$P' = P + \lambda \vec{d}_t \tag{4.83}$$

where λ is the under-relaxation factor. The new point, P', is not on the constraint surface and should be projected on the surface. Point P'' on the constraint surface is found in a way that holds:

$$P'' = \gamma P' \tag{4.84}$$

To this end, the point P' is first transformed into the body-fixed frame and $\gamma P'^b$, is substituted in Equation 4.79. After substitution, the non-linear equation is solved to obtain γ. Having γ, P'' is obtained in the space-fixed frame. If this point is equal or very close to P, the iteration stops, otherwise P is replaced by P'' in the next iteration and this procedure is repeated. This procedure should be followed for both ellipsoids simultaneously until the locations of sliding points on both surfaces do not change in two successive iterations.

4.4 Multi-Sphere Method

The multi-sphere method is another method for generating non-spherical particles. In spite of superellipsoids, which are non-spherical axisymmetric particles, the multi-sphere method can generate particles with irregular shapes. Based on this method, spheres with smaller size are glued to each other to approximately fill the volume/surface of the real particle. These spheres are allowed to overlap to any extent. In this way, any number of spheres, with different sizes and overlaps, can be used. After positioning spheres in the non-spherical particle and producing the final shape, the relative position of spheres within the particle will not change. Thus, the dynamics of the rigid body can be applied to multi-sphere particles.

Figure 4.12 shows a number of different shapes that can be generated by this method. The surface of these particles is rough and bumpy. The particle shape/surface can be more accurately approximated by increasing the number of spheres and reduce the level of bumpiness of the surface. On the other hand, the computational cost of simulation also increases with increasing the number of spheres. Although the accuracy of representation of the surface is improved by using a large number of spheres, the mechanical behavior of the particle is not necessarily improved due to additional dominant errors introduced with increasing the number of spheres in the particle [42]. We are not still definite about an optimal number of spheres as it changes from one case to another.

Consider two colliding non-spherical particles i and j generated by the multi-sphere method as illustrated in Figure 4.13. Each particle has its own body-fixed frame that is oriented relative to the space-fixed frame. Position of the center of each sphere ik in the particle i

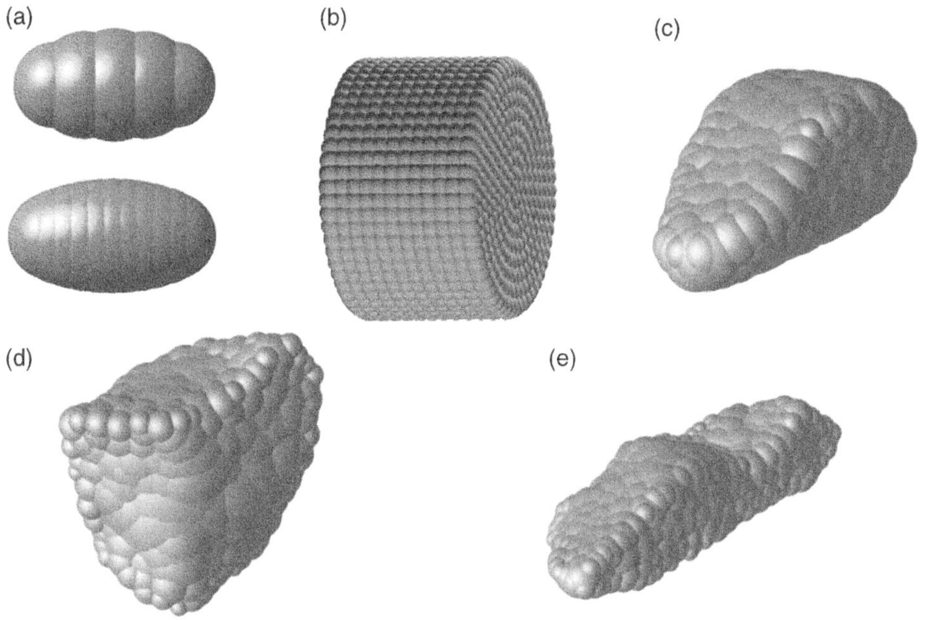

Figure 4.12 Some regular and irregular shapes that can be generated by assembling spheres. (a) Ellipsoidal particles with 5 and 11 spheres (Source: Höhner, D., Wirtz, S., Kruggel-Emden, H., and Scherer, V. (2011) Comparison of the multi-sphere and polyhedral approach to simulate non-spherical particles within the discrete element method: Influence on temporal force evolution for multiple contacts. *Powder Technology*, 208(3), 643–656, from [42] with permission) (b) cylindrical particle (Source: Kodam, M., Bharadwaj, R., Curtis, J., Hancock, B., and Wassgren, C. (2009) Force model considerations for glued-sphere discrete element method simulations. *Chemical Engineering Science*, 64(15), 3466–3475, from [43] with permission), (c) angular particle with 121 sphere, (d) equant particle with 150 spheres, and (e) elongated particles with 247 spheres (Source: Garcia, X., Xiang, J., Latham, J.P., and Harrison, J.P. (2009) A clustered overlapping sphere algorithm to represent real particles in discrete element modelling. *Géotechnique*, 59(9), 779–784, from [44] with permission)

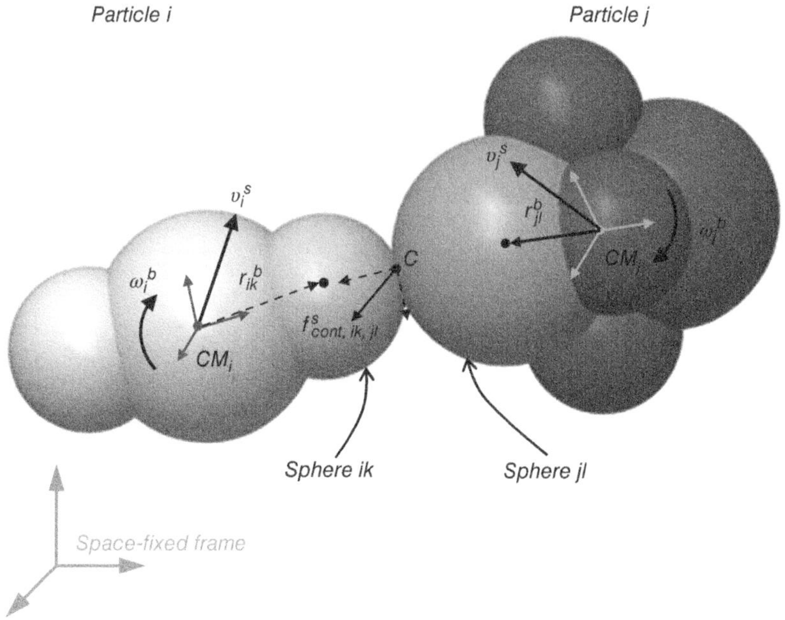

Figure 4.13 Two non-spherical particles *i* and *j* are in contact via their constituent spheres *ik* and *jl*

(in the body-fixed frame of the particle) is defined by a vector pointing from center of mass of particle to the center of sphere and is denoted by $\vec{r}_{ik}^{\,b}$. Therefore, the center position of each sphere in the space-fixed frame is obtained from:

$$\vec{r}_{ik}^{\,s} = \vec{r}_{CM,i}^{\,s} + A^T \vec{r}_{ik}^{\,b} \tag{4.85}$$

where $\vec{r}_{ik}^{\,s}$ is the center position of sphere *ik* in the space-fixed frame and $\vec{r}_{CM,i}^{\,s}$ is the position of the center of mass of particle *i* in the space-fixed frame.[4] In a similar way, the velocity of the center of each sphere in the space-fixed frame is calculated from:

$$\vec{v}_{ik}^{\,s} = \vec{v}_{CM,i}^{\,s} + A^T \left(\vec{\omega}_i^{\,b} \times \vec{r}_{ik}^{\,b}\right) \tag{4.86}$$

As shown in Figure 4.13, contact interactions between colliding particles are governed by contacts between spheres *ik* and *jl*. Having all required variables (positions and velocities) in the space-fixed frame, the contact force between these two spheres can be calculated similar to what we explained for spherical particles.

The net contact force acting the particle, $\vec{f}_{cont,i}^{\,s}$, is sum of contact forces acting on all spheres in that particle:

$$\vec{f}_{cont,i}^{\,s} = \sum_{jl \in CL_i} \vec{f}_{cont,ik,jl}^{\,s} \tag{4.87}$$

where $\vec{f}_{cont,ik,jl}^{\,s}$ is the contact force between sphere *ik* in particle *i* and sphere *jl* in particle *j*. Since the contact force acts on the surface and it is considered on the center of mass of the particle, a torque should be considered around this point. The net contact torque on particle in the space-fixed frame is obtained by:

$$\vec{M}_{cont,i}^{\,s} = \sum_{jl \in CL_i} \vec{R}_{C,ik,jl}^{\,s} \times \vec{f}_{cont,ik,jl}^{\,s} \tag{4.88}$$

It is also possible to consider the rolling resistance by either Equation 2.122 or 2.124. In this case, the rotational velocity of spheres in a particle should be equal to the particle rotational velocity. The total torque (sum of contact torque and rolling torque) acting on the particle in the space-fixed frame should be transferred to the body-fixed of particle to be used in the Euler's equations.

For solving linear and rotational equations of motion, the total mass of particle and principal moments of inertia should be known. It is common to calculate these values from properties of the spheres inside the particle. The following equations are used to obtain the total mass and principal moments of inertia of a multi-sphere particle:

$$m_i = \sum_{k=1}^{NS} m_{ik} \tag{4.89}$$

$$\hat{I}_i = \begin{pmatrix} \sum I_{ik} + \sum m_{ik}(y_{ik}^2 + z_{ik}^2) & 0 & 0 \\ 0 & \sum I_{ik} + \sum m_{ik}(x_{ik}^2 + z_{ik}^2) & 0 \\ 0 & 0 & \sum I_{ik} + \sum m_{ik}(x_{ik}^2 + y_{ik}^2) \end{pmatrix} \tag{4.90}$$

[4]According to the definition of the body-fixed frame, the origin of the body-fixed frame is located on the center of mass of particles, hence, the position of the center of mass of the particle in the body-fixed frame would be (0,0,0).

where m_{ik} and I_{ik} are mass and moment of inertia of sphere k in particle i, x_{ik}, y_{ik}, z_{ik} are the components of vector r_{ik}^b and NS is the total number of spheres in the particle. Using these equations can introduce errors in the simulation. For example, consider a case where we want to generate a long cylinder with spheres. The diameter of this cylinder is 0.05 m, its length 0.25 m, and its density 2500 kg/m^3. Suppose that we generate this cylinder with nine 5-cm spheres. The total mass and principal moments of inertia of this multi-sphere cylinder as calculated by the Equations 4.89 and 4.90 are:

$$m_i = 1.473\text{kg}$$

$$\hat{I}_i = (8.4, 8.4, 2.25) \times 10^{-3}\,\text{kg·m}^2$$

Now if we consider the real geometry of cylinder, the total mass and moments of inertial of cylinder are:

$$m_i = 1.23\text{kg}$$

$$\hat{I}_i = (6.6, 6.6, 0.8) \times 10^{-3}\,\text{kg·m}^2$$

There is a noticeable difference between these values that can affect the dynamics of the particle and introduce error in the simulation. These differences are originated from the overlaps between spheres in the particle, as shown in Figure 4.12. It is recommended that mass and inertias are calculated based on the geometry of the real particle rather than the mass and position of spheres in the particle and using Equations 4.89 and 4.90.

Based on the multi-sphere method, the interaction force between colliding particles is calculated by summing across all interaction forces between their constituent spheres that are in contact. This simple method of calculating the interaction force introduces some error in the macroscopic and microscopic contact parameters, hence, leads to a wrong behavior in the dynamics of particles. We address two issues here: the dissipation rate of kinetic energy in viscoelastic models and existence of multiple contacts.

The first issue is the dissipation rate of kinetic energy in the viscoelastic contact force models. Figure 4.14 shows the position of a single sphere during a vertical free fall. In this case, the normal coefficient of restitution was set to 0.8 and the initial height was 0.6 m. The sphere losses a part of its kinetic energy in each collision with wall, thus travels a shorter height in the rebound. After about 2.6 s, the particle loses a major part of its kinetic energy and rests on the wall. Now, consider a multi-sphere particle which is generated by the similar sphere used in the previous test. All physical properties of the spheres in this particle remain unchanged. We performed similar vertical free fall test with this multi-sphere particle. It should be noted that the orientation of this particle was set in a way that no rotation was observed during its contacts with the wall. Evolution of the position of this particle is illustrated in Figure 4.14. This particle follows a different path from the single sphere. In each contact with the wall, the particle losses its kinetic energy less than expected. We calculated the coefficient of restitution for this particle to be 0.95. The difference between single sphere and multi-sphere particle coefficients of restitution is related to the way the normal damping coefficient is calculated. The restitution coefficient is related to the square root of mass (see e.g., Equation 2.44). When the damping coefficient of a single sphere is also used for the spheres inside the multi-sphere particle, the effect of total mass of particle on this parameter

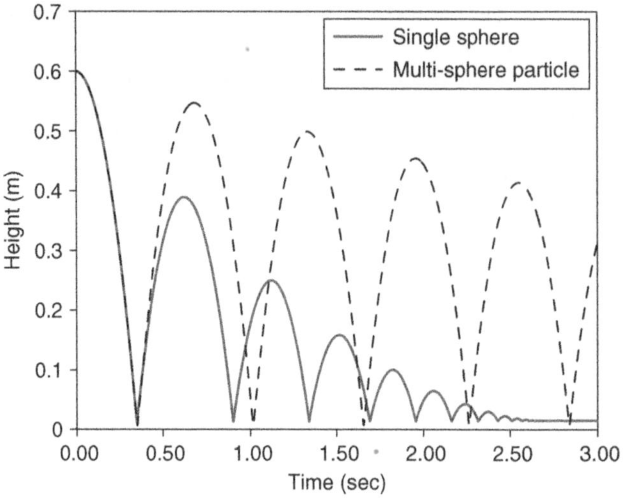

Figure 4.14 The evolution of height of particle in free fall tests with single-sphere and multi-sphere particles. Spheres in the multi-sphere particle are identical to sphere used in the single-sphere free fall

is disregarded. Therefore, the value of damping coefficient would be smaller and less kinetic energy is dissipated during each contact of the multi-sphere particle with the wall. As a result, the damping coefficient of spheres in the multi-sphere particle should be adjusted to obtain a correct coefficient of restitution.

The other issue that may cause deviation in the contact properties of a multi-sphere particle is the possibility of multiple contact points between two particles or a particle and a wall. Consider a multi-sphere particle colliding with a flat wall as shown in Figure 4.15. Depending on the particle orientation and the extent of overlap between wall and particle, one, two, or three contact points may occur. For example, for the case with three contact points, when the overlap of the first colliding sphere increases, other spheres may come into contact with wall, thus, multiple contacts occur. The possibility of multiple contacts increases with increasing the number of spheres in the particle or using very small values of spring stiffness for spheres.

In the case of multiple contacts, combination of parallel springs between spheres and wall increases the final stiffness between particle and wall. This causes a shorter contact time, a greater maximum contact force, and a higher rate of dissipation of kinetic energy. Kruggel-Emden *et al.* [45] examined multiple contacts of a multi-sphere particle with a wall. They suggested that by dividing the total force acting on the particle by the number of contacts, the problem of multiple contacts would be solved. They suggested the following relation for evaluating the contact force:

$$\vec{f}^{s}_{cont,ij} = \frac{1}{NCP_{ij}} \sum \vec{f}^{s}_{cont,ik,jl} \tag{4.91}$$

where $\vec{f}^{s}_{cont,ij}$ is the contact force acting between particles i and j and NCP_{ij} is the number of contact points between particles i and j (here, j is wall). In their tests, all overlaps of spheres with wall at contact points were equal while in reality this rarely occurs. In contrast, Kodam *et al.* [43] used a more sophisticated approach to adjust the total force acting on particle.

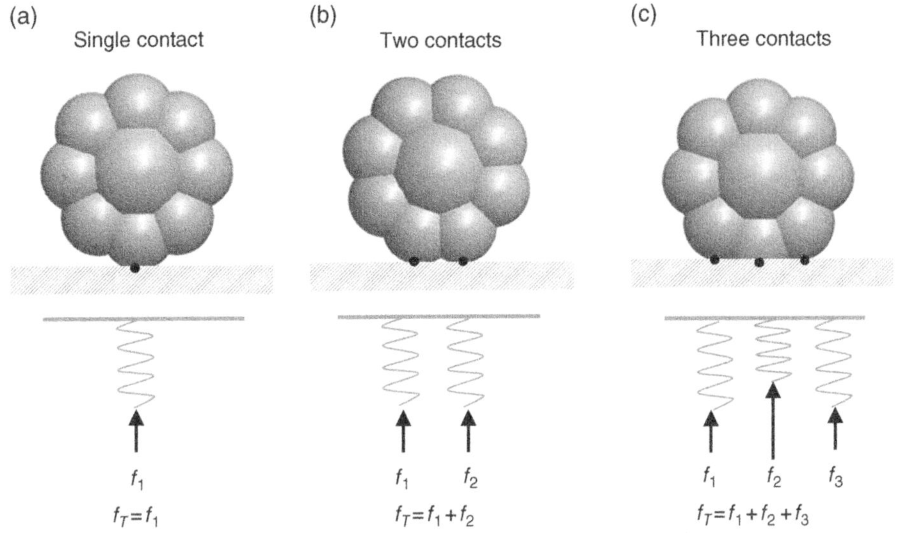

Figure 4.15 The schematic of possible contact conditions between a multi-sphere particle and a flat wall. With more contact points between particle and wall, the equivalent spring would be stiffer

They suggested that the value of spring stiffness used for spheres should be different from the value chosen for the particle as the material property. In addition, the power exponent in the contact force model (1.5 for the Hertizan model and 1 for the linear model) should also be different. They adjusted both the spring stiffness of spheres and the power exponent in the contact force model to minimize the error between the contact force of a multi-sphere particle with multiple contacts and the contact force of a real-shaped particle. However, these optimal values are case sensitive and may vary depending on the particle shape, orientation, degree of overlap, and number of contacts.

Höhner *et al.* [42] proposed a simple approach for adjusting the total contact force between particle and wall (also applicable to two particles). Their approach is based on two facts: first, the overlaps between spheres and wall in the contact points can be unequal and second, the number of contacts can be changed during the collision. They proposed an incremental approach for calculating contact force at each time step. Having the contact force at the previous time step, $\vec{f}^{s,0}_{cont,ij}$, the contact force at the current time step is calculated from:

$$\vec{f}^{s}_{cont,ij} = \vec{f}^{s,0}_{cont,ij} + \Delta\vec{f}^{s}_{cont,ij} \tag{4.92}$$

where $\Delta\vec{f}^{s}_{cont,ij}$ is the incremented contact force from the previous time step to the current time step. This method can be used for all linear and non-linear contact force models. As an example, we describe how to calculate the incremental contact force for a linear model. In normal direction, the incremental contact force is obtained from:

$$\Delta\vec{f}^{s,n}_{cont,ij} = -\frac{k_{n,ij}}{NCP_{ij}}\sum\left(\delta_{n,ik,jl} - \delta^{0}_{n,ik,jl}\right)\vec{n}_{ik,jl} - \frac{\eta_{n,ij}}{NCP_{ij}}\sum\left(v_{rn,ik,jl} - v^{0}_{rn,ik,jl}\right)\vec{n}_{ik,jl} \tag{4.93}$$

where $\delta^0_{n,ik,jl}$ and $\delta_{n,ik,jl}$ are previous and current normal overlaps between sphere *ik* from particle *i* and sphere *jl* from particle *j*, and $v^0_{rn,ik,jl}$ and $v_{rn,ik,jl}$ are previous and current normal relative velocities between spheres *ik* and *jl*. $k_{n,ij}$ and $\eta_{n,ij}$ are normal spring stiffness and normal damping coefficient between particles *i* and *j*, respectively. Note that the summations are done over current contact points between two particles. The incremental contact force for the tangential direction can be obtained using similar methodology. It can be seen that there is no need to adjust the spring stiffness or change the contact force law. Since the incremental approach is used, the number of contacts changes and the overlap of spheres at contact points can be different during a collision. A drawback of this approach is usage of extra memory for storing contact parameters, such as overlap, relative velocity, and so on.

Nomenclature

Symbol	Unit	Description
a	m	Half size of body on the x-axis in a superellipsoid equation
$\vec{a}_i$	m/s^2	Linear acceleration of particle i
A	—	Transformation matrix
A_ϕ, A_θ, A_ψ	—	Rotation matrix for Euler angles
b	m	Half size of body on the y-axis in a superellipsoid equation
$B(x,y)$	—	Beta function
c	m	Half size of body on the z-axis in a superellipsoid equation
C_i	m	Coordinates of the center of particle i
$\vec{C}_i$	—	Principal direction vector of particle i in a body-fixed frame
$\vec{C}_i^s$	—	Principal direction vector of particle i in a space-fixed frame
CL_i	—	Contact list of particle i
$\vec{d}$	m	Vector defined in Equation 4.81
$\vec{d}_t$	m	Tangential component of vector $\vec{d}$
$\tilde{e}_0, \tilde{e}_1, \tilde{e}_2, \tilde{e}_3$	—	Elements of quaternion space
E, F, G	—	Coefficients of the first fundamental form of surface
$\vec{f}^s_{cont,i}$	N	Net contact force acting on the particle i in a space-fixed frame
$\vec{f}^s_{cont,ij}$	N	Contact force acting between particles i and j in a space-fixed frame
$\vec{f}^{s,0}_{cont,ij}$	N	Contact force in previous time step in a space-fixed frame
$\vec{f}^s_{cont,ik,jl}$	N	Contact force between sphere ik in particle i and sphere jl in sphere j
f_i	—	Surface equation of particle i
$\vec{f}_i$	N	Sum of all forces acting on the center of mass of particle i
$\vec{f}_j^s$	N	Contact force in a space-fixed frame with body j
g_j	—	Surface equation of particle j
I	kg·m^2	Inertia tensor of a rigid body
I_{ik}	kg·m^2	Moment of inertia of sphere k in particle i
$\hat{I}_1, \hat{I}_2, \hat{I}_3$	kg·m^2	Principal central moments of inertia about the center mass of body
$k_{n,ij}$	N/m	Spring stiffness between particles i and j
L, M, N	—	Coefficients of the second fundamental form of surface
$\vec{L}$	N·m	Rate of change of angular momentum
$\vec{L}^b$	kg·m^2/s	Angular momentum with respect to the center of mass
m	—	Exponent in Equation 4.30

m_i	kg	Mass of particle i
m_{ik}	kg	Mass of sphere k in particle i
$\vec{M}$	N·m	Torque acting on a particle
$\vec{M}^b$	N·m	Torque acting on a particle in a body-fixed frame
$\vec{M}^s$, $\vec{M}_i^s$	N·m	Torque acting on a particle in a space-fixed frame
$\vec{M}_{cont,i}^s$	N·m	Torque acting on a particle in a space-fixed frame due to contact force
$\vec{M}_{r,i}^s$	N·m	Rolling resistance torque acting on a particle in a space-fixed frame
$\vec{n}$	—	Normal vector at the surface
$\vec{n}_{ik,jl}$	—	Normal vector point from the center of sphere ik to the center of sphere jl
NCP_{ij}	—	Number of contact points between particles i and j
NS	—	Number of spheres in a particle
P	m	Sliding point
q		Quaternion
q^*		Conjugate of q
u	—	Parameter in Equation 4.42
v	—	Parameter in Equation 4.42
$\vec{v}$	—	Vector component of quaternion
$\vec{v}_i$	m/s	Velocity vector of the center mass of particle i
$v_{c,i}^s$	m/s	Velocity of particle i at a contact point in a space-fixed frame
$v_{CM,i}^s$	m/s	Velocity of center of mass of particle i in a space-fixed frame
v_{ik}^s	m/s	Velocity of the center of sphere ik in a space-fixed frame
v_{ij}^s	m/s	Relative velocity of contacting particles i and j in a space-fixed frame
$v_{rn,ik,jl}$	m/s	Current normal relative velocity between spheres ik and jl
$v_{rn,ik,jl}^0$	m/s	Previous normal relative velocity between spheres ik and jl
$\vec{r}$	m	Parametric definition of particle surface in a body-fixed frame
$\vec{r}^b$	—	Vector variable in a body-fixed frame
$\vec{r}^s$	—	Vector variable in a space-fixed frame
$\vec{r}_{CM-j}$	m	Vector pointing from the center of mass of body i to its contact point with body j
$\vec{r}_{CM,i}^s$	m	Position of center of mass of particle i in a space-fixed frame
$\vec{r}_{ik}^b$	m	Vector pointing from the center of mass of a particle to the center of a sphere in a body-fixed frame
$\vec{r}_{ik}^s$	m	Center position of sphere ik in a space-fixed frame
R^*	m	Reduced radius
R_i	m	Radius of particle i
R_i', R_i''	m	Radii of curvature of the surface of a particle i
$\vec{R}_{C,ij}^b$	m	Vector pointing from the center of mass of particle i to a contact point in body-fixed frame
$\vec{R}_{C,ij}^s$	m	Vector pointing from the center of mass of particle i to a contact point in a space-fixed frame
$\vec{R}_{C,ik,jl}^s$	m	Vector pointing from center of mass of particle i to contact point between spheres ik and jl
s	—	Scalar component of quaternion
t	s	Time
x^b, y^b, z^b	m	Coordinates of a point in a body-fixed frame
$\vec{x}^b$		Vector in a body-fixed frame
$\vec{x}^s$	—	Vector in a space-fixed frame
x_{ik}, y_{ik}, z_{ik}	m	Components of vector $\vec{r}_{ik}^b$

Greek symbols

α	rad	Rotation angle around vector u
β	rad	Angle between principal directions of two colliding particles at a contact point in a space-fixed frame
γ	rad	Angle used in Equation 4.32
	—	Correction factor in Equation 4.84
δ	m	Dilation distance of surface
$\delta_{n,ik,jl}$	m	Current overlap between sphere ik from particle i and sphere jl from particle j
$\delta^0_{n,ik,jl}$	m	Previous overlap between sphere ik from particle i and sphere jl from particle j
$\Delta\vec{f}^{s}_{cont,ij}$	N	Incremental change in contact force from previous time step to current time step in a space-fixed frame
Δ, Δ_i, Δ_j	—	Parameters in Equations 4.76–4.78
$\Delta\vec{f}^{s,n}_{cont,ij}$	N	Incremental change in contact force from previous time step to current time step in the normal direction
Δt	S	Time step of integration
ε_1, ε_2	—	Shape exponents of a superellipsoid
$\eta_{n,ij}$	kg/s	Normal damping coefficient between particles i and j
θ	rad	Euler angle, rotation angle around the x'-axis
θ	rad	Angle in Equation 4.67
κ	—	Curvature
λ	rad	Angle in Equation 4.32
λ	—	Direction parameter in Equation 4.49
λ	—	Under-relaxation factor
ϕ	rad	Euler angle, rotation angle around the z-axis
ψ	rad	Euler angle, rotation angle around the z''-axis
$\vec{\omega}^b$	rad/s	Angular velocity in a body-fixed frame
$\vec{\omega}^b_i$	rad/s	Rotational velocity of particle i around its center of mass in a body-fixed frame
$\vec{\omega}^s_i$	rad/s	Rotational velocity of particle i around its center of mass in a space-fixed frame

References

[1] Höhner, D., Wirtz, S., and Scherer, V. (2012) A numerical study on the influence of particle shape on hopper discharge within the polyhedral and multi-sphere discrete element method. *Powder Technology*, **226**, 16–28.

[2] Fraige, F.Y., Langston, P.A., and Chen, G.Z. (2008) Distinct element modelling of cubic particle packing and flow. *Powder Technology*, **186**(3), 224–240.

[3] Cleary, P.W. and Sawley, M.L. (2002) DEM modelling of industrial granular flows: 3D case studies and the effect of particle shape on hopper discharge. *Applied Mathematical Modelling*, **26**(2), 89–111.

[4] Pandey, P. and Turton, R. (2005) Movement of different-shaped particles in a pan-coating device using novel video-imaging techniques. *AAPS PharmSciTech*, **6**(2), E237.

[5] Zhou, Z.Y., Pinson, D., Zou, R.P., and Yu, A.B. (2011) Discrete particle simulation of gas fluidization of ellipsoidal particles. *Chemical Engineering Science*, **66**, 6128–6145.

[6] Favier, J.F., Abbaspour-Fard, M.H., Kremmer, M., and Raji, A.O. (1999) Shape representation of axisymmetrical, non-spherical particles in discrete element simulation using multi-element model particles. *Engineering Computations*, **16**(4), 467–480.

[7] Ting, J.M., Meachum, L., and Rowell, J.D. (1995) Effect of particle shape on the strength and deformation mechanisms of ellipse-shaped granular assemblages. *Engineering Computations*, **12**, 99–108.

[8] Markauskas, D., Kačianauskas, R., Džiugys, A., and Navakas, R. (2009) Investigation of adequacy of multi-sphere approximation of elliptical particles for DEM simulations. *Granular Matter*, **12**(1), 107–123.

[9] Cleary, P.W. (2007) DEM modelling of particulate flow in a screw feeder. *Progress in Computational Fluid Dynamics, an International Journal*, **7**(2–4), 128–138.

[10] Hilton, J.E., Mason, L.R., and P.W. Cleary (2010) Dynamics of gas–solid fluidised beds with non-spherical particle geometry. *Chemical Engineering Science*, **65**, 1584–1596.

[11] Hopkins, M.A. (2004) Discrete element modeling with dilated particles. *Engineering Computations*, **21**(2), 422–430.

[12] Wellmann, C., Lillie, C., and Wriggers, P. (2008) A contact detection algorithm for superellipsoids based on the common-normal concept. *Engineering Computations*, **25**(5), 432–442.

[13] Ouadfel, H. and Rothenburg, L. (1999) An algorithm for detecting inter-ellipsoid contacts. *Computers and Geotechnics*, **24**(4), 245–263.

[14] Lin, X. and Ng, T.-T. (1995) Contact detection algorithms for three-dimensional ellipsoids in discrete element modelling. *International Journal for Numerical and Analytical Methods in Geomechanics*, **19**(9), 653–659.

[15] Song, Y., Turton, R., and Kayihan, F. (2006) Contact detection algorithms for DEM simulations of tablet-shaped particles. *Powder Technology*, **161**(1), 32–40.

[16] Ketterhagen, W.R. (2011) Modeling the motion and orientation of various pharmaceutical tablet shapes in a film coating pan using DEM. *International Journal of Pharmaceutics*, **409**(1–2), 137–149.

[17] Freireich, B., Ketterhagen, W.R., and Wassgren, C. (2011) Intra-tablet coating variability for several pharmaceutical tablet shapes. *Chemical Engineering Science*, **66**, 2535–2544.

[18] Suzzi, D., Toschkoff, G., Radl, S., Machold, D., Fraser, S.D., Glasser, B.J., and Khinast, J.G. (2012) DEM simulation of continuous tablet coating: effects of tablet shape and fill level on inter-tablet coating variability. *Chemical Engineering Science*, **69**(1), 107–121.

[19] Mansourpour, Z., Mostoufi, N., and Sotudeh-Gharebagh, R. (2013) A mechanistic study of agglomeration in fluidised beds at elevated pressures. *The Canadian Journal of Chemical Engineering*, **91**(3), 560–569.

[20] Ren, B., Zhong, W., Chen, Y., Chen, X., Jin, B., Yuan, Z., and Lu, Y. (2012) CFD-DEM simulation of spouting of corn-shaped particles. *Particuology*, **10**(5), 562–572.

[21] Ren, B., Zhong, W., Jin, B., Shao, Y., and Yuan, Z. (2013) Numerical simulation on the mixing behavior of corn-shaped particles in a spouted bed. *Powder Technology*, **234**, 58–66.

[22] Markauskas, D. and Kačianauskas, R. (2010) Investigation of rice grain flow by multi-sphere particle model with rolling resistance. *Granular Matter*, **13**(2), 143–148.

[23] Wu, J. and Aidun, C.K. (2010) A method for direct simulation of flexible fiber suspensions using lattice Boltzmann equation with external boundary force. *International Journal of Multiphase Flow*, **36**, 202–209.

[24] Wang, G., Yu, W., and Zhou, C. (2006) Optimization of the rod chain model to simulate the motions of a long flexible fiber in simple shear flows. *European Journal of Mechanics – B/Fluids* **25**, 337–347.

[25] Lindström, S.B. and Uesaka, T. (2007) Simulation of the motion of flexible fibers in viscous fluid flow. *Physics of Fluids*, **19**(11), 113307.

[26] Chen, J. and Matuttis, H.-G. (2013) Optimization and OpenMP parallelization of a discrete element code for convex polyhedra on multi-core machines. *International Journal of Modern Physics C*, **24**(2), 1350001.

[27] Cundall, P.A. (1988) Formulation of a three-dimensional Distinct Element Model – Part 1. A scheme to detect and represent contacts in a system composed of many polyhedral blocks. *International Journal of Rock Mechanics and Mining Sciences*, **25**(3), 107–116.

[28] Nezami, E.G., Hashash, Y.M.A., Zhao, D., and Ghaboussi, J. (2006) Shortest link method for contact detection in discrete element method. *International Journal for Numerical and Analytical Methods in Geomechanics*, **30**(8), 783–801.

[29] Poschel, T. and Schwager, T. (2005) *Computational Granular Dynamics: Models and Algorithms*, Springer-Verlag Berlin/Heidelberg.

[30] Matuttis, H.-G. and Chen, J. (2014) *Understanding the Discrete Element Method: Simulation of Non-Spherical Particles for Granular and Multi-Body Systems*, John Wiley & Sons, Singapore Pte Ltd., Singapore.

[31] Greenwood, D.T. (2003) *Advanced Dynamics*, Cambridge University Press, New York.

[32] Marghitu, D.B. and Dupac, M. (2012) *Advanced Dynamics: Analytical and Numerical Calculations with MATLAB*, Springer, New York.

[33] Munjiza, A., Latham, J.P., and John, N.W.M. (2003) 3D dynamics of discrete element systems comprising irregular discrete elements-integration solution for finite rotations in 3D. *International Journal for Numerical Methods in Engineering*, **56**(1), 35–55.

[34] Jazar, R.N. (2011) *Advanced Dynamics: Rigid Body, Multibody, and Aerospace Applications*, John Wiley & Sons, Inc., Hoboken, NJ.

[35] Jaklic, A., Leonardis, A., and Solina, F. (2000) *Segmentation and Recovery of Superquadrics*. Vol. **20.** In *Computational Imaging and Vision* (series), Springer Science & Business Media, Dordrecht.

[36] Zheng, Q.J., Zhou, Z.Y., and Yu, A.B. (2013) Contact forces between viscoelastic ellipsoidal particles. *Powder Technology*, **248**, 25–33.

[37] Johnson, K.L. (1985) *Contact Mechanics*, Cambridge University Press, Cambridge.

[38] Hale, L.C. (1999) Principles and techniques for designing precision machines. PhD thesis. MIT.

[39] Harris, W.F. (2006) Curvature of ellipsoids and other surfaces. *Ophthalmic and Physiological Optics*, **26**, 497–501.

[40] Ting, J.M. (1992) Computational laboratory for discrete element geomechanics. *Journal of Computing in Civil Engineering*, **6**(2), 129–146.

[41] Ng, T.T. (1994) Numerical simulations of granular soil using elliptical particles. *Computers and Geotechnics*, **16**(2), 153–169.

[42] Höhner, D., Wirtz, S., Kruggel-Emden, H., and Scherer, V. (2011) Comparison of the multi-sphere and polyhedral approach to simulate non-spherical particles within the discrete element method: Influence on temporal force evolution for multiple contacts. *Powder Technology*, **208**(3), 643–656.

[43] Kodam, M., Bharadwaj, R., Curtis, J., Hancock, B., and Wassgren, C. (2009) Force model considerations for glued-sphere discrete element method simulations. *Chemical Engineering Science*, **64**(15), 3466–3475.

[44] Garcia, X., Xiang, J., Latham, J.P., and Harrison, J.P. (2009) A clustered overlapping sphere algorithm to represent real particles in discrete element modelling. *Geotechnique*, **59**(9), 779–784.

[45] Kruggel-Emden, H., Rickelt, S., Wirtz, S., and Scherer, V. (2008) A study on the validity of the multi-sphere discrete element method. *Powder Technology*, **188**(2), 153–165.

5

DEM Applications to Granular Flows

Granular flow is encountered in many applications such as packing and compression of particles, separation, mixing, agglomeration, excavation and fluid-particle flow, material transport and storage, film coating and granulation, milling and compaction, mixers with impellers, and screening. Of course, the list is far from complete and the interested reader is referred to the literature, as partially addressed here. Discrete element method (DEM) simulations can be used to properly capture dominated phenomena occurring in these applications.

This chapter covers potential applications of DEM in common granular flows in industry (mostly chemical and pharmaceutical). The chapter starts with confined packing, unconfined packing, flexible fibers, and flow in hoppers. In subsequent sections, mixing in drums, tumblers, convective and continuous blenders, solid transport in screw conveyors, and film coating are presented briefly in order to cover the essence of granular flow. The simulation results presented in this chapter are performed by our own numerical program developed with formulations and numerical methods presented in this book. This numerical code accompanies the book. It is important to mention that the same physical properties and simulation conditions for all simulations are performed in this chapter, as listed in Table 5.1, unless stated otherwise.

5.1 Packing of Particles

5.1.1 Confined Packing

Packing of particles is one of the simplest cases investigated in granular material science. Although particles are stationary in the packed state, the packed bed is the outcome of a dynamic process in which many factors can affect it. These factors are particle shape and properties [1, 2], type of particle interaction forces [3, 4] (cohesive and non-cohesive), the

Coupled CFD-DEM Modeling: Formulation, Implementation and Application to Multiphase Flows, First Edition.
Hamid Reza Norouzi, Reza Zarghami, Rahmat Sotudeh-Gharebagh and Navid Mostoufi.

Companion website: www.wiley.com/go/norouzi/CFD-DEM

Table 5.1 Physical properties and operating conditions of simulations

Parameter	Particle	Wall
Properties		
Young's modulus (MPa)	1	1
Poisson's ratio (—)	0.23	0.23
Dynamic friction (—)	0.3	0.4
Rolling friction (—)	0.1	0.1
Restitution coefficient (—)	0.8	0.8
Density (kg/m^3)	2500	—
Simulation		
Time step (s)	0.00001	—
Contact force model	Non-linear with non-limited tangential displacement	
Torque model	Constant torque	
Integration method	Adams–Moulton fifth order	

surrounding medium [5] (gas or liquid), and the packing protocol [6] (continuous charging, step-wise packing, drop height, and inclusion of vibration or tapping). Some properties of the final packed bed, such as force network and interparticle contacts, are of practical interest. Therefore, it is essential to have a tool to consider all these effects in formation of the packed bed. The DEM is a promising tool in this regard.

DEM has been used to investigate various properties of packed beds, such as mean porosity and porosity distribution [7, 8], which are essential to predict hydraulic properties of fixed beds [9] and filter cakes as well as to determine concrete properties, to calculate coordination number (important for estimating thermal properties of packed beds), to estimate particle orientation (has metallurgical and casting applications), and to find pore structure and interconnection [10, 11] (which are applied for permeability studies). DEM is also used for developing/testing correlations or methods of predicting the properties of packed beds [12].

Here, we present results of the numerical code on the formation of packed beds in a cylindrical vessel with an inner diameter of 0.15 m. The physical properties and operating conditions of simulations are listed in Table 5.1. It should be noted that similar properties and conditions have been used for most of the simulations performed in this chapter, unless otherwise stated. Three simulations with different solid mixtures are considered. The first mixture contains mono-sized particles with a diameter of 7 mm, the second mixture contains bidispersed particles of 3.5 and 7 mm (small particle weight fraction is 0.125), and the third mixture contains cylindrical particle (multi-sphere method[1]) with a diameter and length of 5 mm of 25 mm, respectively. To produce the packed bed, an insertion plane (an imaginary plane which is not a physical wall) is considered inside and at the top part of the vessel (see Figure 5.1). Particles with the pre-defined rate of 2.76 kg/s (equivalent to 6150 particles/s for the monodispersed mixture) are randomly inserted into the cylinder through this insertion plane for a period of 1.3 s. Each simulation continues for 2.5 s to ensure that the kinetic energy of all particles is dissipated and the final packing is achieved. For the third simulation with cylindrical particle, the initial orientation of each particle is determined randomly, as well as its initial position on

[1] Multi-sphere method is not a part of the code published with this book and simulation of cylindrical particles was performed by an in-house code.

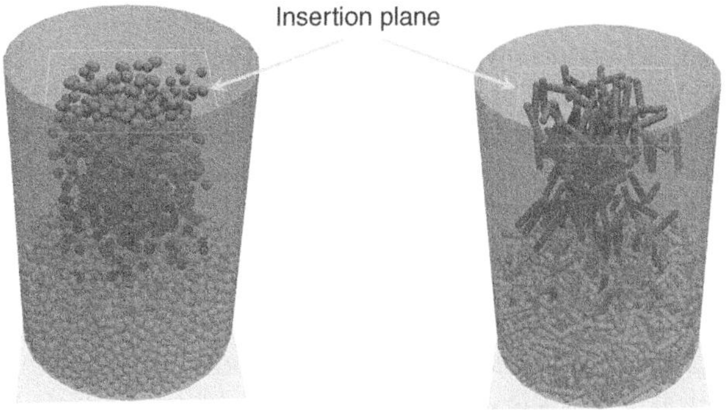

Figure 5.1 Formation of packed bed under gravity. Particles are inserted into the simulation through the insertion plane (rectangle with light shading at top) at a specified rate

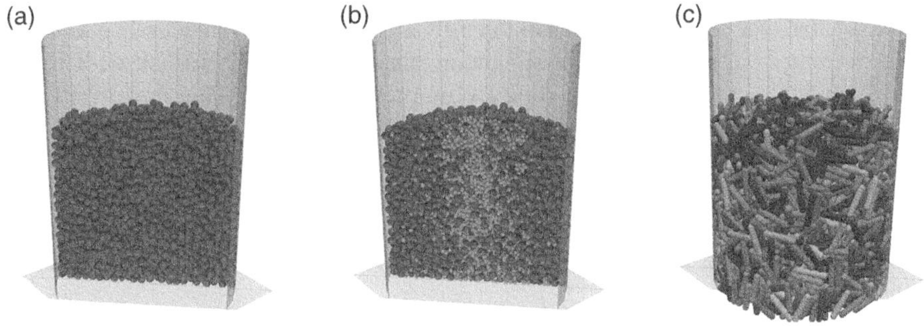

Figure 5.2 Packing of (a) monodispersed spherical particles with the size of 7 mm, (b) binary mixture of spherical particles with the diameters of 7 mm (dark gray) and 3.5 mm (light gray), and (c) cylindrical particles with the aspect ratio of 5. The average porosity in (a) is 0.41, in (b) is 0.38, and in (c) is 0.45

the insertion plane. The code files to setup these simulations and sample animations of them can be found in online content under "Packing" folder.[2]

The produced packed beds are shown in Figure 5.2. Figures 5.2a and b are clipped to make the middle of the packed bed visible. Average porosity of the monodispersed mixture is 0.41 while it is 0.38 for the bidispersed mixture. Small particles fill gaps between large particles and result in a reduction of the overall porosity. The same trend is observed in the corresponding experiments and DEM simulations of Langston and Kennedy [8]. They used a very similar charging protocol (and particle size ratio) to ours and found that the average porosity is 0.42 for the monodispersed powder and decreases to 0.38 when the weight fraction of small particle is increased to 0.125. The minimum average porosity of 0.35 was reached for 0.25 weight fraction in their experiments. The distribution of porosity is also important, since it affects the local phenomena in the packed bed (e.g., fluid bypass). Various studies have shown that the DEM can accurately predict radial porosity distribution, especially its oscillations near the wall [12, 13].

Results of the second simulation show that distribution of small particles is not uniform in the radial direction and these particles are mostly accumulated in the middle of the bed (Figure 5.2b). This is due to the size segregation of particles during their charging. The same

[2] The related files are located in: http://www.wiley.com/go/norouzi/CFD-DEM.

behavior has been observed experimentally in a system with similar particle size ratio and charging protocol [8]. Size distribution of particles can greatly influence the average porosity and coordination number in the packed bed. Simulations showed that when the standard deviation of particle size distribution (lognormal distribution) increases from zero to a moderate value, the average porosity of densely packed beds increases; while, further increase in the standard deviation results in a gradual reduction in the average porosity. The mean coordination number follows an opposite trend in densely packed beds (first decrease and then increase). For loosely packed beds, the coordination number decreases with increasing the standard deviation [14]. It was also found that the shape of size distribution curve does not affect the average porosity and the mean coordination number [15]. The packed bed condition of cylindrical particles is very different from spherical ones (Figure 5.2c). Particles mainly have vertical or near vertical orientation under our simulation conditions. The mean porosity of cylindrical particles in the packed bed is about 0.45, which is higher than the other two mixtures.

Another factor that can affect the packing properties of particles is vibration of the vessel [16]. It was shown that increasing both vibration intensity and frequency in a randomly packed bed of particles decreases the porosity until a minimum value is reached. Nevertheless, at very high vibration amplitudes and frequencies, particles are over-excited and the average void fraction becomes larger (i.e., a loose packing structure is achieved). The particle friction coefficient (dynamic and rolling) and coefficient of restitution also affect packing properties [6, 16, 17]. Simulations show that the average porosity increases with increasing both rolling and dynamic frictions and remains unchanged at high values. The coordination number follows almost the same trend. Increase in the restitution coefficient results in a monotonic decrease in the packing density.

5.1.2 Pile Formation

Formation of piles is another type of packing at unconfined conditions. The angle of repose and the force network are two main properties investigated with DEM. The governing mechanism for the formation of a pile is the friction that may be caused by surface roughness or particle shape irregularity [18]. Other factors, like pile formation history, particle size, and polydispersity, also influence properties of the pile [2]. To better understand the effect of friction and particle size, the following correlation has been proposed [18]:

$$\begin{gathered}\beta = 68.61\mu_{pp}^{0.27}\mu_{pw}^{0.22}\mu_{r,pp}^{0.06}\mu_{r,pw}^{0.12}d_p^{-0.2} \\ 2 \le d_p \le 10\,\text{mm},\ 0 \le \mu_{pp}, \mu_{pw} \le 0.6 \text{ and } 0 \le \mu_{r,pp}, \mu_{r,pw} \le 0.2\,\text{mm}\end{gathered} \tag{5.1}$$

where β is the angle of repose and μ_{pp}, μ_{pw}, $\mu_{r,pp}$, and $\mu_{r,pw}$ are the coefficients of dynamic friction for particle–particle and particle–wall contacts, coefficients of rolling friction for particle–particle and particle–wall contacts, respectively. This correlation suggests that the dynamic angle of repose is directly proportional to coefficients of friction and inversely to the particle diameter. In addition, to have a stable pile in DEM simulations with spherical particles, including and adjusting the value of all coefficients of friction are essential.

Figure 5.3 Pile formation of (a) 9 mm spherical particles, (b) biconvex particles, and (c) 9 mm spherical particles without rolling friction

We performed three DEM simulations to form piles with the code accompanied this book. These piles are formed on a flat plate placed 16 cm below the discharge orifice of a conical hopper (see Section 5.2 for more details on the hopper flow simulations). The first simulation was done with 8500 spherical particles with a diameter of 9 mm and the second with 8000 biconvex particles (multi-sphere method) with the volume equal to the 9-mm sphere. In these simulations, particle–particle and particle-wall dynamic friction coefficients were 0.3 and 0.4, respectively, and the particle–particle and particle–wall rolling friction coefficients were both 0.1. In the third simulation, operating conditions was similar to the first one with the exception that the rolling friction coefficients were set to zero.

Snapshots of the final pile formed in these simulations are shown in Figure 5.3. In the first simulation (spherical particles, Figure 5.3a), the angle of repose was found to be about 24°, which is consistent with Equation 5.1. In the second simulation (biconvex particles, Figure 5.3b) this angle is around 30°, which is greater than that of the spherical particle. However, no stable pile was formed in the third simulation in which no rolling friction was considered (Figure 5.3c).

Experimental measurement of the static angle of repose is relatively easy. Some researchers use experimental angle of repose to calibrate the input parameters of the model, then use this calibrated model for a granular flow that is very different from the pile formation (e.g., in confined dynamic flows such as hopper discharge or granular flow in rotating drums), which is wrong. We showed that no stable pile can be formed when the rolling friction is ignored in the model. We also found that flow pattern and discharge rate of spherical particles from a hopper are not significantly affected by inclusion of rolling friction (results not shown here). The confined dynamic granular flow is less affected by changes in the rolling friction while the formation of pile, which can be categorized as an unconfined granular flow, is greatly influenced by including the rolling friction. Therefore, there is no guarantee that if the model parameters are calibrated with the angle of repose in pile formation, the calibrated model would predict the correct behavior of a dynamic granular flow. As a primary requirement, the flow regime/type in which the model is calibrated should be similar to the flow regime in which the calibrated model is used.

5.1.3 Rigid and Flexible Fibers

The dynamic behavior of a single fiber or a collection of fibers or packing condition of fibers in a non-inertial medium – such as air – or in a medium with noticeable inertial effects – such as a suspension – is of great importance in material science and industries like composite metal and ceramic production, flow of polymeric solutions, manufacturing of fabrics, pulp and paper, membrane science, and tissue engineering [19–23]. For example, in manufacturing of composite materials by pressurized infiltration, packing conditions of fibers (i.e., packing fraction, orientation of fibers, and spatial distribution of pores), dictates the infiltration procedure, and determines the quality of the final product. Another example is the dynamics of fibers and rod-like materials in a liquid such as polymeric suspensions. The modeling of fibers by a discrete approach gives insights into the rheology of suspension, role of fiber–liquid interactions in viscosity of the suspension and the stable shape of fibers in the suspension.

Fibers can be either rigid or flexible. A rigid fiber can be a cylinder [24], a sphero-cylinder, or spheres glued together with or without overlap (multi-sphere method). The kinematics of such a fiber can be fully characterized by rigid body dynamics, described in Chapter 4. A flexible fiber is a chain of rigid segments connected by elastic bonds that allow the fiber to elongates, shortens, bends, or twists. The motion of a flexible fiber can be modeled by tracking each of these segments using proper Newton's and Euler's equations of motion. For example, in a fiber composed of spherical segments, spheres are interconnected through bonds with predefined length, radius, Young's, and shear moduli. Relative motion of two adjacent spheres deforms this bond, which yields bonding forces (constituted of normal and shear forces) and moments (constituted of torsional and bending moments) on the spheres and hinders the deformation of fiber. The particle flexibility can be determined by changing the bond properties, such as Young's and shear moduli, and bond shape, like cross-sectional area and moment of inertia of the bond.

For tracking spheres in each fiber, equations of motion for spherical particles, presented in Chapter 2, can be used with some modifications as follows:

$$m_i \frac{d^2\vec{x}_i}{dt^2} = \sum_{j\in CL_i} \vec{f}_{ij}^{\,p-p} + \sum_{k\in BL_i} \vec{f}_{ik}^{\,bond} + \vec{f}_i^{\,f-p} + \vec{f}_i^{\,ext} \tag{5.2a}$$

$$I_i \frac{d\vec{\omega}_i}{dt} = \sum_{j\in CL_i} \vec{M}_{ij}^{c} + \sum_{k\in BL_i} \vec{M}_{ik}^{bond} \tag{5.2b}$$

In Equation 5.2a, $\vec{f}_{ij}^{\,p-p}$ is the contact force between spheres in different fibers or non-adjacent spheres in the same fiber, $\vec{f}_i^{\,f-p}$ represents the fluid–particle interaction forces acting on the sphere (can be neglected in the absence of significant fluid effects), $\vec{f}_i^{\,ext}$ is the sum of external forces acting on the sphere and $\vec{f}_{ik}^{bond}$ is the bonding force between constituent spheres (i and k) of a bond. The notation k is used here to distinguish between the interaction of spheres in fibers that do not share a bond (where notations i and j are used) and interaction between spheres that are connected to the same bond (where notations i and k are used). In Equation 5.2b, $\vec{M}_{ij}^{c}$ is the contact moment on the sphere and $\vec{M}_{ik}^{bond}$ is the bonding moment. $\vec{f}_{ij}^{p-p}$ and $\vec{M}_{ij}^{c}$ are calculated by relationships presented in Chapter 2. Details for calculating bonding force and moment are presented by Potyondy and Cundall [25]. The outcome of such a model is a flexible fiber that elongates, shortens, bends, and twists and at the same time, and moves as a fiber in space due to various external forces acting on it. Equations 5.2a,b clearly show that the original numerical code used for spherical particles need minor modifications (some modifications in contact search and contact force calculations) to become ready for flexible fiber simulations. This model has been validated by experimental data on bending, twisting, and stretching of fibers and a criterion for selecting integration time step was proposed by Guo *et al.* [26]. It should be noted that if we model the flexible fiber with non-spherical segments, Equations 5.2a and 5.2b cannot be used for tracking fibers (e.g., [22, 23]).

Nan *et al.* [20] studied the packing behavior of straight and curved rigid fibers under gravity. Various packing properties such as porosity (validated against experimental data), coordination number, and fiber orientations were studied. It was shown that the aspect ratio has more pronounced effect on the porosity than the curl index. The porosity increases with increasing the aspect ratio and decreasing the curl index. Langston *et al.* [19] simulated very long flexible fibers (aspect ratios up to 30) using one, three, and five connected sphero-cylinders (see Figure 5.4). Their results showed that increasing the number of segments from one to three influences the packing structure while further increase in the number of segments has a little effect on the structure.

The rod-like particles are also studied in a suspension to characterize the rheology of the suspension as well as the microstructure of fibers in a suspension. In this case, the effect of

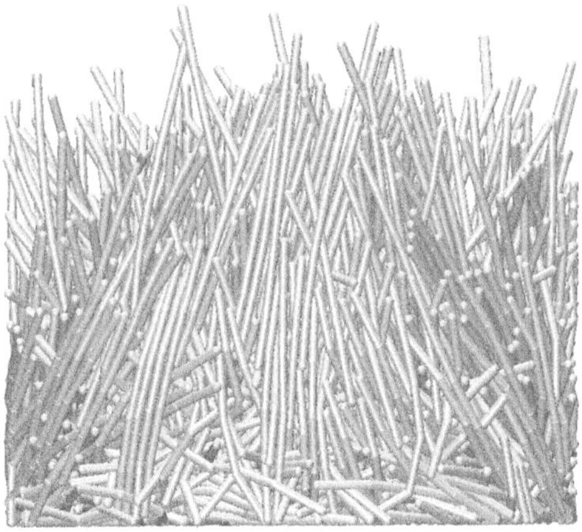

Figure 5.4 Packing of 1200 flexible fibers which comprise of five sphero-cylinders. The length to diameter ratio is 30. Bending that from the joint points between segments and it was shown that the flexible fiber model has better capabilities to predict dynamics and packing condition of long rods. (from [19] with permission)

fluid viscous effects should be considered and proper terms accounting for the solid–fluid interaction forces should be added to the equation of motion of segments (like $\vec{f}_i^{f-p}$ in Equation 5.2a). The fluid motion in the suspension can be obtained by solving incompressible Navier–Stokes equations in a discretized space or incorporating other methods like Lattice–Boltzmann. The equations of motion of fiber segments and fluid are coupled through solid–fluid interaction forces and applying boundary conditions for fluid equations. It was shown that by this coupling, the behavior of fibers in the suspension becomes more consistent to experimental observations. The flexible fiber model in suspensions has been validated against various experimental data and correlations [22, 23].

5.2 Flow in Hoppers

Hoppers/bins/silos (we refer to all of them as hoppers) are used to store bulk solids in any industries dealing with solids. These devices have different designs and are used for various applications. They mostly have circular or rectangular cross section whose area may be constant along the entire height of the hopper or conical at the bottom part. To avoid confusion, we name them as follows:

- *Wedged hopper*: a hopper with rectangular cross section with wedged bottom.
- *Flat-bottomed hopper*: a hopper with rectangular cross section with flat bottom.
- *Conical hopper*: a hopper with circular cross section with conical bottom.
- *Cylindrical hopper*: a hopper with circular cross section with flat bottom.

In addition to these hoppers, other types of hopper can be found. Sometimes, hoppers are used in combination with screw conveyors (screw feeders) for storage and transporting solids at a controlled rate.

Important factors for successful designs of hoppers are flow pattern and discharge rate of solids as well as pressure distribution on walls of the hopper during charge, storage (static pressure distribution), and discharge (dynamic pressure distribution). Although the flow in these containers seems very simple at first glance, many issues have remained unresolved. This complexity arises from the fact that solids with a diverse range of size, shape, and cohesion are processed in these containers. Problems such as arching, solidification, and flow rate fluctuations can be found in hoppers. As a result, many theoretical models have been developed and lots of experimental and computer simulations have been conducted over several decades to study the flow behavior in hoppers.

The DEM simulation is shown to be a powerful tool for predicting flow behavior of dry [27–31], cohesive spherical particles [32, 33], and non-spherical particles [34–38] in different hopper geometries. Flow of solids in special shapes of hoppers were also investigated, like hoppers with eccentric cross sections [39] or the Paul–Wurth hopper [40]. Segregation of particles during filling and discharging is a well-known issue addressed in the literature [40–43].

Wall pressure/stress distribution is another important aspect investigated both experimentally and by modeling with good agreement reported between DEM results and analytical relations or experiments [29, 30, 44, 45]. It was shown that increasing the friction between wall and granules strengthens the force chains on the wall leading to a higher stress on the walls. The pressure profile on the wall is different during charging and discharging and also it depends on the geometry of the hopper. The normal pressure on the wall is less during discharge of the flat-bottomed hopper in comparison to the storage period. In a wedged hopper, the pressure reaches a maximum at the joint point between the wedge and hexahedral sections of the hopper [45]. It was also shown that the pressure distribution along the wall and stress propagation inside the hopper is affected by shape [29, 30] and cohesion [32] of particles. Particles with sharp edges, particles that are able to interlock or cohesive particles have a higher tendency to form an arch above the orifice [35, 46]. The method of representing non-spherical particles (multi-sphere, polyhedral with sharp, and round edges) can also affect the capability of the model to predict arch formation [47].

To get a better insight into the flow of granules in hoppers, we performed a set of simulations with spherical and non-spherical (created by the multi-sphere method) particles. Physical properties and simulation conditions are listed in Table 5.1. Other simulation conditions are as follows:

- Diameter of the hopper is 20 cm and the discharge orifice diameter is 8 cm.
- In conical hoppers, 15° and 45° hopper angles (the angle between its conical section of the hopper and vertical direction, θ_{cone}) are considered.
- Spherical particles have a diameter of 9 mm.
- Biconvex tablets (created with the multi-sphere method) were used as non-spherical particles. The diameter of each tablet in the cross section is 12 mm and its thickness in the center and edges are 6.5 and 4 mm, respectively.

The code files to setup these simulations and sample animations of them can be found in the online content of the book in the "FlowInHopper" folder.

5.2.1 *Flow Patterns*

Figure 5.5 shows snapshots of the particle discharge from various hoppers. In all simulations, after the packing and before discharging, particles were colored based on their initial height in the container. Different layers of particles observed in this figure were formed by this method. Tracking the deformation of these layers during discharge gives qualitative insight into the velocity profile and flow pattern of particles. Two primary flow patterns can be identified in hoppers: funnel flow and mass flow. In the mass flow pattern, all particles move down in the hopper, no stagnant zone can be found and the flowing channel coincides with the walls. The mass flow pattern is possible when the walls are steep or/and friction between particle and wall is low. In the funnel flow pattern, some particles flow toward the outlet of the hopper while the rest are stationary and no intersection between flow channel and walls can be found. This case occurs when there is a high friction between particle and wall and hopper's bottom is near flat or completely flat.

Discharge of spherical particles from a cylindrical hopper is shown in Figure 5.5a. It can be seen in this figure that layers of particles flow with a V-shaped pattern. A stagnant region is formed at the bottom of the hopper. The flow channel coincides with the wall at the top of the bed and we can conclude that it is a funnel flow pattern. The order at which particles exit the hopper is very important. If we divide the hopper into a core (in which particles move) and an annulus, particles in the core start to exit the hopper from the lowest layer. When the top layer

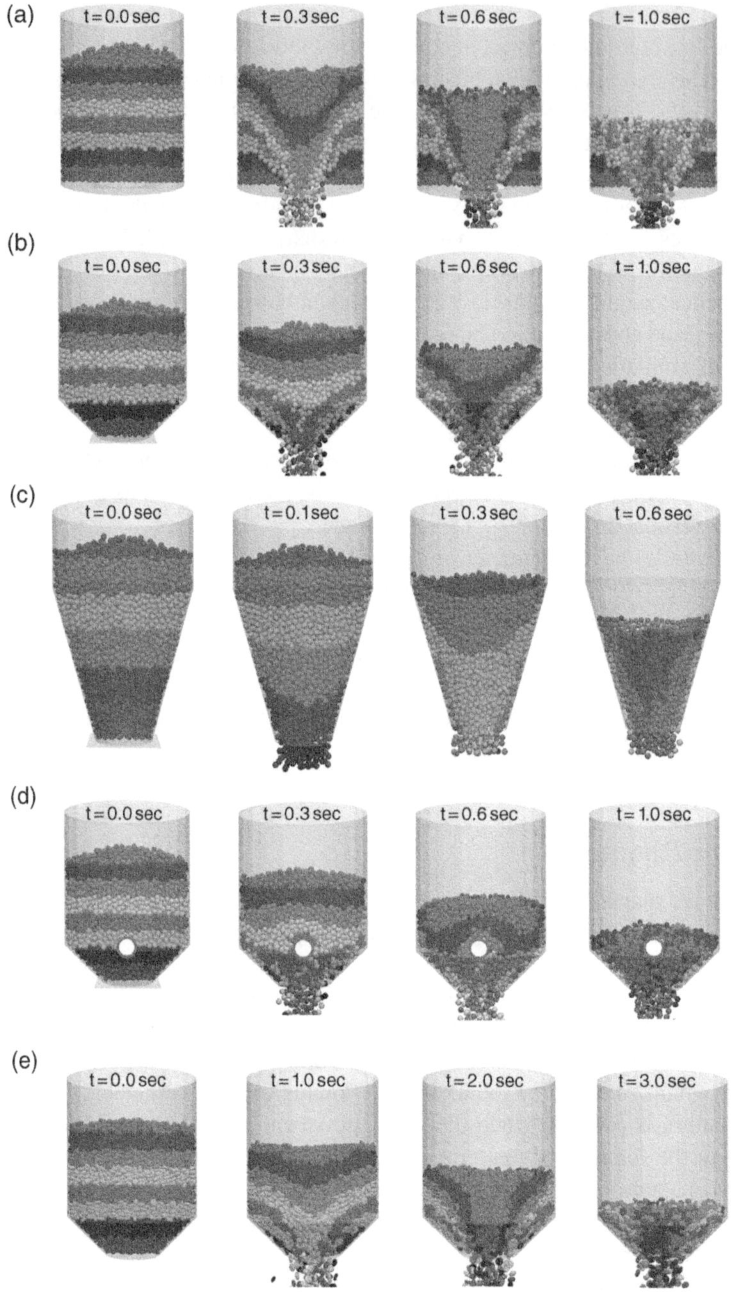

Figure 5.5 Discharge flow patterns of particles in various hoppers. (a) Discharge of spherical particles in a cylindrical hopper, (b) spherical particle in a conical hopper with $\theta_{cone} = 45°$, (c) spherical particle in a conical hopper with $\theta_{cone} = 15°$, (d) spherical particles in a conical hopper with obstacle, and (e) biconvex particles in conical hopper with $\theta_{cone} = 45°$. All snapshots are clipped in the middle to make the inside of the bed visible

of the core exits from orifice, the layers in the annulus start to exit from top to the bottom. This leads to a non-ideal and wide residence time distribution (RTD) of particles in the hopper that deviates from the first-in-first-out flow pattern.

The second row of Figure 5.5b shows the discharge flow pattern of spherical particles from a conical hopper with $\theta_{cone} = 45°$. Here, the flow pattern is a combination of the funnel flow and the mass flow (mixed flow). The flow channel coincides with the wall and the whole bed descend, while layers still do not leave the hoppers in sequence as it is expected in the mass flow pattern. However, the RTD is closer to first-in-first-out pattern in this hopper in comparison to the cylindrical hopper. Figure 5.5c also demonstrates the discharge flow pattern in a hopper with $\theta_{cone} = 15°$. Based on the criterion proposed by Jenike (in [48]), the mass flow pattern should be observed in this hopper. The snapshots also show this behavior and layers of particles leave the hopper in sequence.

In Figure 5.5d, we placed a horizontal tube in the conical hopper near the orifice to act as an obstacle to the flow. We can see that the effect of this obstacle on the flow pattern is significant. It widens the central channel flow in the hopper. Consequently, particles leave the hopper layer by layer. It should be noted that this obstacle does not affect the discharge flow rate in our simulations. More investigations are required to find the optimum location, number, and size of the obstacle to obtain a uniform flow pattern.

Shape of particle is the last parameter studied here (in Figure 5.5e). The discharge flow rate of biconvex particles decreases in comparison to that of spherical particles in the conical hopper (second row). Moreover, the flow pattern becomes more similar to the funnel flow rather than the mass flow.

5.2.2 *Segregation*

Segregation of particles due to size difference is a known phenomenon during charging and discharging of hoppers. As an example, experiments on the funnel flow hopper with binary mixture of particles with size difference and mass composition of 50% have shown that in the first half of discharge, the exit flow contains mostly small particle, while in the second half it contains mostly large particle [49]. We simulated the discharge of two mixtures of small (4.5 mm) and large (9 mm) particles with 11 and 50 wt% of small particle in a cylindrical hopper. Physical properties of particles are listed in Table 5.1. Discharge from this hopper follows the funnel flow pattern. The code files to setup these simulations and sample animations of them can be found in the online content in the "SegregationHopper" folder.

Figure 5.6 shows the snapshots of the case with 11 wt% of small particle. In Figure 5.6a, concentration of small particle in the core section of hopper is high which is due to the method of charging particles. Particles were inserted into the hopper through an insertion plane as we described it in Figure 5.1. During the charging, the concentration of small particles in the core of hopper becomes higher. When the exit gate is removed, the flow contains mostly small particles because they are accumulated in the core section (Figure 5.6a). After formation of the funnel flow pattern (the V-shape surface is not clear in this figure), fine particles percolate into slow moving or stagnant regions (near the walls) and large particles move to the core where velocity of particles is high (Figure 5.6b). Materials in the core (here, large particles) quickly discharge from the hopper. At the final stages of the discharge, a small rich mixture of particles near the wall starts leaving hopper (Figure 5.6c).

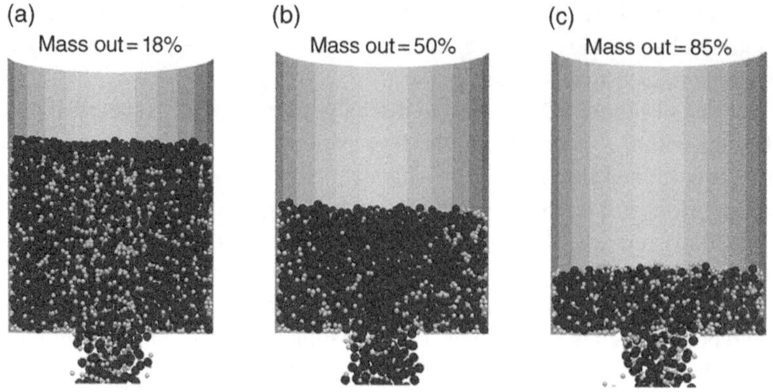

Figure 5.6 Snapshots of discharge of a binary mixture from a cylindrical hopper. Small particles (4.5 mm) are light gray and large particles (9 mm) are black. (a) shows 18% mass out (b) shows 50% mass out, and (c) 85% mass out

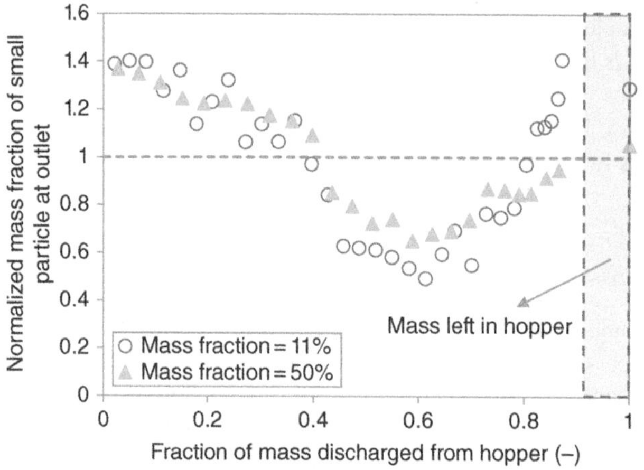

Figure 5.7 Segregation profile during the discharge of two binary mixtures of particles with 11 and 50 wt% of small particle in a cylindrical hopper. Diameters of small and large particles are 4.5 and 9 mm, respectively

Figure 5.7 demonstrates the normalized mass fraction of small particle (instantaneous mass of small particle to initial mass of small particle) in the exit flow versus the fraction of mass discharged from hopper for two simulation cases. In the case of 11% small particles, the exit flow is rich in small particle, then becomes rich in large particle and finally rich in small particle again. For the case of 50% small particles, the same trend is observed while the normalized mass fraction in the exit flow is closer to unity. In addition, the normalized mass fraction at the outlet approaches unity at the end of operation. This indicates that the segregation decreases when the fraction of small particle is increased. DEM simulations showed that other factors, such as particle size ratio, hopper cross section shape, hopper angle (θ_{cone}), and charging method of particle, have significant effects on the segregation pattern [41]. For example, by decreasing the hopper angle, the funnel flow patterns turn into mass flow pattern, and the segregation diminishes.

5.2.3 Discharge Rate

The steady-state mass flow rate of discharge from a hopper can be calculated from the following equation [48]:

$$\dot{W} = \rho_b A \left[\frac{D_0 g}{2(1+m)\tan\theta_{cone}} \right]^{1/2} \tag{5.3}$$

where ρ_b is the bulk density, A and D_0 are area and width of orifice, respectively, and m is a hopper shape factor, which is 1 for a conical hopper and 0 for a wedged hopper. This correlation is valid for free flowing particles with the average diameter of at least 6 mm.

Another correlation, developed by Beverloo *et al.* [50], takes into account the particle diameter. They showed that a straight line is obtained by plotting $\dot{W}^{2/5}$ vs. D_0. The intercept of this line is called Z, which is a parameter related to the mean diameter and shape of particles. $Z = kd_p$, where k is a constant that depends on the particle shape and is ~ 1.5 – 1.9 for spheres. Their correlation for a cylindrical hopper is:

$$\dot{W} = C\rho_b g^{1/2} \left(D_0 - kd_p\right)^{5/2} \tag{5.4}$$

where C is an empirical constant which depends on the material properties and is in the range of $0.55 \le C \le 0.65$. To include the effect of hopper angle in the discharge rate, the following equation was proposed [51]:

$$\dot{W}_{cone} = \dot{W} \frac{1-\cos^{3/2}\theta_{cone}}{\sin^{5/2}\theta_{cone}} \tag{5.5}$$

where $\dot{W}_{cone}$ is the discharge mass flow rate from a conical hopper with the hopper angle θ_{cone}. This correlation clearly shows that the mass flow rate of a conical hopper does not increase noticeably for hopper angles between 90° and 45°; while it rapidly increases at angles smaller than 45°.

The factors affecting the discharge rate are not still very well understood. The DEM can help better understanding the role of shape factor, cohesion, and friction on the discharge rate. Simulation results showed that at small dynamic friction coefficients (less than 0.2 in the operating range investigated), the discharge rate increases with the fill height of hopper [31]. At large values of the dynamic friction coefficient, the discharge rate becomes independent of the fill height. It should be noted that in most practical applications, the friction coefficient is greater than 0.2 and we can safely consider that the discharge rate does not depend on the fill height. Deeper analysis on the discharge rate has shown that the particle–particle friction coefficient has a strong effect on the discharge rate while the particle–wall coefficient has little effect. It has also been shown that the hopper width and the coefficient of restitution of particle do not affect the mass flow rate [31].

Non-sphericity has a significant effect on the discharge flow rate from a hopper. Liu *et al.* [37] simulated the discharge flow rate of super-ellipsoidal particles with aspect ratio in the range

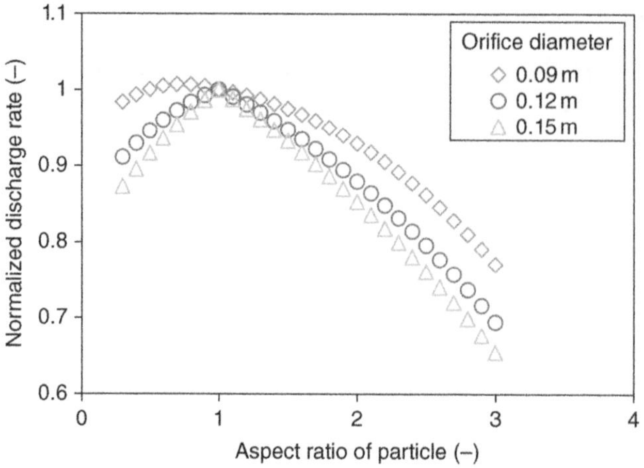

Figure 5.8 Effect of aspect ratio of particles and orifice diameter on the normalized discharge rate (normalized with discharge rate of spherical particle) of 9 mm particles in a cylindrical hopper. Data were created using Equations 5.4, 5.6a, and 5.6b

of $0.3 \le \eta \le 3.0$. They extended the correlation of Beverloo (Equation 5.4) by finding the dependency of parameters C and k to the aspect ratio:

$$C(\eta) = \begin{cases} 0.23\eta + 0.405 & 0.3 \le \eta \le 1.0 \text{ flattened particle} \\ -0.15\eta + 0.785 & 1.0 < \eta \le 3.0 \text{ elongated particle} \end{cases} \tag{5.6a}$$

$$k(\eta) = \begin{cases} 1.37\eta + 0.41 & 0.3 \le \eta \le 1.0 \text{ flattened particle} \\ -0.67\eta + 2.45 & 1.0 < \eta \le 3.0 \text{ elongated particle} \end{cases} \tag{5.6b}$$

When using the Beverloo equation for non-spherical particles, the equivalent diameter of particle should be used, defined as the diameter of a sphere with the same volume. This correlation was successfully tested for the particle diameter to orifice size ratios greater than 10, but its validity is not clear for smaller ratios [37]. Figure 5.8 shows the effect of aspect ratio and orifice diameter on the normalized discharge rate of 9-mm particles in a cylindrical hopper. The discharge rate is normalized with respect to discharge rate of the spherical particle. It can be seen that the aspect ratio has a minor effect on the discharge rate for the small orifice diameters but has a more pronounced effect for larger orifices. The network of normal force between particles becomes stronger (this effect is more pronounced for elongated particles), which leads to stronger interlocking between particles when the shape of particle deviates sphere. This controls the flow of particle in the hopper.

Anand *et al.* [33] investigated the discharge rate of spherical wet particles from a flat-bottomed hopper with a rectangular cross section. They showed that the k parameter in the Beverloo equation should be modified as follows to reflect the effect of humidity on the mass flow rate:

$$k = 1.9e^{0.39Bo} \quad 0 \le Bo \le 3 \tag{5.7}$$

where Bo is the Bond number defined as the maximum cohesive force acting on particle to the gravitational force:

$$Bo = \frac{6\sigma_l}{d_p^2 \rho_l g} \tag{5.8}$$

where σ_l and ρ_l are the surface tension and the density of liquid, respectively. Discharge flow rate is decreased by increasing the Bond number. Although this correlation has been obtained for rectangular hopper, it also can be used for circular hoppers.

5.3 Solid Mixing

Mixing of particles is an important operation in many industries such as chemical, food, pharmaceutical, ceramics, metallurgy, plastics, and cosmetics. The amount of solid processed in the world exceeds billions of tons of solid materials that must be well-mixed in most situations, giving this operation a huge market and lots of opportunities for research and development. Having a well-mixed condition is essential for raw materials or final products. For example, concentration of different components in different packs of fertilizer or concentration of the active ingredient in each drug dose (tablets, capsules, etc.) should be constant. Depending on type and volume of material to be processed and the desired level of mixing, various types of mixers can be employed. In some processes, good mixing of material is essential for a successful operation. Granulation, coating, chemical reactions, and drying of granules are all examples showing that mixing should be done appropriately to make the process viable. In this section, we first look at different mixing and segregation mechanisms. Then we present some conventional methods for quantifying mixing quality. Mixing equipment such as rotating drum, tumbling blenders (V-blender, bin-blender, and tote-blender), convective batch mixers, and continuous mixers are discussed at the end.

5.3.1 Mechanisms of Mixing and Segregation

Different mechanisms for mixing have been proposed, but the following is the most accepted mechanism proposed by Lacey [52]:

- *Convective mixing*: movement of pockets or groups of particles in the macroscopic scale from one location to another due to the action of impeller, rotation of mixer, or inertial effect of fluid. This mixing is rapid and results a high extent of mixing in the very first stages of mixing.
- *Shear mixing*: coexisted of layers/zones of particles with high velocity gradient (shear) that interchange particles with one another and cause mixing. The shearing can be caused by a mechanical agitator or can be created in the bed of particles in a rotating drum (very weak). Shearing is used to break the interparticle bonds in agglomerating particles (cohesive particles) and enhance the homogeneity of the mixture.

- *Diffusive (dispersive) mixing*: When particles are free to move, especially on the surface of the bed, they can have random motion due to contact with other particles. This random motion gives rise to mixing in particle scale. Experimental measurements show that diffusivity of particles in a tumbling drum in the rolling regime is of the order of 10^{-6} to 10^{-5} m^2/s [53, 54]. Cohesion between particles reduces the contribution of diffusive mixing, if any.

All these three mechanisms occur during the mixing, but depending on the type of mixer and particles, one or two of them are prevailing.

Segregation occurs when particles in the mixture have differences in size, shape, density, physical properties, or flowability. The segregation of particles, as an opposing mechanism to mixing, occurs during the mixing process and continues even when the mixer is switched off. Various types and mechanisms of segregation have been identified. The main mechanisms of segregation are:

- *Trajectory segregation*: when a stream of particles with size difference is freely discharged, smaller particles travel a shorter distance than larger particles. In fact, small particles experience greater drag force in a fluid. This leads to a partial segregation, especially when emptying a mixer through a discharge gate in an inclined chute. In another circumstance, when small and large particle move on a curved path (this motion is mostly caused by the geometry of the container), the small particle tends to travel a curved path-line more than the large particle.
- *Percolation/sifting segregation*: in a mixture with size difference, small particles (in the presence of any disturbance) tend to rearrange their position and fill the gaps between the larger particles. This behavior mostly occurs during charging and discharging of hoppers where particle percolation mostly occurs on the free surface and results in an uneven radial distribution of particles and tendency of small particles to be discharged in the last stages of discharging. DEM simulations have predicted the size segregation of particles during discharging [41] and charging of hoppers [43]. Another well-known segregation governed by this mechanism is the "Brazil nut" effect in which small nuts are accumulated at the bottom of the container.
- *Momentum segregation*: when flow of a mixture of solids impacts a stationary heap of particles, large particles tend to collect in outer layers of the heap (and possibly bottom layers) while small particles in the point of impact and preferably on the top of the heap. Combarros *et al.* [55] studied this type of segregation through experiment as well as DEM simulations and showed that the DEM can predict this type of segregation. The size difference has more effect on the segregation than the shape difference.
- *Elutriation segregation*: this type of segregation occurs when the viscous effects of a fluid result in different drag forces on particles with different size or shape. For example, in hopper discharge, the upward movement of air exerts an upward drag force on particles. Particles that experience a greater relative drag force (drag/weight) tend to migrate toward the surface of the bed or even suspend. The difference between drag force on particles is the main motivation for the segregation even in rigorous gas–solid flows like fluidized beds. This effect has been studied both experimentally [56] and numerically [57, 58].

It can be realized from these mechanisms that segregation occurs due to the rearrangement of particles with respect to other particles. For cohesive particles, this rearrangement is hindered by the cohesion, thus, segregation is less problematic. For very fine particles, for

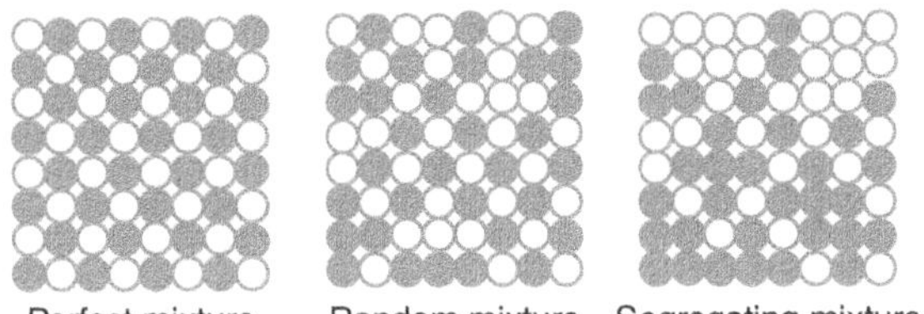

Figure 5.9 Different types of mixtures

which electrostatic or van der Waals forces are dominant, segregation is minimized. However, mixing should be achieved by breaking agglomerates with high shears. The same role is applied to wet particles. The attractive liquid bridge force hinders the segregation.

5.3.2 Mixing Index

The first step in studying the mixing operation is to quantify the mixing quality. Before that, some standard cases, as extremes of mixing, should be defined (see Figure 5.9):

- *Perfect mixture*: consider a mixture of two types of particles such that each sample taken randomly from the mixture has the same composition as the whole mixture. In the real world, such a mixture cannot be found.
- *Random mixture*: a random mixture (as the limit for maximum achievable mixedness) is obtained for non-interacting identical particles whose only difference is their color. The probability of finding particles of a kind in each random sample of the mixture is equal to the composition of that kind of particle in the whole mixture.
- *Segregating mixture*: if a mixture is composed of two interacting constituents, mixing and segregation mechanisms work concurrently in the mixer. In this condition, the probability of finding particles of a kind in one part is different than that in other parts. This is a segregating mixture and is the case achievable in real conditions.

The mixture quality is assessed with respect to the composition of particles in a sample. The question here is what is the proper size of the sample? It is determined by the scale of scrutiny, which was first used by Danckwerts [59]. It is the amount of material in which a certain level of homogeneity is desired. In pharmaceutical processes, the amount of material in a dose of medicine is the maximum scale of scrutiny, since the concentration of active pharmaceutical ingredients should be in an acceptable range. In the mixing process of different components of a fertilizer, the scale of scrutiny is a bag of fertilizer.

To determine the quality of a mixture, some samples should be taken from different parts of the mixture and statistical analyses are applied to determine the extent of mixing. Principles and various methods of sampling are described elsewhere [60–62]. Assume that the mixture is composed of two kinds of particles (say marked and unmarked) and N_s samples are taken from it. Mean and variance of these samples are calculated as:

$$\bar{c} = \frac{1}{N_s}\sum_{i=1}^{N_s} c_i \tag{5.9}$$

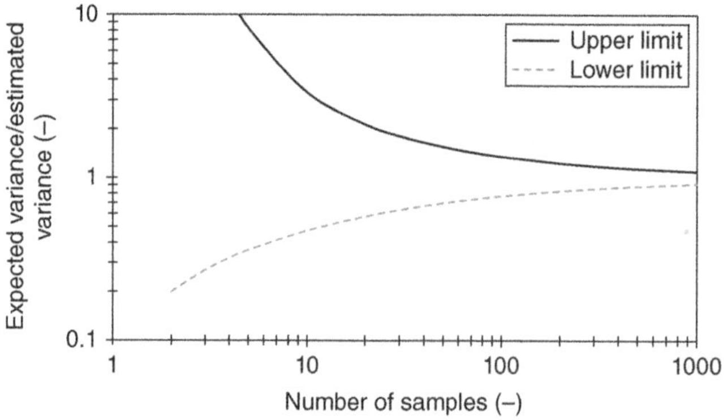

Figure 5.10 Upper and lower limits for expected variance of a sample as a function of number samples

$$S^2 = \frac{\sum_{i=1}^{N_s} (c_i - \bar{c})^2}{N_s - 1} \tag{5.10}$$

where c_i is the concentration of the marked particle in each sample. A large value of variance indicates segregated mixture and a small value corresponds to better mixing quality. This is an estimate for the expected variance of mixture, σ^2, and by increasing the number samples a better estimate of σ^2 can be obtained. The confidence range for σ^2 is obtained from the χ^2 distribution:

$$\frac{S^2 (N_s - 1)}{\chi^2_{\alpha/2}} \leq \sigma^2 \leq \frac{S^2 (N_s - 1)}{\chi^2_{1-\alpha/2}} \tag{5.11}$$

where α is the level of significance ($1-\alpha$ is the confidence level) and χ^2 is the chi-square statistic whose values can be found in standard statistical charts. Figure 5.10 shows upper and lower limits of the estimated variance as a function of number of samples for 95% confidence level. This figure shows that at least 30 samples should be taken from the mixture to have a reasonable estimate of σ^2. For the expected mean of the samples, μ, the confidence range is:

$$-\frac{t_{\alpha/2} S}{\sqrt{N_s}} \leq \mu - \bar{c} \leq \frac{t_{\alpha/2} S}{\sqrt{N_s}} \tag{5.12}$$

where values for $t_{\alpha/2}$ are obtained from standard Student's t-test charts for the desired confidence level.

As we discussed earlier, the random mixture is the maximum achievable mixing state in the real world. In the worst condition, the mixture is completely segregated, which means that marked particles occupy one region of the mixture and the rest is occupied by unmarked particles.

These are considered as two extreme conditions for the mixing quality of a mixture. The variances of these two extreme conditions are [60]:

$$\sigma_R^2 = \frac{w(1-w)}{n} \tag{5.13}$$

$$\sigma_0^2 = w(1-w) \tag{5.14}$$

where w is the composition of marked particle in the whole mixture and n is the average number of particles in samples.

Different statistical-based mixing indices have been proposed to quantify the mixing quality [63, 64]. In this section, we present some of the most frequently used ones. Lacey [52] proposed the following relationship to quantify the mixing quality:

$$MI_{Lacey} = \frac{\sigma_0^2 - \sigma^2}{\sigma_0^2 - \sigma_R^2} \tag{5.15}$$

The numerator shows the achieved level of mixing and the denominator is the maximum achievable level of mixing. The Lacey mixing index varies between one and zero: it is one for a random mixture and zero for a completely segregated mixture. The problem with the Lacey index is that it rapidly reaches to values more than 0.7 and cannot distinguish between mixtures with different qualities very well.

Another mixing index is proposed by Poole *et al.* [65] and reads as:

$$MI_{Poole} = \frac{\sigma}{\sigma_R} \tag{5.16}$$

It varies between a large value (for a segregated mixture) and unity (for a random mixture). Sensitivity of the Poole mixing index to changes in the extent of mixing in a partially mixed condition is greater than that of the Lacey index.

The Kramer mixing index behaves similar to the Poole index (reported in [64]):

$$MI_{Kramer} = \frac{\sigma_0 - \sigma}{\sigma_0 - \sigma_R} \tag{5.17}$$

It varies between zero and one for fully segregated and random mixture conditions, respectively.

The following mixing index is proposed by Valentine (reported in [64]):

$$MI_{Valentine} = \frac{\ln\sigma_0 - \ln\sigma}{\ln\sigma_0 - \ln\sigma_R} \tag{5.18}$$

Since it uses a logarithmic scale for measuring achieved and maximum achievable levels of mixing, it is sensitive to the variation of the extent of mixing from completely segregated to randomly mixed conditions [63].

Contrary to an experiment in which sampling is done after turning the mixer off and letting the bed settle down, in DEM simulations, samples of certain size at desired locations can be taken online. Since the position of all particles is known, it is possible to evaluate the mixing index by considering data of all particles. To this end, the mixer is divided into several cells (usually cubic) and composition of each particle type is computed in each cell. Each non-empty cell is considered as a sample and the mixing index can be calculated by one of the previously mentioned equations. It is possible that some cells contain a few particles in comparison with other non-empty cells. These cells should be excluded from calculations to prevent a biased evaluation.

New definitions of mixing index have been developed that need Lagrangian data of particles in the mixture [66–68]. These definitions are much more robust than the statistical methods. However, they have not been used very frequently yet for evaluating the mixture quality. A big obstacle here is that common experimental techniques cannot provide such data while the DEM simulation can. The generalized mean mixing index (GMMI) is based on evaluating the average height of particles of kind w in the x-, y-, and z-directions [66]:

$$GMMI_w = \frac{1}{3}\left(GMMI_{w,x} + GMMI_{w,y} + GMMI_{w,z}\right) \tag{5.19}$$

where $GMMI_{w,x}$, $GMMI_{w,y}$, and $GMMI_{w,z}$ are the average height of particles of kind w along the x-, y-, and z-axes. As an example, $GMMI_{w,x}$ can be calculated as:

$$GMMI_{w,x} = \frac{N_t}{N_w}\frac{\sum_{i=1}^{N_w}(x_i - x_0)}{\sum_{j=1}^{N_t}(x_j - x_0)} \tag{5.20}$$

where N_w and N_t are the number of particles of kind w and the total number of particles, respectively, and x_0 is the x-coordinate of a reference point. This mixing index is insensitive to the sample size and packing condition of particles and can be used for multi-component mixtures.

In the GMMI method, the average heights of particles are measured and averaged as the mixing index, which shows the mixing intensity along each direction. It is also possible to evaluate the mixing intensity around each particle. Consider a mixture with equal portions of marked and unmarked particles. For each particle, 12 nearest particles (neighbor particles) are determined. Number of neighboring particles that have a different mark than the target particle is then determined. If half of the neighboring particles have a different mark, the mixture is locally (around that particle) mixed and if all neighboring particles have the same mark, the mixture is locally segregated. The nearest neighbor mixing index of the whole mixture is expressed as [68]:

$$MI_{NN} = \frac{1}{N_t}\sum_{N_t}\frac{2n_{diff}}{n_{nb}} \tag{5.21}$$

where n_{nb} and n_{diff} are the number of neighbor particles and the number of neighbor particles with different mark, respectively. The computational complexity of this method is more than the GMMI method. However, it is closer to the definition of the random mixture.

5.3.3 *Rotating Drums*

Tumbling/rotating drums have been studied more extensively by experiment and simulation than other tumbling mixers, which indicates the importance of this equipment. Many parameters affect the flow of solids in these simple drums such that we cannot design and operate them without doing direct experiments or a detailed simulation. Moreover, these drums are used for various processes like coating, granulation, chemical reaction, drying, ball milling, and mixing, which motivate scientists and engineers to study different aspects of particulate flow in these drums rather than only the mixing efficiency. In this section, we do not restrict ourselves to mixing and segregation. We will first discuss different flow regimes and granular flow behavior in these drums and then talk about mixing and segregation phenomena.

5.3.3.1 Flow Regimes and Patterns

Mellmann [69] identified three groups of flow patterns in horizontal rotating drums (slipping, cascading, and cataracting) based on the fill level, Froude number, and particle–wall friction coefficient. The Froude number is defined as the ratio of the centrifugal force to gravity:

$$Fr = \frac{R_D \Omega^2}{g} \tag{5.22}$$

where R_D is the drum radius and Ω is the rotational speed of the drum in radians per second. It should be noted that the transition point between two regimes sometimes depends on additional parameters like particle diameter or particle–wall friction. These regimes can be predicted by DEM simulations [70]. We performed some simulations in a drum with inner diameter of 0.45 m at various rotational speeds to find the flow pattern of spherical particles in different flow regimes. The properties of particles and simulation conditions are listed in Table 5.1. In these simulations, we changed particle–particle and particle–wall coefficients of dynamic friction to 0.5 and 0.7, respectively. The code files to setup these simulations and sample animations of them can be found in online content under “RotatingDrumRegimes” folder. The results of these simulations are shown in Figure 5.11. Different flow regimes are identified in simulations, which will be discussed in the following.

The slipping motion occurs at low Froude numbers ($Fr \leq 10^{-4}$). This type of motion has no practical use in the industry and includes the following regimes:

- *Sliding regime*: when the wall is very smooth, the drum rotates and particles remain stationary in the packed state.
- *Surging regime*: at high fill levels and in a drum with rough walls, the packed bed of particles first rotates with wall and then slides to its original position, periodically.

Simulations of Yang *et al.* [70] showed that the DEM simulation can reproduce the sliding regime if the particle–wall friction coefficient is reduced below the critical friction coefficient proposed by Mellmann [69]:

$$\mu_{p-w,crit} = \frac{2\sin^3 \xi \sin \alpha}{3\pi\varphi\left(1+Fr\right)} \tag{5.23}$$

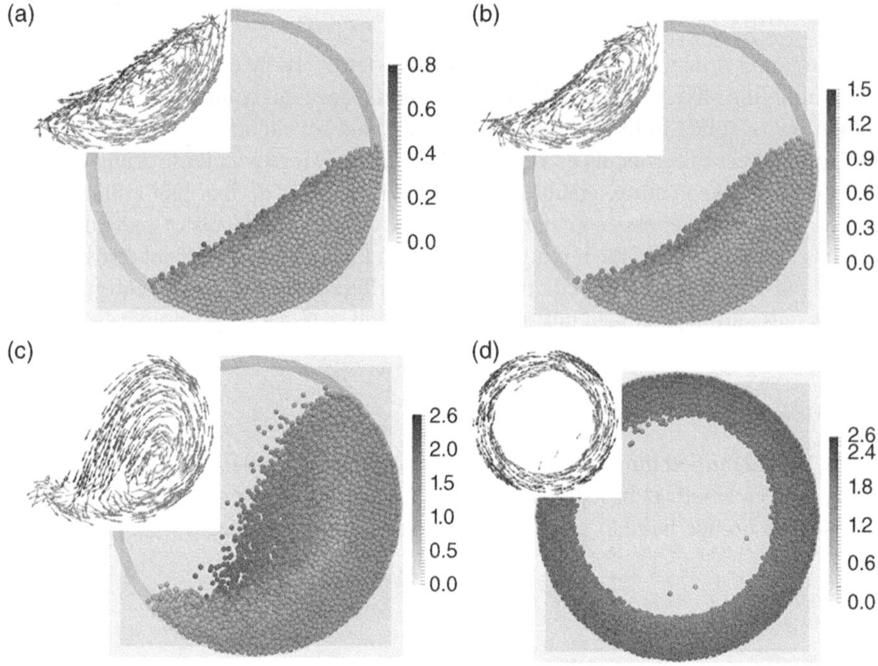

Figure 5.11 Various flow regimes in a rotating drum with inner diameter of 0.45 m, filled with 9 mm spherical particles, (a) rolling regime with $Fr = 5\times10^{-3}$ and $\varphi = 0.25$, (b) cascading regime with $Fr = 5\times10^{-2}$ and $\varphi = 0.25$, (c) cataracting regime with $Fr = 0.63$ and $\varphi = 0.35$, and (d) centrifuging regime with $Fr = 2.52$ and $\varphi = 0.44$. Particles are colored based on their velocity magnitude (the graduated shaded bar shows the velocity magnitude) and small pictures show the instantaneous velocity vectors. Vectors are not scaled and all have the same length

where ξ is the half bed angle of the circular segment filled with particles, α is the dynamic angle of repose, and φ is the fill ratio.

The cascading motion occurs at higher Froude numbers ($10^{-5} \le Fr \le 10^{-1}$), higher fill ratios ($\varphi > 0.1$), and in a drum with a rough wall. This type of motion is observed in coating drums, rotary dryers, and mixing drums. The two main flow regimes here are:

- *Rolling regime*: Figure 5.11a shows the bed of particles in the rolling regime as well as the velocity vector of particles. The whole bed is composed of two layers: active (cascading) layer and passive layer. In the thin active layer, particles avalanche on the bed surface and accelerate from top to middle of the bed surface. In the passive layer (beneath the active layer) a plug of particles is transported upward due to rotation of the drum and enters the active layer. A very thin shearing zone between these two layers can be seen in which the average particle velocity becomes zero. In the rolling regime, the bed surface is almost flat.
- *Cascading regime*: by a further increase in the Froude number, the bed enters the cascading regime (Figure 5.11b). In this regime, the bed surface begins to arch and is transformed into a kidney-shape bed. The flow of particles is similar to the rolling regime in the active and passive layers while the active layer becomes thicker.

Slumping, rolling, cascading, and the slumping-rolling and rolling-cascading transitions have been identified in DEM simulations and their main flow characteristics have been reproduced [70–72]. Simulation results showed that the flat surface of bed in the rolling regime transforms into an S-shape surface (a main characteristics of the cascading regime) when the shape of the particles diverges from spherical. This suggests that the particle shape affects the transition point between rolling and cascading regimes and occurs at a lower Froude number for non-spherical particles.

The cataracting motion can be observed at higher Froude numbers ($Fr > 0.1$) and has some practical applications in the industry (like ball mills). It includes the following flow regimes:

- *Cataracting regime*: as the rotation speed increases, some particles at the top of the bed surface detach from the bed and fall off into the free space of the drum. The height at which this detachment occurs and the portion of particles detached from the bed increase with increasing the rotation speed. The flow pattern considerably changes from cascading regime, that is, the avalanching of particles on the bed surface changes into the free fall of particles in the drum space (Figure 5.11c).
- *Centrifuging regime*: at higher rotational speeds, particles near the drum wall start to adhere to the wall due to the centrifugal force. In an extreme condition, all particles form an annulus in the drum (Figure 5.11d).

The transition between cataracting and centrifuging regimes depends on the fill level according to the correlation proposed by Mellmann [69]:

$$Fr_{crit} = \frac{1}{\sin(\beta_i + \xi - 90)} \tag{5.24}$$

where β_i is the angle of internal friction and ξ is half bed angle of the circular segment filled with particles in degrees. Accordingly for the fill level of 0.44, the transition to centrifuging regime occurs at Froude number around 2. Figure 5.12 shows the particle flow in a drum at different Froude numbers. It can be seen in this figure that when the Froude number is less than 2 (Figure 5.12a), annulus is not completely formed and particles fall down from periphery. As the Froude number increases to around 2 (Figure 5.12b), a few particles can be seen in the

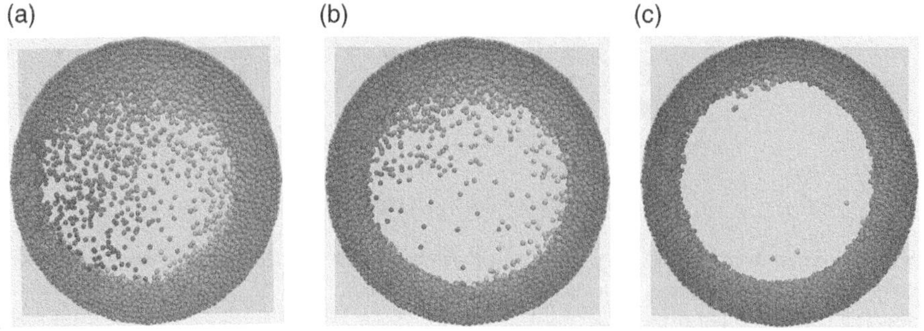

Figure 5.12 Rotating drum at $\varphi = 0.44$ and different Froude numbers, (a) $Fr = 1.61$, (b) $Fr = 2.03$, and (c) $Fr = 2.52$

central part and with further increase in the Froude number, complete centrifuge regime is obtained (Figure 5.12c). This shows that the model can predict the transition between cataracting and centrifuging regimes. As a conclusion, the DEM can predict the flow behavior of different regimes as well as the transition between them. This provides a good potential to produce flow regime diagrams and to obtain criteria for transition between regimes while including more variables, such as shape, size distribution, and properties of particles.

Drums operating in rolling and cascading regimes are used frequently in various industries. In these regimes, a good level of mixing can be achieved with an acceptable rate of renewal of particles on the bed surface with low energy requirement and particle breakage. Velocity profile and flow behavior of solids in these drums have been characterized by experiments [53, 54] and simulations [71, 73, 74]. Also, theoretical and empirical relationships have been developed for predicting velocity profile and the active layer thickness [75–77]. Comparisons between DEM results and experimental measurements have shown good agreements in terms of dynamic angle of repose, mean surface velocity, and velocity profile in the whole drum for both regimes.

We performed 39 simulations with spherical (5–9 mm) and non-spherical particles (biconvex, oblong, and oval) in drums with various fill levels (0.1–0.2), diameters (0.3–0.6 m), and rotational speeds (6–12 rpm) to determine velocity profile in these drums.[3] These simulations cover both rolling and cascading regimes. Figure 5.13 shows the time-averaged velocity profile on the bed surface for particles with different shapes and sizes as a function of chord distance (*X*, the distance from the center of the bed surface to the bottom of the bed surface)

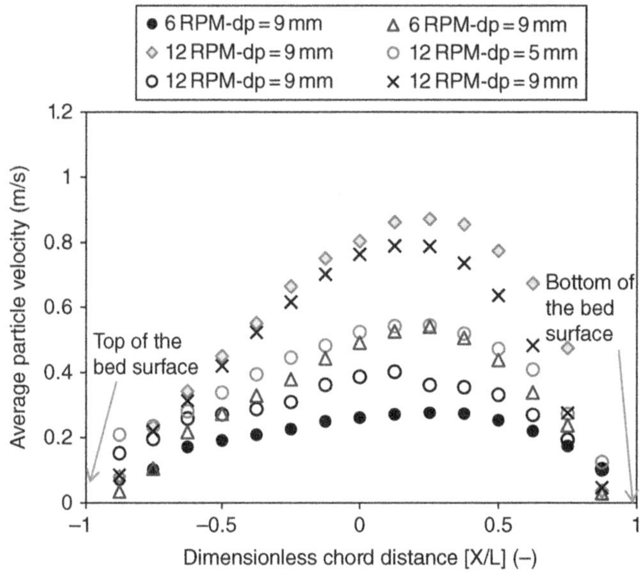

Figure 5.13 Time averaged velocity profile of particles at the surface of the active layer in a drum with inner diameter of 45 cm and fill level of 0.15. (•) Spherical, (Δ) oval, (♦) biconvex, and (×) oblong particles. Data are extracted from [71]

[3] These simulations were performed by an in-house FORTRAN code not published with this book.

in a drum with inner diameter of 0.45 m at various rotational speeds. Particles accelerate from top of the bed due to gravity and reach their peak velocity after the mid-chord point ($X = 0$). Thereafter, the kinetic energy is balanced by frictional and inelastic losses and particles decelerate, leading to an asymmetric profile. The asymmetric profile develops because the kinetic energy of avalanching particles is not balanced by friction and inelastic losses before reaching this point and acceleration continues after the mid-chord point. Khakhar *et al.* [78] assumed the following symmetric profile for particle velocity at the bed surface in the rolling regime:

$$\frac{u_{surf}}{u_{max}} = 1 - \left(\frac{X}{L}\right)^2 \tag{5.25}$$

where L is the mid-chord length and u_{max} is the maximum surface velocity. Alizadeh *et al.* [54] found that this equation fits the experimental data (spherical particles in rolling regime) better when the exponent in the right-hand side of this equation is replaced by 2/3. Nevertheless, an asymmetric profile is observed in experiments [53, 79]. The following equation best describes the velocity profile on the bed surface [71]:

$$\frac{u_{surf}}{u_{max}} = \left[1 - \left(\frac{X}{L}\right)^2\right] \exp\left(\frac{2\beta}{1-\beta^2}\frac{X}{L}\right) \tag{5.26}$$

where β is the asymmetry factor and is defined as $\beta = l/L$, in which l is the distance of the peak velocity from the mid-chord point ($X = 0$). The value of β varies from 0.04 to 0.22 in different cases and on average we found $\beta = 0.12 \pm 0.06$ with 95% confidence level. The only unknown in Equations 5.25 and 5.26 is the maximum surface velocity which should be obtained as a function of operating conditions.

The results in Figure 5.13 show that the maximum velocity increases with increasing the rotational speed and by decreasing the particle size. These trends agree with experimental observations [80–82]. The shape of particles is another effective parameter on the peak velocity. Non-spherical particles show a higher peak velocity (hence, higher surface velocity) compared to spherical particles. This is attributed to higher dynamic angle of repose of non-spherical particles in the drum that converts a higher fraction of the potential energy of particles at the top of the bed into the kinetic energy during avalanching. Different scaling relationships have been proposed for the maximum velocity in drums [73, 77, 83]. These relations are usually either based on some assumptions that are applicable to a narrow range of operating conditions (rolling or cascading) or shape effects are not reflected in them. Following Ding *et al.* [84], we obtained following relation for the maximum velocity from our simulation results:

$$\bar{u}_{max} = \frac{u_{max}}{L\sin(\alpha)\Omega} = 0.198\left(\frac{d_p}{2R_D}\right)^{-0.801} Fr^{-0.300}\varphi^{0.438}\left[\tan(\alpha)\right]^{0.855} \tag{5.27}$$

$$2.5\times10^{-4} \le Fr \le 5\times10^{-2},\ 0.1 \le \varphi \le 0.35,\ 3 \le d_p \le 9\,\text{mm, and } 12 \le R_D \le 30\,\text{cm}$$

where d_p is the particle diameter. This equation can be used for both spherical and non-spherical particles in rolling and cascading regimes. The dynamic angle of repose in this equation connects the maximum velocity of spherical particles to that of non-spherical particles. Equation 5.27 predicts the maximum velocity in the 39 simulations performed here as well as 29 data points from literature with an average absolute error of $10\pm3\%$ [71].

5.3.3.2 Mixing and Segregation

DEM simulations can be used to understand the mixing mechanisms in rotating drums. Consider the simulation of a half-filled rotating drum with the diameter of 0.2 m and the rotation speed of 10 rpm that contains 19,000 3-mm spherical particles. Snapshots taken from this simulation are shown in Figure 5.14. Particles are shown with dots in this figure. As we discussed earlier, the bed in the rolling/cascading regime comprises of active and passive layers which are shown in Figure 5.14a. Since the mean flow rate of particles does not change with time when the drum rotates, the streamlines of particle flow in the passive layer remains unchanged and almost impermeable to the convective flow of particles (Figure 5.14b). Particles have the opportunity to change their streamline in the active layer due to diffusion or their random movement. In general, we can assume that the whole bed is composed of distinct layers of particles with half-circle shape (similar to the shape of the bed). The velocity vectors in Figure 5.14a exactly shows this flow pattern. Note that vectors are not scaled with the velocity magnitude and are not averaged over time.

At steady state operation of the drum, we marked particles in three different parts of the bed with red, green, and blue colored circles to better understand the mixing mechanisms. Red particles are almost inside the active layer and the other two are in the passive layer (see shading in Figure 5.14c). By passing the time, the red circle in the active layer becomes deformed by high velocity shear flow in the active layer, becomes dispersed and then leaves this layer (Figure 5.14d). During the same period, the other two circles remain intact since they are in the passive layer and there is no relative motion between particles (solid-body motion). As the green and blue circles enter the active layer, they also deform and become blurred, then leave the layer (see shading in Figure 5.14e). This process is repeated many times during the rotation of the drum until mixing reaches equilibrium state (Figures 5.14f–h). These results show that, mixing only occurs in the active layer and the passive layer does not contribute to it. The passive layer only transports particles to the top of the bed to reenter the active layer. To better understand this mechanism for mixing of particles in the rolling regime, we recommend the reader to watch the animation of this simulation available in the online content under "RotatingDrumMixing" folder. The code files to setup this simulation are also provided.

Now we can determine effect of various parameters on the mixing mechanism and mixing rate. As the filling fraction increases, up to half-filled bed, the mixing rate decreases due to two reasons. First, the average circulation time of particles increases with filling ratio. The circulation time is the time required for particles to reenter the active layer or, equivalently, the time that particles spend in the passive layer. The particle velocity along the streamline in the passive layer is $r\Omega$, where r is the radial position, and is almost insensitive to the filling fraction and particle properties (size, shape, etc.). By increasing the filling fraction, particles should travel a longer distance to reenter the active layer, hence the circulation time

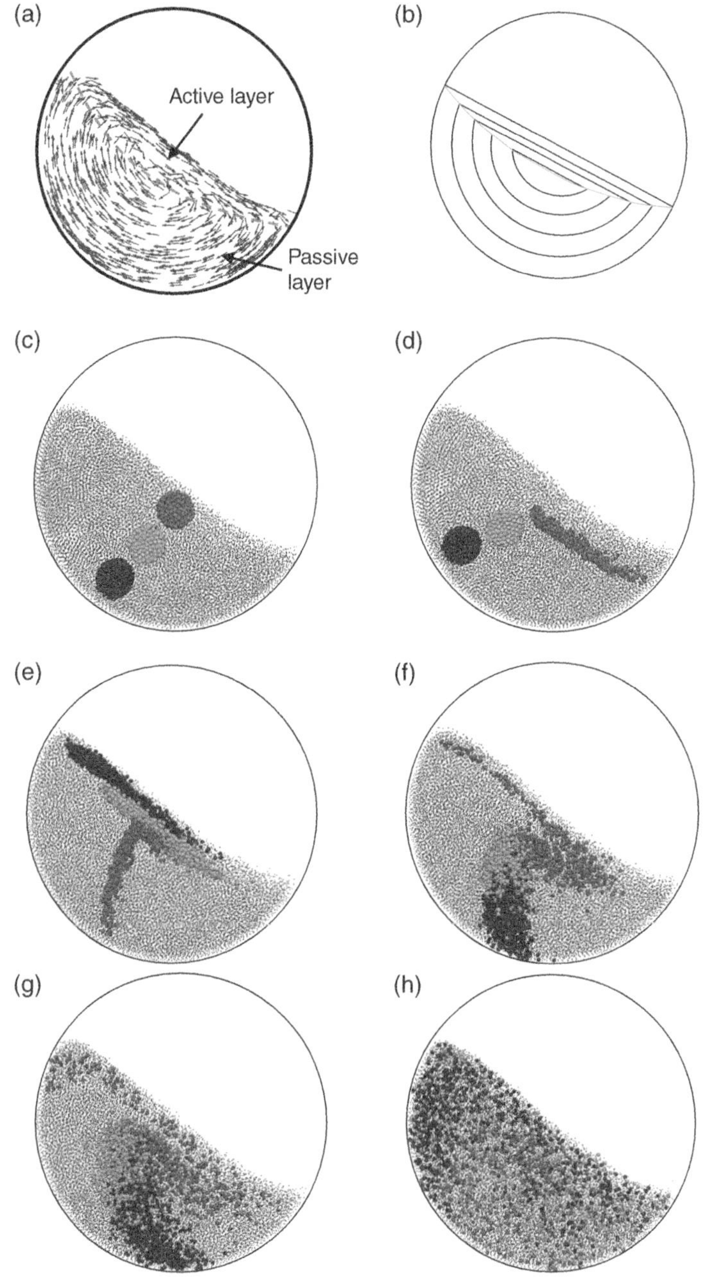

Figure 5.14 Mixing mechanism of 3-mm monodispersed spherical particles in a rotating drum with the diameter of 0.2 m and the rotation speed of 10 rpm (rolling regime), (a) instantaneous non-scaled velocity vectors of particles and (b) the schematic of motion streamlines in active and passive layers. Particles are shown by gray dots in these snapshots and tracers by darker shaded spheres. Snapshots show the position of tracers at the reference time of (c) 0 s, (d) 0.2 s, (e) 0.8 s, (f) 1.6 s, (g) 3.2 s, and (h) 9 s

increases. Second, the volume of active layer (where mixing occurs) decreases with respect to the total volume of the bed. Increasing the rotation speed has a positive effect on the mixing rate for two reasons. The mean circulation time of particles decreases due to increased particle velocity in the passive layer and the shearing effects and diffusion become more pronounced in the active layer. Combining these two parallel effects leads to enhanced mixing rate with increasing the rotation speed.

An extensive amount of experimental and simulation studies can be found on the mixing of particles in rotating drums [85–90]. These studies are focused on three aspects of mixing in rotating drums: the dynamics of mixing and time required to reach the completely mixed condition, traverse/radial mixing, and axial mixing or dispersion coefficient. Using the 3D DEM simulations, Finnie *et al.* [85] showed that the dimensionless axial diffusion coefficient is linearly proportional to the rotation speed while it is insensitive to the filling fraction. In general, the axial diffusion coefficient is proportional to Ω^{λ} in which λ varies between 0.4 and 2. This shows that the diffusion coefficient is a complex function of rotation speed, which requires careful characterization at different operating conditions. It was found that the axial diffusion coefficient is proportional to $d_p^{1.9}$ and is independent of the drum diameter in large drums [91]. In the traverse direction, a better mixture quality is achieved at the same number of revolutions when decreasing the rotation speed and the filling fraction. Kwapinska *et al.* [86] further showed that the number of required revolutions to obtain a perfect mixture is proportional to $Fr_f^{0.22}$, where Fr_f is the modified Froude number that accounts for the filling degree of drum. The mixing rate and mixing mechanism for wet particles is different from those of dry particles [88]. Three-dimensional DEM simulations show that the mixing rate decreases with increasing the fill fraction and cohesion. At the fill fraction more than 50%, poor mixing is achieved for both dry and wet particles due to formation of non-mixing zones in the lower parts of the bed. This was also observed in experiments [92, 93]. Simulation results show that the mean circulation time of particles increases with increasing the wetness of particles and the radial displacement of dry particles becomes more restricted as the wetness of particles increases. These two result in poorer mixing conditions for wet particles in comparison to dry particles.

Different segregation patterns have been observed in rotating drums experimentally [78, 94, 95] and by DEM simulations [96–102]. A very common segregation pattern is radial segregation in which a core of smaller or denser particles is formed in the center of the bed at low rotational speeds. At higher rotational speeds, however, smaller or denser particles are segregated in the periphery of the bed, which indicates the complexity of the segregation in these drums. Radial segregation usually happens after a few revolutions. Another type of segregation in horizontal long rotating drums is axial segregation in which particles with the same size or density (or maybe shape) tend to segregate into axial bands. This type of segregation takes a longer time than axial segregation to become evident (more than 100 revolutions). This type of segregation has been studied experimentally [103, 104] and with DEM simulations [105].

To investigate the mechanism of radial segregation at low rotational speeds, we performed a simulation with a binary mixture of small (3 mm) and large (4.5 mm) particles with volume composition of 50% in a drum with the diameter of 20 cm rotating at 10 rpm. The bed is initially in random mixture state. The code files of this simulation and an animation of the segregation process in this drum can be found in the online content of the book under "RotatingDrumSegregation" folder. Figure 5.15 illustrates snapshots of this simulation.

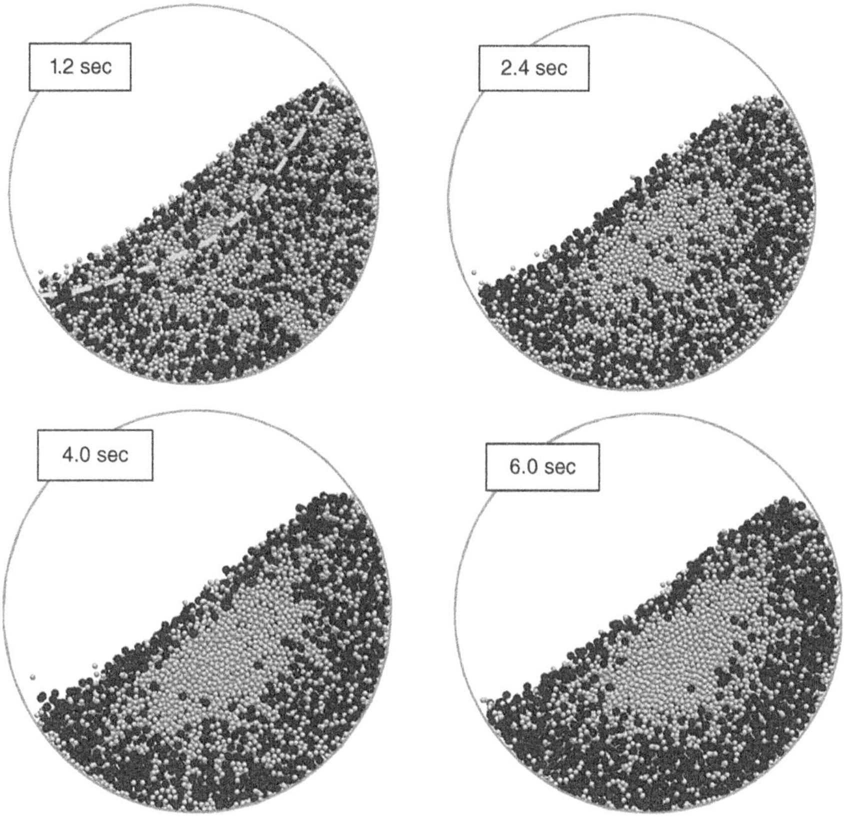

Figure 5.15 Radial segregation process for a 1:1 binary mixture of small (3 mm, light gray) and large (4.5 mm, black) particles in a rotating drum (10 rpm and 0.2 m diameter). Small particles accumulate in the middle of the drum while larger particles accumulate in the periphery of the bed

Particles are shaded based on their size, gray for small particle, and black for large particle. A core of small particles is formed after only 6 s of rotation elapses (one revolution). Shape of the core is similar to the bed suggesting that the streamlines in the passive layer act as impermeable walls to the convective mixing. Gradual formation of the core is evident after only 2.4 s of rotation. As we discussed earlier, particles move along the streamlines as a rigid body in the passive layer. When they reach the bed surface and enter the active layer, smaller particles sink into the lower parts of the active layer due to the percolation mechanism. Consequently, larger particles are partially accumulated in the upper layers and smaller particles in the lower layers of the active layer. In this new arrangement of particles, larger particles enter outermost layers of the passive layer (periphery of the bed) and smaller particles in the innermost layers (the core). No rearrangement occurs in the passive layer. Therefore, these particles reenter the active layer with almost the same configuration as they left this layer. This process is repeated many times and segregation proceeds. The boundary between the core and outer layer is blurred due to diffusion. In the absence of diffusion, a sharp boundary would be obtained. For a binary mixture of particles with identical sizes but different densities, denser particles

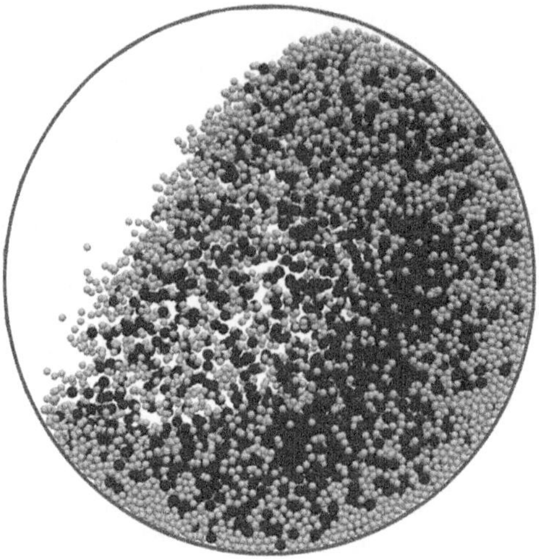

Figure 5.16 Inverted size segregation after five revolutions in the cataracting regime ($Fr = 0.71$) in a drum with the diameter of 0.2 m. Particles are of the same density with diameters 3 mm (gray) and 4.5 mm (black)

accumulate in the core based on a mechanism similar to that described previously. In the active layer denser particles sink into the lower parts due to the buoyant effect and this leads to the accumulation of denser particles in the core of the bed.

DEM simulations showed that the fill level interacts with rotation speed in a complex way that various pattern of segregation can be observed in the drum [99]. At a constant rotation speed, the rate of segregation decreases and the extent of segregation increases with increasing the fill level. The segregation is very pronounced at fill levels greater than 50%, in which we previously observed poor mixing rates due to formation of the non-mixing zone. At a constant fill level (for instance, 50%), by increasing the rotation speed from low to intermediate values the final extent of segregation (hence, the size of the core) decreases and the segregation vanishes. This range of rotation speed covers $10^{-4} \leq Fr \leq 0.56$ and corresponds to full rolling and cascading regimes and partially in cataracting regime. By a further increase in the rotation speed, inverted segregation appears.

As shown in Figure 5.16, in the rolling and cascading regimes, segregation mostly occurs in the active layer in which percolation and shearing/diffusion compete with each other. As the rotation speed increases, the mean residence time of particles on the surface decreases and particles have less time to perform rearrangements. Moreover, DEM results show that the packing density of particles in the active surface decreases, providing more space between particles, which makes percolation less selective to particle size and enhances the diffusion. In the incipient of the cataracting regime, where a very small fraction of particles is thrown off the empty space of the bed, the bed still can be considered to consist of two distinct regions: a lower layer that moves as a rigid body and an upper layer in which mixing of particles with size difference occurs. In this condition, the segregation due to percolation is not as strong as mixing, thereby, segregation vanishes. At higher rotational speeds (cataracting and centrifuging regimes), most of the particles are thrown off the top of the bed and fall in the other side

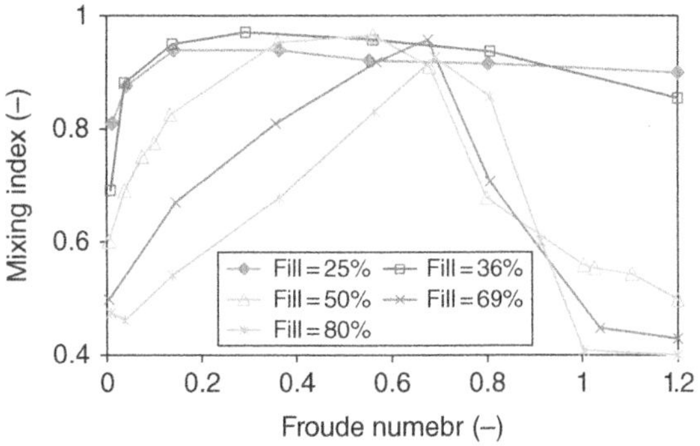

Figure 5.17 Variation of steady state mixing index in a 0.07 m diameter rotating drum filled with small (1 mm) and large (1.5 mm) particles with the volume ratio of 1:1 as a function of Froude number for various fill ratios. (data are taken from [99])

of the bed. In such a flow condition, percolation of small particles gradually occurs in the region where they fall and the direction of percolation is toward the wall. The dominant motion in this region is toward the wall. Therefore, an inverted radial segregation occurs in which small particles are segregated in the periphery of the bed (see Figure 5.16). The same mixing/segregation pattern was observed in systems with varying density and masses [100]. Variation of the final mixing index as a function of Froude number and filling degree in a 7-cm diameter drum is shown in Figure 5.17. At a constant fill level, by increasing the rotation speed, the mixing extent increases to a maximum then decreases. With increasing the filling degree, this maximum occurs at higher Froude numbers.

Various efforts have been made to eliminate or control segregation in drums, such as using cohesive agent, alteration of particle properties, or change in the shape of the drum cross section to introduce chaotic advection/mixing [106]. End plates also affect the segregation extent in drums with short length where wall effects are important [98]. The extent of segregation can be reduced by keeping the end plates stationary (not rotating with the drum) or making their surfaces rough or bumpy. An effective method for controlling the segregation is time modulation of the flow. This can be achieved by changing the direction of flow at a high frequency (faster than the time constant of segregation) in regions where segregation occurs. For example, a partial segregation occurs in the active layer of a rotating drum (rolling and cascading regimes) and this process is repeated many times until the final state of segregation is reached. Consequently, segregation can be effectively controlled if the flow in the active layer is perturbed with a high frequency and at the same time the direction of segregation is changed [107]. Both experimental and DEM simulations showed that using traditional short wall-connected baffles are not effective means of controlling segregation since they cannot perturb the flow in the active layer. However, very long wall-connected baffles or axially located baffles (baffles located in the center of the drum) can effectively control the segregation induced by size or density difference. The characteristic segregation time is related to the required time that a particle reenters the active layer. The circulation time is roughly half the revolution time for a half-filled drum, as shown in simulations. It was found that adding central baffles to the drum broadens the distribution of circulation time, indicating that the flow perturbation is done opposite to the segregation [96].

5.3.4 Tumbling Blenders

A tumbling blender is a closed vessel that rotates about an axis. In standard designs, there is an axis of rotation that produces a symmetric rotation with respect to a plane of symmetry. However, variant designs can be found in which two axes of rotations exist or the axis of rotation is tilted to produce asymmetric rotation. The most common shapes used in industry are double cone blender, V-blender, and bin/tote blender. Some modifications have been performed to enhance mixing and control segregation in these blenders, including changes in the geometry of vessel and installing internal baffles or rotating intensifiers. The main mechanism of the mixing in these drums is diffusion in the axial direction (perpendicular to the plane of rotation) and a combination of the convection and diffusion in the lateral direction. In some operating conditions, evidences of weak shear mixing have also been observed.

Due to more complex geometry of the vessel, in comparison to rotating drums, and various rotation patterns applicable to these drums, less theoretical works have been carried out on tumbling blenders and the granular flow in these blenders is not well understood. The DEM modeling provides a reliable tool to this end. Such a model can predict different mixing and segregation patterns as well as mixing dynamics observed in experiments [108–111] and power requirement for blending [112]. We simulated a V-blender to study mechanisms of mixing of particles with identical size and density. Two loading profiles were used to investigate different mechanisms of mixing in axial and lateral directions: left-right profile for the axial direction and top-bottom loading for the radial direction. Around 11,000 spherical particles with the diameter of 9 mm were used in the simulation. Physical properties and simulation parameters are listed in Table 5.1. This blender rotates at 30 rpm. The code files to simulate this blender and sample animations of flow pattern in it can be found in the online content of the book under "V-blender" folder.

Figure 5.18 illustrates snapshots of the simulations at the beginning of mixing as well as after one, two, and six revolutions. In the first column, the V-blender with the left-right loading profile is shown. With each rotation of the blender, a few white particles migrate to the left arm and a few black particles to the right arm in a way that after six revolutions a very low degree of mixing is obtained. This is due to slow dispersion of particles across the symmetry plane which separates two arms of the blender. The dispersion occurs by random motion of particles on the free surface of the bed during tumbling and redistribution of particles between two arms of the blender rejoining of particles in one revolution. When the blender rotates from its original state to 180°, particles in the blender are redistributed between the two arms and those particles located in the proximity of the symmetry plane should choose between one of the two arms in a random manner. The same random migration between two arms occurs when particles rejoin at the end of one revolution.

The second and third columns in Figure 5.18 show side and front views of the V-blender during the mixing process with the top-bottom loading profile. The two layers of particles are mixed in each revolution and almost the final degree of mixing is attained after six revolutions. The convective motion of particles within each arm of the blender causes immense convective mixing between two layers, which speeds up the mixing process. From the side view, motion of particles is very similar to that in a rotating drum with a high velocity active layer on the top and a slow passive layer beneath it. This will be discussed later in Figure 5.19. Therefore, it can be expected that mixing mechanisms in a V-blender are similar to those in a rotating drum. However, there is a significant difference between these two devices. In the V-blender, the

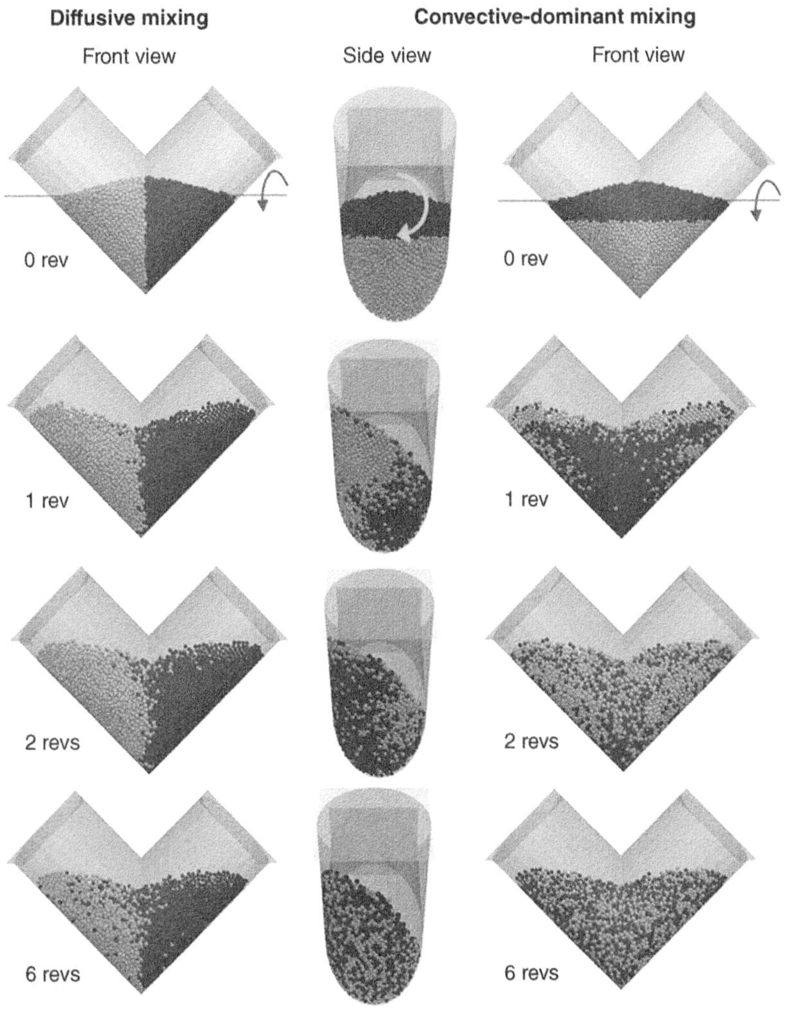

Figure 5.18 Progress of diffusive and convective dominant mixing of identical particles in a V-blender

shape of the cross section of the bed and the motion streamlines constantly change during one revolution, which promotes the chaotic mixing. However, the shape of cross section of the bed in the rotating drum is constant. Slow axial mixing compared to lateral mixing in V-blenders was also observed and confirmed by DEM simulations of Lemieux *et al.* [112].

Different parameters affect the mixing rate and the final degree of mixing in V-blenders. Rotation speed and filling degree are the most important ones. Both mixing rate and final degree of mixing are enhanced by increasing the rotation speed as long as the centrifugal force effect is not dominant [108]. High filling ratio in the blender results in poor mixing. It was also shown that the size distribution of particles has minor effects on the mean velocity distribution and granular temperature in these blenders although the mixing dynamics is altered due to

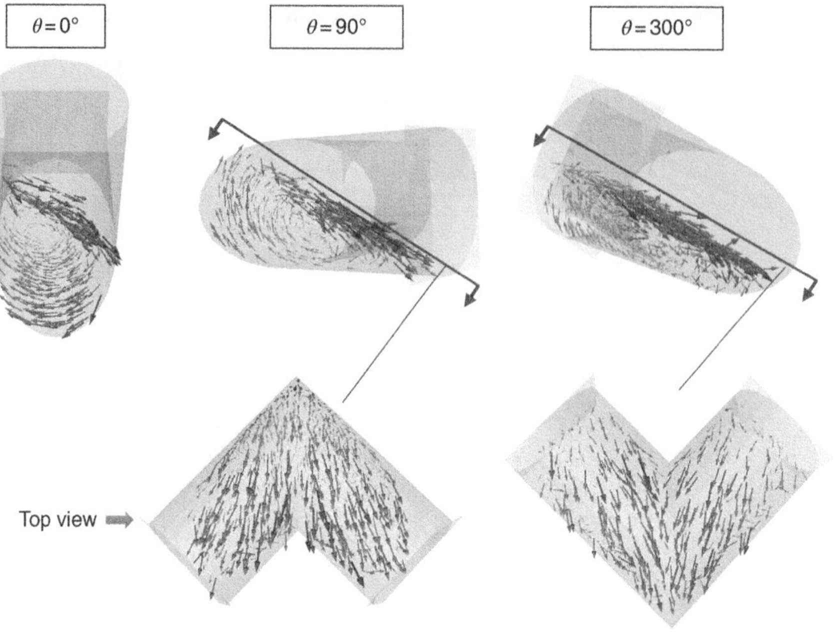

Figure 5.19 Instantaneous velocity vector of particles during one revolution of the V-blender. Vectors are scaled and shaded based on their velocity magnitude

promotion of segregation mechanism [112]. Axial and radial mixing rates in a V-blender and a bin blender were compared and it was concluded that both these blenders have almost similar mixing times in the lateral direction while the V-blender exhibits stronger axial dispersion. Different strategies have been applied to enhance the mixing rate and controlling segregation in these blenders. As an example, Brone *et al.* [113] added a rocking motion to a conventional rotating V-blender to enhance mixing in their experiments. Alizadeh *et al.* [114, 115] used an alternate geometry for the vessel called tetrapodal blender (a pair of V-blenders attached at their bottoms with one twisted by 90°) and compared its performance with a V-blender using DEM simulations and experiential measurements. They showed that in each revolution of the tetrapodal blender, particle distribution between arms and recombination occurs twice, which enhances the mixing, both axially and laterally, and controls patterned segregation in these blenders. Their simulation results showed that the mean radial and axial particle velocities in the tetrapodal blender are higher than those in the V-blender.

The instantaneous particle velocity vectors inside the V-blender for $\theta = 0°$, 90°, and 300° are shown in Figure 5.19. The vectors are scaled and colored according to their velocity magnitude. The side view of the blender shows changes in the shape of the bed and streamlines of particles motion, which steadily alters during each revolution of the blender. Particles at the bed surface avalanche with a high velocity due to gravity and are transported back to the top of the bed through the layer below. Top views of particle velocity at $\theta = 90°$ and $\theta = 300°$ are also shown in this figure. It is clear that particles travel a curved path when being distributed between two arms and rejoining. As stated by Alexander *et al.* [116], movement of small and large particles along this curved path line is the main source of trajectory segregation.

Based on this mechanism, larger particles tend to travel rectilinear paths and smaller particles tend to travel more curved paths. The difference between these tendencies causes a gradual segregation in these vessels, especially at low rotational speeds. However, at high rotational speeds, this mechanism is not dominant since the direction of flow changes very fast and trajectory segregation does not have enough time to develop. Instead, the percolation of small particles through larger ones becomes significant and changes the segregation pattern. Alexander *et al.* [116] identified three different stable segregation patterns: small-out, stripes, and left-right. It should be emphasized that this classification has been suggested based on the operating conditions used in their experiments and cannot be generalized.

The granular flow as well as mixing and segregation dynamics of solids in bin and double cone blenders have been investigated by DEM [110, 117, 118]. Similar to the V-blender, these blenders suffer from segregation and poor axial dispersion. Moakher *et al.* [111] simulated the flow in a double cone blender and observed small-out segregation pattern that was observed in similar experiments. This segregation is governed by trajectory segregation in which small particles tend to travel curved paths near the wall and large particles tend to travel rectilinear path in the middle of the blender. Sudah *et al.* [118] showed that by tilting the axis of rotation of a bin blender, an asymmetric particle flow is created which produces periodic particle flow in each revolution (left to right and right to left) of the blender. Manickam *et al.* [110] simulated a double cone blender with one axis of rotation and a double cone blender with two perpendicular axes of rotation. They studied a range of operating conditions, such as filling profile, rotation speed, particle size, and filling ratio and concluded that the mixing rate is increased by 60–90% in the blender with two axes of rotations when compared with the blender with one axis of rotation at the same operating conditions.

In new designs of blenders, a combination of mixing techniques may be used. For example, in the blender studied by Alian *et al.* [119], a combination of tumbling blending in a slant cone blender and agitation caused by an internal agitator was used. The DEM was validated by experimental data in terms of kinetics of mixing in this blender. It was shown that increasing the filling ratio reduces the rate and the final degree of mixing. The efficiency of agitator is high when the blender works with full capacity and the increase in the rotation speed of the agitator enhances mixing. The co-rotating agitator (in which the rotation direction of the agitator is the same as the vessel) performs better than the counter-rotating agitator at the same rotation speed. The mixing/segregation dynamics of polydispersed particles was also investigated in this blender [120]. The simulation results showed that the mixing characteristics of polydispersed particles are very complex. For a bidispersed mixture, the best mixing degree was attained at high rotation speed of the drum (45 rpm) and mixing was improved by the use of agitation. However, for polydispersed mixtures the best mixing quality was obtained at a low rotation speed (15 rpm). It was then concluded that polydispersed mixtures have a lower tendency to segregation than bidispersed mixtures.

5.3.5 Shaft Batch Mixers

In the rotary and tumbling blenders considered in the previous sections, the vessel rotates and the solid material tumbles inside due to wall effects. However, in shaft mixers, the vessel is stationary and ribbons or blades, fitted around an axial shaft, rotate inside the vessel and produce convective and shear motion of particles. Various designs of shaft mixers can be found, which are different in orientation of the shaft (horizontal or vertical) and shape of the blade or ribbon. Examples of shaft mixers are horizontal paddle mixer, plough blade mixer, horizontal ribbon mixer, vertical ribbon mixer, vertical shear mixer, Nauta mixer, and draught tube mixer. Shaft mixers can operate in both batch and continuous modes. We focus on the batch mode in this section and on continuous mode in the next section.

The DEM has been used for studying the mixing of solids in plough blade mixers [121–123]. Laurent and Cleary [121] simulated a single plow mixer with spherical particles and compared the simulation results with experimental observations in the same mixer with non-spherical rice grains. They found qualitative agreements between simulation and experiment in terms of flow pattern and mixing behavior. They found that the angle of repose of the plough trench is much lower in the simulation than in the experiment and attributed this to using spherical particles in the simulation rather than their real shape. Later, Cleary [122] simulated the same mixer with non-spherical particles (superquadric) and obtained closer results to the experiments. It was shown that the mixing rate of non-spherical particles is inferior to that of spherical particles due to lower mobility of non-spherical particles. Alian *et al.* [123] studied the mixing kinetics in a blender with six plows. They showed that the mixing rate improves when the rotation speed of the shaft increases, using larger particles and decreasing the filling fraction of the vessel.

Velocity distribution of particles has been investigated in paddle mixers experimentally as well as with DEM simulations [124, 125]. It was shown that the model can qualitatively predict the flow pattern in the mixer, while a good quantitative agreement was observed for velocity distribution of particles. When using larger particles, they move faster and their velocity distribution becomes wider. Mixing in other types of mixers, such as vertical cylindrical vessel with flat blades [126] and vertical vessel with ribbon blade [127] were investigated. Analysis of results in the vertical ribbon mixer showed that the filling height of particles is the most influential factor on flow pattern and mixing rate in this type of mixer. When the filling height is shorter than the blade height, particles are transported to the top of the blade by the wedge action of ribbon, then fall into the core section of the mixer, giving it a regular particle circulation and rapid mixing. On the other hand, when the filling height is longer than the blade height, a semi-stagnant zone of particles is formed in the core section of the mixer, giving it an irregular particle circulation and a slow mixing rate [127].

A diverse range of flow patterns and mixing mechanisms have been identified in the shaft mixers. This diversity arises from the variation in the blade design and rotor configuration. Therefore, we cannot propose a general flow pattern and mixing mechanism for this type of mixers. The DEM simulation enables us to perform a detailed study on the flow pattern and mixing kinetics of particles in shaft mixers. As an example, we simulated the granular flow in a Nauta mixer.

We performed a series of DEM simulations to investigate flow pattern and mixing kinetics of particles in a Nauta-type blender. These simulations were performed using open source code LIGGGHTS [128] and the script and stl files to setup them can be found in the online content of the book under "NautaMixer" folder. This blender consists of a stationary vessel and a rotating screw. The vessel is a frustum cone with radii 0.56 and 0.11 m and the height of 0.62 m, giving it a capacity of about 60 l. The screw diameter is 0.108 m and its pitch is 0.07 m. The screw rotates around its axis (the primary rotation) with the rotational speed of Ω_p and around the axis of the vessel (the sweeping rotation) with the rotational speed of Ω_s. In the sweeping rotation, the screw sweeps the whole side wall of the vessel. Around 90,000 spherical

Table 5.2 Properties of the particle and the walls used in simulation of the Nauta mixer

Parameter	Particle	Wall
Properties		
Young's modulus (MPa)	40	40
Poisson's ratio (—)	0.2	0.2
Dynamic friction (—)	0.25	0.2
Rolling friction (—)	0.07	0.05
Restitution coefficient (—)	0.5	0.5
Diameter (m)	0.006	—
Density (kg/m^3)	900	—
Simulation		
Time step (s)	1×10^{-6}	—
Number of particles	90,000	—
Simulation time (s)	60	—

particles with a diameter of 6 mm were used to fill the blender up to 0.34 m. Time step for the integration was 1×10^{-6} seconds and all simulations were continued for 60 seconds of real time. Material properties of the particle and the walls are listed Table 5.2. Based on these simulations, effects of primary and sweeping rotation speeds on the mixing kinetics are described.

To investigate the mixing dynamics of mono-sized particles, we post-processed the simulation results. Particles were marked based on their initial position before the screw starts moving. The initial loading profile of particles with light and dark shading is shown in Figure 5.20 which demonstrates that dark and bright particles are filled vertically side by side. When the screw starts its rotation, it produces two distinct convective motions: bottom-top-bottom motion and tangential motion along the sweeping direction. In the bottom-top-bottom motion, particles are drawn toward the screw near the bottom and are lifted by the primary rotation of the screw to the bed surface, where particles exit the screw region and avalanche on the inclined surface of the bed. Our deeper observations of the simulation results showed that the particles start exiting the screw region a few flights (one to three) before the last flight on the bed surface. To compensate the particles deficiency at the bottom, particles far from the screw descend in the mixer due to their gravity. As a result, the primary rotation of screw produces a convective layered flow of particles in which particles are lifted to the bed surface and then descend at the opposite side of the screw. A semi-stagnant zone is formed in the middle of the bed near the cone axis in which particles move slowly. In the tangential motion, the sweeping rotation of screw pushes particles forward and induces a tangential flow behind the screw to compensate the particles deficiency in that region. The bottom-top-bottom and the tangential flow are combined and produce a coherent 3D flow in the Nauta mixer. Since the semi-stagnant zone is off-axis, the screw reaches this region and destroys it during its sweeping rotation. Therefore, no permanent stagnant zone can be formed in this mixer. Location of this region, as well as particles inside it, changes during the operation of mixer. The animation of this simulation, is available in the online content of the book ("NautaMixer" folder), better shows the flow pattern and mixing progress of particles.

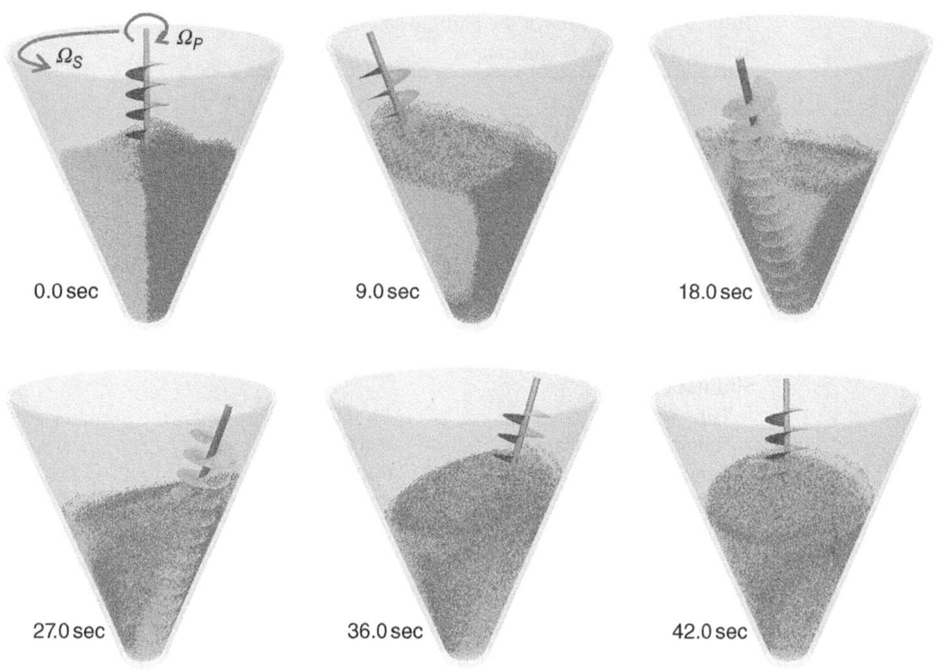

Figure 5.20 Snapshots of mixing evolution in the Nauta mixer

The mixing process shown in Figure 5.20 is in agreement with the flow pattern described above. Dark and bright particles are drawn into the screw at the bottom and become mixed when being lifted to the surface. This pattern produces a good particle mixing in a hypothetical vertical plane crossing the screw axis. This vertical plane rotates with the sweeping rotation of the screw and almost the whole mixer undergoes such a mixing. After completion of one sweeping rotation, the mixing is almost complete.

It should be noted that the mixing occurs due to both convection and dispersion in this mixer. The granular temperature is a measure of local particle fluctuations with respect to the local average flow. A high granular temperature indicates intense random motion/agitation of particles while a low value shows non-random advective flow of particles. Contour of the granular temperature of particles is shown in Figure 5.21. This contour belongs to a vertical plane crossing the screw and cone axes. The impeller region and the bed surface have the highest granular temperature indicating that the dispersive mixing is dominant in these regions and the convective mixing is dominant in the remaining parts of the mixer.

To support the mixing mechanism proposed above, we further processed the results and marked seven spherical groups of particles (each containing 100–110 particles) in various locations in the mixer and tracked their shapes and positions during the operation of the mixer. Particles in these groups were marked when the mixer was in the steady state condition. Figure 5.22 illustrates the initial position of groups as well as their locations during the operation of the mixer. Four of the marked groups are in the impeller region (group 1), one in the semi-stagnant zone (group 2), and two in the convective part (group 3) of the mixer. It can be seen in this figure that the tracers in the impeller region are lifted to the bed surface and

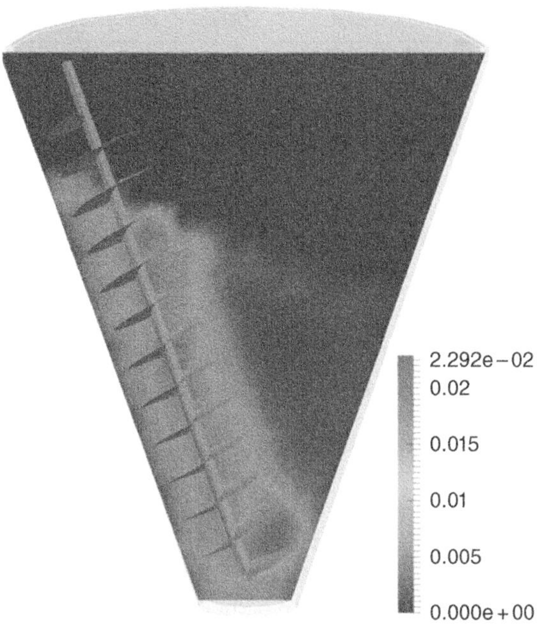

Figure 5.21 Granular temperature distribution on a vertical plane crossing the screw axis in the Nauta mixer. The shaded bar shows the granular temperature in m^2/s^2

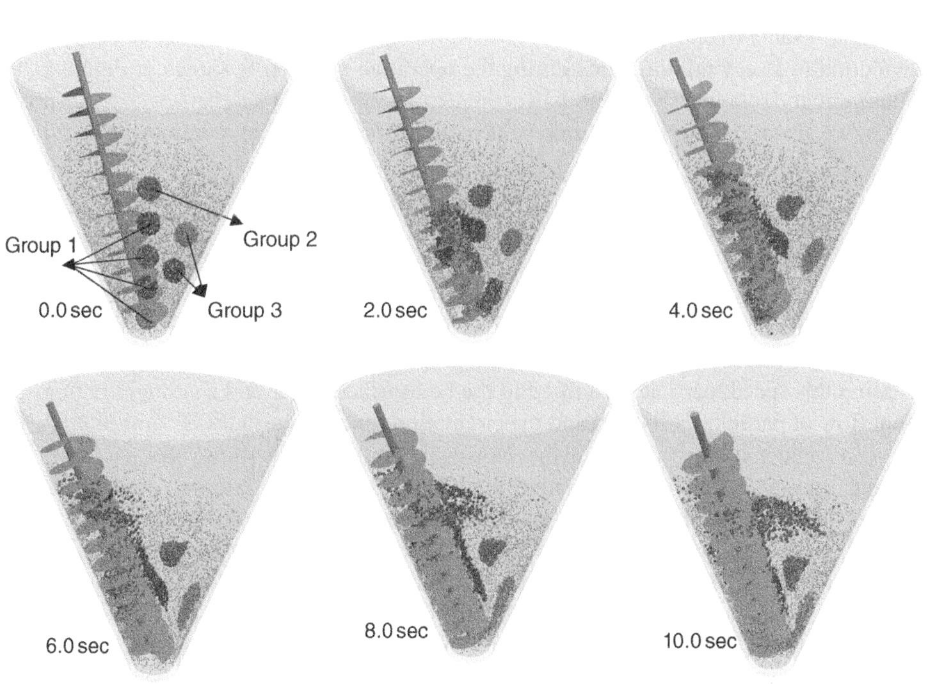

Figure 5.22 Position of marked groups during the operation of the Nauta mixer

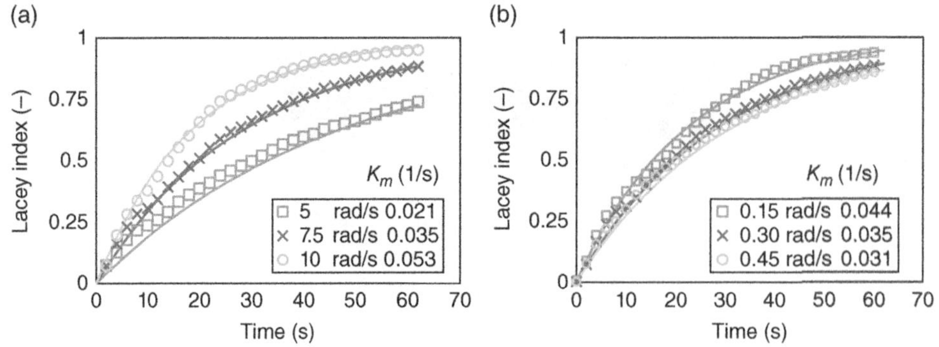

Figure 5.23 Evolutions of Lacey mixing index during the operation of the Nauta mixer at (a) various primary rotation speeds and (b) various sweeping rotation speeds. Symbols are obtained from DEM simulations and lines are the fitted exponential decay model (Equation 5.28)

dispersed during this transport, then exit the screw zone at the bed surface. These tracers are completely deformed and transformed into layers of particles at the surface. The tracers in the convective part are partially deformed and descend toward the screw at the bottom of the mixer due to the bottom-top-bottom motion of particles. Shape of the tracer group in the semi-stagnant region remains almost intact and it slowly descends in this region. When the screw hits this group during the course of sweeping rotation, it becomes deformed and sucked it in the screw flights (this is not shown in Figure 5.22). We recommend the reader to watch the animation created from the motions of these tracers. It is available in the online content of the book under "NautaMixer" folder.

Evolution of Lacey mixing index during the operation of mixer is shown in Figure 5.23 at various primary rotation speeds and sweeping rotation speeds. To characterize the mixing kinetics, we also fitted the simulation results to an exponential decay relationship as follows:

$$MI_{Lacey} = \frac{\sigma_0^2 - \sigma^2}{\sigma_0^2 - \sigma_R^2} = 1 - e^{-K_m t} \tag{5.28}$$

where K_m is the variance decay mixing rate constant. The mixing rate constants are also reported on this figure. It can be seen in Figure 5.23a that the mixing rate constant linearly increases with increasing the primary rotation speed of the screw from 5 to 10 rad/s. By increasing this speed, particles are lifted to the bed surface faster and a stronger bottom-top-bottom flow of particles is formed and the circulation times of particles become shorter. This leads to a rapider mixing in the mixer. However, Figure 5.23b shows that increasing the sweeping rotation speed has a negative impact on the mixing rate and the mixing rate decreases as the sweeping rotation speed is increased. When the sweeping rotation is 0.15 rad/s, most of the particles are drawn toward the screw at the bottom and leave the screw flights at the surface. Therefore, these particles undergo a complete circulation in the mixer. While at the sweeping rotation speeds of 0.3 and 0.45 rad/s, a large number of particles leave the screw flights at the rear before they reach the bed surface due to higher sweeping speed of the screw. Therefore, incomplete circulation of particles is promoted in the bed, leading to a lower mixing rate.

5.3.6 Continuous Mixers

Continuous mixers can handle a large amount of powders when compared with the analogous batch mixers. Although they offer advantages over batch mixers, only a few experimental and numerical studies can be found in literature about continuous mixers. Perhaps the most important obstacle in using continuous mixers is the difficulty associated with their operation. Controlling the feed flow rate of each component and handling and storage of feeds and product are the most important ones. Moreover, risks and costs of failures of these mixers are higher than those of batch mixers. A limited number of literature can be found on DEM simulation of continuous tumbling blenders [129, 130] and continuous convective mixers [131–138]. In simulation of a continuous convective mixer, usually a slice of the mixer with the length of one blade set is considered to reduce the computational cost of simulation. To this end, the periodic boundary conditions are applied to the faces perpendicular to the rotation shaft of the mixer.

The rate of material transport along the mixer length and the performance of local mixing inside the mixer (called cross-sectional mixing) per one rotation of the shaft or drum (or equivalently, per unit length of the mixer) are of great importance in continuous mixers. The rate of the material transport along the mixer can be characterized by the analysis of RTD, which reflects effects of mean axial velocity and axial dispersion [139]. The mixing performance inside the mixer is evaluated by studying the mixing in a periodic section of the mixer by DEM [133, 137]. A qualitative analogy can be established between the performance of the periodic section of a continuous mixer and the similar batch mixer. For example, in a convective batch mixer, a low fill level or a high blade rotation leads to high rate of mixing and the same can be observed in a continuous mixer. However, the performance of the batch mixer cannot be used for quantitative evaluation of the cross-sectional mixing in a continuous mixer due to pronounced end walls effects on the flow pattern in batch mixers [133]. The overall performance of mixer, in terms of homogeneity of the outflow, is affected by the rate of material transport and quality of cross-sectional mixing. Very long residence times of materials and good cross-sectional mixing are essential to achieve good mixing in the mixer.

In most situations, change in the operating condition has positive effects on the residence time and negative effects on the cross-sectional mixing or vice versa. As an example, Gao *et al.* [135] optimized the mixer performance at high rotational velocities and low fill levels by DEM simulation. At high rotation speeds and low fill levels, cross-sectional mixing is good while the residence time of material in the mixer is short (due to high axial velocities). They used two methods to keep the axial velocity constant (constant residence time) while increasing the rotation speed (increasing cross-sectional mixing) to enhance the mixer performance. In the first method, decreasing the blade angle (with respect to the axial shaft) and increasing the shaft speed were assessed. In the second method, increasing the weir height at the end of the periodic section (baffle) and increasing the shaft speed were considered. It was shown that these two methods can improve the mixing rate by five and nine times, respectively.

The size difference in a mixture is another important factor which influences the RTD of particles in the mixer. To investigate this effect, we simulated a full length bladed continuous mixer of Figure 5.24. In this mixer, two separate streams of small (3 mm) and large (5 mm) particles are poured into the mixer through the inlet gate. The inner diameter and the length of the mixer are 0.22 and 0.4 m, respectively, and dimensions of blades are $3 \times 9 \times 1\ cm^3$. The clearance between the blade tip and the wall is 2 cm. Simulations were performed for

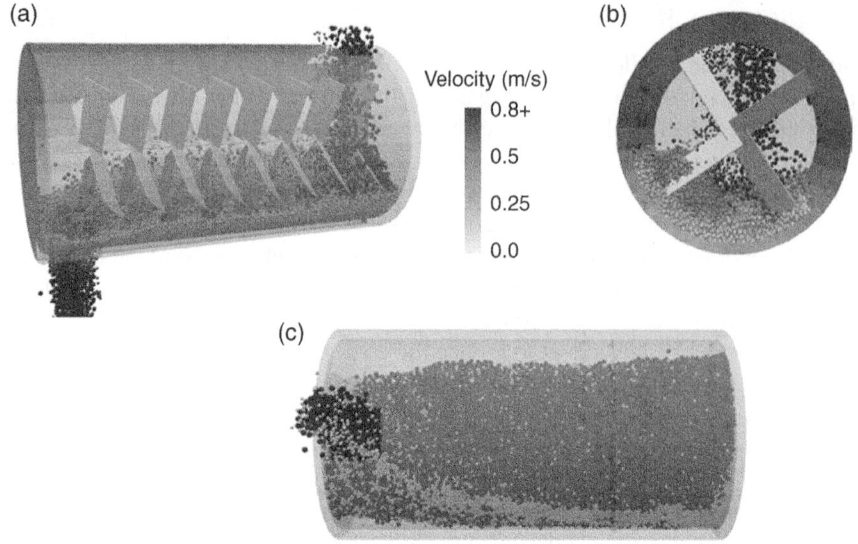

Figure 5.24 DEM simulation of a continuous mixer. Two separate streams of small (3 mm) and large (5 mm) particles are fed into the mixer ($\Omega = 4\pi$ rad/s). Particles are shaded based on their velocity magnitudes in (a) and (b) and particles are shaded based on their sizes in (c): light gray for small particles and black for large particles

more than 40 s. These simulations were performed using open source code LIGGGHTS [128]. Results of this simulation are shown in Figure 5.24. It can be seen that the fill level of the blender is almost constant along the blender length (Figure 5.24a). The cross-sectional view of the bed shows that the particles near the bed surface, where they are in contact with the blade, have a higher velocity than particles in other parts (Figure 5.24b). We observed that small and large particles accumulate in two regions of the mixer (Figure 5.24c). Small particles are accumulated at the end of the blender near the exit and large particles near the inlet where the inlet flow impacts the bed. The bottom view of the mixer also shows that a strip of small particle is formed along the mixer length. This is mainly due to the clearance between blade and drum. In the absence of pronounced convective mixing caused by blade rotation in the lower most layers of particle (near the wall), segregation proceeds in those layers.

In a continuous system, the RTD can be measured by a pulse or step injection of the tracer. Here, we describe the procedure of calculating the RTD of particles in a continuous mixer based on the pulse injection of the tracer. For this purpose, a certain amount of tracer is instantaneously injected into the mixer and its concentration is monitored at the outlet. The tracer should be injected in an infinitely short time and it should not alter the flow condition inside the mixer. Since we have all particles' data at each instance of simulation, we can post process results to mimic a perfect pulse injection. When the simulation of the mixer is reached the steady condition, a certain number of particles are marked in the input stream and tracked until they leave the mixer. Marking of particles is simply done by recording their identification number in the input stream. It is clear that this type of tracer injection does not alter the flow

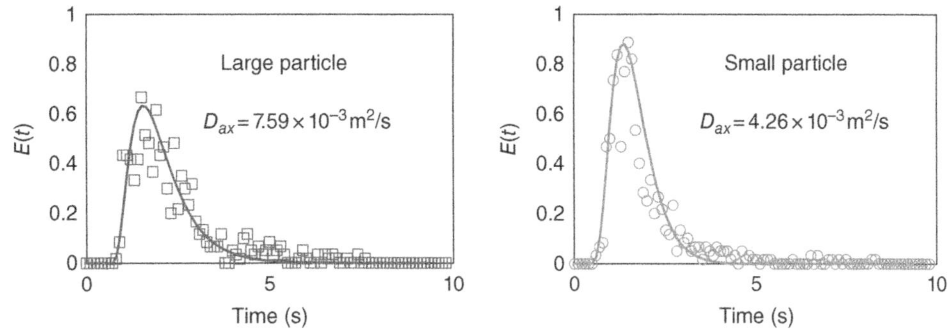

Figure 5.25 The residence time distribution of large and small particles in a continuous mixer. Markers show the pulse injection obtained from post processing DEM results and curves show the fitted dispersion model on the simulation results

condition of the mixer and is performed in a very short time. Concentration of marked particles in the output stream, $C(t)$, is then recorded and the RTD can be evaluated from:

$$E(t) = \frac{C(t)}{\int_0^\infty C(t)\,dt} \tag{5.29}$$

The mean residence time of particles in the mixer can be calculated from the first moment of RTD:

$$\tau = \int_0^\infty tE(t)\,dt \tag{5.30}$$

Typical RTD curves obtained from the simulation of continues mixer is shown in Figure 5.25. This figure demonstrates that after the tracer injection, it takes a while for the tracer to appear in the outlet. A narrow curve indicates a flow with low dispersion and a wide curve a flow with high dispersion. Figure 5.25 shows that the RTD of small particles is narrower than that of large particles, which indicates that the dispersion of small particle is less than that of large particle in the mixer. Moreover, larger particles stay longer in the mixer.

To better understand the flow pattern in a continuous mixer, various theoretical models have been proposed, including tanks-in-series and dispersion models. In the tanks-in-series model, it is assumed that the whole mixer is composed of N_{tnk} identical sequentially connected stirred tanks. The RTD at the mixer outlet based on this model is [140]:

$$E(\theta) = \frac{N_{tnk}\left(N_{tnk}\theta\right)^{N_{tnk}-1}}{\left(N_{tnk}-1\right)!} e^{(-N_{tnk}\theta)} \tag{5.31}$$

where $\theta = t/\tau$ is the dimensionless time and τ is the mean residence time of the mixer. This is a one-parameter model. The number of tanks can be obtained by fitting this model to the RTD data. When N_{tnk} is large, the flow becomes similar to a plug flow model (low dispersion) while a small N_{tnk} indicates that the flow is close to the stirred tank (high dispersion).

Another model is the Taylor dispersion model that considers a plug flow in the mixer and superimposes dispersion on it. The mass balance equation in the mixer based on this model can be written as follows [140]:

$$\frac{\partial C}{\partial t} + \bar{u}_z \frac{\partial C}{\partial z} = D_{ax} \frac{\partial^2 C}{\partial z^2} \tag{5.32}$$

where $\bar{u}_z$ is the mean axial velocity and D_{ax} is the axial dispersion coefficient which characterizes the degree of random motion and back mixing in the axial direction. In fact, all deviations from the plug flow are lumped in the axial dispersion coefficient. When D_{ax} is small, the flow in the mixer behaves like a plug flow while a large D_{ax} indicates that the flow in the mixer is similar to the stirred tank. The solution to Equation 5.32 with open boundary conditions is [140]:

$$E\left(\bar{z},\bar{\theta}\right) = \frac{1}{\left(4\pi\bar{\theta}/Pe\right)^{1/2}} e^{-Pe\left(\bar{z}-\bar{\theta}\right)^2/\left(4\bar{\theta}\right)} \tag{5.33}$$

where $\bar{\theta} = (t - t_o)/\tau$ is the dimensionless time, $\bar{z} = z/L_z$ is the dimensionless location, t_o is the time that takes the tracer to first appear at the exit stream, L_z is the longitudinal length between injection and detection points, and $Pe = \bar{u}_z L_z / D_{ax}$ is the Peclet number which is the ratio of the rate of transport by advection to the rate of transport by dispersion.

When $\bar{z} = 1$, Equation 5.33 represents the RTD of the mixer. This equation was fitted to the RTD data of the mixer and the calculated dispersion coefficients are shown in Figure 5.25. The curves in this figure are the results of this fitting. It can be seen in this figure that this model properly matches the RTD data, which indicates that the dispersion model can adequately describe the flow in this mixer. We also performed three sets of simulations to study the effect of inlet flow rate and blade design (number of blade sets and angle between the blade and the axial shaft) on the RTD of solids in the mixer. The dispersion model was fitted to the RTD of the mixer and the corresponding model parameters are listed in Table 5.3. We also calculated the mean and the variance of axial velocity of particles and listed them in this table.

It can be seen in Table 5.3 that the mean axial velocities of the small and large particle increase with increasing the number of blade sets in the mixer. With a greater number of blade sets, more kinetic energy is transferred to particles in each rotation of the shaft, which increases the mean axial velocity. The same trend is observed when the inlet mass flow rate of particles is increased. When the inlet flow rate increases, so does the mean axial velocity. At the same time, the fill level of the mixer increases from 11 to 16%. Trend of changes of axial dispersion coefficients calculated from Peclet numbers show that the axial dispersion is sensitive to both number of blade sets and inlet flow rate. Axial dispersion of small particles is less than that of large particles in all cases. By comparing the axial dispersions and variances of axial velocity it can be seen that there is a direct relationship between these two quantities. The correlation coefficient between axial dispersion and its variance was estimated to be 0.6, which suggests a moderate linear relationship between these quantities. Dispersion coefficients of small and large particles increase with an increase in the number of blade sets and the inlet flow rate.

Table 5.3 Parameters of the dispersion model and other variables extracted from DEM simulations

	$D_{ax} \times 10^4$ (m²/s)		$\bar{u}_z$ (m/s)		Variance of $u_z \times 10^4$(m²/s²)		Model parameters					
							Pe		τ(s)		t_o(s)	
	Small	Large	Small	Large	Small	Large	Small	Large	Small	Large	Small	Large
Blades = 6[a]	31.2	41.3	0.166	0.150	1.08	5.78	14.9	10.1	1.20	1.58	0.27	0.41
Blades = 7	39.1	64.5	0.202	0.175	1.66	5.33	14.5	7.6	1.23	1.11	0.14	0.41
Blades = 8	42.6	75.9	0.219	0.190	5.43	8.58	14.4	7.0	1.35	1.20	0.11	0.50
Flow = 0.31[b]	27.5	68.3	0.192	0.176	5.24	8.64	19.5	7.2	1.25	0.84	0.12	0.55
Flow = 0.38	42.6	75.9	0.219	0.190	5.43	8.58	14.4	7.0	1.35	1.2	0.11	0.50
Flow = 0.44	66.2	121.2	0.234	0.199	5.71	8.58	9.9	4.6	1.36	0.84	0.20	0.55
Angle = 30°[c]	19.4	64.1	0.153	0.183	2.92	1.23	22	8	1.31	0.81	0.12	0.8
Angle = 45°	42.6	75.9	0.219	0.190	5.43	8.58	14.4	7.0	1.35	1.20	0.11	0.50
Angle = 60°	10.9	34.8	0.175	0.143	2.56	6.56	45	11.5	1.01	1.20	0.45	0.65

[a] Number of blade sets.

[b] Inlet flow rate in kg/s.

[c] Angle between the blades and the axial shaft.

In bladed mixers (convective mixers), most of the velocity fluctuation of particles occurs near the blades [141]. By increasing the number of blade sets, the number of blade-particle collisions increases and so does the velocity fluctuations. When the inlet flow rate increases, the blade becomes more immersed, which results in more pronounced particle-blade contacts and more intense axial dispersion.

The effect of blade angle (the angle between the blade and the axial shaft) on the mean axial velocity and dispersion coefficient is rather complex (see Table 5.3). The maximum mean axial velocity occurs at the blade angle of 45° while it decreases for both larger and smaller angles. The same trend is also reported by Gao *et al.* [135] using DEM simulations. When the blade angle increases from 0° to 90°, the mechanism of momentum transfer from blades to particles changes from pure normal impulse to pure tangential impulse restricted by the friction. At 0°, no tangential component exists in the direction of blade rotation and impulse is normal. By increasing the blade angle, a tangential component appears that leads to a decline in the normal component and a growth in the tangential component of the blade-particle impulse. However, according to the Coulomb's friction law, the tangential impulse is restricted by the blade-particle friction. This limits the tangential impulse, hence the amount of kinetic energy transferred to particles. Therefore, the total energy transfer to particles decreases monotonously by increasing the blade angle. On the other hand, the direction of energy transfer changes from the lateral to the axial direction. At 30° blade angle, a large amount of kinetic energy is transferred to particles in the radial direction (ineffective direction) and at 60° blade angle, a small amount of kinetic energy is transferred in the axial direction (effective direction). However, at a 45° blade angle, a sufficient amount of kinetic energy is transferred to particles in the effective direction.

5.4 Screw Conveying

Screw conveyors are commonly used for transporting different feedstock at a controlled and stable flow rate in industry. They can be used for feeding biomass, coal, or other solid fuels to the combustion, gasification, or pyrolysis reactor [142], transporting corn or wheat seeds in harvesting machines, discharging the excavated soil from front face of excavating machine, moving carrot, green beans, and so on in a steam blancher, dosing/mixing of fine materials (like pigments) with bulk powders [143]. They are also used for drying of wet solids since they have a high surface-to-volume ratio and provide a good medium for drying of materials. Steam, hot oil, or water is passed through the central screw blade to act as an immersed heat transfer surface through which the heat is transferred to materials. Due to continuous rotation of screw and high renewal rate of particles on the surface, a uniform heating is attained in this drying device.

Different designs of the screw conveyor exist, depending on the required mass flow rate, elevation discharge point, type of the feed stock and intended use of the conveyor. A standard screw conveyor consists of a stationary cylindrical casing (sometimes it is U-shape), a rotating screw, and a small hopper that acts as a feeder (see Figure 5.26a). The hopper discharge opens into the casing and the screw rotates in this casing by a motor driver mounted at one end. The length of the hopper opening into the casing is almost one to three times the screw pitch. Solid particles are drawn to the screw flight and pushed forward along the casing by rotation of screw due to

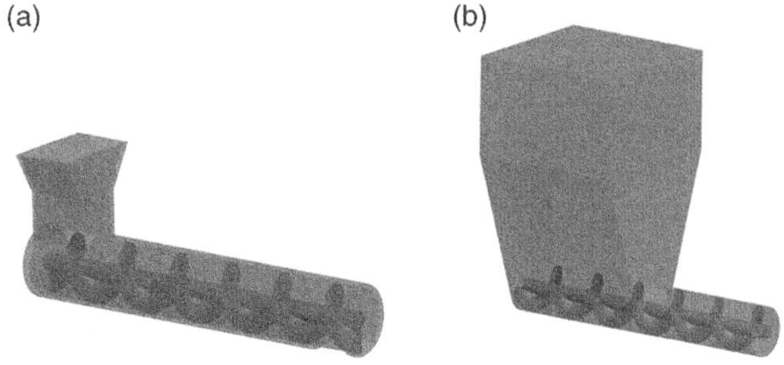

Figure 5.26 Schematic of (a) a horizontal screw conveyor and (b) a screw feeder with its wedge hopper

the wedge action of screw. Two major types of these conveyors can be found, enclosed and U-shape conveyors. Enclosed or auger conveyors operate at high rotation speeds (200–2000 rpm) and can transport lightweight, non-cohesive, free-flowing solids to long distances in horizontal, inclined, or vertical directions. The U-shaped conveyors operate at low rotation speed and can transport heavy and cohesive solids to short distances horizontally or slightly inclined.

The screw feeder is another variant design of the screw conveying system, which is used for drawing solid particles from storage hoppers or bins and transporting them at a controlled rate. As shown in Figure 5.26b, the screw is mounted at the bottom of a wedge hopper (or a bin) and the bulk of solids rests on the top of the screw and fills its flights. A part of the screw is inside a tubular casing through which solid particles are to be transported. When the screw rotates, solid particles inside the screw flights move forward along the screw and are transported out of the hopper. Solids on the top of the screw are drawn down to fill the empty space in the flights (mostly two or three first flights of the screw, depending on the screw design). Since a long height of solids is available on the top of the screw, the flights are always full of particles and the screw operates with full capacity. However, the whole free volume of a screw conveyor may not be filled by solids. A very important characteristic of the screw feeder and container configuration is to obtain uniform drawdown of particles along the screw in which the surface of particle bed descends uniformly and remains horizontal. Uneven drawdown of particles always involves a severe recirculation of particles in the bed in which particles that enter the screw region recirculate back into the container. These phenomena lead to a large variation in the residence time of particles in the container, a variable composition in the output of the feeder and high power requirements, all of which should be minimized.

Although mechanical operation of these devices is very simple, the flow behavior of solids inside them is very complex. Therefore, their design and optimization for different applications require a deep understanding of solid flow behavior inside them. The solid flow pattern, power requirement, and throughput are affected by screw design, geometry of casing, rotational speed, particle shape, presence of cohesion between particles, shape of particles, particle-particle and particle-wall frictions, and so on. In addition, some problems, like temporal or permanent blockage of material may occur and stable flow of material cannot be provided. Bridge/arch formation on the intake of screw (top of the screw where particles are drawn down) is another problem that can be initiated by large and long particles, cohesive (wet) particles, and poor design of screw-casing configuration. All these factors motivate the need for vast understanding of flow pattern of solids in these devices for their design and optimization.

Table 5.4 Experimental and theoretical studies on the screw conveyors

Reference	Type of study	Type of device	Main effect
[142, 149, 150]	Experiment/ theoretical	Horizontal biomass screw feeder	Mechanism of blockage and effect of particle size, distribution, particle shape, moisture content, and different screws on the feeding and models for predicting torque requirement and onset of blockage
[151]	Experiment	Horizontal screw feeder	Degree of mixing with various screw designs and rotational speed
[152]	Experiment	Horizontal screw conveyor	Effect of screw geometry, speed, filling, and material on conveying performance
[153]	Experiment/ theoretical	Inclined screw conveyor	Effect of rotation speed, inclination angle feed intake length on the conveyor volumetric efficiency
[154]	Experiment/ theoretical	Horizontal screw feeder	Experimental measurement and theoretical model on torque requirement of screw feeders
[155]	Experiment	Horizontal screw feeder	Flow pattern in the hopper and feeder throughput with different screw designs to obtain uniform drawdown of particles
[156]	Experiment	Horizontal screw feeder	Flow pattern of solids in the screw
[157]	Experiment	Biomass screw feeder	Using automatic shaking mechanism in the hopper to mitigate feeding problems of biomass solids and enhancing flow rate of feeder
[158–160]	Experiment	Screw conveyor dryer	Residence time distribution of sand particles as a function of screw speed and solid flow rate. Heat transfer coefficient, thermal efficiency, and power consumption of screw dryer in drying of fine crystalline solids
[161]	Experiment	Horizontal screw feeder	Effect of screw speed and moisture content of starch on residence time distribution and flow pattern of starch granules
[162]	Theoretical	Screw conveyor blancher	Simulating energy consumption and optimization of the process

Many experiments have been conducted to characterize the flow pattern of solids, power requirement, throughput, and solid holdup in conveyors (as cited in Table 5.4). These experiments answer a wide range of questions associated with screw conveyors and provide enough information for their design. However, they are applicable only to a specific geometry and feedstock. Beside the experimental works, some theoretical models have been developed to characterize the performance of screw conveyors and their power requirement. These models are established based on some simplifying assumptions that limit their applicability to very simple geometries and coarse particles. These studies provide a good basis for design and optimization of these devices but they cannot answer some basic questions. For example, effect of particle shape and size distribution, interparticle forces, and mutual interactions of particle and surfaces have not been investigated yet. In addition, using these devices for a new feedstock or application requires additional experiments and theoretical models. Finding mechanism of particle breakage due to high mechanical forces and places where it occurs and proposing methods to mitigate this effect cannot be answered easily via experiments.

The framework of DEM allows us to model solid flow in screw conveyors with any level of geometric complexity and considering particle shape, size distribution, and interparticle force (like liquid bridge force for wet particles or van der Waals force for fine particles) in various operational conditions at very low cost. DEM simulations of horizontal screw conveyor [144, 145], inclined and vertical conveyors [146, 147] can predict the experimental data and theoretical relationships on the solid flow rate and solid hold up [148]. With such numerical experiments we can obtain detailed information about flow pattern in the casing and container, determine the role of different operating conditions on the flow rate, the power requirement, the wear condition and mechanical forces experienced by solid particles in different parts of the equipment. In addition, they can be used for optimizing case-specific conveying systems to obtain the desired flow pattern and flow rate.

5.4.1 Simulation of Screw Conveyor

We set up a simulation for a simple horizontal screw conveyor with a small hopper as the particle collector. The casing is tubular with an inner diameter of 0.125 m and length of 0.75 m. The length of opening of hopper to the casing is 0.11 m and the length of exit at the end of casing is 0.1 m. Screw flight diameter is 0.125 m (the clearance is zero) and the screw shaft diameter is 0.035 m. This leads to six flights along the screw. Surfaces of different parts of the screw conveyor were constructed with a drawing tool, then triangulated and converted into STL[4] format and saved. Three files were created: "screw.stl" for the helical blade, "inner.stl" for the cylindrical shaft, and "shell.stl" for the casing and small hopper mounted on it. While performing the simulation, these STL files were loaded into the program using subroutine `add_stl_file` defined in the `Geometry` type (see Chapter 3). Code and STL files to setup this simulation can be found in the online content of the book under "ScrewConveyor" folder.

A mixture of spherical particles with three different sizes (uniform number distribution) was used as the feed. Table 5.5 lists the physical properties of particles and walls. Solids with the rate of 1000 particles per second (equivalent to 0.7 kg/s) were inserted into the simulation form the top of the hopper. Three simulations with different shaft rotational speeds (30, 45, and 60 rpm) and constant feed rate were performed. Simulations lasted for 20 s, which was long enough to get the steady state operation of the conveyor.

The time step for simulation was 0.00001 s. According to the physical properties in Table 5.5 and the Rayleigh analysis (see Chapter 3), the critical time step for the small particles is 1.2×10^{-4} seconds. The selected time step for simulation is 8% the critical time step, which is acceptable. Tip velocity of the screw blade is 0.4 m/s for the rotation speed of 60 rpm. We assumed the maximum relative velocities between two particles and between particle and

[4] A file with STL (standard tessellation language) format contains representation of a surface in 3D space. The surface is tessellated into oriented triangles, each of which is described by a unit normal vector and three vertex points.

Table 5.5 Physical properties of particles and walls used in the screw conveyor simulation

Property		Particle	Wall
Young's modulus (MPa)		10	10
Poisson's ratio (—)		0.23	0.23
Dynamic friction (—)		0.3	0.4
Rolling friction (—)		0.1	0.1
Restitution coefficient (—)		0.7	0.7
Density (kg/m^3)		2500	—
Diameter (mm)	Small	4.5	—
	Medium	7.0	
	Large	10.5	

wall to be 1 m/s. Accordingly, the maximum overlap becomes 3% the diameter of the smallest particle, which is reasonable for this application. The time step resolution for contact between two small particles is 46, which is high enough to assure accurate and stable integration.

5.4.2 Results of the Simulations

Figure 5.27 shows the results of three simulations. Particles are shaded according to their size with black for large, blue for medium, and light gray for small particles. The casing wall is also clipped to make inside it visible. In each simulation, a snapshot from the cross section of casing is also provided to better picture the flow pattern of particles. This figure shows the effect of rotation speed at constant solid flow rate of 0.7 kg/s on the flow pattern and solids holdup. At high rotation speed (60 rpm), once particles are collected in the first flight of the screw under the feeder, they are pushed forward due to rotation of the screw. A heap of particles is formed in each flight where particles avalanche from top toward the bottom of the heap. Due to the friction between particles and screw, particles in contact or near the wall rise up the heap. These two mechanisms form a circulation of particle in the heap formed in each flight of the screw. A few particles near the shaft have the chance to roll over it and fall into the previous flight. This results in a back mixing of particles and increases the residence time of particles in the casing. The cross section view of the conveyor clearly shows that some particles roll over the shaft. The animations of the particle flow in this simulation and the next two are available online. This animation better shows the flow behavior explained previously.

At 45 rpm, the particle flow is similar to that which is observed above at 60 rpm. However, the fill ratio of the conveyor increases, which is a result of slow rotation of the shaft. As the rotation speed decreases, the first flight has more time to be filled under the feeder. Due to lower velocity of particles along the conveyor and increased holdup, the mean residence time increases in the conveyor. Looking at the cross-sectional view of the conveyor, it can be seen that more particles roll over the shaft and fall into the previous flight resulting in a broader RTD due to promoted back mixing [148]. At 30 rpm, the whole space in each flight is filled with particles and the free surface vanishes, hence, no particle avalanches of on the surface.

Size segregation occurs in the screw conveyor. We would not quantify the extent of segregation here. Therefore, discussions are qualitative and based on the snapshots presented in Figure 5.27.

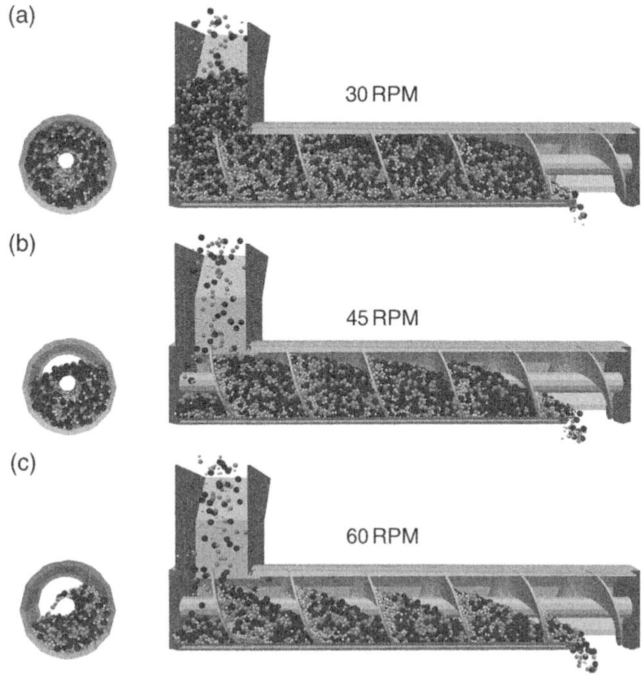

Figure 5.27 Snapshots of the solids flow in a screw conveyor at various rotational speed with constant solid flow rate (0.7 kg/s). Left-hand images (a–c) show the cross-sectional views and the right-hand side shows the side views of the conveyor

At 60 rpm, small particles are collected almost near the rear surface of the heap (left side of the heap in each flight) and larger particles are concentrated in the leading part (right side of the heap). The main mechanisms of segregation are percolation of small particles into larger particles on the free surface and shear effect of the screw wall. DEM simulation results show that the extent of segregation decreases with decreasing the rotation speed. It is expected that the extent of segregation is decreased by the elimination of the segregation on the surface due to elimination of the free surface of the heap. As the rotation speed decreases to 45 rpm, the fraction of particles which contribute to free surface motion decreases. By further decrease of the speed to 30 rpm, the free surface completely vanishes. Therefore, the extent of size segregation decreases with decreasing the rotation speed as predicted by the DEM simulation.

5.4.3 Literature

Using DEM simulations, effects of rotation speed and feed flow rate on the performance of horizontal, inclined, and vertical screw conveyors [146, 163, 164] and screw feeders [165] were studied. Other factors like screw design [166–168], cohesion between particles [144], and particle shape [165] on the performance of screw feeders were also studied by DEM. In addition to conventional screw conveyors, the performance of alternative designs of horizontal [169] and vertical conveyors [147] were also studied.

Simulations on horizontal, inclined, and vertical screw conveyors, using a single-pitch screw conveyor and applying periodic boundary condition, showed two extreme flow regimes in conveyors [146, 163]. The first regime, called the circulating heap, is similar to that observed in Figure 5.27c in which particles avalanche on the surface of the particle bed and are transported up to the surface of the particle bed due to the motion of the screw blade. This regime occurs in horizontal to low inclined conveyors at low fill levels. At high inclination angles and vertical configuration, a layer of particles is formed on the leading baled in which particles are collected near the casing wall and away from the screw shaft. This is called the bed type regime. Results revealed that with increase in inclination, the axial velocity of particles decreases while the swirling velocity increases due to the action of gravity. These trends lead to a transition from the circulating heap regime to the bed type regime. It was also shown that inclination of the conveyor reduces the mass flow rate and increases the power requirement, which was found to be related to motivated particle–particle and particle–wall collisions, which in turn cause higher kinetic energy dissipation [146].

The RTD of particles is an important characteristic in conveyors with heat transfer. DEM was used to investigate the RTD of high density polyethylene particles in an extruder with open and restricted discharge [145]. Simulations of spherical particle flow in a screw conveyor dryer [148] showed that the RTD of particles has a normal distribution at low flow rates while it is skewed toward high values at high flow rates. Inclination of the screw (0–15°) and decreasing the rotation speed increases the mean residence time and broaden the RTD.

Cohesion between particles exists in some applications. For example, it can be caused by surface wetness, reduced elasticity in heated material, electrostatic force due to tribo-electrification of mixtures of particles with different properties or the van der Waals force in fine particles. Hou *et al.* [144] considered a simple and general cohesive model in their DEM simulation to study this effect on the screw feeder performance. They identified three flow regimes with increasing the strength of cohesive force: continuous, intermittent, and stable arch. At low levels of cohesion, a continuous flow from bin to conveyor discharge can be observed. At intermediate cohesive force, instability in the flow rate occurs due the competition between cohesive force network and perturbation effects of the screw blade. Finally, at high cohesive forces, the cohesive force network wins the competition and a stable arch is formed around the screw and no particle is drawn into the flights. With the aid of the DEM simulation, they also showed that by varying the length of screw in the bin, the drawdown pattern of solids in the bin can be controlled and a uniform drawdown profile can be obtained at a cost of a little decline in the flow rate of particles. They also showed that geometry of the container (bin or hopper) of the feeder plays an important role in controlling the solid flow rate of the feeder and using bin as the container is better for conveying cohesive particles, since the converging walls in the hopper container enhances the interparticle cohesion between particles.

It was experimentally shown that the screw design can affect the drawdown and flow pattern of particles in the container of a screw feeder [155]. However, a DEM simulation can provide more than what we can observe and perform in experiments at very low cost. Fernandez *et al.* [166] compared six different screw designs to find the best screw in terms of drawdown profile, power consumption, and wear conditions. These designs included changes in the screw flight diameter, shaft diameter, and pitch length along the screw. The screw with the tapered shaft diameter and variable pitch showed the best performance considering uniform drawdown, power requirement, and moderate wear condition. Changes in the design of screw may lead to a significant increase in the power consumption with no increase in the flow rate. With the aid of detailed analysis on the force distribution and particle compaction along the screw, provided by the DEM, it was found that particle compaction is essential for uniform drawdown. Moreover, the power consumption can be reduced, with a weak decline in the solid flow rate, by decreasing the particle-wall friction. This can be achieved by regular polishing the surface or selecting a proper material for the conveyor.

The framework of the DEM allows simulating alternate designs of screw conveying systems for new applications in the industry. Landry *et al.* [169] simulated a conveyer composed of a top-open conveying system and a screw spreader for distributing manure over the farm land. They studied flow rate, spreading quality, and power computation of the system at different operating conditions. McBride and Cleary [147] simulated a so called OLDS vertical conveyor in which the screw is stationary and the casing rotates with a feeding cutter at the bottom of the container. Two different flow patterns were identified in simulations depending on the height of particle bed in the container and the height of feeding cutter. Both experimental and simulation results showed that the mass flow rate varies linearly with the rotation speed in the full range of operational conditions, suggesting that the throughput rate can be controlled by adjusting the rotation speed. The feeding cutter helps mobilization of particles in the container and facilitates feeding into the casing. Tall cutters mobilize particles more, which can be suitable for hard-to-flow particles (such as irregularly shaped particles with interlocks or particles with cohesion) and short cutters impose less mobilization particles on the feed intake, which is suitable for fragile particles. Force analysis showed that most of the particle damage occurs before particles enter the casing.

5.5 Film Coating

Coating is used to cover surface of particles with a coating material in industries such as pharmaceutical, food, and agrochemicals. Coating is applied for various reasons: changing or masking the smell or taste of the product, increasing the lifetime of the product, changing the functionality of the product like controlling the release rate of drug or determining the site of drug release. Various types of coating equipment are used, which are categorized mainly into pan coaters and fluidized bed coaters. Particles smaller than 6 mm and without attrition problems can be coated in fluidized beds while larger particles are coated in pan coaters [170]. In general, fluidized bed coaters result in a more uniform product than pan coaters. Here, we focus on pan coaters since this can be considered as a granular flow in a rotating pan or a vessel with rotating blades.

Two important factors in coating operations are interparticle variability and intrarparticle variability. Interparticle variability refers to the variation in the mass of coating solution that particles receive during the coating process. The mass of coating solution on particles has a distribution that is statistically defined as:

$$CoV_{inter} = \frac{\sigma_{coat}}{\mu_{coat}} \tag{5.34}$$

where σ_{coat} and μ_{coat} are the standard deviation and mean of coating mass distribution on particles, respectively. A smaller CoV_{inter} indicates a more uniform coating. Intrarparticle

Figure 5.28 A typical pan coater with spray zone

variability is defined as the variability of coating thickness on the surface of one particle. Uniform coating of a particle is important in coating of tablets with functional materials, for example, for controlling the release rate of the drug. In this section, we mainly focus on interparticle variability.

A simple pan coater can be a rotating drum in which a nozzle sprays the coating solution on the bed surface (see Figure 5.28). Pan coaters operate in either rolling or cascading regime. During rotation of the drum, a continuous flow of particles is formed in the drum. Particles enter the spraying zone, receive coating material, and then leave it. This is called a pass. This process is repeated for a number of times for each particle. The total coating mass deposited on a particle is proportional to the number of passes and the mass deposited on the surface in each pass. The mass deposited on the surface is proportional to the rate of spraying, the surface area exposed toward the spray, and the time spent in the spraying zone. Variation of the coating in each pass originates from distribution of residence time of particles and variable exposed area of particles toward the spray in the spraying zone (distribution in mass coating in each pass). It is possible that particles bypass the spraying zone and do not appear on the bed surface when they circulate in the pan (wider distribution in the circulation time).

Since information at the particle scale is readily available in a DEM simulation, it can be a perfect tool to be used for predicting interparticle (and also intrarparticle mass) variability. It is also possible to couple it with a simple or comprehensive spray model [171] to make the simulation sensitive to spray parameters. However, there is a major limitation in directly using it for predicting the interparticle variability. Most practical coating operations contain thousands of particles and need several minutes to become complete. Simulating such a case takes a long CPU time, which makes this approach unfeasible. In contrast, there are other models for calculating the interparticle coating variability, which are:

- Phenomenological models.
- The Monte-Carlo method.

In this part, we briefly introduce them and explain how a DEM simulation can be used in conjunction with these methods.

5.5.1 Phenomenological Models

As we discussed previously, the interparticle variability is caused due to variability in two factors: circulation time and mass coating received in each pass. Mann [172] obtained the following equation for interparticle variability based on the renewal theory and assuming that the circulation time is independent of the spraying zone residence time:

$$CoV_{inter} = \sqrt{\frac{\mu_{CT}}{t_{coat}}\left[\left(\frac{\sigma_{m,pass}}{\mu_{m,pass}}\right)^2 + \left(\frac{\sigma_{CT}}{\mu_{CT}}\right)^2\right]} \tag{5.35}$$

where μ_{CT} and σ_{CT} represent the mean and standard deviation of circulation times, and $\mu_{m,pass}$ and $\sigma_{m,pass}$ represent the mean and standard deviation of mass coating received in each pass, respectively. Assuming that the mass deposited on a particle is only proportional to the total time that it spends in the spray zone, the mass deposited on the particle in each pass becomes proportional to the residence time in the spray zone. Therefore, we can write $\mu_{m,coat} = k\mu_{ST}$ and $\sigma_{m,caot} = k\sigma_{ST}$, where k is a proportionality constant and μ_{ST} and σ_{ST} are the mean and standard deviation of time that particle spends in the spray zone in each pass, respectively [173]. Accordingly, Equation 5.35 can be written as:

$$CoV_{inter} = \sqrt{\frac{\mu_{CT}}{t_{coat}}\left[\left(\frac{\sigma_{ST}}{\mu_{ST}}\right)^2 + \left(\frac{\sigma_{CT}}{\mu_{CT}}\right)^2\right]} \tag{5.36}$$

It should be noted that this assumption (proportionality of mass deposition on particle to spray zone time) can be held for a system with a narrow size distribution or near-spherical particles in which the exposed surface of particle to the spray has a narrow distribution. This equation shows that the interparticle variability decreases with square root of the coating time and decreases the standard deviation of spray zone time and circulation time.

Distributions of circulation time and spray zone time depend on the operating condition and geometry of the pan. Therefore, we must measure these distributions to obtain CoV_{inter}. For example, Sandadi *et al.* [81] installed a camera in place of the nozzle and captured images of the bed surface during rotation of the drum. Using an image analysis technique, a tablet with a different color to the others was tracked in the spray zone to obtain the circulation time and spray time distributions. They discussed some critical factors, such as the threshold for time between successive sights of the tablet tracer on the spray zone and the threshold for exposure area.

An alternative approach to experiment is using the DEM. After carrying out simulations at the desired operating conditions, the position data and orientation of particles should be processed to calculate the desired distributions [174, 175]. To this end, a number of particles are considered as the tracer and are tracked for a certain period. This period should be at least 15–20 times the average circulation time of particles to ensure that we have included very long circulation times in our analysis. A pre-defined spray zone is considered in the pan and the exposure area of particle toward the spray nozzle and spray zone time are measured using a proper numerical method. The time and area thresholds should also be considered in the measurements [82, 176]. Figure 5.29a–d shows distributions of circulation time, spray zone time,

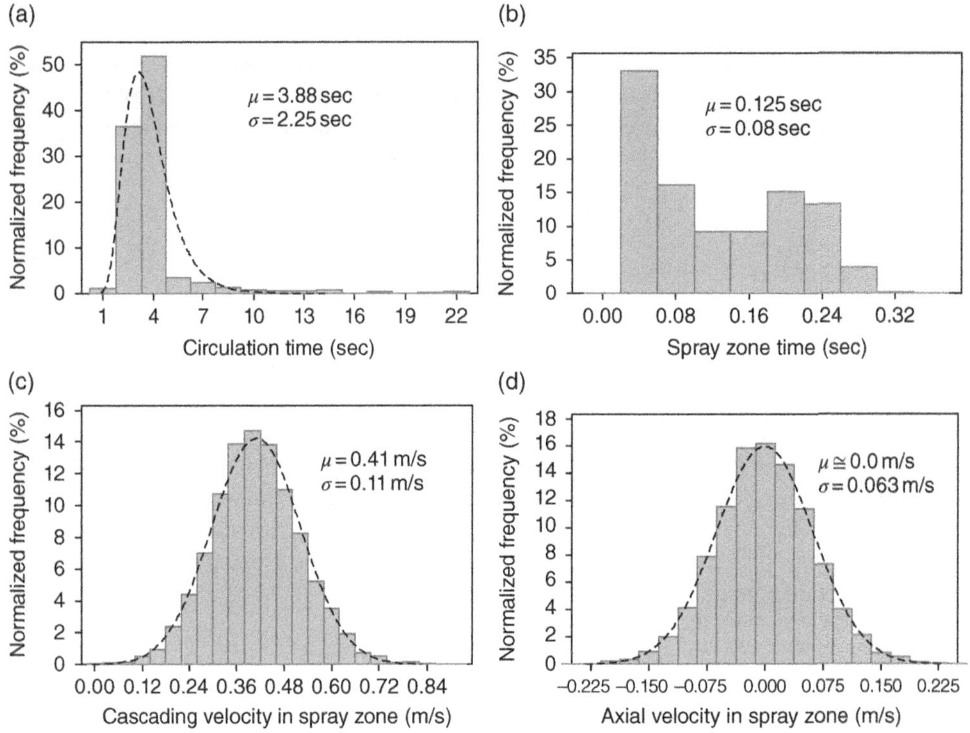

Figure 5.29 Distribution of various parameters in a pan coater with a 0.45 m drum at 9 rpm rotation speed (a) circulation time, (b) spray zone time, (c) cascading velocity, and (d) axial velocity

axial velocity, and cascading velocity of spherical particles in a rotating drum with the diameter of 0.45 cm and the rotation speed of 9 rpm, respectively. These values were obtained by post processing the particle data obtained from the DEM simulation performed by the code that accompanies this book. Using values reported in Figure 5.29, one can calculate the variation of interparticle variability as a function of time by Equation 5.36.

5.5.2 *Monte-Carlo Method*

In the Monte-Carlo method, each coating pass is discretized into sub-events. The discretization is done when the particle is in the spray zone. To evaluate the motion of particle in the spray zone, true distributions of particle velocities (such as that shown in Figure 5.29) are used. Since the residence of particle in the spray zone is discretized into sub-events, more parameters such as surface area of particle toward spray and spray parameters like distribution of spray flux and shape of the spray zone can be included in the calculation of coating mass deposited on the particle. A very simplified flowchart of the Monte-Carlo method used to calculate interparticle variability in a pan coater is presented in Figure 5.30 [170].

A spray zone with predefined size and location is selected. To start calculations, a random starting position in the spray zone is selected for the particle. It can be provided by DEM

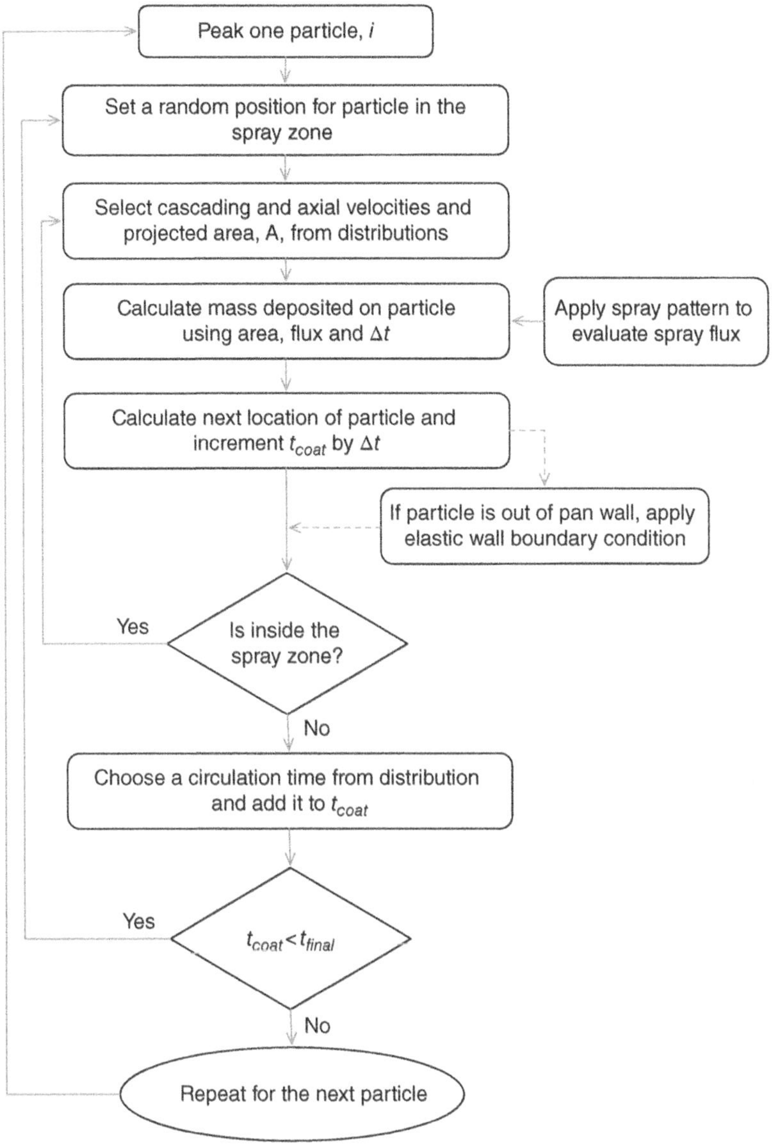

Figure 5.30 The flowchart of the Monte-Carlo method for calculating inter-particle variability [170]

simulation results since it not uniformly distributed. To evaluate the motion of particle in the spray zone, axial velocity (V_X), and cascading velocity (V_Y) are randomly selected from their true distributions and the new position of particle in the spray zone is calculated from:

$$X = X + V_X \Delta t \tag{5.37a}$$

$$Y = Y + V_Y \Delta t \tag{5.37b}$$

where X (in the direction perpendicular to cascading layer flow in the plane of this layer) and Y (in the direction parallel to cascading layer) are coordinates of the particle center in the spray zone and Δt is time step. Using spray flux at the particle location, S_{flux}, and randomly selected exposed area of particle toward the spray, A_{exp}, the coating mass deposited on the particle during one event is calculated and added to its previous coating mass:

$$m_{i,coat} = m_{i,coat} + A_{exp} S_{flux} \Delta t \tag{5.38}$$

Once the particle goes out of the spray zone, one coating event is complete. A random circulation time is selected from the true distribution of circulation time and added to the coating time of particle. If the current coating time of particle is larger than the final coating time (t_{final}), calculations for the selected particle stop. Otherwise, calculations for the next coating event start. A certain number of particles are coated according to this procedure. Then, the inter-particle mass variability is calculated by Equation 5.34. Pandey *et al.* [177] used this method to investigate effect of operating conditions and spray dynamics on the coating mass variability on tablets. They showed that the spray shape does not affect CoV_{inter}, while CoV_{inter} is decreased by increasing the spray area. They also obtained the following equation in the operating condition of their work:

$$CoV_{inter} \propto \frac{d_p^{1.2} N_{tablet}^{0.5}}{\Omega^{0.4} t_{coat}^{0.5}} \tag{5.39}$$

where N_{tablet} is the number of tablets in the pan coater. Kalbag and Wassgren [175] used DEM data to obtain the true distributions of system variables and calculated CoV_{inter}. They obtained the same results as reported by Pandey *et al.* [177]. Kandela *et al.* [178] used the same model and found that CoV_{inter} is proportional to $\phi^{-0.0127}$, where ϕ is the sphericity of tablets.

Although the experimental measurement is a way of obtaining distribution of various parameters required in the Monte-Carlo method, the DEM can easily provide these data since position, orientation, and velocity of particles are readily available in the simulation. Good agreements between simulations and experiments were reported in terms of dynamic angle of repose and velocity distributions in the spray zone [73]. Analysis of the spray zone RTD and circulation time of spherical particles were performed by Kalbag *et al.* [174] who found that mean and standard deviation of spray zone time decrease by increasing the rotation speed while dependence of these parameters on fill level is complex due to the use of anti-slip baffles in the coater. Kumar *et al.* [176] compared CoV_{inter} obtained from DEM with that obtained from a population balance model and found a good agreement between these two. They finally concluded that using DEM for estimating inputs of the population balance model is more feasible than purely using DEM simulation for obtaining CoV_{inter}.

The Monte-Carlo method was used in conjunction with the DEM to investigate the intraparticle coating variability, CoV_{intra}, for spherical and non-spherical particles [179, 180]. It was shown that increasing fill level and drum rotation speed decrease CoV_{intra} due to increasing the surface velocity. Particles in the spray zone may have no preferred orientation, have preferred rotation direction (like spheres), or have preferred orientation (like many non-spherical particles). In the case of random orientation, CoV_{intra} is inversely proportional to the square root of number of coating passes (or time) as also observed for CoV_{inter}. However, in the case of particle with a preferred rotation direction, the decrease in CoV_{intra} is delayed and for particles with a preferred orientation, CoV_{intra} decreases with time slower and approaches a non-zero asymptotic value, which means that increasing the coating time would not improve the intraparticle mass uniformity. Other operating conditions like, fill level (negative effect), and drum speed (positive effect) affect the intraparticle mass uniformity but not as strong as the particle shape [181]. Orientation index of particles, which is a measure of variability of tablet orientation in the spray zone, was shown to be inversely proportional to the sphericity of particle. CoV_{intra} is directly (not necessarily linearly) related to the orientation index. This trend is confirmed by Kandela *et al.* [178] who showed the intraparticle mass variability is proportional to $\phi^{-0.65}$.

Nomenclature

Symbol	Unit	Description
A	m^2	Area of orifice
A_{exp}	m^2	Exposed area of particle toward spray
BL_i	—	List of particles which have flexible bond with particle i
Bo	—	Bond number
$\bar{c}$	kg/kg	Mean concentration of marked particles in samples
c_i	kg/kg	Concentration of marked particles in sample i
C	—	Empirical constant in Equation 5.4
C	kg/kg	Concentration of marked particles in outlet flow
CL_i	—	Contact list of particle i
CoV_{inter}	—	Interparticle coating variability
CoV_{intra}	—	Intraparticle coating variability
d_p	m	Particle diameter
D_{ax}	m^2/s	Axial dispersion
D_0	m	Width of orifice
E	1/s	Residence time distribution
$\vec{f}_{ik}^{bond}$	N	Bonding force between constituent spheres of a bond
$\vec{f}_i^{ext}$	N	Sum of external forces acting on sphere i
$\vec{f}_i^{f-p}$	N	Fluid-particle interaction forces acting on sphere
$\vec{f}_{ij}^{p-p}$	N	Contact force between spheres in different fibers or non-adjacent spheres
Fr	—	Froude number
Fr_{crit}	—	Critical Froude number
Fr_f	—	Modified Froude number that accounts for filling degree of drum
g	m/s^2	Acceleration of gravity
$GMMI_w$	—	Generalized mean mixing index of white particle
$GMMI_{w,x}$	—	Weighted average height of white particle along the x-axis
$GMMI_{w,y}$	—	Weighted average height of white particle along the y-axis
$GMMI_{w,z}$	—	Weighted average height of white particle along the z-axis
I_i	$kg \cdot m^2$	Moment of inertia of sphere i
k		Particle shape constant in Equation 5.4
K_m	1/s	Variance decay mixing rate constant
L	m	Mid-chord length in a drum
L_z	m	Longitudinal length between injection and detection points

m	—	Shape factor
m_i	kg	Mass of sphere i
$m_{i,coat}$	kg	Mass of coating material deposited on particle i
$\vec{M}_{ik}^{bond}$	N·m	Bonding moment between constituent spheres (i and k) of a bond
$\vec{M}_{ij}^{c}$	N·m	Moment on the sphere i due to contact with sphere j
MI_{Kramer}	—	Kramer mixing index
MI_{Lacey}	—	Lacey mixing index
MI_{NN}	—	Nearest neighbor mixing index
MI_{Poole}	—	Poole mixing index
$MI_{Valentine}$	—	Valentine mixing index
n	—	Average number of particles in samples
n_{diff}	—	Number of neighbor particles with different marks
n_{nb}	—	Number of neighbor particles
N_s	—	Number of samples
N_t	—	Total number of particles
N_{tablet}	—	Number of tablets
N_{tnk}	—	Number of tanks
N_w	—	Number of particles of kind w
Pe	—	Peclet number
r	m	Radial position
R_D	m	Drum radius
S_{flux}	kg/(m^2.s)	Local spray flux
S^2	(kg/kg)^2	Variance of concentration
t	s	Time
t_{coat}	s	Coating time
t_{final}	s	Final coating time
t_O	s	First appearance time
u_{max}	m/s	Maximum surface velocity
$\bar{u}_{max}$	—	Dimensionless maximum surface velocity
u_{surf}	m/s	Surface velocity of particles
$\bar{u}_z$	m/s	Mean axial velocity of particles
V_X	m/s	Axial velocity
V_Y	m/s	Cascading velocity
$\vec{x}_i$	m	Center position of sphere i
x_i	m	x-coordinate of center of particle i
x_0	m	x-coordinate of a reference point
X	m	Distance
Y	m	Distance
w	kg/kg	Composition of white particle
$\dot{W}$	kg/s	Discharge flow rate
$\dot{W}_{cone}$	kg/s	Discharge flow rate
z	m	Axial location
$\bar{z}$	—	Dimensionless axial location

Greek symbols

α	°	Dynamic angle of repose
β	°	Angle of repose
β	—	Asymmetric factor
β_i	°	Angle of internal friction

Δt	s	Time step
η	—	Aspect ratio of super-ellipsoids
θ	—	Dimensionless time
$\bar{\theta}$	—	Dimensionless time
θ_{cone}	°	Cone angle
μ		Expected mean
μ_{coat}	kg	Mean of coating material deposited on particles
μ_{CT}	s	Mean of circulation times
$\mu_{m,pass}$	kg	Mean of mass coating per pass
μ_{pp}	—	Coefficient of dynamic friction between two particles
μ_{pw}	—	Coefficient of dynamic friction between particle and wall
$\mu_{p-w,crit}$	—	Critical dynamic friction between particle and wall below which sliding regime occurs
$\mu_{r,pp}$	m	Coefficient of rolling friction between two particles
$\mu_{r,pw}$	m	Coefficient of rolling friction between particle and wall
μ_{ST}	s	Mean of time that particle spends in the spray zone in each pass
ξ	°	Half bed angle of the circular segment filled with particles
ρ_b	kg/m^3	Bulk density of powder
ρ_l	kg/m^3	Density of liquid
σ^2	(kg/kg)^2	Expected variance of mixture
σ_{coat}	kg	Standard deviation of coating material deposited on particles
σ_{CT}	s	Standard deviation of circulation times of particles
$\sigma_{m,pass}$	kg	Standard deviation of mass coating received in each pass
σ_l	N/m	Surface tension of liquid
σ_R^2	(kg/kg)^2	Variance of random mixture
σ_{ST}	s	Standard deviation of time that particle spends in spray zone in each pass
σ_0^2	(kg/kg)^2	Variance of segregated mixture
τ	s	Mean residence time
φ	—	Fill ratio
ϕ	—	Sphericity
χ^2	—	Chi-square statistic
$\vec{\omega}_i$	rad/s	Rotational velocity of sphere i
Ω	rad/s	Rotational speed of drum
Ω_p	rad/s	Primary rotational speed of screw
Ω_s	rad/s	Sweeping rotational speed of screw

References

[1] He, H., Guo, Z., Stroeven, P., Stroeven, M., and Sluys, L.J. (2009) Characterization of the packing of aggregate in concrete by a discrete element approach. *Materials Characterization*, **60**(10), 1082–1087.

[2] Matuttis, H.G., Luding, S., and Herrmann, H.J. (2000) Discrete element simulations of dense packings and heaps made of spherical and non-spherical particles. *Powder Technology*, **109**, 278–292.

[3] Yang, R.Y., Zou, R.P., Yu, A.B., and Choi, S.K. (2006) Pore structure of the packing of fine particles. *Journal of Colloid and Interface Science*, **299**(2), 719–725.

[4] Yang, R.Y., Zou, R.P., and Yu, A.B. (2003) Numerical study of the packing of wet coarse uniform spheres. *AIChE Journal*, **49**(7), 1656–1666.

[5] Dong, K.J., Zou, R.P., Yang, R.Y., Yu, A.B., and Roach, G. (2009) DEM simulation of cake formation in sedimentation and filtration. *Minerals Engineering*, **22**(11), 921–930.

[6] Zhang, Z.P., Liu, L.F., Yuan, Y.D., and Yu, A.B. (2001) A simulation study of the effects of dynamic variables on the packing of spheres. *Powder Technology*, **116**, 23–32.

[7] Mostofinejad, D. and Reisi, M. (2012) A new DEM-based method to predict packing density of coarse aggregates considering their grading and shapes. *Construction and Building Materials*, **35**, 414–420.
[8] Langston, P. and Kennedy, A.R. (2014) Discrete element modelling of the packing of spheres and its application to the structure of porous metals made by infiltration of packed beds of NaCl beads. *Powder Technology*, **268**, 210–218.
[9] Dorai, F., Moura Teixeira, C., Rolland, M., Climent, E., Marcoux, M., and Wachs, A. (2015) Fully resolved simulations of the flow through a packed bed of cylinders: effect of size distribution. *Chemical Engineering Science*, **129**, 180–192.
[10] Kang, D.H., Yun, T.S., and Matthew Evans, T. (2014) Pore orientation of granular materials during biaxial compression. *Computers and Geotechnics*, **59**, 1–11.
[11] Augier, F., Idoux, F., and Delenne, J.Y. (2010) Numerical simulations of transfer and transport properties inside packed beds of spherical particles. *Chemical Engineering Science*, **65**(3), 1055–1064.
[12] Theuerkauf, J., Witt, P., and Schwesig, D. (2006) Analysis of particle porosity distribution in fixed beds using the discrete element method. *Powder Technology*, **165**(2), 92–99.
[13] Fu, X., Dutt, M., Bentham, A.C., Hancock, B.C., Cameron, R.E., and Elliott, J.A. (2006) Investigation of particle packing in model pharmaceutical powders using X-ray microtomography and discrete element method. *Powder Technology*, **167**(3), 134–140.
[14] Ueda, T., Matsushima, T., and Yamada, Y. (2012) Micro structures of granular materials with various grain size distributions. *Powder Technology*, **217**, 533–539.
[15] Wiącek, J. and Molenda, M. (2014) Microstructure and micromechanics of polydisperse granular materials: effect of the shape of particle size distribution. *Powder Technology*, **268**, 237–243.
[16] An, X.Z., Yang, R.Y., Zou, R.P., and Yu, A.B. (2008) Effect of vibration condition and inter-particle frictions on the packing of uniform spheres. *Powder Technology*, **188**(2), 102–109.
[17] Suikkanen, H., Ritvanen, J., Jalali, P., and Kyrki-Rajamäki, R. (2014) Discrete element modelling of pebble packing in pebble bed reactors. *Nuclear Engineering and Design*, **273**, 24–32.
[18] Zhou, Y.C., Xu, B.H., Yu, A.B., and Zulli, P. (2002) An experimental and numerical study of the angle of repose of coarse spheres. *Powder Technology*, **125**, 45–52.
[19] Langston, P., Kennedy, A.R., and Constantin, H. (2015) Discrete element modelling of flexible fibre packing. *Computational Materials Science*, **96**, 108–116.
[20] Nan, W., Wang, Y., Ge, Y., and Wang, J. (2014) Effect of shape parameters of fiber on the packing structure. *Powder Technology*, **261**, 210–218.
[21] Wang, G., Yu, W., and Zhou, C. (2006) Optimization of the rod chain model to simulate the motions of a long flexible fiber in simple shear flows. *European Journal of Mechanics – B/Fluids* **25**, 337–347.
[22] Wu, J. and Aidun, C.K. (2010) A method for direct simulation of flexible fiber suspensions using lattice Boltzmann equation with external boundary force. *International Journal of Multiphase Flow*, **36**, 202–209.
[23] Lindström, S.B. and Uesaka, T. (2007) Simulation of the motion of flexible fibers in viscous fluid flow. *Physics of Fluids*, **19**(11), 113307.
[24] Guo, Y., Wassgren, C., Hancock, B., Ketterhagen, W., and Curtis, J. (2013) Granular shear flows of flat disks and elongated rods without and with friction. *Physics of Fluids*, **25**(6), 063304.
[25] Potyondy, D.O. and Cundall, P.A. (2004) A bonded-particle model for rock. *International Journal of Rock Mechanics and Mining Sciences*, **41**(8), 1329–1364.
[26] Guo, Y., Wassgren, C., Hancock, B., Ketterhagen, W., and Curtis, J. (2013) Validation and time step determination of discrete element modeling of flexible fibers. *Powder Technology*, **249**, 386–395.
[27] Yu, Y. and Saxén, H. (2011) Discrete element method simulation of properties of a 3D conical hopper with mono-sized spheres. *Advanced Powder Technology*, **22**(3), 324–331.
[28] Balevičius, R., Kačianauskas, R., Mróz, Z., and Sielamowicz, I. (2011) Analysis and DEM simulation of granular material flow patterns in hopper models of different shapes. *Advanced Powder Technology*, **22**(2), 226–235.
[29] Balevičius, R., Sielamowicz, I., Mróz, Z., and Kačianauskas, R. (2011) Investigation of wall stress and outflow rate in a flat-bottomed bin: a comparison of the DEM model results with the experimental measurements. *Powder Technology*, **214**(3), 322–336.
[30] Balevičius, R., Kačianauskas, R., Mróz, Z., and Sielamowicz, I. (2008) Discrete-particle investigation of friction effect in filling and unsteady/steady discharge in three-dimensional wedge-shaped hopper. *Powder Technology*, **187**(2), 159–174.
[31] Anand, A., Curtis, J.S., Wassgren, C.R., Hancock, B.C., and Ketterhagen, W.R. (2008) Predicting discharge dynamics from a rectangular hopper using the discrete element method (DEM). *Chemical Engineering Science*, **63**(24), 5821–5830.

[32] Faqih, A.N., Chaudhuri, B., Mehrotra, A., Tomassone, M.S., and Muzzio, F. (2010) Constitutive model to predict flow of cohesive powders in bench scale hoppers. *Chemical Engineering Science*, **65**(10), 3341–3351.
[33] Anand, A., Curtis, J.S., Wassgren, C.R., Hancock, B.C., and Ketterhagen, W.R. (2009) Predicting discharge dynamics of wet cohesive particles from a rectangular hopper using the discrete element method (DEM). *Chemical Engineering Science*, **64**(24), 5268–5275.
[34] Höhner, D., Wirtz, S., and Scherer, V. (2015) A study on the influence of particle shape on the mechanical interactions of granular media in a hopper using the Discrete Element Method. *Powder Technology*, **278**, 286–305.
[35] Höhner, D., Wirtz, S., and Scherer, V. (2013) Experimental and numerical investigation on the influence of particle shape and shape approximation on hopper discharge using the discrete element method. *Powder Technology*, **235**, 614–627.
[36] Höhner, D., Wirtz, S., and Scherer, V. (2012) A numerical study on the influence of particle shape on hopper discharge within the polyhedral and multi-sphere discrete element method. *Powder Technology*, **226**, 16–28.
[37] Liu, S.D., Zhou, Z.Y., Zou, R.P., Pinson, D., and Yu, A.B. (2014) Flow characteristics and discharge rate of ellipsoidal particles in a flat bottom hopper. *Powder Technology*, **253**, 70–79.
[38] Tao, H., Jin, B., Zhong, W., Wang, X., Ren, B., Zhang, Y., and Xiao, R. (2010) Discrete element method modeling of non-spherical granular flow in rectangular hopper. *Chemical Engineering and Processing: Process Intensification*, **49**(2), 151–158.
[39] Ketterhagen, W.R. and Hancock, B.C. (2010) Optimizing the design of eccentric feed hoppers for tablet presses using DEM. *Computers & Chemical Engineering*, **34**(7), 1072–1081.
[40] Yu, Y. and Saxen, H. (2014) Segregation behavior of particles in a top hopper of a blast furnace. *Powder Technology*, **262**, 233–241.
[41] Ketterhagen, W.R., Curtis, J.S., Wassgren, C.R., Kong, A., Narayan, P.J., and Hancock, B.C. (2007) Granular segregation in discharging cylindrical hoppers: a discrete element and experimental study. *Chemical Engineering Science*, **62**(22), 6423–6439.
[42] Ketterhagen, W.R., Curtis, J.S., Wassgren, C.R., and Hancock, B.C. (2008) Modeling granular segregation in flow from quasi-three-dimensional, wedge-shaped hoppers. *Powder Technology*, **179**, 126–143.
[43] Wu, S., Kou, M., Xu, J., Guo, X., Du, K., Shen, W., and Sun, J. (2013) DEM simulation of particle size segregation behavior during charging into and discharging from a Paul–Wurth type hopper. *Chemical Engineering Science*, **99**, 314–323.
[44] Xu, P., Duan, X., Qian, G., and Zhou, X.-G. (2015) Dependence of wall stress ratio on wall friction coefficient during the discharging of a 3D rectangular hopper. *Powder Technology*, **284**, 326–335.
[45] Goda, T.J. and Ebert, F. (2005) Three-dimensional discrete element simulations in hoppers and silos. *Powder Technology*, **158**(1–3), 58–68.
[46] Li, J., Langston, P.A., Webb, C., and Dyakowski, T. (2004) Flow of sphero-disc particles in rectangular hoppers – a DEM and experimental comparison in 3D. *Chemical Engineering Science*, **59**(24), 5917–5929.
[47] Mack, S., Langston, P., Webb, C., and York, T. (2011) Experimental validation of polyhedral discrete element model. *Powder Technology*, **214**(3), 431–442.
[48] Carson, J.W. (2008) Hopper/Bin design. In *Bulk Solids Handling: Equipment Selection and Operation* (ed. D. McGlinchey), Blackwell Publishing Ltd., Singapore.
[49] Sleppy, J.A. and Puri, V.M. (1996) Size-segregation of granulated sugar during flow. *Transactions of the ASABE*, **39**(4), 1433–1439.
[50] Beverloo, W.A., Leniger, H.A., and van de Velde, J. (1961) The flow of granular solids through orifices. *Chemical Engineering Science*, **15**(3–4), 260–269.
[51] Brown, R.L. and Richards, J.C. (1965) Kinematics of the flow of dry powders and bulk solids. *Rheologica Acta*, **4**(3), 153–165.
[52] Lacey, P. (1954) Developments in theory of particulate mixing. *Journal of Applied Chemistry*, **4**, 257.
[53] Dubé, O., Alizadeh, E., Chaouki, J., and Bertrand, F. (2013) Dynamics of non-spherical particles in a rotating drum. *Chemical Engineering Science*, **101**, 486–502.
[54] Alizadeh, E., Dubé, O., Bertrand, F., and Chaouki, J. (2013) Characterization of mixing and size segregation in a rotating drum by a particle tracking method. *AIChE Journal*, **59**(6), 1894–1905.
[55] Combarros, M., Feise, H.J., Zetzener, H., and Kwade, A. (2014) Segregation of particulate solids: experiments and DEM simulations. *Particuology*, **12**, 25–32.
[56] Goldschmidt, M.J.V., Link, J.M., Mellema, S., and Kuipers, J.A.M. (2003) Digital image analysis measurements of bed expansion and segregation dynamics in dense gas-fluidised beds. *Powder Technology*, **138**(2–3), 135–159.

[57] Feng, Y.Q., Xu, B.H., Zhang, S.J., Yu, A.B., and Zulli, P. (2004) Discrete particle simulation of gas fluidization of particle mixtures. *AIChE Journal*, **50**(8), 1713–1728.

[58] Norouzi, H.R., Mostoufi, N., and Sotudeh-Gharebagh, R. (2012) Segregation behaviour of particles in gas solid fluidized beds at elevated pressure. *Journal of Chemical and Petroleum Engineering, University of Tehran*, **46**(2), 111–121.

[59] Danckwerts, P.V. (1952) The definition and measurement of some characteristics of mixtures. *Applied Scientific Research, Section A*, **3**(4), 279–296.

[60] Schofield, C. (1976) The definition and assessment of mixture quality in mixtures of particulate solids. *Powder Technology*, **15**, 169–180.

[61] Muzzio, F.J., Alexander, A., Goodier, C., Shen, E., Shinbrot, T., Manjunath, K., Dhodapkar, S., and Jacob, K. (2004) Solids mixing. in *Handbook of Industrial Mixing: Science and Practice* (eds E.L. Paul, V.A. Atiemo-Obeng, and S.M. Kresta), John Wiley & Sons, Inc., Hoboken, NJ.

[62] Rhodes, M. (2008) *Introduction to Particle Technology*, 2nd edn, John Wiley & Sons, Ltd, Chichester.

[63] Fan, L.T. and Wang, R.H. (1975) On mixing indices. *Powder Technology*, **11**, 27–32.

[64] Poux, M., Fayolle, P., Bertrand, J., Bridoux, D., and Bousquet, J. (1991) Powder mixing: some practical rules applied to agitated systems. *Powder Technology*, **68**(3), 213–234.

[65] Poole, K.R., Taylor, R.F., and Wall, G.P. (1964) Mixing powders to fine-scale homogeneity: studies of batch mixing. *Transactions of the Institution of Chemical Engineers*, **41**, T305–T315.

[66] Asmar, B.N., Langston, P.A., and Matchett, A.J. (2002) A generalised mixing index in distinct element method simulation of vibrated particulate beds. *Granular Matter*, **4**(3), 129–138.

[67] Doucet, J., Bertrand, F., and Chaouki, J. (2008) A measure of mixing from Lagrangian tracking and its application to granular and fluid flow systems. *Chemical Engineering Research and Design*, **86**(12), 1313–1321.

[68] Deen, N.G., Willem, G., Sander, G., and Kuipers, J.A.M. (2010) Numerical analysis of solids mixing in pressurized fluidized beds. *Industrial & Engineering Chemistry Research*, **49**, 5246–5253.

[69] Mellmann, J. (2001) The transverse motion of solids in rotating cylinders—forms of motion and transition behavior. *Powder Technology*, **118**(3), 251–270.

[70] Yang, R.Y., Yu, A.B., McElroy, L., and Bao, J. (2008) Numerical simulation of particle dynamics in different flow regimes in a rotating drum. *Powder Technology*, **188**(2), 170–177.

[71] Norouzi, H.R., Zarghami, R., and Mostoufi, N. (2015) Insights into the granular flow in rotating drums. *Chemical Engineering Research and Design*, **102**, 12–25.

[72] Wachs, A., Girolami, L., Vinay, G., and Ferrer, G. (2012) Grains3D, a flexible DEM approach for particles of arbitrary convex shape — Part I: numerical model and validations. *Powder Technology*, **224**, 374–389.

[73] Pandey, P., Song, Y., Kayihan, F., and Turton, R. (2006) Simulation of particle movement in a pan coating device using discrete element modeling and its comparison with video-imaging experiments. *Powder Technology*, **161**(2), 79–88.

[74] Alizadeh, E., Bertrand, F., and Chaouki, J. (2014) Comparison of DEM results and Lagrangian experimental data for the flow and mixing of granules in a rotating drum. *AIChE Journal*, **60**(1), 60–75.

[75] Orpe, A.V. and Khakhar, D.V. (2001) Scaling relations for granular flow in quasi two-dimensional rotating cylinders. *Physical Review E*, **64**, 031302.

[76] Ding, Y.L., Seville, J.P.K., Forster, R.N., and Parker, D.J. (2001) Solids motion in rolling mode rotating drums operated at low to medium rotational speeds. *Chemical Engineering Science*, **56**, 1769–1780.

[77] Liu, X.Y., Specht, E., Gonzalez, O.G., and Walzel, P. (2006) Analytical solution for the rolling-mode granular motion in rotary kilns. *Chemical Engineering and Processing: Process Intensification*, **45**(6), 515–521.

[78] Khakhar, D., McCarthy, J., and Ottino, J. (1997) Radial segregation of granular mixtures in rotating cylinders. *Physics of Fluids*, **9**(12), 3600–3614.

[79] Alexander, A., Shinbrot, T., and Muzzio, F.J. (2002) Scaling surface velocities in rotating cylinders as a function of vessel radius, rotation rate, and particle size. *Powder Technology*, **126**(2), 174–190.

[80] Jain, N., Ottino, J.M., and Lueptow, R.M. (2002) An experimental study of the flowing granular layer in a rotating tumbler. *Physics of Fluids*, **14**(2), 572.

[81] Sandadi, S., Pandey, P., and Turton, R. (2004) In situ, near real-time acquisition of particle motion in rotating pan coating equipment using imaging techniques. *Chemical Engineering Science*, **59**, 5807–5817.

[82] Pandey, P. and Turton, R. (2005) Movement of different-shaped particles in a pan-coating device using novel video-imaging techniques. *AAPS PharmSciTech*, **6**(2), E237.

[83] Mueller, R. and Kleinebudde, P. (2007) Prediction of tablet velocity in pan coaters for scale-up. *Powder Technology*, **173**(1), 51–58.

[84] Ding, Y.L., Forster, R.N., Seville, J.P.K., and Parker, D.J. (2001) Scaling relationships for rotating drums. *Chemical Engineering Science*, **56**, 3737–3750.

[85] Finnie, G.J., Kruyt, N.P., Ye, M., Zeilstra, C., and Kuipers, J.A.M. (2005) Longitudinal and transverse mixing in rotary kilns: a discrete element method approach. *Chemical Engineering Science*, **60**(15), 4083–4091.

[86] Kwapinska, M., Saage, G., and Tsotsas, E. (2006) Mixing of particles in rotary drums: a comparison of discrete element simulations with experimental results and penetration models for thermal processes. *Powder Technology*, **161**(1), 69–78.

[87] Xu, Y., Xu, C., Zhou, Z., Du, J., and Hu, D. (2010) 2D DEM simulation of particle mixing in rotating drum: a parametric study. *Particuology*, **8**(2), 141–149.

[88] Liu, P.Y., Yang, R.Y., and Yu, A.B. (2013) DEM study of the transverse mixing of wet particles in rotating drums. *Chemical Engineering Science*, **86**, 99–107.

[89] van Puyvelde, D.R., Young, B.R., Wilson, M.A., and Schmidt, S.J. (1999) Experimental determination of transverse mixing kinetics in a rolling drum by image analysis. *Powder Technology*, **106**, 183–191.

[90] McCarthy, J.J., Khakhar, D.V., and Ottino, J.M. (2000) Computational studies of granular mixing. *Powder Technology*, **109**(1–3), 72–82.

[91] Third, J.R., Scott, D.M., and Scott, S.A. (2010) Axial dispersion of granular material in horizontal rotating cylinders. *Powder Technology*, **203**(3), 510–517.

[92] Metcalfe, G., Graham, L., Zhou, J., and Liffman, K. (1999) Measurement of particle motions within tumbling granular flows. *Chaos*, **9**, 581–593.

[93] Metcalfe, G., Shinbrot, T., McCarthy, J.J., and Ottino, J.M. (1995) Avalanche mixing of granular solids. *Nature*, **374**, 39–41.

[94] Hill, K.M., Gioia, G., and Amaravadi, D. (2004) Radial segregation patterns in rotating granular mixtures: waviness selection. *Physical Review Letters*, **93**(22), 224301.

[95] Santomaso, A.C., Artoni, R., and Canu, P. (2013) Controlling axial segregation in drum mixers through wall friction: cellular automata simulations and experiments. *Chemical Engineering Science*, **90**, 151–160.

[96] Bhattacharya, T., Hajra, S.K., and McCarthy, J.J. (2014) A design heuristic for optimizing segregation avoidance practices in horizontal drum mixers. *Powder Technology*, **253**, 107–115.

[97] Ayeni, O.O., Wu, C.L., Joshi, J.B., and Nandakumar, K. (2015) A discrete element method study of granular segregation in non-circular rotating drums. *Powder Technology*, **283**, 549–560.

[98] Chand, R., Khaskheli, M.A., Qadir, A., Ge, B., and Shi, Q. (2012) Discrete particle simulation of radial segregation in horizontally rotating drum: effects of drum-length and non-rotating end-plates. *Physica A: Statistical Mechanics and its Applications*, **391**(20), 4590–4596.

[99] Arntz, M.M.H.D., den Otter, W.K., Briels, W.J., Bussmann, P.J.T., Beeftink, H.H., and Boom, R.M. (2008) Granular mixing and segregation in a horizontal rotating drum: a simulation study on the impact of rotational speed and fill level. *AIChE Journal*, **54**(12), 3133–3146.

[100] Arntz, M.M.H.D., Beeftink, H.H., den Otter, W.K., Briels, W.J., and Boom, R.M. (2014) Segregation of granular particles by mass, radius, and density in a horizontal rotating drum. *AIChE Journal*, **60**(1), 50–59.

[101] Jiang, M., Zhao, Y., Liu, G., and Zheng, J. (2011) Enhancing mixing of particles by baffles in a rotating drum mixer. *Particuology*, **9**(3), 270–278.

[102] Yada, H., Kawaguchi, T., and Tanaka, T. (2010) Relation between segregation patterns and granular flow modes in conical rotating drum. *Flow Measurement and Instrumentation*, **21**(3), 207–211.

[103] Zik, O., Levine, D., Lipson, S.G., Shtrikman, S., and Stavans, J. (1994) Rotationally induced segregation of granular materials. *Physical Review Letters*, **73**(5), 644–647.

[104] Huang, A.-N., Liu, L.-C., and Kuo, H.-P. (2013) The role of end wall shearing in the drum segregation band formation. *Powder Technology*, **239**, 98–104.

[105] Cui, Z., Zhao, Y., Chen, Y., Liu, X., Hua, Z., Zhou, C., and Zheng, J. (2014) Transition of axial segregation patterns in a long rotating drum. *Particuology*, **13**, 128–133.

[106] Jain, N., Ottino, J.M., and Lueptow, R.M. (2005) Combined size and density segregation and mixing in noncircular tumblers. *Physical Review E*, **71**(5), 051301.

[107] Shi, D., Abatan, A.A., Vargas, W.L., and McCarthy, J.J. (2007) Eliminating segregation in free-surface flows of particles. *Physical Review Letters*, **99**(14), 148001.

[108] Lemieux, M., Bertrand, F., Chaouki, J., and Gosselin, P. (2007) Comparative study of the mixing of free-flowing particles in a V-blender and a bin-blender. *Chemical Engineering Science*, **62**(6), 1783–1802.

[109] Kuo, H.P., Knight, P.C., Parker, D.J., Tsuji, Y., Adams, M.J., and Seville, J.P.K. (2002) The influence of DEM simulation parameters on the particle behaviour in a V-mixer. *Chemical Engineering Science*, **57**(17), 3621–3638.

[110] Manickam, S.S., Shah, R., Tomei, J., Bergman, T.L., and Chaudhuri, B. (2010) Investigating mixing in a multi-dimensional rotary mixer: experiments and simulations. *Powder Technology*, **201**(1), 83–92.

[111] Moakher, M., Shinbrot, T., and Muzzio, F.J. (2000) Experimentally validated computations of flow, mixing and segregation of non-cohesive grains in 3D tumbling blenders. *Powder Technology*, **109**, 58–71.

[112] Lemieux, M., Léonard, G., Doucet, J., Leclaire, L.A., Viens, F., Chaouki, J., and Bertrand, F. (2008) Large-scale numerical investigation of solids mixing in a V-blender using the discrete element method. *Powder Technology*, **181**(2), 205–216.

[113] Brone, D., Wightman, C., Connor, K., Alexander, A., Muzzio, F.J., and Robinson, P. (1997) Using flow perturbations to enhance mixing of dry powders in V-blenders. *Powder Technology*, **91**(3), 165–172.

[114] Alizadeh, E., Hajhashemi, H., Bertrand, F., and Chaouki, J. (2013) Experimental investigation of solid mixing and segregation in a tetrapodal blender. *Chemical Engineering Science*, **97**, 354–365.

[115] Alizadeh, E., Bertrand, F., and Chaouki, J. (2014) Discrete element simulation of particle mixing and segregation in a tetrapodal blender. *Computers & Chemical Engineering*, **64**, 1–12.

[116] Alexander, A., Muzzio, F.J., and Shinbrot, T. (2003) Segregation patterns in V-blenders. *Chemical Engineering Science*, **58**, 487–496.

[117] Arratia, P.E., Duong, N.-h., Muzzio, F.J., Godbole, P., and Reynolds, S. (2006) A study of the mixing and segregation mechanisms in the Bohle Tote blender via DEM simulations. *Powder Technology*, **164**, 50–57.

[118] Sudah, O.S., Arratia, P.E., Alexander, A., and Muzzio, F.J. (2005) Simulation and experiments of mixing and segregation in a tote blender. *AIChE Journal*, **51**(3), 836–844.

[119] Alian, M., Ein-Mozaffari, F., Upreti, S.R., and Wu, J. (2015) Using discrete element method to analyze the mixing of the solid particles in a slant cone mixer. *Chemical Engineering Research and Design*, **93**, 318–329.

[120] Alchikh-Sulaiman, B., Ein-Mozaffari, F., and Lohi, A. (2015) Evaluation of poly-disperse solid particles mixing in a slant cone mixer using discrete element method. *Chemical Engineering Research and Design*, **96**, 196–213.

[121] Laurent, B.F.C. and Cleary, P.W. (2012) Comparative study by PEPT and DEM for flow and mixing in a ploughshare mixer. *Powder Technology*, **228**, 171–186.

[122] Cleary, P.W. (2013) Particulate mixing in a plough share mixer using DEM with realistic shaped particles. *Powder Technology*, **248**, 103–120.

[123] Alian, M., Ein-Mozaffari, F., and Upreti, S.R. (2015) Analysis of the mixing of solid particles in a plowshare mixer via discrete element method (DEM). *Powder Technology*, **274**, 77–87.

[124] Hassanpour, A., Tan, H., Bayly, A., Gopalkrishnan, P., Ng, B., and Ghadiri, M. (2011) Analysis of particle motion in a paddle mixer using Discrete Element Method (DEM). *Powder Technology*, **206**(1–2), 189–194.

[125] Pasha, M., Hassanpour, A., Ahmadian, H., Tan, H.S., Bayly, A., and Ghadiri, M. (2015) A comparative analysis of particle tracking in a mixer by discrete element method and positron emission particle tracking. *Powder Technology*, **270**, 569–574.

[126] Chandratilleke, G.R., Zhou, Y.C., Yu, A.B., and Bridgwater, J. (2010) Effect of blade speed on granular flow and mixing in a cylindrical mixer. *Industrial & Engineering Chemistry Research*, **49**(11), 5467–5478.

[127] Kaneko, Y., Shiojima, T., and Horio, M. (2000) Numerical analysis of particle mixing characteristics in a single helical ribbon agitator using DEM simulation. *Powder Technology*, **108**, 55–64.

[128] Kloss, C., Goniva, C., Hager, A., Amberger, S., and Pirker, S. (2012) Models, algorithms and validation for opensource DEM and CFD–DEM. *Progress in Computational Fluid Dynamics*, **12**(2–3), 140–152.

[129] Florian-Algarin, M. and Méndez, R. (2015) Effect of material properties and design parameters on the final blend uniformity using experimental and simulation results. *Powder Technology*, **276**, 156–165.

[130] Florian-Algarin, M. and Méndez, R. (2015) Blend uniformity and powder phenomena inside the continuous tumble mixer using DEM simulations. *AIChE Journal*, **61**(3), 792–801.

[131] Sarkar, A. and Wassgren, C.R. (2012) Comparison of flow microdynamics for a continuous granular mixer with predictions from periodic slice DEM simulations. *Powder Technology*, **221**, 325–336.

[132] Sarkar, A. and Wassgren, C. (2010) Continuous blending of cohesive granular material. *Chemical Engineering Science*, **65**(21), 5687–5698.

[133] Gao, Y., Ierapetritou, M., and Muzzio, F. (2011) Investigation on the effect of blade patterns on continuous solid mixing performance. *The Canadian Journal of Chemical Engineering*, **89**(5), 969–984.

[134] Gao, Y., Ierapetritou, M., and Muzzio, F. (2012) Periodic section modeling of convective continuous powder mixing processes. *AIChE Journal*, **58**(1), 69–78.

[135] Gao, Y., Muzzio, F.J., and Ierapetritou, M.G. (2012) Optimizing continuous powder mixing processes using periodic section modeling. *Chemical Engineering Science*, **80**, 70–80.

[136] Yijie, G., Muzzio, F.J., and Ierapetritou, M.G. (2013) Scale-up strategy for continuous powder blending process. *Powder Technology*, **235**, 55–69.
[137] Sarkar, A. and Wassgren, C.R. (2009) Simulation of a continuous granular mixer: effect of operating conditions on flow and mixing. *Chemical Engineering Science*, **64**(11), 2672–2682.
[138] Dubey, A., Sarkar, A., Ierapetritou, M., Wassgren, C.R., and Muzzio, F.J. (2011) Computational approaches for studying the granular dynamics of continuous blending processes, 1 – DEM based methods. *Macromolecular Materials and Engineering*, **296**(3–4), 290–307.
[139] Gao, Y., Vanarase, A., Muzzio, F., and Ierapetritou, M. (2011) Characterizing continuous powder mixing using residence time distribution. *Chemical Engineering Science*, **66**(3), 417–425.
[140] Levenspiel, O. (1999) *Chemical Reaction Engineering*, John Wiley & Sons, Inc., New York.
[141] Chandratilleke, G., Yu, A.B., and Bridgwater, J. (2012) A DEM study of the mixing of particles induced by a flat blade. *Chemical Engineering Science*, **79**, 54–74.
[142] Dai, J., Cui, H., and Grace, J.R. (2012) Biomass feeding for thermochemical reactors. *Progress in Energy and Combustion Science*, **38**(5), 716–736.
[143] Tardos, G.I. and Lu, Q. (1996) Precision dosing of powders by vibratory and screw feeders: an experimental study. *Advanced Powder Technology*, **7**(1), 51–58.
[144] Hou, Q.F., Dong, K.J., and Yu, A.B. (2014) DEM study of the flow of cohesive particles in a screw feeder. *Powder Technology*, **256**, 529–539.
[145] Moysey, P.A. and Thompson, M.R. (2005) Modelling the solids inflow and solids conveying of single-screw extruders using the discrete element method. *Powder Technology*, **153**(2), 95–107.
[146] Owen, P.J. and Cleary, P.W. (2009) Prediction of screw conveyor performance using the discrete element method (DEM). *Powder Technology*, **193**(3), 274–288.
[147] McBride, W. and Cleary, P.W. (2009) An investigation and optimization of the "OLDS" elevator using discrete element modeling. *Powder Technology*, **193**(3), 216–234.
[148] Osman, H. (2012) Granular flow and heat transfer in a screw conveyor heater: a discrete element modeling study. Master of Engineering thesis. National University of Singapore.
[149] Dai, J. and Grace, J.R. (2011) Biomass granular screw feeding: an experimental investigation. *Biomass and Bioenergy*, **35**(2), 942–955.
[150] Dai, J. and Grace, J.R. (2008) A model for biomass screw feeding. *Powder Technology*, **186**(1), 40–55.
[151] Tsai, W.-R. and Lin, C.-I. (1994) On the mixing of granular materials in a screw feeder. *Powder Technology*, **80**(2), 119–126.
[152] Carleton, A., Miles, J., and Valentin, F. (1969) A study of factors affecting the performance of screw conveyers and feeders. *Journal of Manufacturing Science and Engineering*, **91**(2), 329–333.
[153] Roberts, A.W. and Willis, A.H. (1962) Performance of grain augers. *Proceedings of the Institution of Mechanical Engineers*, **176**(1), 165–194.
[154] Yu, Y. and Arnold, P.C. (1997) Theoretical modelling of torque requirements for single screw feeders. *Powder Technology*, **93**(2), 151–162.
[155] Yu, Y. and Arnold, P.C. (1996) The influence of screw feeders on bin flow patterns. *Powder Technology*, **88**(1), 81–87.
[156] Bates, L. (1969) Entrainment patterns of screw hopper dischargers. *Journal of Engineering for Industry*, **91**(2), 295–302.
[157] Zhang, J., Yuan, Z.-w., and Wei, X.-l. (2011) An automatic shake mechanism for the biomass pyrolysis feeding system. *Powder Technology*, **207**(1–3), 348–352.
[158] Waje, S.S., Thorat, B.N., and Mujumdar, A.S. (2006) An experimental study of the thermal performance of a screw conveyor dryer. *Drying Technology*, **24**(3), 293–301.
[159] Waje, S.S., Patel, A.K., Thorat, B.N., and Mujumdar, A.S. (2007) Study of residence time distribution in a pilot-scale screw conveyor dryer. *Drying Technology*, **25**(1), 249–259.
[160] Waje, S.S., Thorat, B.N., and Mujumdar, A.S. (2007) Hydrodynamic characteristics of a pilot-scale screw conveyor dryer. *Drying Technology*, **25**(4), 609–616.
[161] Seker, M. (2005) Residence time distributions of starch with high moisture content in a single-screw extruder. *Journal of Food Engineering*, **67**(3), 317–324.
[162] Togeby, M., Hansen, N., Mosekilde, E., and Poulsen, K.P. (1986) Modelling energy consumption, loss of firmness and enzyme inactivation in an industrial blanching process. *Journal of Food Engineering*, **5**(4), 251–267.

[163] Owen, P.J. and Cleary, P.W. (2009) Screw conveyor performance: comparison of discrete element modeling with laboratory experiment. Seventh International Conference on CFD in the Minerals and Process Industries, Melbourne, Australia.

[164] Shimizu, Y. and Cundall, P.A. (2001) Three-dimensional DEM simulation of bulk handling by screw conveyors. *Journal of Engineering Mechanics*, **127**(9), 864–872.

[165] Cleary, P.W. (2007) DEM modelling of particulate flow in a screw feeder. *Progress in Computational Fluid Dynamics*, **7**(2–4), 128–138.

[166] Fernandez, J.W., Cleary, P.W., and McBride, W. (2011) Effect of screw design on hopper drawdown of spherical particles in a horizontal screw feeder. *Chemical Engineering Science*, **66**(22), 5585–5601.

[167] Fernandez, J.W., Cleary, P.W., and McBride, W. (2009) Effect of screw design on hopper draw down by a horizontal screw feeder. Seventh International Conference on CFD in the Minerals and Process Industries, Melbourne, Australia.

[168] Jovanović, A., Pezo, L., Stanojlović, S., Kosanić, N., and Lević, L. (2015) Discrete element modelling of screw conveyor-mixers. *Hemijska industrija*, **69**(1), 95–101.

[169] Landry, H., Laguë, C., and Roberge, M. (2006) Discrete element modeling of machine–manure interactions. *Computers and Electronics in Agriculture*, **52**(1–2), 90–106.

[170] Pandey, P., Song, Y., and Turton, R. (2007) Modelling of pan-coating processes for pharmaceutical dosage forms. In *Granulation* (eds A.D. Salman, M.J. Hounslow, and J.P.K. Seville), Elsevier.

[171] Suzzi, D., Radl, S., and Khinast, J.G. (2010) Local analysis of the tablet coating process: impact of operation conditions on film quality. *Chemical Engineering Science*, **65**(21), 5699–5715.

[172] Mann, U. (1983) Analysis of spouted-bed coating and granulation. 1. Batch operation. *Industrial & Engineering Chemistry Process Design and Development*, **22**(2), 288–292.

[173] Freireich, B. and Li, J. (2013) A renewal theory approach to understanding interparticle coating variability. *Powder Technology*, **249**, 330–338.

[174] Kalbag, A., Wassgren, C., Sumana Penumetcha, S., and Pérez-Ramos, J.D. (2008) Inter-tablet coating variability: residence times in a horizontal pan coater. *Chemical Engineering Science*, **63**(11), 2881–2894.

[175] Kalbag, A. and Wassgren, C. (2009) Inter-tablet coating variability: tablet residence time variability. *Chemical Engineering Science*, **64**(11), 2705–2717.

[176] Kumar, R., Freireich, B., and Wassgren, C. (2015) DEM–compartment–population balance model for particle coating in a horizontal rotating drum. *Chemical Engineering Science*, **125**, 144–157.

[177] Pandey, P., Katakdaunde, M., and Turton, R. (2006) Modeling weight variability in a pan coating process using Monte Carlo simulations. *AAPS PharmSciTech*, **7**(4), Article 83.

[178] Kandela, B., Sheorey, U., Banerjee, A., and Bellare, J. (2010) Study of tablet-coating parameters for a pan coater through video imaging and Monte Carlo simulation. *Powder Technology*, **204**(1), 103–112.

[179] Freireich, B. and Wassgren, C. (2010) Intra-particle coating variability: analysis and Monte-Carlo simulations. *Chemical Engineering Science*, **65**, 1117–1124.

[180] Freireich, B., Ketterhagen, W.R., and Wassgren, C. (2011) Intra-tablet coating variability for several pharmaceutical tablet shapes. *Chemical Engineering Science*, **66**, 2535–2544.

[181] Ketterhagen, W.R. (2011) Modeling the motion and orientation of various pharmaceutical tablet shapes in a film coating pan using DEM. *International Journal of Pharmaceutics*, **409**(1–2), 137–149.

Part II

CFD-DEM

6

CFD-DEM Formulation and Coupling

Multiphase flows are encountered in many industrial processes such as fluidized bed combustion, gasification and drying, spouted bed drying, pneumatic conveying, and coating. A detailed knowledge of these flows is crucial for design, scale-up, optimization, de-bottlenecking, and troubleshooting of such processes. This knowledge can be acquired by experimental measurements and mathematical modeling at any desired level. Modeling enables us to understand different phenomena occurring in these flows, to perform sensitivity analysis and case studies on the effect of different operating conditions, and to examine different geometric configurations at lower expenses.

To model fluid–particle flows, Euler–Euler and Euler–Lagrange approaches are used as most common combinations. In the Euler–Euler approach, both fluid and solid phases are considered as continuum phases. This has been successfully utilized to obtain the flow pattern of various non-reacting and reacting flows. The Euler–Lagrange approach considers a combination of computational fluid dynamics (CFD) and the discrete element method (DEM), referred as CFD-DEM, which is the focus of this chapter. Inclusion of constitutive equations relating to fluids and solids and proper coupling strategy in CFD-DEM allows us to study momentum, heat and mass transports, and chemical reactions in almost any detail on the particle scale.

In this chapter, the CFD-DEM formulation and coupling of momentum, energy, and mass between phases are presented. Different coupling strategies and interaction parameters among phases and fluid flow field are also discussed in detail. This is followed by detailing the solid phase flow, interphase coupling, coupling framework, fluid volume fraction, and mapping from the Eulerian to Lagrangian domain. Analytical or empirical fluid–particle interaction forces are presented with details to represent the fluid forces on the surface of particle. The heat and mass transfer formulations with and without chemical reaction for both solid phases as well as coupling strategies are also presented.

Coupled CFD-DEM Modeling: Formulation, Implementation and Application to Multiphase Flows, First Edition.
Hamid Reza Norouzi, Reza Zarghami, Rahmat Sotudeh-Gharebagh and Navid Mostoufi.

Companion website: www.wiley.com/go/norouzi/CFD-DEM

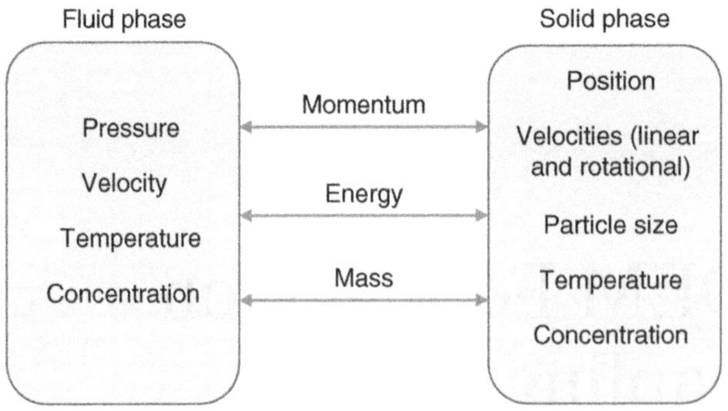

Figure 6.1 Exchange of momentum, energy, and mass between phases

6.1 Multiphase Coupling

Phase coupling is a basic concept in description of any multiphase flows. As shown in Figure 6.1, coupling can occur through exchange of momentum (e.g., interaction forces between phases, momentum addition, or depletion due to mass transfer), energy, and mass (e.g., evaporation and condensation) among phases. In principle, the fluid phase is usually described by defining distributions of pressure, velocity, temperature, and species concentration in the flow field. The dispersed solid phase is also described by size, position (and orientation if particles are non-spherical), linear and rotational velocities, temperature, and species concentration of each particle.

A few fundamental questions should be answered before choosing or developing a model for the fluid–particle flow:

- What types of interphase transfers occur between phases?
- What are the mutual interactions of fluid and solid phases? Is there any intra or interphase interaction to be included in the model?
- What is the effect of particle concentration on the coupling, especially momentum coupling?
- What are the proper length scales for fluid and solid phases to capture all the intended phenomena in the system?

In this section, we try to give some guidelines for answering these questions and choosing the right model.

6.1.1 Coupling Strategies

Interactions between phases in a two-phase flow may be grouped into four phase coupling categories [1, 2] as shown in Figure 6.2 and described next. The fourth category of coupling includes particle collisions while the first three categories include only particle–fluid interactions. It should be noted that these categories are described for momentum coupling next, but can be extended to heat and mass couplings.

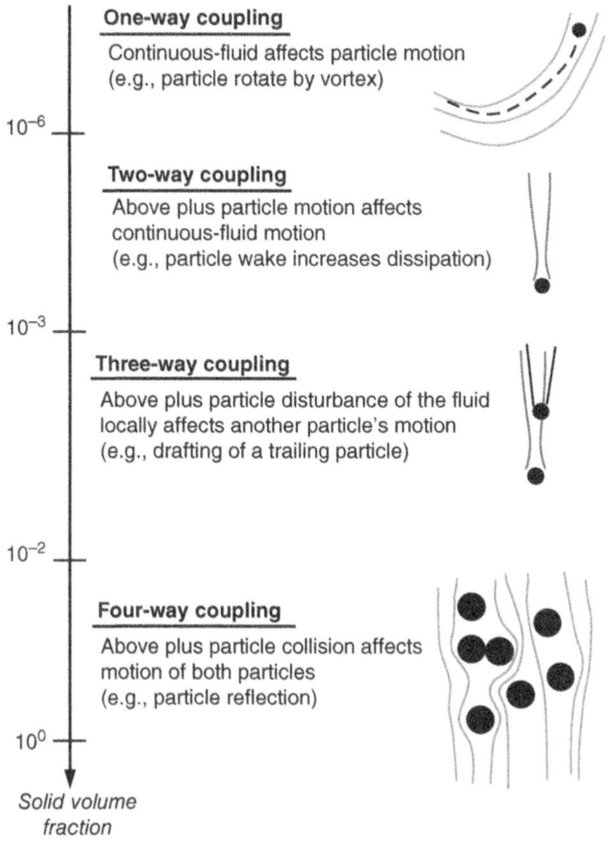

Figure 6.2 Various coupling strategies

6.1.1.1 One-Way Coupling

If the motion of dispersed solid phase is primarily affected by the motion of the continuous fluid phase while motion of particles has a negligible influence on the motion of continuous phase, the flow is one-way coupled. In this situation, equations of the fluid phase can be solved independently from equations of particles whereas equations of particles need some information from equations of the fluid – fluid–particle force – for solution. The dispersed dilute flow of particles is an example of one-way coupling.

6.1.1.2 Two-Way Coupling

If there is a mutual effect between flows of both phases, the flow is said to be two-way coupled. In this case, the motion of particles also affects the motion of fluid phase and vice versa. Particle-induced fluid disturbances (such as wakes behind a particle) in a particle-laden turbulent flow without particle collision is a typical example of two-way coupling.

6.1.1.3 Three-Way Coupling

As shown in Figure 6.2, the continuous phase disturbances (e.g., eddies and wakes) can affect the motion of particles. This flow is three-way coupled. For example, the wakes caused by particles affect the motion of nearby particles. Three-way coupling is usually important when a particle is drafting behind another particle.

6.1.1.4 Four-Way Coupling

Four-way coupling is encountered in situations where, in addition to interactions between solid and fluid phases, particle–particle collisions also influence the overall motion of phases. For example, the turbulent flow in a channel with colliding particles, like that which can be observed in pneumatic conveying [3], is categorized as four-way coupling. Fluidization of particles is another example of four-way coupling. Dense flows generally include four-way coupling where the frequency of particle–particle collisions is high. In some practical applications of granular flow, the fluid–particle interactions are weak and can be neglected. For example, the fluid effects can be neglected in most granular flows discussed in Chapter 5.

While this phenomenological description of phase interactions can be used to classify the type of coupling, it can also be used to determine its appropriate numerical solution. For example, in one-way coupling, governing equations of the continuous phase can be solved independently from the solid phase equations. However, for example, in four-way coupling, the computation sequences should be modified to include evaluation of inter-phase exchanges in equations of both phases. In general, the modeling complexity associated with these coupling categories, to capture all structures within both phases, increases as more effects associated with different time and length scales are included in the model.

6.1.2 Types of Coupling

Not all types of coupling should necessarily be included in the model. Different types of coupling parameters have been defined in the literature for exchange of momentum, energy, and mass between phases [1, 4]. Using these parameters, we can decide which type of coupling should be considered in the model at a given flow condition.

6.1.2.1 Momentum Coupling

Momentum coupling between phases within a control volume can be addressed by comparing the total drag force due to solid phase (F_d) with the convective momentum flux of fluid phase (Mom_f). The momentum coupling number is defined as:

$$\prod_{mom} = \frac{F_d}{Mom_f} \tag{6.1}$$

If $\Pi_{mom} \ll 1$, the momentum coupling could be carried out as one-way coupling. The momentum coupling number can be related to the momentum Stokes number [2]:

$$\Pi_{mom} = \frac{C}{1 + St_{mom}} \tag{6.2}$$

where, the momentum Stokes number is defined as:

$$St_{mom} = \frac{\tau_p}{\tau_f} \tag{6.3}$$

In this equation, τ_f, is the characteristic time of the flow field:

$$\tau_f = \frac{l}{|\vec{u}|} \tag{6.4}$$

where l and $\vec{u}$ are the length scale and the velocity associated with the fluid phase flow, respectively. The particle momentum response time, τ_p, is the time required for a particle to respond a velocity change in the fluid. At very low fluid Reynolds numbers, this momentum response time of particle is [2]:

$$\tau_p = \frac{\rho_i d_i^2}{18\mu_f} \tag{6.5}$$

where μ_f is the fluid viscosity and ρ_i and d_i are the density and diameter of particle i, respectively. This time can be considered as the time required for a particle to start movement from rest to achieve 63% of the free stream velocity [1]. If $St_{mom} \ll 1$, the particles would have enough time to respond to changes in the flow velocity and the flow would be at velocity equilibrium, that is, both phases have nearly the same velocity. However, at $St_{mom} \gg 1$, the particles won't have enough time to respond to changes in fluid velocity and velocity of particles is not affected by changes in the fluid velocity. As the momentum Stokes number approaches zero, particle velocity approaches the fluid velocity. For example, very small particles of low density move with the same velocity as the fluid. On the other hand, as the momentum Stokes number becomes very large, the particle velocity approaches zero. For instance, velocity of a large and high density particle is unaffected by the motion of fluid. The momentum coupling becomes less important for flows with low particles concentrations and large momentum Stokes numbers.

6.1.2.2 Energy Coupling

The energy coupling can be assessed by the energy coupling number through comparing the total rate of heat transfer to/from the solid phase ($\dot{Q}_p$) and the rate of energy transported by the fluid phase within the control volume ($\dot{E}_f$):

$$\Pi_{ener} = \frac{\dot{Q}_p}{\dot{E}_f} \tag{6.6}$$

If $\prod_{ener} \ll 1$, the energy coupling would be one-way. The energy coupling number can be also expressed in terms of energy Stokes number (St_{ener}):

$$\prod_{ener} = \frac{C}{1+St_{ener}} \tag{6.7}$$

The energy Stokes number is defined as:

$$St_{ener} = \frac{\tau_T}{\tau_f} \tag{6.8}$$

where the particle thermal response time, τ_T, is the time required for a particle to achieve 63% of a step change in temperature of the fluid phase. At very low fluid Reynolds numbers, where Nusselt number is about 2, the particle thermal response time is [2]:

$$\tau_T = \frac{\rho_i c_{p,i} d_i^2}{12k_f} \tag{6.9}$$

where k_f is the fluid thermal conductivity and $c_{p,i}$ is the specific heat of a particle. Similar to the momentum coupling number, effect of the energy coupling becomes less important at low particle concentrations and large energy Stokes numbers. An example of such a situation is the movement of large particles with high density and high specific heat in a fluid.

6.1.2.3 Mass Coupling

Exchange of mass between phases within a control volume can be defined by mass coupling number:

$$\prod_{mass} = \frac{\dot{M}_p}{\dot{M}_f} \tag{6.10}$$

where, $\dot{M}_p$ is the rate of mass generated or consumed by the solid phase and $\dot{M}_f$ is the rate of mass of the fluid passing through the control volume. Generation or consumption of mass by the solid phase is caused by adsorption or condensation on the surface of particles, chemical reaction in/on particles, and desorption of material or evaporation from the surface of particles. If $\Pi_{mass} \ll 1$, the effect of mass change of the solid phase on the fluid phase would be insignificant and could be neglected. Therefore, the mass coupling can be performed as the one-way coupling in this condition.

The mass coupling number can be related to the mass exchange rate of the solid phase as [2]:

$$\prod_{mass} = \frac{C}{St_{mass}} \tag{6.11}$$

In this equation, C is the solid phase fraction and the mass Stokes number (St_{mass}) is defined as:

$$St_{mass} = \frac{\tau_m}{\tau_f} \tag{6.12}$$

in which τ_m is the time associated with mass transfer and can be defined as:

$$\tau_m = \frac{m}{\dot{m}} \tag{6.13}$$

where $\dot{m}$ is the mass exchange rate between particle and fluid (e.g., evaporation rate) and m is the mass of each particle. At low solid phase concentration or low mass exchange rate (i.e., large mass Stokes number), mass coupling is insignificant and can be omitted. An example of such a situation is the drying of particles in the dilute phase flow at low ambient temperature where the evaporation rate is low.

6.1.3 Interphase Interactions

When dealing with fluid–particle interphase coupling, the concentration of the solid phase and required resolutions of time and length scales in each phase affects the choice of the model to be used for describing the motion of phases and interphase coupling. In the following, we address these effects and introduce different modeling approaches for fluid–particle interphase coupling.

6.1.3.1 Effect of Particle Concentration

Tsuji [5] classified the gas-particle flow based on the extent of particle–particle interaction and developed a multi-scale modeling approach that covers gas–particle flows from dilute to dense phase (see Figure 6.3). In this approach, the gas–particle flow is classified into following cases based on particle–particle interaction:

- *Collision free flow* corresponds to the dilute phase flow in which the particle–particle interaction can be neglected. Examples for this type of flow are dust collection, dust flow in the room, and dilute phase pneumatic transport of particles.
- *Collision dominated flow* is encountered when particles collide with each other while still dispersed. This type of flow is observed in particle concentrations greater than that in the collision free flow. Examples of collision dominated flow are the upper region of circulating fluidized beds and pneumatic transport of particles at intermediate concentrations.
- *Contact dominated flow* can be observed at even higher particle concentrations in which particles keep on their contact and are not dispersed. While in collision free and collision dominated flows existence of a fluid is required for establishing the motion of particles, in the contact dominated flow the effect of motion of fluid on the dynamics of particles can be neglected in some circumstances. Thus, the contact dominated flow can be divided into two categories, with and without fluid effect. Many unit operations for powder handling are examples of contact dominated flows without the fluid effect. Examples of this type of flow are hoppers, particle mixers, rotating drums, screw feeders, and granulators. Fluidized bed and dense phase transport, on the other hand, are examples of contact dominated flow with the fluid effect.

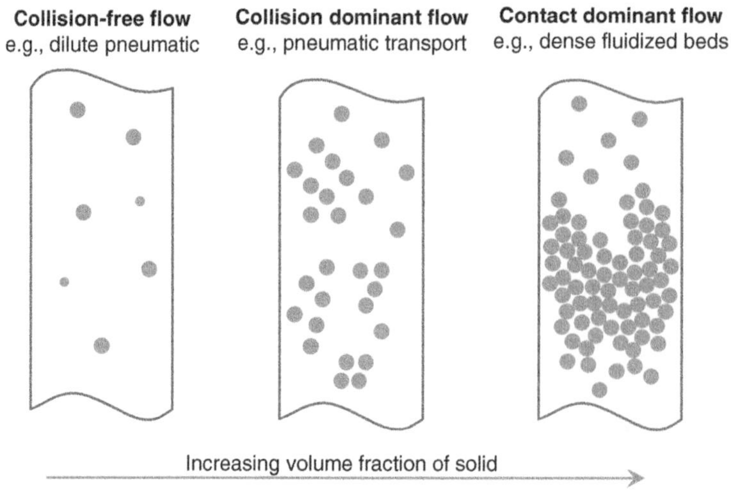

Figure 6.3 Various flow conditions based on the particle–particle interaction

The fluid–particle interaction coupling depends on the volume fraction of particles. The flow can be classified into dilute flow and dense flow based on the particle volume fraction. The difference between dilute and dense flows can be distinguished by the ratio of particle momentum response time (τ_p) to average time between particle collisions (τ_c). The flow is considered dilute if $\tau_p/\tau_c < 1$ and it is considered dense if this ratio is greater than 1. In the dilute flow, the overall motion of particles is affected by interactions between particle and fluid while in the dense flow, the particle–particle interaction is dominant. Particle–particle interaction refers to two separate phenomena: particle–particle collision and particle–particle fluid dynamic interaction (where the closeness of particles can affect the dynamic forces on the particle). Fluid-particle flow coupling can be classified based on the type of particle–particle interaction. Isolated particles in an unrestricted flow can be considered as the collision free flow in which particles have a negligible effect on the motion of fluid while for a slightly higher particles concentration, the interactions between the two phases become so complicated that the influence of the motion of particles begins to affect the motion of the fluid phase. Further increase in the volume fraction of particles can lead to increased particle–particle interactions. In this condition, the particle motion is controlled by fluid forces and particles have a sufficient time to respond to local fluid forces before the next collision. At even greater volume fractions, up to a value close to the volume fraction of the packed bed, the motion of particles is controlled by collisions and particles do not have enough time to respond to changes in the local fluid forces before the next collision.

6.1.3.2 Multi-Length Scale and Multi-Time Scale Interactions

Based on different time and length scales of the natural phenomena in either solid or fluid phase, different numerical models for solving the governing equations of these flows and corresponding coupling approaches can be developed [6]. As mentioned earlier, the DEM method is based on the estimation of motion of each individual particle according to the

Newton's equation of motion. Depending on the required level of time and length scales (in either laminar or turbulent flow), different approaches from discrete Lagrangian (e.g., molecular dynamics simulation (MDS), lattice–Boltzmann (LB), pseudo-particle modelling (PPM), smoothed particle hydrodynamics (SPH)) to continuum conventional Eulerian CFD techniques (e.g., Reynolds averaged Navier–Stokes (RANS), large eddy simulation (LES), direct numerical simulation (DNS)) have been developed for modeling the flow of fluids. The MDS is a computer based simulation method for modeling the physical movements of atoms and molecules which are allowed to interact. In the LBM, the discrete Boltzmann equation is solved instead of directly solving Navier–Stokes (NS) equations for modeling the flow of a Newtonian fluid with proper collision models. The PPM resolves the fluid into the pseudo particle s (PPs), which are much smaller than solid particles. The SPH is a mesh free Lagrangian method that works by dividing the fluid into a set of discrete elements [7].

In principle, all fluid phase modeling approaches can be combined with the previously mentioned solid phase models to describe the particle-fluid coupling approach. As summarized by Zhu *et al.* [7], many of these coupling approaches have been applied by researchers. Examples of these combinations are TFM, CFD-DEM, DNS-DEM [8, 9], and LES-DEM [10–13] as well as non-continuum discrete fluid flow-DEM coupling approaches such as LB-DEM [14, 15], PPM-DEM [16, 17], SPH-DEM [18], and FT-DEM (front tracking–discrete-element method) [19, 20].

In addition to conventional DEM models that use a deterministic approach for treatment of particles collisions (either hard-sphere collision model for low volume fraction or soft-sphere collision model for high volume fraction), other discrete particle models exist, such as direct simulation Monte Carlo (DSMC) [21, 22] that treat the collisions statistically. The DSMC uses the probabilistic Monte-Carlo simulation to solve the Boltzmann equation of the fluid flow at a finite Knudsen number.

6.2 Momentum Coupling

In the CFD-DEM, the motion of fluid phase is described by the volume averaged NS equation, while the translational and rotational motions of discrete particles are described by Newton's and Euler's second laws of motion, respectively. Forces acting on particles are gravitational force, contact forces between colliding particles, fluid–particle interaction forces as well as interparticle forces such as van der Waals and liquid and solid bridge forces. Depending on the particle operating condition and volume fraction of particles in the flow, one of the momentum coupling strategies can be used. We start it with a very simple case of one-way coupling and extend it to complex four-way coupling.

6.2.1 Single Phase Flow of Fluids

Let us consider a single phase flow and ignore the effect of solid phase on the flow of continuous phase (i.e., one-way coupling, for simplicity). The equations of change (continuity and momentum) in this flow can be written as [23]:

$$\frac{\partial \rho_f}{\partial t} + \nabla.\left(\rho_f \vec{u}\right) = 0 \tag{6.14a}$$

$$\frac{\partial(\rho_f\vec{u})}{\partial t}+\nabla.(\rho_f\vec{u}\vec{u})=-\nabla.\vec{\pi}_f+\rho_f\vec{g}=-\nabla p-\nabla.\vec{\tau}_f+\rho_f\vec{g} \tag{6.14b}$$

where $\vec{\pi}_f = p\vec{\delta}+\vec{\tau}_f$ is the tensor of molecular momentum flux of fluid (usually referred to as the fluid stress tensor), ∇ is differential operator vector, p is pressure, $\vec{u}$ is the velocity, $\vec{g}$ the vector of gravity acceleration, and $\vec{\tau}_f$ is the viscous stress tensor that can be determined from its generalized Newtonian law of viscosity:

$$\vec{\tau}_f=-\mu_f\left(\nabla\vec{u}+(\nabla\vec{u})'\right)+\left(\frac{2}{3}\mu_f-k\right)(\nabla.\vec{u})\vec{\delta} \tag{6.15}$$

here, $\vec{\delta}$ is the unit tensor with components δ_{ij}[1], μ_f, and k are dynamic and the dilatational viscosities of the fluid. It should be noted that for an ideal monoatomic gas (extended to ideal gases), k is zero and for an incompressible fluid, $\nabla.\vec{u}=0$. Therefore, knowing the value of k is not necessary in many applications. The convective momentum flux tensor, $\nabla.(\rho_f\vec{u}\vec{u})$, exhibits hyperbolic behavior of the transport partial differential equation (PDE) and the viscous diffusion term, $\vec{\tau}_f=-\mu_f(\nabla\vec{u}+(\nabla\vec{u})')$, yields its elliptic character.

By substituting the Newtonian expression for $\vec{\tau}_f$ from Equation 6.15 into Equation 6.14b and considering an incompressible, Newtonian fluid (constant ρ_f and μ_f), the following equation, called the Navier–Stokes equation, is obtained:

$$\rho_f\frac{\partial\vec{u}}{\partial t}+\rho_f\nabla.(\vec{u}\vec{u})=-\nabla p+\mu_f\nabla^2\vec{u}+\rho_f\vec{g} \tag{6.16}$$

With increasing the Reynolds number (i.e., increasing the inertia force compared to the viscous force) the flow regime changes from creeping (where the momentum convection terms can be neglected), to laminar, to transitional, and finally to turbulent. Numerical treatment of NS equation can change if the flow regime is changed. The Reynolds number in a macroscopic scale is defined as:

$$\mathrm{Re}_l=\frac{\rho_f|\vec{u}|l}{\mu_f} \tag{6.17}$$

6.2.1.1 Laminar Flow

At very low Reynolds numbers ($Re_l \ll 1$), the inertia (convection) force is small compared with the viscous force and the momentum convection terms can be neglected. In this situation, the flow is highly laminar and is called creeping flow or Stokes flow. The NS equation at steady state reduces to the following in this case [23]:

$$0=-\nabla p-\nabla.\vec{\tau}_f+\rho_f\vec{g} \tag{6.18}$$

---

[1] The value of the Kronecker delta is 1 if $i=j$ and zero if $i\neq j$.

The creeping flow regime yields a strong elliptic PDE character and should be solved by an implicit numerical method. At unsteady conditions, however, a parabolic approach may be employed.

At higher Reynolds numbers, the flow is still laminar. The convection terms in the NS equation are important in this case and cannot be neglected. Therefore, this equation can be presented as:

$$\rho_f \nabla.(\vec{u}\vec{u}) = -\nabla p - \nabla.\vec{\tau}_f + \rho_f \vec{g} \tag{6.19}$$

The elliptic characteristic of this PDE remains strong and a direct or iterative numerical approach is required to solve this equation. This flow persists until $Re_l < Re_{crit}$, where Re_{crit} demarks the onset of transition flow.

In an unsteady flow, the domain discretization in space and time should be considered and a parabolic approach may be employed in the direction that the viscous diffusion can be neglected. However, at higher Reynolds numbers, where the convective terms begin to become dominant and the elliptic effect of the viscous terms can be neglected compared with the hyperbolic character of the convective term, an explicit scheme is needed for solving the NS equation. Basically, a hybrid implicit-explicit scheme is needed to capture complex features of the incompressible laminar flow.

6.2.1.2 Transitional Flow

For transitional flow, where the intermittent flow behavior yields a difficult flow field to simulate, the flow instabilities become important [1]. This flow regime exists in the range of $Re_{crit} < Re_l < Re_{turb}$, where Re_{turb} is the Reynolds number at onset of fully turbulent regime. Continuity and NS equations are valid for description of the transitional flow. While transitional flow may be initially two dimensional in a macroscopic scale; the instabilities are typically significant in small scales with 3D features (e.g., vortex).

6.2.1.3 Turbulent Flow

While the laminar flow is orderly, the flow is significantly nonlinear and instable in the turbulent regime where $Re_l > Re_{turb}$. Turbulence instability occurs when the time scale of the viscous damping of velocity fluctuations is much larger than the time scale of the convection. This happens at high Reynolds numbers where the nonlinear convective term becomes more significant than the diffusion term in the NS equation. In other words, tendency of the instability to become damped by viscosity decreases with increasing the Reynolds number [24]. Despite initial impression about irregular and stochastic behavior of the turbulent flow, turbulent fluctuations have coherent regular structures, callededdies, with a wide range of length and time scales which highly persist on their dynamic features. In addition, eddies can stretch, rotate, and break up into smaller eddies or coalesce into larger ones. The turbulence consists of three-dimensional eddies with wide ranges of size and shape. These eddies contain the turbulence energy and change over the time. They are bounded by the largest length scale (with the order of flow domain scale, l) and the smallest Kolmogorov length scale (λ_K), which can be related to the turbulent kinetic energy dissipation rate (ε) as [24]:

$$\lambda_K = \left(\nu_f^3 / \varepsilon\right)^{1/4} \tag{6.20}$$

in which ν_f is the fluid kinematic viscosity. The intermediate length scales between these two bounded scales, also referred to as Taylor micro scales, form the inertial sub-range. While the small scale turbulence is isotropic and homogeneous at sufficiently large Reynolds numbers, most of the flow domain scale structures (e.g., sufficiently large eddies) are anisotropic and have highly distorted dimensions. The range of required spatial resolution to capture the whole flow structures is based on the ratio of these two length scales, which increase with Reynolds number as [24]:

$$l/\lambda_K \propto \mathrm{Re}_l^{3/4} \tag{6.21}$$

In addition, in all turbulent flows there is a flux of energy (usually referred to as the energy cascade) from the largest turbulent scales (eddies obtaining energy from the main flow and from each other) to the smallest Kolmogorov length scales (that form the viscous sub-layer, where the turbulent kinetic energy is dissipated into heat due to viscous stress). The inertial sub-range scales transfer the energy only from the largest to the smallest eddies without any dissipation.

Since the smallest turbulent eddies are several orders of magnitude larger than the mean free path of molecules of the fluid, the turbulence can be considered a continuum phenomenon and equations of change can still be applicable to this regime. Although studying the details of turbulent coherent structures is possible through numerical solution of these nonlinear equations, such computationally expensive detailed information in some cases is not required. In practical engineering applications, time averaged PDEs can be used for simulation of turbulent flows. Therefore, two different categories of models, time-smoothed simulation (no structure resolved) and eddy-resolved simulation (some of or all structures resolved), have been developed for modeling of turbulence. Passages of large and small eddies through a measurement point induce velocity and pressure fluctuations. This kind of velocity fluctuation and the fluctuation predicted by some techniques of turbulent flow simulation are schematically represented in Figure 6.4. It is worth noting that in RANS an integral length scale is considered, in LES larger eddies are considered and in DNS all eddies are considered in the model. It should be noted that there is no single source for the fluctuations and they are the

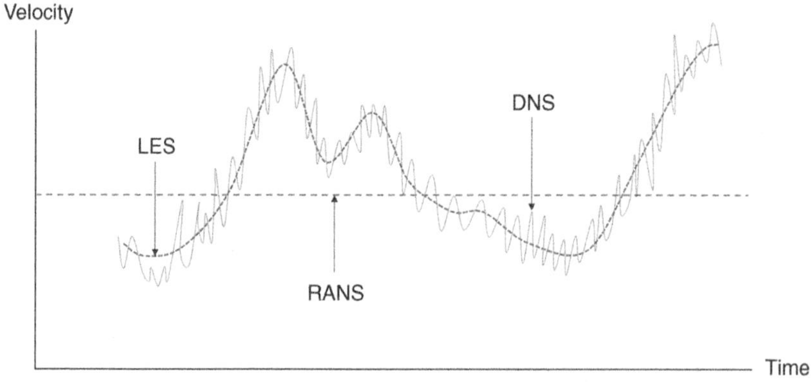

Figure 6.4 Velocity fluctuations and different resolved turbulence scales in a steady turbulent flow. While RANS resolves integral scales, LES resolves only the largest scales, and DNS resolves all scales

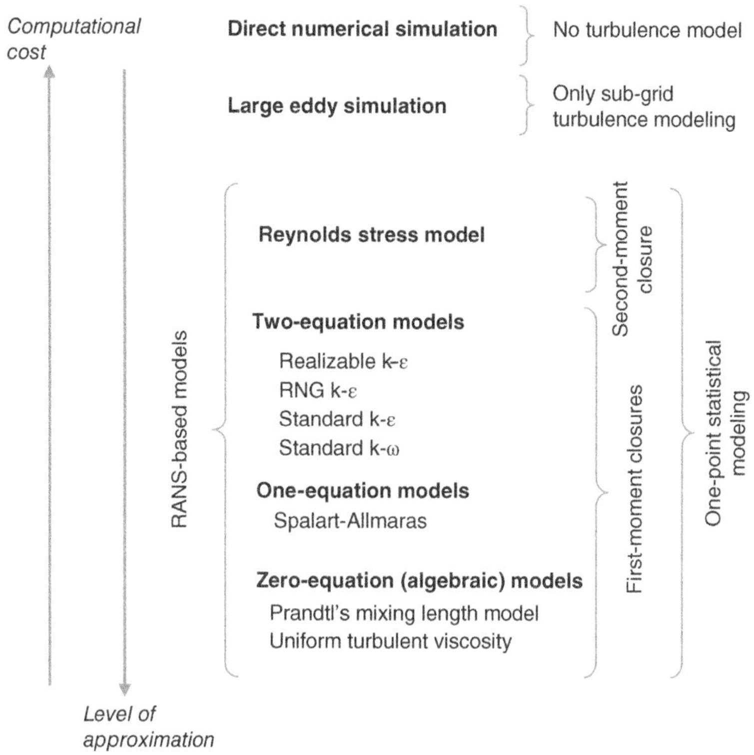

Figure 6.5 An overview of the different turbulent models. (Source: Andersson, B., Andersson, R., Hakansson, L., Mortensen, M., Sudiyo, R., Wachem, B.v., 2012. *Computational Fluid Dynamics for Engineers*. Cambridge University Press, with permission)

result of a number of phenomena within the turbulent flow, such as passage, coalescence, and breakage of eddies. Large eddies induce fluctuations with larger amplitudes and lower frequencies; whereas small eddies induce small amplitude fluctuations at higher frequencies.

Figure 6.5 shows a general view of different turbulence models which will be discussed in the following sections. As addressed in this figure, there is a trade-off between accuracy of the model (time and length resolution) and its computational cost [24].

Time-Smoothed Models

Time-smoothed models (sometimes referred to as statistical models) focus on the average flow and effects of turbulence on averaged properties of the flow. Therefore, only a mean turbulence length and time scales can be estimated and these models cannot predict structure or dynamics of eddies. In these approaches, all flow variables are individually decomposed into their smoothed and fluctuating components. For example:

$$\vec{u} = \overline{\vec{u}} + \vec{u}' \tag{6.22a}$$

$$p = \overline{p} + p' \tag{6.22b}$$

where $(\bar{\ })$ is time-smoothed value and $(.')$ is the fluctuating part. This is sometimes called the Reynolds decomposition. The time-smoothed equation of motion for a fluid with constant density and viscosity then can be derived from the NS equations by changing all $\vec{u}$ to $\bar{\vec{u}}$, p to $\bar{p}$, and $\vec{\tau}_f$ to $\bar{\vec{\tau}}_f$:

$$\nabla.\left(\rho_f \bar{\vec{u}}\right) = 0 \tag{6.23a}$$

$$\rho_f \nabla.\left(\bar{\vec{u}}\bar{\vec{u}}\right) = -\nabla\bar{p} - \nabla.\bar{\vec{\tau}}_f + \rho_f \vec{g} \tag{6.23b}$$

The shear stress tensor consists of viscous and turbulent momentum flux tensors:

$$\bar{\vec{\tau}}_f = \bar{\vec{\tau}}_f^{v} + \bar{\vec{\tau}}_f^{t} \tag{6.24}$$

where

$$\bar{\vec{\tau}}_f^{v} = -\mu_f \left(\nabla\bar{\vec{u}} + \left(\nabla\bar{\vec{u}}\right)' \right) \tag{6.25a}$$

$$\bar{\vec{\tau}}_f^{t} = \rho_f \overline{\vec{u}'\vec{u}'} \tag{6.25b}$$

The components of turbulent momentum flux are commonly referred to as Reynolds stresses. An equation analogous to the Newton's law of viscosity can be adopted for expressing the Reynolds stress tensor:

$$\bar{\vec{\tau}}_f^{t} = -\mu_f^t \nabla\bar{\vec{u}} \tag{6.26}$$

where μ_f^t is the turbulent viscosity, also called eddy viscosity. Generally, the Reynolds stresses can be estimated from experimental data, empirical correlations of turbulent viscosity, or solving equations of change for the Reynolds stresses. Therefore, as shown in Figure 6.5, the time-smoothed turbulence models are generally classified into the traditional turbulent viscosity models (first-moment closures or first-order theory) and the more advanced Reynolds-stress models (second-moment closure or second-order theory) [23].

The turbulent viscosity models are usually classified based on the required number of partial differential transport equations that should be solved to estimate μ_f^t, in addition to those for the RANS equations. These models generally ranged from zero-equation models (e.g., the Prandtl mixing length model), one-equation models (e.g., Spalart–Allmaras), and two-equation models (e.g., k-ε). As mentioned earlier, the Reynolds-stress models employ individual equations of change for the Reynolds stresses. However, equations of change for the Reynolds stresses contain quantities like $\overline{\vec{u}'\vec{u}'\vec{u}'}$, which should be again estimated based on experimental information [23].

Eddy-Resolved Models

Eddy-resolved models estimate those features of eddy structures that can be resolved within the computational grid. These models include DNS, in which the turbulence is attributed to eddies with all wave numbers, and LES, where the turbulence is only resolved for larger eddies.

Direct numerical simulation: This turbulent model computes the mean flow as well as all turbulent velocity fluctuations directly from the NS equation. In this method, the 3D unsteady NS Equation 6.16 is solved on a small enough spatial grids to resolve the Kolmogorov length scale and with a time step sufficiently small to resolve the fastest fluctuations. The computational cost of the DNS is high, thus, it is not suitable for industrial applications. Moreover, the hybrid RANS–LES technique has been used to treat a part of the computational domain by the RANS model and in the other part with the LES approach. Other eddy resolved techniques, which can predict specific features of the turbulence, include detached eddy simulation s (DES s) and proper orthogonal decomposition (POD).

Large eddy simulation: This is a form of eddy resolved turbulent model that models the behavior of large eddies. It applies a low-pass space filter on the unsteady NS equation with a specific cutoff frequency to keep large eddies and eliminate small eddies. With this filtering, all velocities are decomposed into their resolved (unfiltered) and unresolved (filtered) components:

$$\vec{u} = \tilde{\vec{u}} + \vec{u}' \tag{6.27}$$

where (˜) is the filtered quantity from which the effects of small eddies are removed. The spatial-filtered equation of motion for a fluid with constant density and viscosity can be derived from the NS equation by changing all $\vec{u}$ to $\tilde{\vec{u}}$, p to $\tilde{p}$, and $\vec{\tau}_f$ to $\tilde{\vec{\tau}}_f$:

$$\frac{\partial \rho_f}{\partial t} + \nabla.\left(\rho_f \tilde{\vec{u}}\right) = 0 \tag{6.28a}$$

$$\rho_f \frac{\partial \tilde{\vec{u}}}{\partial t} + \rho_f \nabla.\left(\tilde{\vec{u}}\,\tilde{\vec{u}}\right) = -\nabla\tilde{p} - \nabla.\tilde{\vec{\tau}}_f + \rho_f \vec{g} \tag{6.28b}$$

Here, the shear stress tensor is:

$$\tilde{\vec{\tau}}_f = \tilde{\vec{\tau}}_f^{\,v} + \tilde{\vec{\tau}}_f^{\,t} \tag{6.29}$$

where

$$\tilde{\vec{\tau}}_f^{\,v} = -\mu_f\left(\nabla\tilde{\vec{u}} + \left(\nabla\tilde{\vec{u}}\right)'\right) \tag{6.30a}$$

$$\tilde{\vec{\tau}}_f^{\,t} = \rho_f \widetilde{\vec{u}'\vec{u}'} \tag{6.30b}$$

The value $\widetilde{\vec{u}'\vec{u}'}$ represents the velocity tensor of unresolved fluctuations. The main advantage of the LES method is that anisotropic distorted large eddies, which cannot be modeled easily, are directly simulated and the effect of unresolved scales, which cannot be neglected, are included in the model by means of $\widetilde{\vec{u}'\vec{u}'}$. In the LES model, the turbulence scales are usually resolved by filtering up to the inertial sub-range.

6.2.2 Fluid Resolution in CFD-DEM

In the CFD-DEM, the solid phase is fully resolved and each particle is tracked in the system. The size of particle may be in the range of a few microns to a few centimeters. Therefore, the length scale of discrete solid particles may change by 3 orders of magnitude. This raises some challenges regarding the type of particle–particle and particle–fluid interactions. For example, when dealing with very small particles, interparticle forces may become dominant and should be included in the equation of motion of particles (see Chapter 8). However, apart from such challenges, the coupling of particles (DEM) with the fluid phase (CFD) strongly depends on the desired resolution in the fluid phase. This length scale is defined relative to the particle size. Let's consider two extreme cases: size of the fluid cells is much smaller or much larger than the particle size. The former is known as the resolved surface CFD-DEM and the latter as the unresolved surface CFD-DEM.

Figure 6.6 schematically shows the difference between resolved and unresolved surface treatments. In the CFD, fluid properties are defined in the fluid cells of size Δx, which cover the whole simulation domain. Therefore, fluid properties are defined at a length scale as large as the size of fluid cells. Since the size of fluid cell is much smaller than the size of particle, fluid flow is fully resolved over the entire surface of particles and detailed profiles of fluid fields are obtained around all particles. These profiles are then integrated to obtain the total hydrodynamic forces and forces acting on individual particles. Since the computational cost of solving the fluid phase equations over the entire grid cells is high, this treatment is only feasible for systems containing no more than a few hundred particles.

In many practical applications, we are interested in studying the dynamics of fluid-particle flow in the length scales much larger than the particle size. This makes the use of resolved surface treatment unfeasible. In the unresolved surface treatment, the size of fluid cell is larger than the size of particle. This makes studying of systems containing thousands or even millions of particles possible. The coupling between solid phase at particle scale and fluid phase at fluid cell scale is done via drag force and with empirical or theoretical correlations developed for the flow across a single particle or an assembly of particles. Effect of the presence of particles on changes of flow profile inside the fluid cell is neglected. The unresolved surface

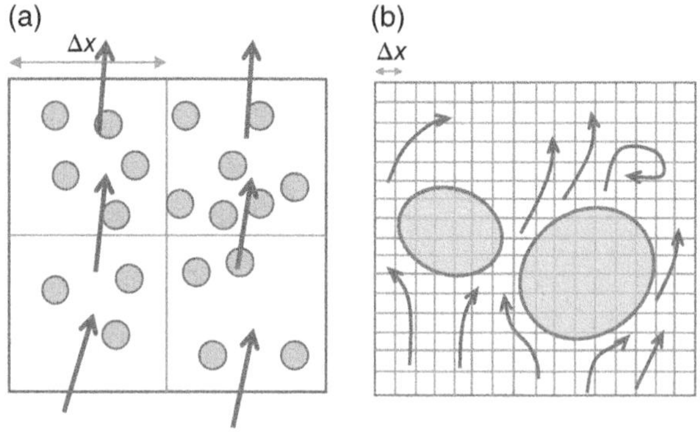

Figure 6.6 Different representations of the particle treatment: (a) unresolved surface method ($\Delta x > d_i$) and (b) resolved surface method ($\Delta x \ll d_i$). Arrows show the local fluid velocity vectors

CFD-DEM will be discussed in the following section for which we need to consider the effect of motion of fluid on the motion of particles and vice versa.

In some conditions we may encounter a situation in which the size of fluid cell is comparable to the size of particles for which neither unresolved nor resolved treatment can be utilized. In this case we need a resolution in the fluid flow higher than that in the unresolved treatment. As a result, a fluid cell may be covered by a particle with a few cells touching surface of the particle or not enough number of particles can reside in each cells. Therefore, the averaged Navier–Stokes equation cannot be used due to sharp changes in the fluid porosity in the adjacent cells. Some solutions to this condition have been proposed. For example, Link *et al.* [25] proposed the porous cubes method to prevent such problems. In this method, each particle is presented by a porous cube with the diameter of $d_{c,i} = \alpha d_i$, where α is the ratio of the cube side to particle diameter. In contrast to real particles, porous cubes are allowed to overlap. Volume of the cube should be larger than or equal to the volume of the particle:

$$\alpha \geq \left(\frac{\pi}{6}\right)^{1/3} \tag{6.31}$$

6.2.3 *Unresolved Surface CFD-DEM*

As mentioned before, unresolved surface CFD-DEM models consider a fluid cell larger than the particle size. Hence, it is always possible for a cell to contain a number of particles. Therefore, the volume averaged Navier–Stokes equation can be used for the fluid phase through the local averaging of solid phase properties. The averaged properties of the solid phase are fluid–particle interaction force and solid phase volume fraction. In the following sections, we describe the method of calculating these properties as well as the coupling scheme and different model formulations of the unresolved CFD-DEM model.

The coupling scheme of the CFD-DEM model is quite different from that of the TFM. In the TFM, the information of solid and fluid phases is available at the length-scale of fluid cell, thus, the coupling is straightforward. In the CFD-DEM model, however, the solid phase information is available at the particle scale while the fluid phase information is available at the fluid cell scale. Therefore, a special attention is required when particle scale information is used to calculate a fluid cell property (e.g., porosity) and the fluid cell information is used to calculate the particle scale property (e.g., Reynolds number of particle).

Two sets of formulations, referred to as model A and model B, have been proposed for describing the governing equations of fluid and solid phases [26]. These models have been used by various researchers with different coupling scheme in both TFM and CFD-DEM. If the pressure is shared between fluid and solid phases, it is referred to as model A and if the pressure is considered for the fluid phase alone, it is referred to as model B. Bouillard *et al.* [27] demonstrated that there is little difference between models A and B in TFM simulations for monosized particles, which is also the case in CFD-DEM simulations [28, 29]. However, Feng and Yu [30] compared the results of models A and B with experimental data and concluded that there is a significant difference between these two formulations in terms of solid flow pattern and mixing kinetics. Di Renzo and Di Maio [31] compared these two models and showed that the model B is valid at the minimum fluidization velocity. Their results were different from those reported by other researchers [28, 32]. Recently, Zhou *et al.* [33] discussed the origins of the difference between these formulations in both TFM and CFD-DEM and compared three sets of formulations (called sets I, II, and III) with the conventional models A and B. They found that model A and B are analogous to sets II and III, respectively. In addition, they showed that set III (model B) is a simplified form of set I that is in fact the original model. In the following we describe the CFD-DEM model and its coupling scheme following the work of Zhou *et al.* [33].

6.2.3.1 Equations of Motion for Solid Phase

Based on the soft-sphere model described in Chapter 2, a particle may have two types of translational and rotational motions. The equations of motion for individual spherical particle i with the mass m_i and the moment of inertia I_i can be written as:

$$m_i \frac{d\vec{v}_i}{dt} = \vec{f}_i^{f-p} + \sum_{j \in CL_i} \left(\vec{f}_{ij}^{c} + \vec{f}_{ij}^{nc} \right) + \vec{f}_i^{g} \tag{6.32}$$

$$I_i \frac{d\vec{\omega}_i}{dt} = \sum_{j \in CL_i} \vec{M}_{ij}^{c} + \vec{M}_i^{f-p} \tag{6.33}$$

where $\vec{v}_i$ and $\vec{\omega}_i$ are the translational and angular velocities of particle i, respectively, $\vec{f}_{ij}^{c}$ and $\vec{M}_{ij}^{c}$ are the contact force and torque (tangential and rolling torques) acting on particle i by particle j or wall, $\vec{f}_{ij}^{nc}$ is the noncontact force acting on particle i by particle j or other sources, k is the number of particles or wall in interaction with the particle i, and $\vec{f}_i^{g} = m_i \vec{g}$ is the gravitational body force. The interaction force between fluid and particle is $\vec{f}_i^{f-p}$. There is also another interaction between fluid and particle which opposes the rotation of particle in the fluid, known as fluid rotational resistance torque, $\vec{M}_i^{f-p}$. It includes the rotational drag on the particle ($\vec{M}_i^{d}$) and an unsteady Basset term in rectilinear acceleration (which is not addressed here).

The fluid-particle interaction force represents the fluid forces on the surface of particle and can be treated with two fundamentally different methods, that is, resolved-surface and unresolved-surface. Knowing velocity and pressure fields around particle i (which is the case in the resolved-surface method), the fluid-particle interaction force can be directly evaluated by integrating the fluid stress tensor (both pressure and shear) over the surface of particle. However, in the unresolved surface CFD-DEM, where the size of the fluid cell is larger than the size of the particle, the fluid–particle interaction force cannot be integrated over the particle surface and an averaged force is employed instead. This interaction force is obtained from either analytical or empirical expressions of the fluid properties in the vicinity of the particle. In the resolved-surface approach, no specific decomposition of the surface force effects is needed, whereas the unresolved-surface method can be represented as a linear decomposition of specific independent forces (e.g., quasi steady drag force, unsteady drag (Basset or history) force, lift (Saffman and Magnus) force, added mass force, etc.). It should be mentioned that various forces in the resolved surface method have a nonlinear interaction. However, this linear combination of forces in the unresolved-surface method is assumed to be valid.

6.2.3.2 Fluid–Particle Interaction Force Decomposition

According to the resolved surface method [23], the fluid–particle interaction force is obtained by:

$$\vec{f}_i^{f-p} = \oint_{A_i} -\vec{\pi}_f.\vec{n}ds = \oint_{A_i} -\left(p\vec{\delta} + \vec{\tau}_f\right).\vec{n}ds \tag{6.34}$$

where A_i is the particle surface area and $\vec{n}$ is the normal unit vector outward from the control surface of particle (ds).

The fluid–particle interaction force in the unresolved surface method can be obtained from [2]:

$$\vec{f}_i^{f-p} = \vec{f}_i^{d} + \vec{f}_i^{u} + \vec{f}_i^{\nabla p} + \vec{f}_i^{\nabla.\vec{\tau}} + \vec{f}_i^{l} \tag{6.35}$$

where $\vec{f}_i^{d}$ is the steady drag force. The unsteady force includes unsteady drag (history Basset) force ($\vec{f}_i^{ud}$) and added mass force ($\vec{f}_i^{a}$):

$$\vec{f}_i^{u} = \vec{f}_i^{ud} + \vec{f}_i^{a} \tag{6.36}$$

The pressure gradient force is obtained from:

$$\vec{f}_i^{\nabla p} = -V_i \nabla p \tag{6.37}$$

where V_i is the volume of particle i. The viscous force due to the fluid shear stress or deviatoric stress tensor can be calculated from:

$$\vec{f}_i^{\nabla.\vec{\tau}} = -V_i\left(\nabla.\vec{\tau}_f\right) \tag{6.38}$$

The lift force includes Saffman and Magnus forces:

$$\vec{f}_i^{l} = \vec{f}_i^{Saffman} + \vec{f}_i^{Magnus} \tag{6.39}$$

As we mentioned when introducing the unresolved-surface method, the fluid–particle interaction force acting on the particle, $\vec{f}_i^{f-p}$, is estimated at the particle scale while the analogous term in the fluid phase momentum equation, which is the volumetric fluid–particle interaction force acting on the fluid, should be calculated in each fluid cell. This difference in the length scale of fluid interaction forces between fluid and solid phases as resulted in development of different schemes, as reviewed by Feng and Yu [30]:

Scheme 1: The average force acting on the fluid phase due to the motion of particles is estimated by the average value of variables in each fluid cell (e.g., average gas and particle velocities and average particle diameter) like that we see in the TFM, while the force acing on a particle from the fluid phase is determined based on the velocity of each particle [34–42].

Scheme 2: The force acting on the fluid phase due to the motion of particles is first evaluated as in Scheme 1, then it is distributed among individual particles based on a certain averaging method (e.g., surface area or volume of particles) [43–50].

Scheme 3: The fluid–particle interaction force acting on individual particles is first estimated in a fluid cell, then this force is summed on all particles to estimate the fluid–particle interaction force in the fluid cell [32, 51–64].

As summarized by Zhu *et al.* [7] and Zhou *et al.* [33], Scheme 1 does not satisfy Newton's third law of motion. In fact, the force of the fluid phase acting on the solid phase should be equal to the force of the solid phase acting on the fluid phase in the opposite direction. Although Newton's third law of motion is satisfied in Scheme 2, it has a drawback regarding uniformly distributing the interaction force among particles in a computational cell. This is not correct in a real case where the fluid–particle interaction force acting on particles in the cell is different to non-uniform fluid-particle flow. In addition, Scheme 2 needs an extra calculation of the mean particle velocity for evaluating the fluid–particle interaction forces, which may not be correct, especially in polydipersed systems. Scheme 3, however, does not have the problems mentioned for Schemes 1 and 2. Therefore, calculation of the volumetric fluid interaction force term will be explained by this scheme in the following sections.

6.2.3.3 Formulation of Governing Equations

Different model formulations for the motion of phases in a fluid–particle system were first developed using the continuum–continuum approach. Therefore, we start our discussion with the TFM equations. The following volume-averaged momentum conservation equations have been derived by Anderson and Jackson [65] for fluid and solid phases in each fluid cell:

$$\frac{\partial\left(\rho_f\varepsilon_f\vec{u}\right)}{\partial t}+\nabla.\left(\rho_f\varepsilon_f\vec{u}\vec{u}\right)=-\nabla.\vec{\pi}_f-\vec{F}+\rho_f\varepsilon_f\vec{g} \tag{6.40a}$$

$$\frac{\partial\left(\rho_p\varepsilon_p\vec{v}\right)}{\partial t}+\nabla.\left(\rho_p\varepsilon_p\vec{v}\vec{v}\right)=-\nabla.\vec{\pi}_p+\vec{F}+\rho_p\varepsilon_p\vec{g} \tag{6.40b}$$

where $\vec{v}$ and ρ_p are the average velocity and density of the solid phase, respectively. Parameters ε_p and ε_f are solid and fluid phase volumetric fractions, respectively. $\vec{\pi}_p$ is the solid phase stress tensor (defined with the same analogy as the fluid stress tensor). $\vec{F}$ is the volumetric mean value of forces acting on particles by surrounding fluid. These equations show that the true coupling can be achieved when the same source terms are used for fluid and solid phases. The force $\vec{F}$ is comprised of two parts: macroscopic variations in the fluid stress tensor and detailed variations in the point stress tensor around a particle. The latter is the drag force exerted on particles as the fluid passes through the assembly of particle and the former is related to the fluid stress tensor, $\vec{\pi}_f$, which includes fluid pressure and viscous stress tensors.

Original Model

Although the governing equations in the CFD-DEM model are quite different from those in the TFM, we can use the same methodology to establish true coupling relations between both phases. Therefore, the equation of momentum conservation for the fluid phase, Equation 6.40a, can be rewritten as:

$$\frac{\partial\left(\rho_f\varepsilon_f\vec{u}\right)}{\partial t}+\nabla.\left(\rho_f\varepsilon_f\vec{u}\vec{u}\right)=-\nabla p-\nabla.\vec{\tau}_f-\vec{F}^{OM}+\rho_f\varepsilon_f\vec{g} \tag{6.41}$$

where $\vec{F}^{OM}$ is the volumetric average of all interaction forces acting on the solid phase due to motion of fluid in each fluid cell. These forces are fluid drag force, fluid pressure force, shear stress force, and other possible fluid-particle interaction forces. Since the solid phase is considered discrete in the CFD-DEM model, this volumetric mean can be obtained from:

$$\vec{F}^{OM} = \frac{1}{V_{cell}} \sum_{i=1}^{k_v} \vec{f}_i^{f-p} = \frac{1}{V_{cell}} \sum_{i=1}^{k_v} \left(\vec{f}_i^{d} + \vec{f}_i^{u} + \vec{f}_i^{\nabla p} + \vec{f}_i^{\nabla.\vec{\tau}_f} + \vec{f}_i^{l} \right) \tag{6.42}$$

where k_v is the number of particles in each fluid cell of volume V_{cell} and the summation is performed on all particles in the fluid cell.

Model A

The second formulation is derived directly from the original model by decomposing the volumetric fluid–particle interaction, $\vec{F}^{OM}$, into the two parts. The first part is the fluid stress tensor and the second part is the drag force and other remaining forces:

$$\vec{F}^{OM} = \frac{1}{V_{cell}} \sum_{i=1}^{k_v} \left(\vec{f}_i^{\nabla p} + \vec{f}_i^{\nabla.\vec{\tau}_f} \right) + \frac{1}{V_{cell}} \sum_{i=1}^{k_v} \left(\vec{f}_i^{d} + \vec{f}_i^{u} + \vec{f}_i^{l} \right) \tag{6.43}$$

The first summation on the right-hand side can be rewritten as follows:

$$\frac{1}{V_{cell}} \sum_{i=1}^{k_v} \left(\vec{f}_i^{\nabla p} + \vec{f}_i^{\nabla.\vec{\tau}_f} \right) = -\frac{1}{V_{cell}} \sum_{i=1}^{k_v} \left(V_i \nabla p + V_i \nabla.\vec{\tau}_f \right) = -\varepsilon_p \nabla p - \varepsilon_p \nabla.\vec{\tau}_f \tag{6.44}$$

The second summation on the right-hand side of Equation 6.43 is defined as the volumetric fluid–particle interaction force for model A, $\vec{F}^{A}$. Therefore, Equation 6.43 can be rewritten as:

$$\vec{F}^{OM} = -\varepsilon_p \nabla p - \varepsilon_p \nabla.\vec{\tau}_f + \vec{F}^{A} \tag{6.45}$$

Substituting Equation 6.45 into Equation 6.41 yields the main equation of model A:

$$\frac{\partial \left(\rho_f \varepsilon_f \vec{u} \right)}{\partial t} + \nabla.\left(\rho_f \varepsilon_f \vec{u}\vec{u} \right) = -\varepsilon_f \nabla p - \varepsilon_f \nabla.\vec{\tau}_f - \vec{F}^{A} + \rho_f \varepsilon_f \vec{g} \tag{6.46}$$

where

$$\vec{F}^{A} = \frac{1}{V_{cell}} \sum_{i=1}^{k_v} \left(\vec{f}_i^{d} + \vec{f}_i^{u} + \vec{f}_i^{l} \right) \tag{6.47}$$

As can be seen from the derivation of model A, these two are similar with the exception that in model A, the fluid stress tensor interaction force on particles is separated from the fluid–particle interaction force. Therefore, although these two models are named differently in literature, they are physically the same.

Model B

Derivation of model B from the original model is based on following important assumption:

$$\frac{\partial\left(\rho_f\varepsilon_p\vec{u}\right)}{\partial t}+\nabla.\left(\rho_f\varepsilon_p\vec{u}\vec{u}\right)\cong 0 \tag{6.48}$$

This means that the fluid flow through the solid phase is steady and uniform [26, 65]. Thus, the fluid force due to the stress tensor is reduced to the pressure gradient force:

$$\vec{f}_i^{\nabla p}+\vec{f}_i^{\nabla.\vec{\tau}_p}\approx\vec{f}_i^{\nabla p}=-V_i\nabla p \tag{6.49}$$

In this equation, ∇p is comprised of hydrostatic pressure gradient due to the fluid gravity and dynamic pressure (p_d) drop due to the relative motion of fluid and particles. The following relationship is obtained for a bed of mono-sized particles at minimum fluidization condition [66]:

$$-V_i\nabla p=-V_i\nabla p_d-V_i\rho_f\vec{g}\approx\frac{\varepsilon_p}{\varepsilon_f}\vec{f}_i^{d}-V_i\rho_f\vec{g} \tag{6.50}$$

Substituting Equations 6.49 and 6.50 into the expression of fluid–particle force results in the main equation of model B:

$$\frac{\partial\left(\rho_f\varepsilon_f\vec{u}\right)}{\partial t}+\nabla.\left(\rho_f\varepsilon_f\vec{u}\vec{u}\right)=-\nabla p-\nabla.\vec{\tau}_f-\vec{F}^B+\rho_f\varepsilon_f\vec{g} \tag{6.51}$$

where [30]

$$\vec{F}^B=\frac{1}{\varepsilon_f V_{cell}}\sum_{i=1}^{k_v}\left(\vec{f}_i^{d}+\vec{f}_i^{u}+\vec{f}_i^{l}\right)-\frac{1}{V_{cell}}\sum_{i=1}^{k_v}\left(\rho_f V_i\vec{g}\right)=\frac{\vec{F}^A}{\varepsilon_f}-\rho_f\varepsilon_p\vec{g} \tag{6.52}$$

We should talk about two points here. First, it should be mentioned that Equation 6.35 is valid for original and A models. However, in order to guarantee the Newton's third law of motion in model B, the particle–fluid interaction force acting on particles ($\vec{f}_i^{f-p}$) should be obtained from the following equation instead of Equation 6.35:

$$\vec{f}_i^{f-p}=\frac{1}{\varepsilon_f}\left(\vec{f}_i^{d}+\vec{f}_i^{u}+\vec{f}_i^{l}\right)-\rho_f V_i\vec{g} \tag{6.53}$$

Another point is that the assumption of steady and uniform flow of fluid, given by Equation 6.48, is held exactly at the minimum fluidization condition [31]. This assumption leads to erroneous results as the fluid–particle flow deviates from the minimum fluidization condition, like in bubbling or turbulent fluidization. Nevertheless, it was shown that model B can be applied in bubbling fluidization [30, 32]. Zhou *et al.* [33] assessed the applicability of these three models in different fluid-particle flows, namely fluidization, pneumatic conveying, and hydro-cyclones. They showed that these models are valid in the fluidization, while the applicability of model B is somehow questionable for complex flow systems like hydro-cyclones. Bokkers *et al.* [63] experimentally studied injection of single bubbles into a monodispersed and bidispersed fluidized bed using particle image velocimetry. They adopted model A as the CFD-DEM model and two drag force correlations, Ergun-Wen and Yu [67, 68] and Koch–Hill–Ladd (KHL) [69]. Their simulation showed that when KHL drag correlation is used, this model can satisfactorily predict the experimental data. In conclusion, the original model and model A are applicable to all type of fluid–solid flows. While model B is only applicable to simple gas-solid flows like bubbling fluidization and pneumatic conveying and its validity is not known in more complex situations like polydispersed systems, hydro-cyclones, and turbulent or fast fluidizations. The governing equations of three CFD-DEM models are summarized in Table 6.1.

Table 6.1 Governing equations of three types of CFD-DEM models

Formulation	Equation no.
Mass conservation for all models	
$\frac{\partial(\rho_f \varepsilon_f)}{\partial t} + \nabla.(\rho_f \varepsilon_f \vec{u}) = 0$	(6.54)
Original model	
$\frac{\partial(\rho_f \varepsilon_f \vec{u})}{\partial t} + \nabla.(\rho_f \varepsilon_f \vec{u}\vec{u}) = -\nabla p - \nabla.\vec{\tau}_f - \vec{F}^{OM} + \rho_f \varepsilon_f \vec{g}$	(6.41)
$\vec{F}^{OM} = \frac{1}{V_{cell}} \sum_{i=1}^{k_v} \vec{f}_i^{f-p} = \frac{1}{V_{cell}} \sum_{i=1}^{k_v} \left(\vec{f}_i^d + \vec{f}_i^u + \vec{f}_i^{\nabla p} + \vec{f}_i^{\nabla.\vec{\tau}_f} + \vec{f}_i^l\right)$	(6.42)
$\vec{f}_i^{f-p} = \vec{f}_i^d + \vec{f}_i^u + \vec{f}_i^{\nabla p} + \vec{f}_i^{\nabla.\vec{\tau}_f} + \vec{f}_i^l$	(6.35)
Model A	
$\frac{\partial(\rho_f \varepsilon_f \vec{u})}{\partial t} + \nabla.(\rho_f \varepsilon_f \vec{u}\vec{u}) = -\varepsilon_f \nabla p - \varepsilon_f \nabla.\vec{\tau}_f - \vec{F}^A + \rho_f \varepsilon_f \vec{g}$	(6.46)
$\vec{F}^A = \frac{1}{V_{cell}} \sum_{i=1}^{k_v} \left(\vec{f}_i^d + \vec{f}_i^u + \vec{f}_i^l\right)$	(6.47)
$\vec{f}_i^{f-p} = \vec{f}_i^d + \vec{f}_i^u + \vec{f}_i^{\nabla p} + \vec{f}_i^{\nabla.\vec{\tau}_f} + \vec{f}_i^l$	(6.35)
Model B	
$\frac{\partial(\rho_f \varepsilon_f \vec{u})}{\partial t} + \nabla.(\rho_f \varepsilon_f \vec{u}\vec{u}) = -\nabla p - \nabla.\vec{\tau}_f - \vec{F}^B + \rho_f \varepsilon_f \vec{g}$	(6.51)
$\vec{F}^B = \frac{1}{\varepsilon_f V_{cell}} \sum_{i=1}^{k_v} \left(\vec{f}_i^d + \vec{f}_i^u + \vec{f}_i^l\right) - \frac{1}{V_{cell}} \sum_{i=1}^{k_v} \left(\rho_f V_i \vec{g}\right)$	(6.52)
$\vec{f}_i^{f-p} = \frac{1}{\varepsilon_f}\left(\vec{f}_i^d + \vec{f}_i^u + \vec{f}_i^l\right) - \rho_f V_i \vec{g}$	(6.53)

6.2.3.4 Fluid Volume Fraction

The local fluid volume fraction (ε_f) is defined as the fraction of fluid volume in the fluid cell with volume V_{cell}. The fluid volume fraction is also called porosity or void fraction. An accurate estimation of the fluid volume fraction in each cell is required [70]. Local and fluid cell porosities appear in mass and momentum conservation equations and are used in the fluid-particles interaction force formulas. Hence, this parameter significantly affects the results of a CFD-DEM simulation. In addition, governing equations of the CFD-DEM include time derivative of the fluid volume fraction. Sudden changes in its estimation in two sequential time steps would add erroneous pressure spikes into the solution.

The exact fluid volume fraction in a computational cell (Figure 6.7a) can be estimated through the equation of Hoomans *et al.* [34]:

$$\varepsilon_f = 1 - \frac{1}{V_{cell}} \sum_{i=1}^{k_V} \varphi_i V_i \tag{6.55}$$

where k_V is the number of particles partially located within the fluid cell and $\varphi_i \in [0,1]$ denotes the volume fraction of particle i that belongs to the fluid cell as estimated through geometrical techniques. The efficiency of this method mainly depends on accurate estimation of φ_i.

In general, the fluid volume fraction in a cell is calculated by two different approaches: exact analytical and approximate non-analytical methods. Analytical approaches use exact geometrical methods for calculating the actual volume of each particle in the cell [72–74]. Wu *et al.* [72] used a 3D analytical approach for spherical particles and typical unstructured

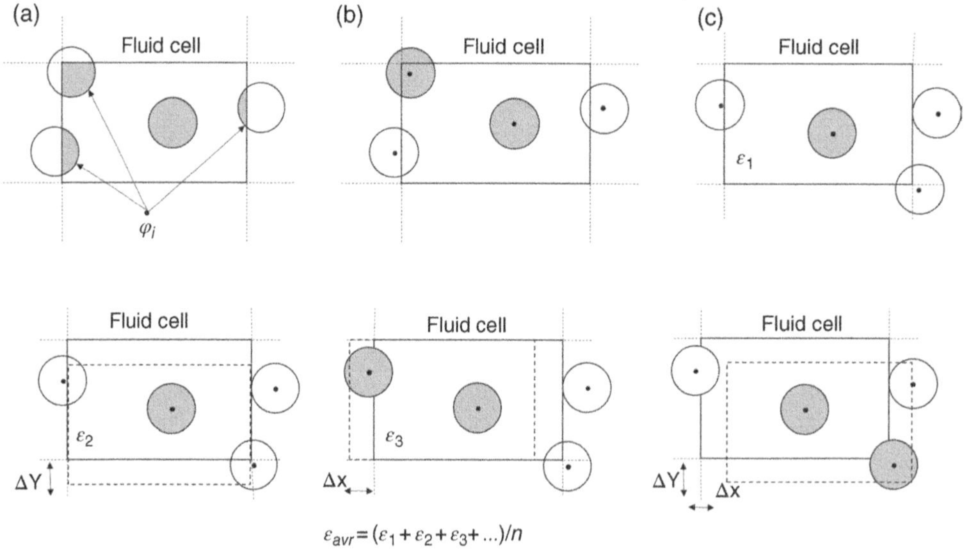

Figure 6.7 Estimation of fluid volume fraction by (a) exact method, (b) PCM, and (c) PCM with the offset method. (Source: Alobaid, F. and B. Epple, Improvement, validation and application of CFD/DEM model to dense gas–solid flow in a fluidized bed. *Particuology*, 2013. **11**(5): p. 514–526 with permission)

meshes (e.g., tetrahedral, wedged, hexahedral). Peng *et al.* [73], on the other hand, used a 3D analytical approach for regular structural Cartesian meshes. Although this method is accurate, it is difficult to use it for irregular meshes and/or particles with irregular shapes [73].

The non-analytical approaches are simpler and can almost be generalized to all types of mesh and particle shape. They include a wide range of methods, from the simple particle center method (PCM) or improved PCM [53, 75–77], to more complex and accurate methods, such as porous cubes [6, 25, 78], statistical approaches [79], sub-element [13, 70], and spherical control volume [80]. As shown in Figure 6.7b, in the PCM, if a particle center is located inside a computational cell, all the volume of that particle is assumed to be in the cell [53]:

$$\varepsilon_f = 1 - \frac{1}{V_{cell}} \sum_{i=1}^{k_c} V_i \tag{6.56}$$

where k_c is the number of particles within the fluid cell if their centers are located within the cell boundaries. This method works when volume of the computational cell is much greater than total volume of particles. This approach may cause a relative error of up to 50% in the estimation of fluid volume fraction if the particle center is near the cell boundaries [72, 73, 76]. Recently, Peng *et al.* [73] compared the accuracy of classical PCM and analytical methods for estimation of fluid-particle flow characteristics in a bubbling fluidized bed. They found that there is a critical computational cell size (3.82 particle diameter) where beyond that the PCM has the same numerical stability and accuracy of the analytical solution.

It was tried to improve the accuracy of the conventional PCM, for example, by applying the offset method [71]. In this method, as shown in Figure 6.7c, the computational fluid cell is adjusted in all directions with $\Delta x, \Delta y, \Delta z \in [-d_i/2, 0, d_i/2]$. Then, the fluid volume fraction is estimated by averaging all determined fluid volume fractions with regard to the number of displacements. Accuracy of the PCM was compared with and without the offset method and it was shown that the mean error of both methods increases as the volume of the computational cell is decreased relative to the particles. Nevertheless, the offset method can improve the accuracy up to 1 order of magnitude.

Another approach that is very similar to the PCM, but can improve its accuracy, is the sub-element method. In this method, the particle is sub-divided into N_e equal parts with center points $C_{i,j}$. For each particle, if the center of any of these sub-elements resides in a fluid cell, the whole volume of that sub-element is assumed to be in that cell. As an example, we illustrated a schematic of this method, shown in 2D for clarity, for calculating the fluid porosity of a spherical particle in Figure 6.8. Number of sub-elements is five in this example. The volume of sub-element 5 for particle 1 would not be added to the total solid volume in the target fluid cell but it would be added to the total solid volume of the adjacent cell instead. For particle 2, volumes of sub-elements 1 and 4 would be added to the total solid volume in the target cell and volumes of other sub-elements would be added to the total solid volume in the adjacent cells. It can be seen that this method is as simple as the PCM with a higher accuracy in calculation of porosity and can be applied to regular and irregular structural meshes. Moreover, it can be applied to non-spherical particles with slight modifications. Accuracy of this method increases with increasing N_e at the expense of more computational costs.

In two-dimensional simulations, the porosity is estimated based on the area of particles that occupies the fluid cell. The calculated 2D porosity is not consistent with empirical drag formulas that are derived for real 3D cases. To solve this problem, several researchers have

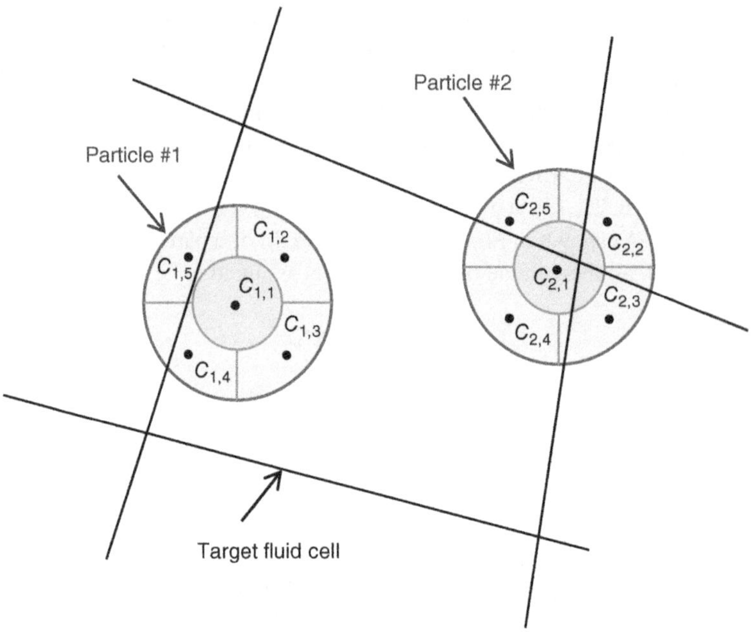

Figure 6.8 The sub-element method for calculating the fluid volume fraction

attempted to obtain more comparable results by relating the 2D fluid volume fraction (ε_f^{2D}) to the 3D fluid volume fraction (ε_f^{3D}). For this purpose, two methods have been developed to map the 2D fluid volume fraction to the 3D one. The first method was proposed by Hoomans *et al.* [34], which employs the following equation:

$$\varepsilon_f^{3D} = 1 - \frac{2}{\sqrt{\pi\sqrt{3}}}\left(1 - \varepsilon_f^{2D}\right)^{3/2} \tag{6.57}$$

This equation was developed based on equal inter-particle distances for a 2D hexagonal lattice and a 3D cubic lattice. The second method was suggested by Xu and Yu [53], which employs a simple pseudo 3D concept as:

$$\varepsilon_f^{3D} = 1 - \frac{1}{V_{cell}} \sum_{i=1}^{k_v} V_i \tag{6.58}$$

where

$$V_{cell} = \Delta x \, \Delta y d_i \tag{6.59}$$

in which Δx and Δy are sides of the 2D computational fluid cell in the x and y directions, respectively.

6.2.3.5 Mapping from Eulerian Variables to Lagrangian Values

For calculating the fluid–particle interaction force, Lagrangian values of fluid velocity or fluid volume fraction in a cell at the particle position are required. A smooth transfer of properties from the Eulerian cell to the Lagrangian position of a particle can be achieved through a mapping method. As described by many researchers [6, 25, 73, 78], an area (in 2D) or volume (in 3D) weighted nodal averaging method, introduced by Hoomans *et al.* [34], can be used as the mapping function. This nodal averaging method is difficult to implement for unstructured schemes [73]. Lagrangian values of Eulerian variables can be estimated based on their spatial distribution [74]:

$$\phi_p = \phi_c + \left(x_p - x_c\right)\frac{\sum_i^{n_f} \phi_{f,i} A_{f,i}}{V_{cell}} \tag{6.60}$$

where ϕ_p and ϕ_c are the fluid variables stored at the particle position and cell center. $\phi_{f,i}$ is the face fluid value interpolated by adjacent cells following a certain numerical scheme (e.g., central-differenced, second-order upwind). $A_{f,i}$ is area of the cell face i and n_f is the number of cell faces. Parameters x_c and x_p are the cell centroid and the particle positions, respectively.

6.2.3.6 Sequence of Calculations

Figure 6.9 shows the general framework of coupling in the unresolved CFD-DEM model. Initially, all components of simulation, DEM, CFD, and coupling, should be initialized. The coupling starts with calculating the fluid porosity in each fluid cell based on the position of particles and fluid mesh information. Thereafter, velocity of particles, velocity of fluid as well as the pressure and stress tensor at the current fluid time step are used to calculate the fluid–particle interaction force acting on each particle, $\vec{f}_i^{f-p}$. To hold Newton's third law of motion, the volumetric fluid–particle interaction force in each fluid cell is calculated with a proper formula, depending on the type of model formulation (see Table 6.1).

The next step is the iteration loop of the DEM. The value of $\vec{f}_i^{f-p}$ calculated in the coupling step is used in the equation of motion of each particle. Details of the iteration loop of the DEM and different methods of integrating the equation of motion are presented in Chapter 3. The time step for integrating the equation of motion of particles is Δt_p and iteration loop of DEM is repeated m times. After the DEM loop is completed, the new position and translational and rotational velocities of all particles in the next fluid time step are obtained.

For solving mass and momentum conservation equations of the fluid phase, the calculated porosity, and volumetric fluid-particle interaction force in each fluid cell are used. Any standard or advanced numerical method, which is generally used for solving coupled mass and momentum equations of fluid phase, can be employed in this step. For example, the SIMPLE (Semi-Implicit Method for Pressure-Linked Equation) [81] or PISO (Pressure implicit with splitting operator) [82] can be used for this purpose. To reduce the pressure checkerboard oscillation problem in the numerical solution, a staggered grid should be employed. Thus, scalar properties, fluid pressure, and fluid volume fraction are determined at the center of each fluid cell, while velocities and forces are obtained at the faces of the fluid cell. A standard first order upwind or the second order QUICK scheme may be used to discretize the convective

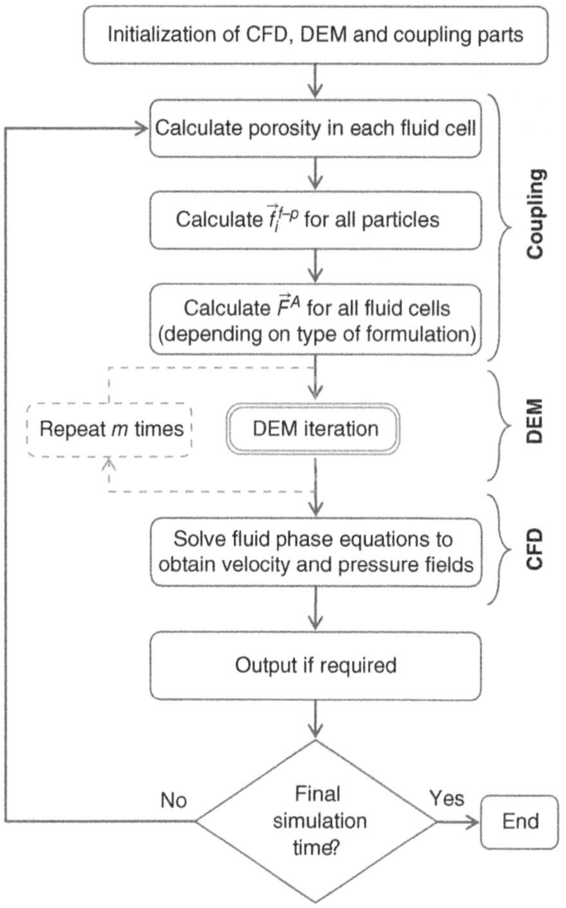

Figure 6.9 Framework of explicit coupling in the CFD-DEM model

terms and the divergence and gradient terms are calculated using the semi-finite difference in the selected staggered grid. The fluid time step for integration is Δt_f. Results of this step are pressure and velocity fields of the fluid phase in the next fluid time step.

The maximum allowable time step for solving the fluid phase equations is estimated from the Courant limit and the viscosity stability criterion (see Chapter 1). However, the time step in DEM is determined by the Rayleigh criteria described in Chapter 3. In most CFD-DEM simulations, the selected time step for DEM is much smaller than that for the CFD. We defined the ratio of these time steps as:

$$m = \frac{\Delta t_f}{\Delta t_p} \tag{6.61}$$

Suppose that $\Delta t_p = 10^{-5}$ seconds for DEM and $\Delta t_f = 10^{-4}$ seconds. This results in $m = 10$ and it suggests than the program must perform 10 iterations (10 particle time steps) in the DEM part and 1 iteration of the CFD (1 fluid time step) in each coupling step.

The reader should note that the proposed sequence of calculations is arbitrary and some variants can be found in literature. In the proposed sequence here, we assumed a fully explicit coupling scheme, meaning that the coupling terms are calculated with the information at the current time step. Then, equations of motion of both phases are solved to advance one step in time. Results of calculations are velocity and pressure fields of the fluid phase and position and velocity of all particles in the next time step. Moreover, the Newton's third law is held since $\vec{f}_i^{f-p}$ and the volumetric fluid-particle interaction force in each fluid cell (either $\vec{F}^{OM}$, $\vec{F}^{A}$, or $\vec{F}^{B}$) are calculated at the same time.

6.2.4 *Surface Force Decomposition*

6.2.4.1 Steady Drag Force

In all fluid-particle flows, drag and pressure gradient forces together play an important role in the movement of particles. Thus, an accurate calculation of drag force is important for successful simulation of fluid-particle flows. With detailed velocity and pressure distributions around a single sphere, the force exerted by the surrounding fluid on a particle can be determined from Equation 6.34 by integrating normal and tangential stresses over the particle surface.

Drag Force on an Isolated Particle

From the theoretical point of view, the only exact relation that computes the drag force on a spherical particle is the of Stokes–Einstein relationship [23], which is valid for an isolated sphere in the creeping flow regime. At creeping flow conditions, where Reynolds number is less than 0.1, the resultant force exerted on the surface of a single sphere i with diameter d_i can be analytically derived as:

$$\vec{f}_i^{f-p} = \vec{f}_i^{(n)} + \vec{f}_i^{(t)} = \left(-V_i \nabla p + \pi \mu_f d_i \vec{w}_i\right) + 2\pi \mu_f d_i \vec{w}_i \tag{6.62}$$

where $\vec{f}_i^{(n)}$ and $\vec{f}_i^{(t)}$ are normal and tangential forces exerted by the surrounding fluid on a particle, respectively. The first term in the parentheses is the pressure gradient force for a fluid at rest. The second term in the parentheses, which containing the relative velocity $\vec{w}_i = \vec{u} - \vec{v}_i$, is the form drag. The term outside the parentheses is the friction drag. It is convenient to rearrange these forces into two general parts: the force exerted by the fluid even if it was stationary and the additional force associated with the motion of the fluid kinetic force, referred to as the Stokes–Einstein drag force ($\vec{f}_i^{d}$):

$$\vec{f}_i^{f-p} = -V_i \nabla p + 3\pi \mu_f d_i \vec{w}_i = \vec{f}_i^{\nabla p} + \vec{f}_i^{d} \tag{6.63}$$

In many practical applications, it is also convenient to express the drag force based on a characteristic area (e.g., projected area of particle i in the direction of motion, A_i^p) and a characteristic kinetic energy per unit volume, K_i:

$$\left|\vec{f}_i^{d}\right| = C_d A_i^p K_i \tag{6.64}$$

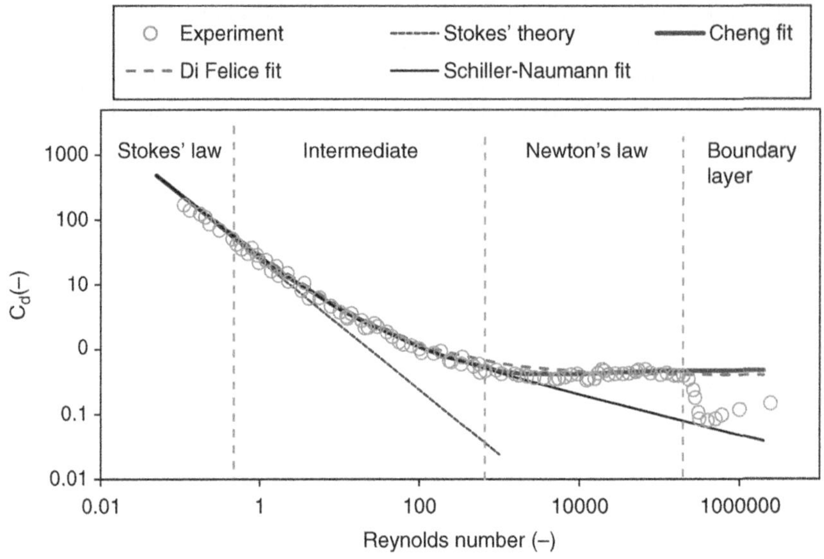

Figure 6.10 Drag coefficient for an incompressible flow over a smooth sphere as a function of Reynolds number obtained from experimental data [83], Stokes' theory [23], Cheng's correlation [84], Di Felice's correlation [85], and Schiller–Naumann's correlation [86]

in which the proportionality dimensionless constant, C_d, is called the drag coefficient. The drag coefficient can be obtained as a relatively simple function of Reynolds number of particle i [23]. For example, for the motion of a smooth sphere in the Stokes regime, the drag coefficient is:

$$C_d = \frac{24}{\mathrm{Re}_i} \tag{6.65}$$

Figure 6.10 shows the drag coefficient of an isolated sphere against the Reynolds number. Four regions can be identified in this figure: The Stokes' region where drag force decreases inversely proportional to the Reynolds number, the Newton's region where the drag coefficient is independent of Reynolds number ($C_d \approx 0.44$), the intermediate region between Stokes and Newton regions and the boundary layer separation region at very Reynolds numbers. Many correlations have been developed for evaluating the drag coefficient shown in Figure 6.10 as a function of Reynolds number. However, a systematic study of all these correlations is beyond the scope of this book.

Drag Force on a Particle in an Assembly of Particles

The correlation of drag coefficient for an isolated particle is not suitable for practical problems to be solved by the CFD-DEM since in a swarm of particles the drag force exerted on a particle must be determined in the presence of other particles. Correct theoretical formula for estimating fluid–particle interactions in such systems is not available yet. Thus, empirical correlations have been developed for this purpose [8, 67–69, 84–92]. Care must be taken when using the correlations in literature because sometimes the total force exerted on a particle ($\vec{f}_i^{f-p}$) is also referred to as the drag force. These two forces are related to each other by the following equation for monodispersed system of spherical particles:

$$\vec{f}_i^{d} = \varepsilon_f \vec{f}_i^{f-p} \tag{6.66}$$

These two forces cannot be linked by such a simple formula in a polydispersed system [91, 93]. In this chapter, all correlations are presented for the drag force on a particle rather than the total force. It is most convenient to use the drag force in its normalized dimensionless form:

$$\hat{f}_i^{d}\left(\varepsilon_f, \mathrm{Re}_i\right) = \frac{\vec{f}_i^{d}\left(\varepsilon_f, \mathrm{Re}_i\right)}{3\pi\mu_f \varepsilon_f d_i \vec{w}_i} \tag{6.67}$$

where

$$\vec{w}_i = \vec{u} - \vec{v}_i \tag{6.68}$$

$$\mathrm{Re}_i = \frac{\rho_f \varepsilon_f d_i \left|\vec{w}_i\right|}{\mu_f} \tag{6.69}$$

It should be noted that for an isolated particle in the Stokes regime, the normalized dimensionless drag force is 1.

Generally, two types of correlations have been developed to determine the normalized dimensionless drag force in packed and dilute conditions. The first type of relation is based on the well-known relation of the Kozeny and Carman equations, which comprises a term representing the drag force on an array of particles in the Stokes regime, $\hat{f}_i^{d}(\varepsilon_f, 0)$, and a term that accounts for the inertial effects, $\alpha(\varepsilon_f)\mathrm{Re}_i$; that is, a linear function of Reynolds number [66]:

$$\hat{f}_i^{d}\left(\varepsilon_f, \mathrm{Re}_i\right) = \hat{f}_i^{d}\left(\varepsilon_f, 0\right) + \alpha\left(\varepsilon_f\right)\mathrm{Re}_i \tag{6.70}$$

The Ergun equation [67] is a typical example of such correlations:

$$\hat{f}_i^{d}\left(\varepsilon_f, \mathrm{Re}_i\right) = \frac{A\left(1-\varepsilon_f\right)}{18\varepsilon_f^2} + \frac{B}{18\varepsilon_f^2}\mathrm{Re}_i \tag{6.71}$$

where A and B are constants that vary from case to case. Constants of the Ergun equation for a packed bed are $A = 150$ and $B = 1.75$.

The second type of correlation is based on the expression for an isolated particle for the full range of Reynolds number, $\hat{f}_i^{d}(1, \mathrm{Re}_i)$ in which the influence of other surrounding particles is taken into account by introducing a term that is a function of porosity, $f(\varepsilon_f)$:

$$\hat{f}_i^{d}\left(\varepsilon_f, \mathrm{Re}_i\right) = \hat{f}_i^{d}\left(1, Re_i\right) f\left(\varepsilon_f\right) \tag{6.72}$$

It is more convenient to express $\hat{f}_i^d(1, \mathrm{Re}_i)$ in terms of drag coefficient. By considering the relation between normalized dimensionless drag force and drag coefficient, C_d, the following equation can be obtained:

$$\hat{f}_i^d\left(1, Re_i\right) = \frac{C_d}{24} Re_i \tag{6.73}$$

Correlations of Wen and Yu [68] and Di Felice [85] are categorized as the second type. The frequently used Wen and Yu correlation [68] reads as:

$$\hat{f}_i^d\left(\varepsilon_f, Re_i\right) = \frac{C_d}{24} Re_i \varepsilon_f^{-3.65} \tag{6.74}$$

$$C_d = \begin{cases} \dfrac{24}{Re_i}\left(1 + 0.15 Re_i^{0.687}\right) & Re_i \le 1000 \\ 0.44 & Re_i > 1000 \end{cases} \tag{6.75}$$

This correlation is only valid for dilute flow of particles, that is, $\varepsilon_f > 0.8$. The Di Felice correlation [85], which is valid for both dense and dilute flow of particles, is as follows:

$$\hat{f}_i^d\left(\varepsilon_f, Re_i\right) = \frac{C_d}{24} Re_i \varepsilon_f^{-\chi} \tag{6.76}$$

$$\chi = 3.7 - 0.65 e^{\left(-0.5\left(1.5 - \log_{10} Re_i\right)^2\right)} \tag{6.77}$$

$$C_d = \left(0.63 + 4.8 Re_i^{-0.5}\right)^2 \tag{6.78}$$

There are also many correlations available in the literature for evaluating the drag coefficient. Recently, Cheng reviewed available correlations for C_d and presented the following correlation based on the experimental data [84]:

$$C_d = \frac{24}{\mathrm{Re}_i}\left(1 + 0.27 \mathrm{Re}_i\right)^{0.43} + 0.47\left(1 - e^{\left(-0.04 \mathrm{Re}_i^{0.38}\right)}\right) \tag{6.79}$$

It was shown that this correlation is valid for Reynolds numbers up to 2×10^5 and gives more accurate values compared to previously reported correlations.

These correlations, as well as similar ones in literature, were obtained based on experimental data. However, studying the effects of different arrangements of particles and measuring the drag force exerted on each particle (rather than averaged drag force on all particles) in a monodispersed or polydispersed system in the full range of Reynolds numbers and porosity is not experimentally possible. Apart from empirical approaches, fortunately, accurate and efficient numerical techniques, such as LB and DNS have been developed since the beginning of this millennium [88, 94]. These *in-silico* experiments[2] provide detailed information about the fluid-particle interactions and make it possible to consider different arrangements of poly-sized particles with various packing fractions at different Reynolds numbers.

Correlations for Monodispersed

A number of correlations have been developed for evaluating the drag force in a monodispersed system based on either experimental data or numerical studies. The correlation of Ergun [67] is valid for dense systems and cannot be applied to dilute systems while the correlation of Wen and Yu works suits well for the dilute system [68]. Gidaspow [26] combined both these correlations and provided a general drag force correlation for fluid–particle systems. In this method, the Ergun equation is used for fluid volume fractions less than 0.8 and the Wen and Yu equation for fluid volume fractions greater than 0.8. The Ergun–Wen and Yu correlation reads as:

$$\hat{f}_i^d\left(\varepsilon_f, Re_i\right) = \begin{cases} \dfrac{150\left(1-\varepsilon_f\right)}{18\varepsilon_f^2} + \dfrac{1.75}{18\varepsilon_f^2} Re_i & \varepsilon_f \leq 0.8 \\ \dfrac{C_d}{24} Re_i \varepsilon_f^{-3.65} & \varepsilon_f > 0.8 \end{cases} \tag{6.80}$$

$$C_d = \begin{cases} \dfrac{24}{Re_i}\left(1+0.15 Re_i^{0.687}\right) & Re_i \leq 1000 \\ 0.44 & Re_i > 1000 \end{cases} \tag{6.81}$$

Kafui *et al.* [56] compared the Ergun–Wen and Yu correlation with the Di Felice equation [85] at various slip velocities and porosities and showed that there is a step change in the drag force evaluated by the Ergun–Wen and Yu correlation at a fluid porosity of 0.8, which is not consistent with reality. Nonetheless, the Ergun–Wen and Yu correlation has been widely used by many researchers in multiphase flows. The Gibilaro *et al.* correlation [87] is:

$$\hat{f}_i^d\left(\varepsilon_f, Re_i\right) = \left(\frac{17.3}{18} + \frac{0.336}{18} Re_i\right) \varepsilon_f^{-3.8} \tag{6.82}$$

This correlation does not have the problem of the Ergun–Wen and Yu correlation but has not been used so often by researchers.

The Di Felice correlation [85] is a monotonic function of Reynolds number and porosity and does not have the problem of the Ergun–Wen and Yu correlation. This correlation is applicable to the full range of Reynolds numbers and porosity, and its capabilities in predicting the

[2] Experiments performed by a computer simulation.

flow characteristics of gas-solid fluidized beds have been tested against experimental data [31, 32, 56]. The correlation by Hill–Koch–Ladd (HKL) was the first to be derived based on LB simulations [69]:

$$\hat{f}_i^d\left(\varepsilon_f, Re_i\right) = \frac{A\varepsilon_p}{18\varepsilon_f^2} + \frac{B}{18\varepsilon_f^2} Re_i \tag{6.83}$$

$$A = \begin{cases} 180 & \varepsilon_f \le 0.6 \\ \left(\dfrac{18\varepsilon_f^3}{\varepsilon_p}\right)\left(\dfrac{1 + \dfrac{3}{\sqrt{2}}\sqrt{\varepsilon_p} + \dfrac{135}{64}\varepsilon_p \ln\varepsilon_p + 16.14\varepsilon_p}{1 + 0.681\varepsilon_p - 8.48\varepsilon_p^2 + 8.16\varepsilon_p^3}\right) & \varepsilon_f > 0.6 \end{cases} \tag{6.84}$$

$$B = \left(0.6057 + 1.908\varepsilon_s\right)\varepsilon_f^3 + 0.209\varepsilon_f^{-2} \tag{6.85}$$

This correlation is only valid for Reynolds numbers between 40 and 120. Thus, although it cannot be considered a general correlation, it can be applied in simulations of fluidization beds for which the Reynolds number falls in this range. Moreover, the calculated drag force may have a step change at $\varepsilon_f \le 0.6$. Later, Benyahia *et al.* [89] extended this correlation to the full range of Reynolds numbers and fluid porosity.

Beetstra *et al.* [95] followed Hill *et al.* [69] and derived a monotonic correlation that is applicable to a wider range of Reynolds numbers (up to 1000) and porosities ($\varepsilon_f \le 0.9$):

$$\hat{f}_i^d\left(\varepsilon_f, Re_i\right) = \left(\frac{180\varepsilon_p}{18\varepsilon_f^2}\right) + \varepsilon_f^2\left(1 + 1.5\sqrt{\varepsilon_p}\right) + \left(\frac{0.413}{24\varepsilon_f^2}\right)\left(\frac{\varepsilon_f^{-1} + 3\varepsilon_p\varepsilon_f + 8.4Re_i^{-0.343}}{1 + 10^{3\varepsilon_p} Re_i^{-(1+4\varepsilon_p)/2}}\right) Re_i \tag{6.86}$$

This correlation was later extended to bidispersed and polydispersed systems, which will be discussed in the next section [8, 88, 91, 95].

Cello *et al.* [92] derived a complex but accurate correlation for the whole range of fluid porosity ($0.35 \le \varepsilon_f \le 1$) and Reynolds numbers up to 1000:

$$\hat{f}_i^d\left(\varepsilon_f, Re_i\right) = k_1 + k_2\varepsilon_f^4 + k_3\left(1 - \varepsilon_f^4\right) \tag{6.87}$$

$$k_0 = \frac{\left(1 - \varepsilon_f\right)}{1 + 3\varepsilon_f} \tag{6.88a}$$

$$k_1 = \frac{1 + 128k_0 + 715k_0^2}{\varepsilon_f^2\left(1 + 49.5k_0\right)} \tag{6.88b}$$

$$k_2 = \frac{1 + 0.13Re_i + 6.66\times10^{-4} Re_i^2}{1 + 3.42\times10^{-2} Re_i + 6.92\times10^{-6} Re_i^2} - 1 \tag{6.88c}$$

$$k_3 = \left(\frac{2Re_i^2}{1+Re_i}\right)\left(\frac{-410\varepsilon_f + 9.2\times10^7 Re_i k_0^{20} + 1900\varepsilon_f^2 - 6.6\times10^{-2} Re_i}{6600\varepsilon_f + 4.92\times10^{-4} Re_i - 4.3\times10^4\varepsilon_f^2 - 1.31\times10^{-4} Re_i^2 + 7.38\times10^4\varepsilon_f^3}\right) \tag{6.88d}$$

They also showed that this complex equation requires less computational time than the Beetstra correlation due to the fact that Equation 6.87 involves only polynomials with integer powers [8, 88, 91, 95].

Sometimes other forms of interphase momentum transfer coefficients are used in the literature. The following relations are useful to convert these equations to other forms:

$$18\mu_f\varepsilon_f\varepsilon_p\hat{f}_i^d\left(\varepsilon_f,\mathrm{Re}_i\right) = \beta d_i^2 \tag{6.89}$$

$$\vec{f}_i^d\left(\varepsilon_f,\mathrm{Re}_i\right) = V_i\frac{\beta}{\varepsilon_p}\left(\vec{u}-\vec{v}_i\right) \tag{6.90}$$

where β is the fluid friction coefficient.

Extension to Polydispersed Systems

Drag force correlations given in the previous section were derived for monodispersed systems. However, many researchers have used these correlations for bidispersed and polydispersed systems based on slip velocity and local porosity. That is, the drag force exerted on particle i is calculated using slip velocity and diameter of the particle and its local porosity. In this way, porosity is the only variable that accounts for the effect of surrounding particles on the drag force. We call this approach the *ad hoc* procedure. However, LB simulations showed that the variation of mean diameter of particles around the particle i also affects the drag force on particle i. The results showed that in this way, smaller normalized fluid-particle force $\hat{f}_i^d$ $(\varepsilon_f, \mathrm{Re}_i)$ would be exerted on smaller particles than in a monodispersed system at the same porosity [91]. To remedy this problem associated with the *ad hoc* procedure, van der Hoef *et al.* [88] developed another method in which monodispersed correlations can be used in a polydispersity system by taking into account the effect of mean diameter of surrounding particles. This procedure is explained in the following.

The normalized total drag force exerted on particle i is plotted in Figure 6.11 as a function of dimensionless particle diameter at various mean Reynolds numbers and the fluid volume fraction of $\varepsilon_f = 0.6$ in a bidispersed system. In this figure, $\langle d\rangle$ is the mean diameter of particles. The symbols are results of LB simulations [8]. The dashed lines represent normalized drag force predicted from the Beetstra *et al.* correlation [95] using the *ad hoc* procedure in which individual particle Reynolds numbers were used in the calculation. The solid lines show the results of the procedure proposed by van der Hoef *et al.* [88] and Beetstra *et al.* [95] for bidispersed and polydispersed systems. As can be seen in Figure 6.11, the normalized drag force estimated form the *ad hoc* procedure deviates largely from the results of LB simulations at low Reynolds numbers and moderately at higher Reynolds numbers. Thus, effect of the degree of polydispersion must be considered to obtain a better estimation of drag force exerted on each particle.

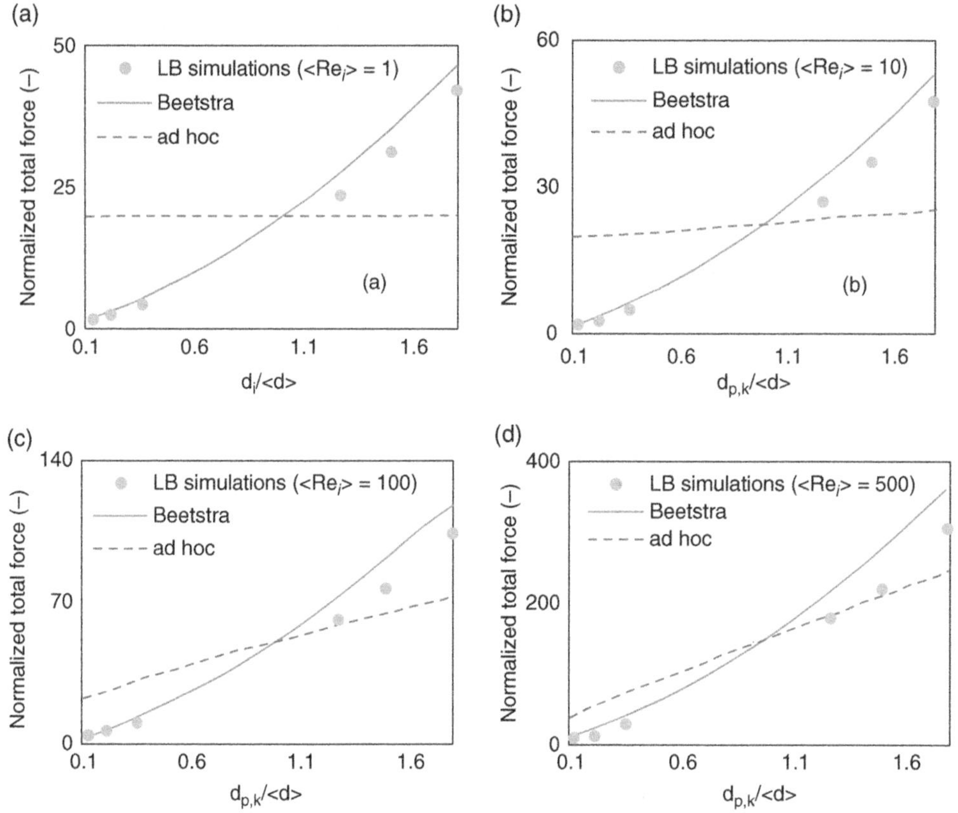

Figure 6.11 Comparison of normalized drag force calculated from LB simulations and correlations versus particle diameter for bidispersed system at $\varepsilon_f = 0.6$ and (a) $\mathrm{Re}_i = 1$, (b) $\mathrm{Re}_i = 10$, (c) $\mathrm{Re}_i = 100$, and (d) $\mathrm{Re}_i = 500$. The size ratio of the smallest to largest particle was varied between 1 : 5 and 1 : 10. Symbols represent the results of LB simulations [8], dashed lines represent the *ad hoc* procedure, and solid lines represent the correlation of Beetstra *et al.* [93, 95]. (Source: Sarkar, S., et al., Gas–solid interaction force from direct numerical simulation (DNS) of binary systems with extreme diameter ratios. *Particuology*, 2009. **7**: p. 233–237, with permission)

Consider a mixture of particles with different sizes that are classified into n_c groups of particles according to their size. Each group of particles contains N_k particles with diameter of $d_{i,k}$. For calculating the drag force on particle i, which belongs to the group k in this polydispersed system, the following step-by-step procedure should be followed:

Step I: Consider a region around the particle that contains particles of different sizes. Dimensionless parameters x_k and y_k are defined as:

$$x_k = \frac{\varepsilon_{p,k}}{\varepsilon_p} \tag{6.91}$$

$$y_k = \frac{d_{i,k}}{\langle d \rangle} \tag{6.92}$$

where ε_p and $\varepsilon_{p,k}$ are the particle volume fraction of all particles and the volume fraction of particles of type k, respectively. Here, $\langle d \rangle$ is the mean Sauter diameter:

$$\langle d \rangle = \left(\sum_{k=1}^{n_c} \frac{x_k}{d_{i,k}} \right)^{-1} \tag{6.93}$$

Using the local porosity and mean diameter, the mean Reynolds number is defined as:

$$\langle \mathrm{Re}_i \rangle = \frac{\rho_f \langle d \rangle \varepsilon_f |\vec{u} - \vec{v}_i|}{\mu_f} \tag{6.94}$$

Step II: The mean normalized drag force is calculated using the mean Reynolds number and porosity from the correlation of Beetstra *et al.* [95] as follows:

$$\langle \hat{f}_i^d \rangle = \left(\frac{180\varepsilon_p}{18\varepsilon_f^2} \right) + \varepsilon_f^2 \left(1 + 1.5\sqrt{\varepsilon_p}\right) + \left(\frac{0.413}{24\varepsilon_f^2} \right) \left(\frac{\varepsilon_f^{-1} + 3\varepsilon_p\varepsilon_f + 8.4 \langle Re \rangle_i^{-0.343}}{1 + 10^{3\varepsilon_p} \langle Re \rangle_i^{-(1+4\varepsilon_p)/2}} \right) \langle Re_i \rangle \tag{6.95}$$

Step III: The normalized drag force exerting on particle i is calculated from:

$$\hat{f}_i^d \left(\varepsilon_f, \mathrm{Re}_i\right) = \beta_k \left(\varepsilon_f, y_k\right) \langle \hat{f}_i^d \rangle \tag{6.96}$$

where β_k is a correction factor that accounts for the effects of the degree of polydispersion around particle i and $\mathrm{Re}_i = y_k \langle \mathrm{Re}_i \rangle$. Note that the mean normalized drag force is calculated from a correlation for the monodispersed system. Based on an earlier expression addressed by van der Hoef *et al.* [88], Sarkar *et al.* [91] obtained the following expression for β_k, which is valid for a bidispersed system as well as polydispersed systems with Reynolds numbers up to 500 and y_k ranging from 0.1 to 1.8 [91, 93, 95]:

$$\beta_k = \varepsilon_f y_k + 0.064 y_k^3 \tag{6.97}$$

Again, it should be mentioned that many correlations reported in literature are developed based on the normalized total force exerted on particle rather than the normalized drag force, as used here. For polydispersed systems, these two normalized interaction forces are related to each other as [91, 93]:

$$\varepsilon_f \sum_k \frac{x_k}{y_k^2} \hat{f}_i^{f-p} \left(\varepsilon_f, \mathrm{Re}_i\right) = \sum_k \frac{x_k}{y_k^2} \hat{f}_i^d \left(\varepsilon_f, \mathrm{Re}_i\right) \tag{6.98}$$

where $\hat{f}_i^{f-p}$ is normalized total force exerted on a particle i. Cello *et al.* [92] followed the approach of van der Hoef *et al.* [8, 88, 91] and proposed the following correction factor for polydispersed systems:

$$\beta_k = y_k + \frac{1-\varepsilon_f}{\varepsilon_f} \left(\frac{1 - \varepsilon_f - 0.27}{1 - 0.27} \right) \frac{y_k^2 - y_k}{\sum_{j=1}^{n_c} x_j y_j} \tag{6.99}$$

Note that in this case, the mean normalized drag force should be calculated by Equation 6.98. This correlation was validated against the experimental data and LB simulation for Reynolds numbers up to 500 and y_k ranging from 0.36 to 3.25.

This averaging procedure should be performed in a region with a certain size. Selecting a value for this size is very critical. For now, there is no definite suggestion for it. We suggest that the fluid cell, in which the particle i resides, is selected as the region. In this way, the calculated porosity that is available for each fluid cell can be directly used in calculations.

Extension to Irregular Shaped Particles

The drag force correlations were derived for spherical particles. However, many researchers have extended these correlations to particles of irregular shapes by introducing a proper correction factor to take into account the effect of non-sphericity of particles on the drag coefficient. One simple approach is to describe the shape of a particle in terms of its shape factor. Different shape factors, such as equal-volume sphere diameter, equal-surface sphere diameter, surface to volume equivalent sphere diameter,[3] and sphericity can be found in the literature to describe the effect of particle shape on the drag coefficient. Figure 6.12 shows experimental and numerical drag coefficients for regular and irregular particles.

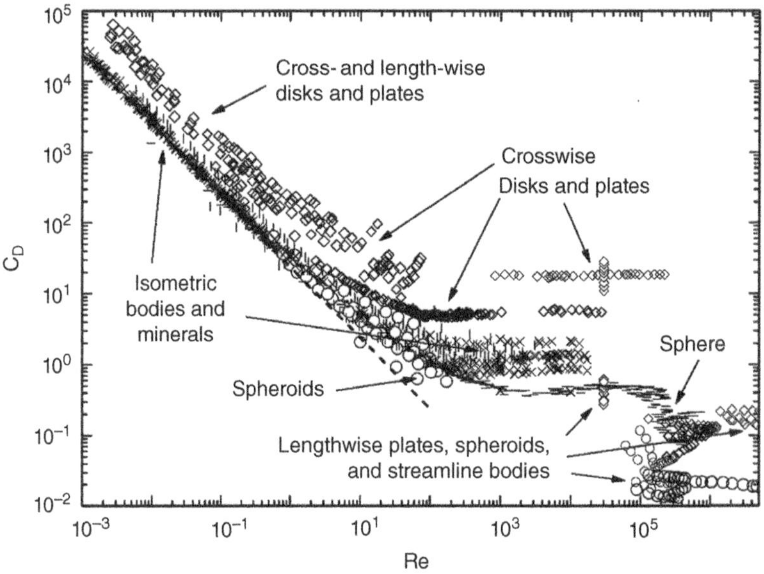

Figure 6.12 Drag coefficient of particles with different shapes as a function of Reynolds number of particle. (----) Stokes; (–) sphere; (◊) disk and plate; (×) isometric bodies; (|) minerals; and (o) spheroids and streamline bodies. (Source: Hölzer, A., Sommerfeld, M., 2009. Lattice–Boltzmann simulations to determine drag, lift and torque acting on non-spherical particles. *Computers & Fluids* **38**, 572–589, with permission)

[3] Diameter of a sphere having the same surface-to-volume ratio as the irregular particle.

By introducing the sphericity into the Ergun equation [67], the effect of shape on the drag force is considered as [97]:

$$\hat{f}_i^d\left(\varepsilon_f, \mathrm{Re}_i^V\right) = \frac{k_1 \varepsilon_p}{\varepsilon_f^2} + \frac{k_2 \psi_i}{\varepsilon_f^2} \mathrm{Re}_i^V \tag{6.100}$$

where k_1 and k_2 are functions of the particle sphericity and

$$\mathrm{Re}_i^V = \frac{\rho_f \varepsilon_f d_i^V \left|\vec{w}_i\right|}{\mu_f} \tag{6.101}$$

in which d_i^V is the diameter of the sphere with the same volume as that of the particle and particle sphericity is defined as:

$$\psi_i = \frac{\pi\left(d_i^V\right)^2}{S_i} \tag{6.102}$$

where S_i is actual surface area of the particle. Rhodes [98] stated that the Ergun correlation [67] can be used for non-spherical particles with its original constants. Some researchers demonstrated that there is a large deviation from the correlation of Ergun [67] when applied to non-spherical particles and proposed some modifications [99, 100].

Among the many correlations that have been modified to estimate the fluid drag on non-spherical particles, the correlation proposed by Di Felice [85] has been widely used because it is the product of two terms: the first term reflects the drag on an isolated particle and the second term is the effect of surrounding particles. Since calculating the drag force on an isolated non-spherical particle from the Di Felice correlation is relatively easy, this correlation is widely used for non-spherical particles. The Di Felice [85] correlation is also used in the following from:

$$\hat{f}_i^d\left(\varepsilon_f, \mathrm{Re}_i^V\right) = \frac{C_d^{ns}}{24} \mathrm{Re}_i^V \varepsilon_f^{-\chi} \tag{6.103}$$

$$\chi = 3.7 - 0.65 e^{\left(-0.5\left(1.5 - \log_{10} \mathrm{Re}_i^V\right)^2\right)} \tag{6.104}$$

where C_d^{ns} is the drag coefficient of an isolated non-spherical particle, which can be determined by different correlations, either based on macroscopic experimental observations [97, 100–102] or the microscopic lattice Boltzmann method simulations [96].

Haider and Levenspiel [101] introduced the following general correlation for evaluating the drag coefficient on an isolated non-spherical particle:

$$C_d^{ns} = \frac{24}{\mathrm{Re}_i^V}\left[1 + A_1\left(\mathrm{Re}_i^V\right)^{A_2}\right] + \frac{A_3}{1 + A_4/\mathrm{Re}_i^V} \tag{6.105}$$

$$A_1 = e^{\left(2.3288 - 6.4581\psi_i + 2.4486\psi_i^2\right)} \tag{6.106a}$$

$$A_2 = 0.0964 + 0.5565\psi_i \tag{6.106b}$$

$$A_3 = 73.69e^{(-5.0748\psi_i)} \tag{6.106c}$$

$$A_4 = 5.378\, e^{(6.2122\psi_i)} \tag{6.106d}$$

Ganser [100] considered both shape and orientation of the non-spherical particle and proposed the following drag correlation for $k_1 k_2 Re_i^V \leq 10^5$:

$$\frac{C_d^{ns}}{k_2} = \frac{24}{k_1 k_2 Re_i^V}\left(1 + 0.1118\left(k_1 k_2 Re_i^V\right)^{0.65657}\right) + \frac{0.4305}{1 + 3305/k_1 k_2 Re_i^V} \tag{6.107}$$

where k_1 and k_2 are Stokes and Newton shape factors, respectively. Shape of particle affects the drag coefficient more in the intermediate and Newton regions than in the Stokes–Einstein region [98]. However, Holzer and Sommerfeld [96] proved that the drag force, especially at high Reynolds numbers, strongly depends on particle shape, angle of incidence, and particle rotation.

Tran-Cong *et al.* [103] proposed another correlation by taking both shape and orientation of the non-spherical particle into account:

$$C_d^{ns} = \frac{24}{Re_i^V}\frac{d_i^{A^P}}{d_i^V}\left(1 + \frac{0.15}{\sqrt{c_i}}\left(\frac{d_i^{A^P}}{d_i^V}Re_i^V\right)^{0.687}\right) + \frac{0.42\left(\frac{d_i^{A^P}}{d_i^V}\right)^2}{\sqrt{c_i}\left(1 + 4.25\times 10^4\left(\frac{d_i^{A^P}}{d_i^V}Re_i^V\right)^{-1.16}\right)} \tag{6.108}$$

$$0.15 < Re_i^V < 1500, \quad 0.8 < d_i^{A^P}/d_i^V < 1.5, \quad 0.4 < c_i < 1$$

where the diameter of the sphere having the same projected area of the particle (A_i^P) is:

$$d_i^{A^P} = \sqrt{4A_i^P/\pi} \tag{6.109}$$

and the particle circularity (also called surface sphericity) is defined as:

$$c_i = \frac{\pi d_i^{A^P}}{P_i^P} \tag{6.110}$$

where P_i^P is the projected perimeter of particle in the direction of motion.

Based on Ganser [100] and Tran-Cong *et al.* [103], Holzer and Sommerfeld [102] presented the following drag coefficient for an isolated non-spherical particle, which is valid over the full practical range of Re_i^V:

$$C_d^{ns} = \frac{8}{Re_i^V}\frac{1}{\sqrt{\psi_i^{\parallel}}} + \frac{16}{Re_i^V}\frac{1}{\sqrt{\psi_i}} + \frac{3}{\sqrt{Re_i^V}}\frac{1}{\psi_i^{3/4}} + 0.42\times 10^{0.4(-\log_{10}\psi_i)^2}\frac{1}{\psi_i^{\perp}} \tag{6.111}$$

where the crosswise sphericity is the ratio of the projected area of the volume equivalent sphere to the projected area of the particle perpendicular to the flow and is defined by:

$$\psi_i^{\perp} = \frac{\pi \left(d_i^V\right)^2}{4A_i^p} \tag{6.112}$$

and $\psi_i^{\parallel}$ is lengthwise sphericity is the ratio of the cross-sectional area of the volume equivalent sphere to the difference between half the surface area and the average projected area parallel to the flow. Since the calculation of lengthwise sphericity is complex, an approximate correlation of Holzer and Sommerfeld [102] has been introduced by replacing the complex lengthwise sphericity with crosswise sphericity:

$$C_d^{ns} = \frac{8}{Re_i^V}\frac{1}{\sqrt{\psi_i^{\perp}}} + \frac{16}{Re_i^V}\frac{1}{\sqrt{\psi_i}} + \frac{3}{\sqrt{Re_i^V}}\frac{1}{\psi_i^{3/4}} + 0.42\times 10^{0.4(-\log_{10}\psi_i)^2}\frac{1}{\psi_i^{\perp}} \tag{6.113}$$

This approximate correlation has been widely used and validated over the full range of Re_i^V [70, 104].

6.2.4.2 Unsteady Forces

Unsteady drag and added mass forces are both unsteady forces that are generated by the non-stationary motion of particle and fluid.

Unsteady Drag Force

Unsteady drag force is caused by the lagging time response of the boundary layer (the time required for penetration of the generated momentum into the body of fluid) to accelerate particles. In other words, the unsteady drag force is generated by temporal development of the viscous region near particles and provides a prompt flow resistance. Particles, therefore, experience an unsteady force due to the instability of the boundary layer. The unsteady drag force is also referred to as Basset force or Boussinesq–Basset force [1]. Since this force depends on the penetration history [23], it is called history force and is defined based on the integral of all past particle accelerations. The Basset force is negligible in gases and at low particle accelerations [105]. It has the following form for an isolated spherical particle at low Reynolds numbers in a fluid at rest [1, 4]:

$$\vec{f}_i^{ud} = \frac{3}{2}d_i^2\sqrt{\pi\rho_f\mu_f}\int_0^{\tau_p}\frac{d\vec{v}_i/dt}{\sqrt{\tau_p - t}}dt \tag{6.114}$$

in which τ_p is the particle momentum response time, defined from Equation 6.5 at low fluid Reynolds number.

Equation 6.114 was then extended to include the effect of convective acceleration of fluid around particles at higher Reynolds numbers by introducing a correction factor C_B [106] and replacing $d\vec{v}_i/dt$ by derivative of the relative velocity $d\vec{w}_i/dt$ [107–109]:

$$\vec{f}_i^{ud} = \frac{3}{2}d_i^2\sqrt{\pi\rho_f\mu_f}C_B\int_0^{\tau_p}\frac{d\vec{w}_i/dt}{\sqrt{\tau_p - t}}dt \tag{6.115}$$

$$C_B = 0.48 + \frac{0.52\left(A_i^n\right)^3}{\left(1 + A_i^n\right)^3} \tag{6.116}$$

$$A_i^n = \frac{\left|d\vec{w}_i / dt\right|}{\vec{w}_i^2} d_i \tag{6.117}$$

Reeks and McKee [110] introduced the initial relative velocity ($\vec{w}_{i,0}$) into Equation 6.115 and proposed the following formula:

$$\vec{f}_i^{ud} = \frac{3}{2} d_i^2 \sqrt{\pi \rho_f \mu_f} \left(\frac{\vec{w}_{i,0}}{\sqrt{\tau_p}} + \int_0^{\tau_p} \frac{d\vec{w}_i / dt}{\sqrt{\tau_p - t}} dt \right) \tag{6.118}$$

This equation has been used by many researchers [108, 109].

Added Mass Force

When a particle is accelerated in a fluid, it can also accelerate a part of its surrounding fluid. The corresponding force of displaced fluid is called added or virtual mass force. This is due to the fact that it is equivalent to adding a virtual mass to the sphere. According to Hjelmfel and Mockros [105], the added mass is insignificant at small density ratios ($\rho_f / \rho_p \sim 10^{-3}$). For a particle moving in an inviscid fluid at rest, the added mass force can be analytically derived as [2, 111]:

$$\vec{f}_i^a = \frac{\pi}{12} d_i^3 \rho_f \frac{d\vec{w}_i}{dt} \tag{6.119}$$

This formula has been used by researchers in classical Eulerian-Lagrangian modeling (e.g., Zhu *et al.* [109]). The following equation was developed by Odar and Hamilton [106]:

$$\vec{f}_i^a = \frac{1}{2} C_A \rho_f V_i \left(\left(\frac{\partial \vec{u}}{\partial t} + \vec{u}(\nabla . \vec{u}) \right) - \frac{d\vec{v}_i}{dt} \right) \tag{6.120}$$

$$C_A = 2.1 - \frac{0.132}{\left(0.12 + \left(A_i^n\right)^{-2}\right)} \tag{6.121}$$

More recently, Michaelides and Roig [112] reinterpreted the data of the Odar and Hamilton [106] and found that the added mass coefficient (C_A) is not a function of the particle acceleration and should be equal to unity.

6.2.4.3 Lift Force

Lift force is generally normal to the relative motion of fluid and particle and is due to particle rotation and fluid shear stress. This force involves the shear-dependent Saffman lift force, caused by the fluid velocity gradient, and the rotational-dependent Magnus lift force, imposed by particle contact and rebound from a surface.

Saffman Lift Force

The Saffman force is caused by the pressure difference on a non-rotating particle under a non-uniform shear velocity field, as shown in Figure 6.13. Saffman [113] calculated this force as follows for an isolated particle at low Reynolds number [111]:

$$\vec{f}_i^{Saffman} = 1.61 d_i^2 \sqrt{\mu_f \rho_f}\, |\vec{\omega}|^{-0.5} \left(\vec{w}_i \times \vec{\omega}\right) \tag{6.122}$$

$$\mathrm{Re}_i \ll \mathrm{Re}_i^s \ll 1 \tag{6.123}$$

where $\vec{\omega}$ is the curl of velocity vector:

$$\vec{\omega} = \nabla \times \vec{u} \tag{6.124}$$

and Re_i^s is the shear Reynolds number based on the velocity difference between bottom and top of the particle:

$$\mathrm{Re}_i^s = \frac{\rho_f d_i^2 |\nabla \vec{u}|}{\mu_f} \tag{6.125}$$

Drew [114] proposed following correlation for evaluating the Saffman lift force:

$$\vec{f}_i^{Saffman} = 1.615 C_{lS} \vec{w}_i d_i^2 \sqrt{\mu_f \rho_f}\, |\vec{\omega}|^{-0.5} \left(\vec{w}_i \times \vec{\omega}\right) \tag{6.126}$$

where C_{lS} is the Saffman lift coefficient given by Mei [115]:

$$C_{lS} = \begin{cases} e^{-0.1\mathrm{Re}_i} + 0.3314\sqrt{\alpha_i}\left(1 - e^{-0.1\mathrm{Re}_i}\right) & \mathrm{Re}_i \le 40 \\ 0.0524\sqrt{\alpha_i \mathrm{Re}_i} & \mathrm{Re}_i > 40 \end{cases} \tag{6.127}$$

where

$$\alpha_i = \frac{|\vec{\omega}| d_i}{2|\vec{w}_i|} \tag{6.128}$$

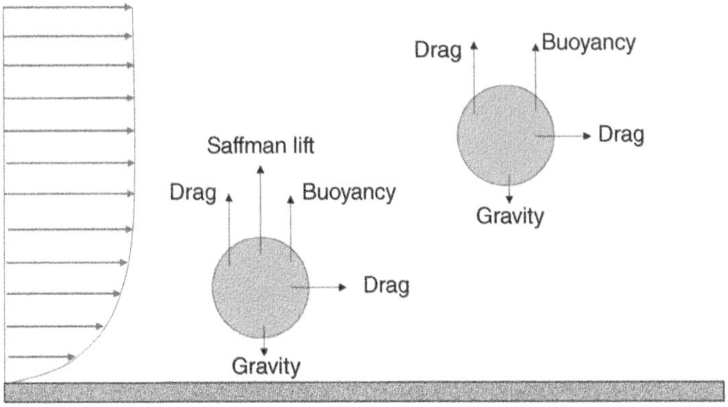

Figure 6.13 Saffman lift force on a particle in a non-uniform flow field

Magnus Lift Force

Magnus lift force is related to the particle rotation in a uniform flow, as shown in Figure 6.14. The Magnus lift force is caused by asymmetric pressure distribution around the particle due to velocity difference between bottom and top of the particle. Particle rotation is caused by sources other than the velocity gradient, for example, particle contact or rebound of particle from a wall.

The Magnus lift force is given by Rubinow and Keller [116]:

$$\vec{f}_i^{Magnus} = \frac{1}{8} C_{lM} \vec{w}_i^2 \pi d_i^2 \rho_f \sqrt{\mu_f \rho_f} \frac{(\vec{\omega}_r \times \vec{w}_i)}{|\vec{\omega}_r||\vec{w}_i|} \tag{6.129}$$

where C_{lM} is the Magnus lift coefficient given by Lun and Liu [117]:

$$C_{lM} = d_i \frac{|\vec{\omega}_r|}{|\vec{w}_i|} \begin{cases} 1 & \mathrm{Re}_i \leq 1 \\ \left(0.178 + 0.822 \mathrm{Re}_i^{-0.522}\right) & \mathrm{Re}_i > 1 \end{cases} \tag{6.130}$$

and

$$\vec{\omega}_r = \frac{1}{2} \vec{\omega} - \vec{\omega}_i \tag{6.131}$$

Holzer and Sommerfeld [96] showed that the lift force for spheres and spheroid particles is affected by fluid shear and particle rotation. Hilton and Cleary [104] used Saffman and Magnus lift forces in their simulation for modeling the pneumatic conveying of particles with different shapes. However, they argued that there is no clear approach to modify these expressions to account for the particle shape and presence of nearby particles. Many researchers [118–121] have considered the previously mentioned Saffman and Magnus lift forces acting

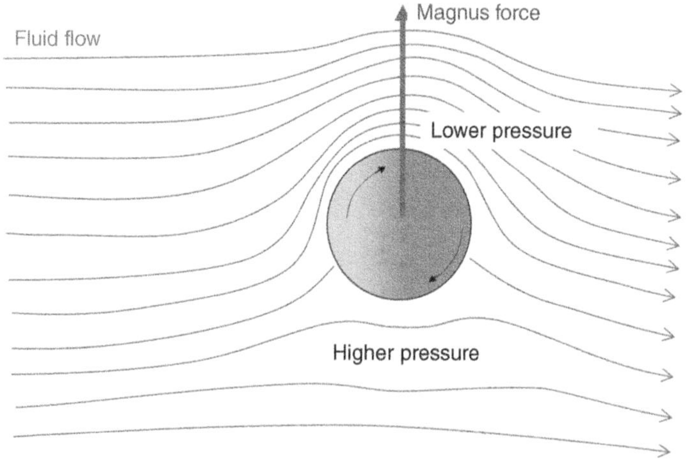

Figure 6.14 Magnus lift force on a particle rotating in a uniform flow field

on an individual particle in their simulations. For example, Zhong *et al.* [118] and Ren *et al.* [119, 121] used Saffman and Magnus lift forces in their three-dimensional CFD-DEM simulations of a conical-cylindrical spouted bed. They showed that values of Saffman and Magnus lift forces are much smaller than drag, contact, and gravitational forces. The latter forces are greater in the spout region as well as the interface between spout and annulus due to the significant gas velocity gradient. Zhong *et al.* [118] showed that the shear induced Saffman lift force and rotation induced Magnus lift force contribute to the entrainment of particles from the annular region into the spout region.

6.2.4.4 Rotational Drag

Rotation drag incorporates the drag on a rotating particle from the surrounding fluid and expresses the rotational resistance due to inertia of fluid. The rotation Reynolds number is defined as:

$$\mathrm{Re}_i^{\vec{\omega}} = \frac{\rho_f \left|\vec{\omega}_r\right| d_i^2}{4\mu_f} \tag{6.132}$$

The rotation drag (also called Stokesian rotational drag) at low rotating Reynolds number is given by [2]:

$$\vec{M}_i^d = \pi d_i^3 \mu_f \vec{\omega}_r \tag{6.133}$$

Dennis *et al.* [122] performed an analytical study on the torque required to rotate a sphere in a viscous fluid and found the following formula for estimating the rotation drag at rotation Reynolds numbers between 20 and 1000 [2]:

$$\vec{M}_i^d = -2.01 d_i^3 \mu_f \vec{\omega}_i \left(1 + 0.201\left(\mathrm{Re}_i^{\vec{\omega}}\right)^{0.5}\right) \tag{6.134}$$

Holzer and Sommerfeld [96] argued that the rotational drag on spheroid particles is strongly affected by particle rotation and fluid shear. Hilton and Cleary [104] incorporated the rotational drag on spinning particles caused by the surrounding gas into their model. However, it is not clear how much this force has affected their simulation results.

6.3 Energy Coupling

In many occasions, the granular flow is not isothermal. In this case, it is necessary to obtain the temperature of particles as well as the temperature distribution in the fluid phase. In order to calculate these temperatures during the simulation, it is necessary to consider energy conservation equations for both particles and fluid. Obviously, equations of the Newton's second law for motion of particles (described in Chapter 2) and equation of motion of fluid phase (described earlier in this chapter) remain unchanged for both phases and equations of energy for these phases should be solved simultaneously to obtain temperature profiles.

6.3.1 Governing Equations

Governing equations for obtaining temperatures of particles and fluid are described in the following. In all these equations, we consider the following as the positive direction of heat transfer:

- from fluid to particle
- from wall to particle
- from particle j to particle i
- heat generated by an exothermic reaction or an exothermic physical change.

6.3.1.1 Governing Equation for Particles

The unsteady equation of conservation of energy for a particle can be written as:

$$m_i c_{p,i} \frac{dT_i}{dt} = S_{p,h} \tag{6.135}$$

In this equation, $S_{p,h}$ is the net rate of heat transferred to particle i, m_i, $c_{p,i}$, and T_i are the mass, heat capacity, and temperature of particle i, respectively. The differential equation (6.135) should be written for all particles in the system and the set of equations should be solved by knowing the initial temperature for all particles:

$$T_i = T_{0,i} \tag{6.136}$$

where $T_{0,i}$ is the temperature of particle i at time $t = 0$.

An important assumption in Equation 6.135 is that the inner thermal resistance of particles can be neglected and the temperature is uniform within the particle. This is true when

$$Bi = \frac{h_i d_i}{k_i} \ll 1 \tag{6.137}$$

where h_i is the heat transfer coefficient corresponding to i, d_i is the diameter of particle i, and k_i is the thermal conductivity of particle i. In fact, the lumped formulation, given as Equation 6.135, is valid when $Bi < 0.1$.

The energy source term in Equation 6.135, $S_{p,h}$, is the net rate of heat transferred to the particle and is comprised of several terms, including:

- the rate of heat exchanged between particle i and neighboring particles due to conduction ($q_{i,j}$),
- the rate of particle-wall heat exchanged ($q_{i,w}$),
- the rate of the heat exchanged between particle and surrounding fluid due to convection ($q_{i,f}$),
- radiation heat transfer to the particle from surrounding environment ($q_{i,r}$), and
- the rate of heat generated in/on the particle due to chemical reaction or phase change ($q_{i,c}$).

In other words,

$$S_{p,h} = q_{i,j} + q_{i,w} + q_{i,f} + q_{i,r} + q_{i,c} \tag{6.138}$$

It is worth noting that not all these heat transfer mechanisms may exist in a specific problem. When solving a problem, only rates of those mechanisms that exist in the problem should be kept on the right-hand side of Equation 6.138.

6.3.1.2 Governing Equation for Fluid

The volume-averaged equation of fluid phase energy is:

$$\frac{\partial(\varepsilon_f \rho_f E_f)}{\partial t} + \nabla.\left[\vec{u}\varepsilon_f(\rho_f E_f + p)\right] = \nabla.\left(\varepsilon_f k_f^{eff} \nabla T_f\right) + S_{f,h} \tag{6.139}$$

in which T_f is the temperature of fluid. Energy of fluid in this equation is given by:

$$E_f = h_f - \frac{p}{\rho_f} + \frac{\vec{u}.\vec{u}}{2} \tag{6.140}$$

where h_f is the specific enthalpy of fluid. Therefore, in terms of heat capacities, Equation 6.139 can be written as:

$$\frac{\partial}{\partial t}\left[\varepsilon_f \rho_f \left(c_{v,f} T_f + \frac{\vec{u}.\vec{u}}{2}\right)\right] + \nabla.\left[\vec{u}\varepsilon_f \rho_f \left(c_{p,f} T_f + \frac{\vec{u}.\vec{u}}{2}\right)\right] = \nabla.\left(\varepsilon_f k_f^{eff} \nabla T_f\right) + S_{f,h} \tag{6.141}$$

where $c_{v,f}$ and $c_{p,f}$ are the heat capacities of fluid at constant volume and constant pressure, respectively. Subtracting the mechanical energy equation from this equation and assuming that the pressure is constant throughout the process, Equation 6.141 can be simplified to [23]:

$$\frac{\partial}{\partial t}\left(\varepsilon_f \rho_f c_{p,f} T_f\right) + \nabla.\left(\vec{u}\varepsilon_f \rho_f c_{p,f} T_f\right) = \nabla.\left(\varepsilon_f k_f^{eff} \nabla T_f\right) + S_{f,h} \tag{6.142}$$

Note that in the unsteady term, $c_{v,f}$ in Equation 6.141 is changed to $c_{p,f}$ in Equation 6.142 after this change. Effective fluid thermal conductivity in these equations includes molecular and turbulent thermal conductivities:

$$k_f^{eff} = k_f + k_{f,t} \tag{6.143}$$

where k_f and $k_{f,t}$ are molecular and turbulent thermal conductivity of fluid.

It should be noted that the correct formulation of the conductive heat transfer in the heat balance Equations 6.142 (the first term on the right-hand side) is the one shown here. This term is sometimes used as $\nabla.\left(k_f^{eff} \nabla T_f\right)$, that is, without multiplication of porosity. However, since this term corresponds to the heat conduction of the fluid phase only, rather than the whole mixture, it should be multiplied by porosity, as shown in Equation 6.142. Effect of this mis-formulation cannot be identified in some simulation results since the rate of conductive heat transfer is negligible in many cases, especially when the fluid is a gas (low conductivity) and the rate of conductivity is negligible compared to other terms in Equation 6.142.

The energy source term, $S_{f,h}$, in these equations is the net rate of heat transferred to the fluid per unit volume of fluid by various mechanisms, including:

- the rate of the heat exchanged between fluid and particles per unit volume of fluid ($Q_{f,p}$),
- the rate of the heat exchanged between fluid and wall per unit volume of fluid ($Q_{f,w}$),
- the rate of the heat generated due to chemical reaction per unit volume of fluid ($Q_{f,R}$),
- the rate of the mechanical work done by friction at wall ($W_{friction}$), and
- the rate of work done on fluid by viscous forces ($W_{viscous}$).

In other words,

$$S_{f,h} = Q_{f,p} + Q_{f,w} + Q_{f,R} + W_{friction} + W_{viscous} \tag{6.144}$$

Viscous and friction forces are significant at high shear flows and where the fluid velocity is very high. Examples of such cases are the pneumatic transport of solid particles and high shear fluid–solid mixers. However, these two terms can be neglected in most applications, such as fluidized beds. Therefore, these two terms are not discussed next.

6.3.1.3 Coupling between Particle and Fluid Equations

Some terms in the heat source term of the particle equation of heat conservation, $S_{p,h}$ in Equation 6.135, is related to the energy source term of the fluid equation of energy, $S_{f,h}$ in Equation 6.142. In general, there are four kinds of heat sources in these two terms: exchange between fluid and particle, exchange between particles, exchange between wall and fluid/particle, and heat generation in fluid/particle. These heat sources are illustrated in Figure 6.15. Equations of energy for particle and fluid, that is, Equations 6.135 and 6.142, are coupled with each other through the first kind, that is, exchange between fluid and particle. Here, we briefly describe how these equations are coupled.

Equations of energy for particle and fluid are coupled with each other through $Q_{f,p}$ and $q_{i,f}$ in these equations. The volume-averaged equation of energy conservation for fluid is obtained by

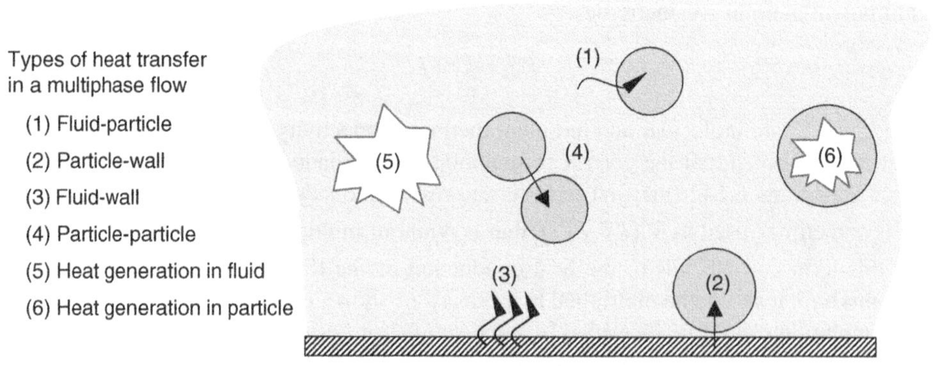

Figure 6.15 Various mechanisms of heat exchange between/in phases

holding energy balance over a fluid cell in the fluid–particle system. Therefore, when we solve this equation for each fluid cell, the term $Q_{f,p}$ should be calculated based on the volume of the fluid cell and the particles that it contains.

In Figure 6.16, we illustrated an arbitrary fluid cell and particles in it. In order to evaluate $Q_{f,p}$ convective heat transfer rate of all particles in the fluid computational cell should be taken into account. Accordingly, we have:

$$Q_{f,p} = -\frac{1}{V_{cell}} \sum_{i=1}^{k_c} q_{i,f} \tag{6.145}$$

where $q_{i,f}$ is the rate of heat transferred from fluid to the particle i. Since the positive direction of $q_{i,f}$ is toward the particle, the minus sign is used in this equation to emphasize that it is a heat sink for the fluid.

Applying Equation 6.145 is straightforward for particles that are completely inside the fluid cell for which the heat transfer area is the surface area of the particle. However, in many cases, there are particles that are not entirely located inside the fluid cell and a portion of such particles may fall inside a neighboring cell. This situation is sketched in Figure 6.16. Particles A and B are partially in the fluid cell and a part is inside a neighboring cell. We can handle these particles in two ways. The first way is to consider that the particle belongs to the fluid cell if its center is inside the cell. In this way, particle A (whose center is within the boundaries of the cell) is considered among the particles of this cell and its heat transfer rate in the summation of Equation 6.145 should be evaluated based on the total surface area of the particle. However, particle B (whose center is outside the boundaries of the cell) is excluded from this summation.

The second approach is to consider only a portion of the surface of the particle in the summation of fluid–particle heat exchanged inside this fluid cell. In this manner, both particles

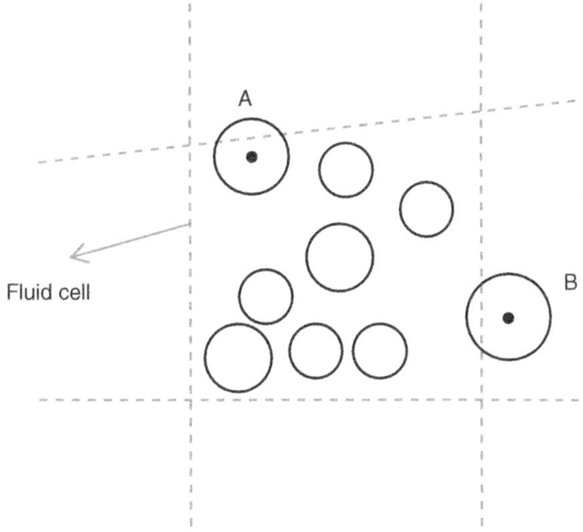

Figure 6.16 The fluid computational cell and corresponding particles for calculating the fluid–particle convective heat transfer

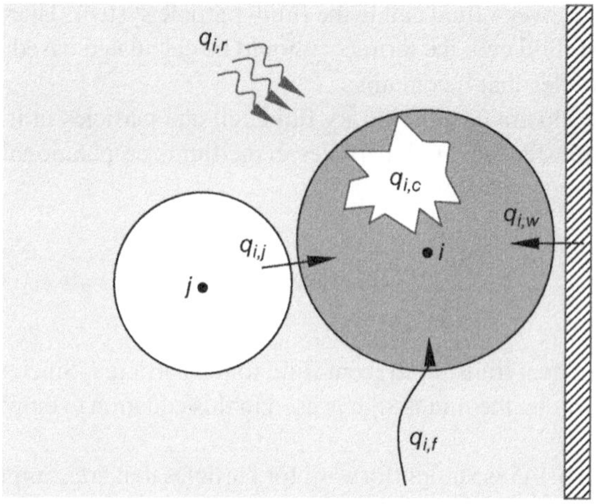

Figure 6.17 Different heat exchange mechanisms between particles and surrounding objects

A and B appear in the summation or Equation 6.145, but the corresponding heat transfer rate should be evaluated based on the transfer area equal to the portion of particle that is within the fluid cell. Therefore, Equation 6.145 should be rewritten as:

$$Q_{f,p} = -\frac{1}{V_{cell}} \sum_{i=1}^{k_V} f_{A,i}\, q_{i,f} \tag{6.146}$$

where $f_{A,i}$ is the fraction of particle surface that resides in the fluid cell. When the size of particle is very small compared to the size of the fluid cell, we expect that both approaches lead to the same temperature distribution of solids and fluid.

It is worth mentioning that $q_{i,f}$ in Equations 6.145 and 6.146 should be estimated from the rate of convective heat transfer between particle in a fluid cell and the surrounding fluid. Of course, the fluid radiation between particle and fluid can be ignored due to low fluid emissivity, especially in case of gas. Also, note that physical properties, heat of reaction, and rate constants are functions of temperature. Therefore, they can also be considered as parameters that couple momentum and energy equations. However, this type of coupling is implicit and we do not discuss the coupling for these terms in this chapter.

6.3.2 Rates of Heat Transfer for Particles

In this section, we describe methods and formulas necessary to calculate various heat sources for particles, given in Equation 6.138. Different mechanisms of heat exchange among a particle and other objects are demonstrated in Figure 6.17. We remind the reader again that the positive direction of heat transfer is always considered to be "toward particle *i*."

6.3.2.1 Conductive Heat Transfer

Particles can exchange heat when in contact or when they are in proximity of each other. Relative position of two colliding particles with overlap and two non-colliding particles are shown schematically in Figure 6.18. Heat can be transferred to the particle by two main

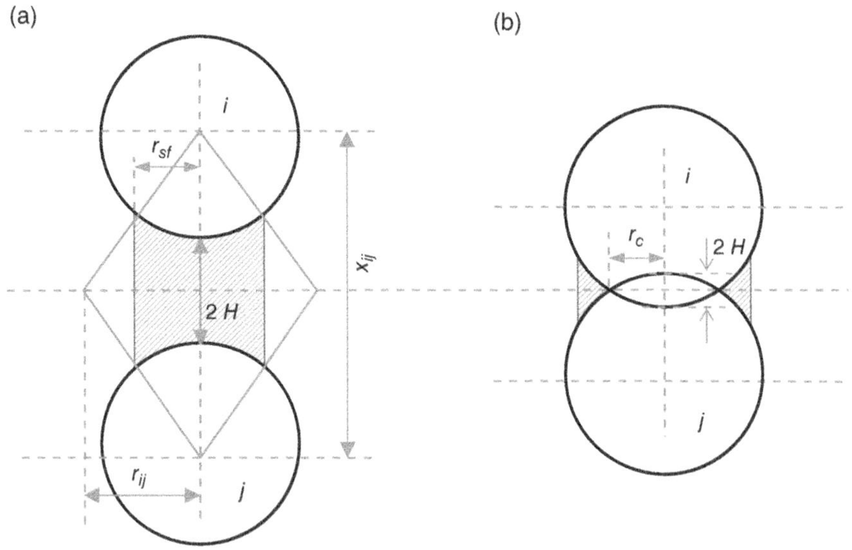

Figure 6.18 Relative position of two spherical particles: (a) non-colliding and (b) colliding with overlap. (Source: Zhou, Z.Y., A.B. Yu, and P. Zulli, Particle Scale Study of Heat Transfer in Packed and Bubbling Fluidized Beds. *AIChE Journal*, 2009. **55**: p. 868–884, with permission)

mechanisms: particle–fluid–particle conduction (Figure 6.18a) and direct particle-to-particle conduction (Figure 6.18b). Therefore, we have:

$$q_{i,j} = q_{i,j}^{p-p} + q_{i,j}^{p-f-p} \tag{6.147}$$

There are two kinds of particle–particle heat transfer due to conduction: due to collision and due to static contact. When two particles collide, they exchange heat by conduction through their contact area. The exchanged heat during the collision is calledcollisional heat transfer. Zhou *et al.* [124] simulated the transient heat transfer between two colliding particles and determined the rate of this exchange as follows:

$$q_{i,j}^{p-p} = c' \frac{\pi r_c^2 t_c^{-1/2}}{\left(\rho_i c_{p,i} k_i\right)^{-1/2} + \left(\rho_j c_{p,j} k_j\right)^{-1/2}} \left(T_j - T_i\right) \tag{6.148}$$

where

$$c' = \frac{0.435}{c_1}\left(\sqrt{c_2^2 - 4c_1\left(c_3 - Fo\right)} - c_2\right) \tag{6.149}$$

$$c_1 = -2.3\left(\frac{\rho_i c_{p,i}}{\rho_j c_{p,j}}\right)^2 + 8.909\left(\frac{\rho_i c_{p,i}}{\rho_j c_{p,j}}\right) - 4.235 \tag{6.150}$$

$$c_2 = 8.169\left(\frac{\rho_i c_{p,i}}{\rho_j c_{p,j}}\right)^2 - 33.77\left(\frac{\rho_i c_{p,i}}{\rho_j c_{p,j}}\right) + 24.885 \tag{6.151}$$

$$c_3 = -5.785\left(\frac{\rho_i c_{p,i}}{\rho_j c_{p,j}}\right)^2 + 24.464\left(\frac{\rho_i c_{p,i}}{\rho_j c_{p,j}}\right) - 20.511 \tag{6.152}$$

$$Fo = \frac{\alpha t_c}{r_c^2} = \frac{k_i t_c}{\rho_i c_{p,i} r_c^2} \tag{6.153}$$

In Equation 6.148, r_c is the contact radius (as shown in Figure 6.18), which is calculated from:

$$r_c = \left(\frac{5m_{ij}}{4E_{ij}} R_{ij}^2 \left|\vec{v}_{ij}^{\,n}\right|^2\right)^{1/5} \tag{6.154}$$

and t_c is the contact time that can be obtained from:

$$t_c = 2.94\left(\frac{5m_{ij}}{4E_{ij} R_{ij}^2 \left|\vec{v}_{ij}^{\,n}\right|^2}\right)^{2/5} \tag{6.155}$$

in which $\vec{v}_{ij}^{\,n}$ is the normal relative velocity vector between particles i and j and:

$$m_{ij} = \frac{m_i m_j}{m_i + m_j} \tag{6.156}$$

$$R_{ij} = \frac{R_i R_j}{R_i + R_j} \tag{6.157}$$

$$E_{ij} = \frac{4/3}{\dfrac{1-\nu_i^2}{E_i} + \dfrac{1-\nu_j^2}{E_j}} \tag{6.158}$$

where ν_i and E_i are the Poisson ratio and Young's modulus of particle i, respectively.

Equation 6.148 provides the heat transfer rate during the collision of two particles. In situations where particles are in static contact, that is, there is a permanent contact between particles (like in fixed beds) or they stick together (like in granulators), the conduction heat transfer rate should be obtained from [125]:

$$q_{i,j}^{p-p} = \frac{4r_c}{\left(1/k_i + 1/k_j\right)}\left(T_j - T_i\right) \tag{6.159}$$

It is important to distinguish between two conductive heat transfer mechanisms characterized by Equations 6.148 and 6.159. When particles are always in static contact (which is the case in fixed beds, agglomerates of sticky particles, and granules formed in granulators), the heat transfer rate should be obtained from Equation 6.148. In a system in which particles are moving, but may stick together, if the contact lasts longer than the contact time given by Equation 6.155, heat transfer rate after the contact time should be evaluated first from Equation 6.148 up to the contact time and from Equation 6.159 after that.

As mentioned previously, conductive heat transfer takes place through particle-fluid-particle conduction. Here, we present the model developed by Cheng *et al.* [125] as modified by Zhou *et al.* [123]. According to this model, particle–fluid–particle heat transfer rate due to conduction can be obtained from:

$$q_{i,j}^{p-f-p} = \left(T_j - T_i\right) \int_{r_{sij}}^{r_{sf}} \frac{2\pi r dr}{\left[\sqrt{R_i^2 - r^2} - \frac{r}{r_{ij}}\left(R_i + H\right)\right]\left(\frac{1}{k_i} + \frac{1}{k_j}\right) + \frac{2}{k_f}\left[\left(R_i + H\right) - \sqrt{R_i^2 - r^2}\right]} \tag{6.160}$$

in which

$$H = \frac{x_{ij}}{2} - R_i \tag{6.161}$$

$$r_{sf} = \frac{R_i r_{ij}}{\sqrt{r_{ij}^2 + \left(R_i + H\right)^2}} \tag{6.162}$$

$$r_{ij} = \sqrt{\frac{3V_{ij}}{\pi x_{ij}}} \tag{6.163}$$

In Equation 6.160, the lower limit of the integral, r_{sij}, is 0 when the two particles are not in physical contact (non-colliding condition) and should be set to r_c after the collision (colliding condition). In Equation 6.163, V_{ij} is the volume of Voronoi polyhedra between particles. Voronoi volume changes when particles move and calculation of this transient volume becomes time consuming. Therefore, instead of Equation 6.163, a simplified model is usually used for evaluating r_{ij}. According to Yang *et al.* [126], the value of r_{ij} can be estimated based on the face area of Voronoi polyhedra as follows:

$$r_{ij} = 0.56 R_i \left(1 - \varepsilon_i\right)^{-1/3} \tag{6.164}$$

here, ε_i is the local porosity around particle i, which can be considered as the average porosity of the cell in which the particle resides. The particle–fluid–particle conductive heat transfer rate decreases sharply when the distance between two non-contacting particles becomes larger. Zhou *et al.* [123] showed that heat transfer due to this mechanism can be ignored when H/R_i is 0.5 or greater. Therefore, $H/R_i = 0.5$ can be considered as the cut-off value, beyond which particle–fluid–particle conductive heat transfer can be ignored.

Although Equation 6.160 has been derived for a bed of single size particles, we suggest using the same formula for estimating particle–fluid–particle heat transfer rate due to conduction in the case of two particles of different sizes by replacing the radius of particle, R_i, in this formula with the following value:

$$R^* = \frac{x_{ij} - 2H}{2} \tag{6.165}$$

where x_{ij} is the center-to-center distance of particles and $2H$ is the face to face distance between particles.

In many soft-sphere DEM simulations, a very low value of Young's modulus is considered in the simulation in order to increase the particle time step and perform the calculations faster. This of course leads to a larger contact radius between colliding particles, as shown in Equation 6.154. The conductive particle–particle heat transfer is directly related to the contact radius. Therefore, the calculated particle–particle heat exchanged rate would become greater than that in the reality. To mitigate this problem, Zhou *et al.* [127] suggested using a corrected contact radius in their thermal model. Considering Equation 6.154, the contact radius is related to Young's modulus as $r_c \propto E_{ij}^{1/5}$. A correction factor for the contact radius is introduced based on the real value of the Young's modulus, $E_{ij,0}$ and the selected value in the simulation, E_{ij}:

$$c = \left(\frac{E_{ij}}{E_{ij,0}} \right)^{1/5} \tag{6.166}$$

where r_c' is the corrected contact radius for the particle-particle heat transfer. The corrected contact radius should be used in Equations 6.148 and 6.159 in place of r_c. In this way, DEM calculations can become faster while obtaining more realistic heat transfer characteristics.

$$r_c' = c\, r_c \tag{6.167}$$

6.3.2.2 Particle–Wall Heat Transfer

All mechanisms and equations introduced for particle–particle heat transfer due to conduction in the previous section can be applied to particle–wall heat transfer. In fact, there are collisional and static contacts between a particle and wall, thus, the same equations can be used for heat transfer in a particle-wall contact. The difference here is that the wall should be considered as a particle (particle *j*) with an infinite radius and mass.

6.3.2.3 Convective Heat Transfer

The convective heat flow between the particle *i* and surrounding fluid can be obtained from:

$$q_{i,f} = h_i A_i \left(T_f - T_i \right) \tag{6.168}$$

where T_f is the fluid temperature of the fluid cell that contains particle *i*, and h_i and A_i are the convective heat transfer coefficient and surface area of particle *i*, respectively. There are many correlations for calculating the heat transfer coefficient for an isolated particle in terms of Nusselt number as a function of Reynolds and Prandtl numbers. Such a correlation changes by increasing the Reynolds number. For Reynolds numbers up to 200, the Ranz and Marshall equation can be used [128]:

$$Nu_i = \frac{h_i d_i}{k_f} = 2 + 0.6 Re_i^{1/2} Pr^{1/3} \quad Re_i < 200 \tag{6.169}$$

In which

$$Re_i = \frac{\rho_f d_i \varepsilon_f \left|\vec{u} - \vec{v}_i\right|}{\mu_f} \tag{6.170}$$

$$\mathrm{Pr} = \frac{c_{p,f}\mu_f}{k_f} \tag{6.171}$$

At higher Reynolds numbers, this equation cannot predict the heat transfer coefficient with acceptable precision and the following correlation should be used instead for Reynolds numbers up to 1500 [129]:

$$Nu_i = 2 + 0.5 Re_i^{1/2} Pr^{1/3} + 0.02 Re_i^{0.8} Pr^{1/3} \quad 200 < Re_i < 1500 \tag{6.172}$$

At even higher Reynolds numbers, dependence of the heat transfer coefficient to Reynolds number increases and also becomes independent of Prandtl number [129]:

$$Nu_i = 2 + 0.000045 Re_i^{1.8} \quad Re_i > 1500 \tag{6.173}$$

Equations 6.170–6.173 are developed for the movement of an isolated particle in an infinite fluid. The heat transfer coefficient is affected by the presence of other particles. In order to take concentration of particles into account, these equations should be modified as follows [129]:

$$Nu_i = \begin{cases} 2 + 0.6\varepsilon_f^n Re_i^{1/2} Pr^{1/3} & Re_i < 200 \\ 2 + \varepsilon_f^n \left(0.5 Re_i^{1/2} + 0.02 Re_i^{0.8}\right) Pr^{1/3} & 200 < Re_{i\leq} < 1500 \\ 2 + 0.000045\varepsilon_f^n Re_i^{1.8} & Re_i > 1500 \end{cases} \tag{6.174}$$

The exponent n was found to be 3.5 for dilute granular flows. However, this value should be changed for dense flows [129]. In the absence of sufficient experimental data and a correlation for the exponent n, this should be adjusted in the specific case of interest. Alternatively, the correlation by Gunn [130] can be used:

$$\begin{gathered} Nu_i = \left(7 - 10\varepsilon_f + 5\varepsilon_f^2\right)\left(1 + 0.7\mathrm{Re}_i^{0.2}\,\mathrm{Pr}^{1/3}\right) + \left(1.33 - 2.4\varepsilon_f + 1.2\varepsilon_f^2\right)\mathrm{Re}_i^{0.7}\,\mathrm{Pr}^{1/3} \\ \mathrm{Re}_i < 10^5, \quad 0.35 < \varepsilon_f < 1 \end{gathered} \tag{6.175}$$

Recently, the fluid–particle heat transfer coefficient was evaluated by conducting DNSs of flow of gas through an assembly of particles. Based on this approach, Sun *et al.* [131] proposed the following correlation for spherical particle:

$$Nu_i = \left(\frac{-0.46 + 1.77\varepsilon_f + 0.69\varepsilon_f^2}{\varepsilon_f^3} \right) + \left(1.37 - 2.4\varepsilon_f + 1.2\varepsilon_f^2\right) \mathrm{Re}_i^{0.7} \mathrm{Pr}^{1/3}$$
$$\mathrm{Re}_i < 100, \quad 0.5 < \varepsilon_f < 1 \qquad (6.176)$$

Tavassoli *et al.* [132] derived the following correlation:

$$Nu_i = \left(7 - 10\varepsilon_f + 5\varepsilon_f^2\right)\left(1 + 0.1\mathrm{Re}_i^{0.2} \mathrm{Pr}^{1/3}\right) + \left(1.33 - 2.19\varepsilon_f + 1.15\varepsilon_f^2\right) \mathrm{Re}_i^{0.7} \mathrm{Pr}^{1/3}$$
$$\mathrm{Re}_i < 100, \quad 0.35 < \varepsilon_f < 1 \qquad (6.177)$$

A comparison among various correlations given in this section is made in Figure 6.19.[4] It can be seen in this figure that various correlations provide different values for heat transfer coefficient. Using an inappropriate correlation can provide erroneous results when simulating a specific case. Therefore, it is important to employ the correct correlation in the problem. However, we cannot recommend any of the correlations shown in Figure 6.19 as the ultimate one and the reader is encouraged to choose one of them based on its agreement with experimental

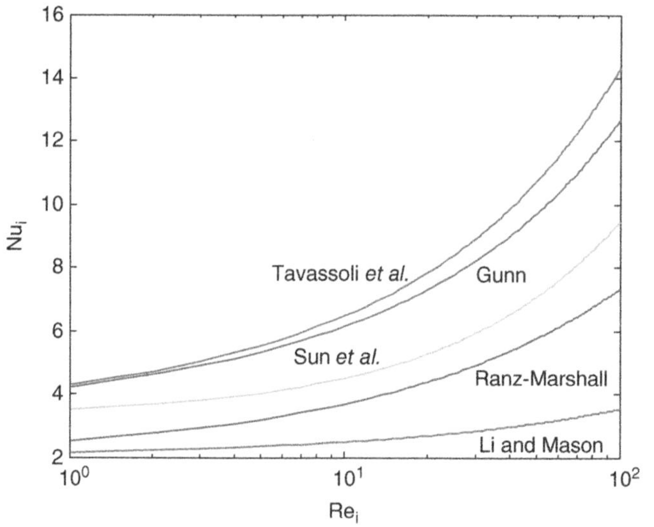

Figure 6.19 Comparison between various correlations for convective heat transfer coefficient. Pr = 0.7, ε_f = 0.7 obtained from correlations of Ranz and Marshall [128], Li and Mason [129], Gunn [130], Sun *et al.* [131], and Tavassoli *et al.* [132]

[4] These correlations are evaluated for Reynolds numbers up to 100 in this figure. This range is applicable to many conventional systems, such as air fluidization of 1 mm particles in which the velocity of particles is of the order of 1 m/s.

data in a particular problem. It should be noticed that even if the inaccurate heat transfer coefficient is employed, it is still possible to obtain good results from a DEM model that agrees well with the experimental values. This arises from the fact that there are a bunch of parameters and physical properties in the DEM that can be altered such that the overall result will become consistent with the experimental data. Of course, we strongly recommend using the exact values and proper correlations in the simulation and finding a phenomenological reason for the deviations if the agreement between the simulation output and reality is not attained.

6.3.2.4 Radiative Heat Transfer

In a high temperature granular flow, particles can exchange heat through radiation. The rate of heat transfer by this mechanism becomes significant when the temperature is at least 400 °C and can be ignored in cooler beds. For evaluating the rate of heat transfer due to radiation, the particle i is assumed to exchange heat with other particles and fluid in its surrounding environment:

$$q_{i,r} = \sigma \epsilon_i A_i \left(T_{se,i}^4 - T_i^4\right) \tag{6.178}$$

where σ is the Stefan–Boltzmann constant (5.67×10^{-8} W/m^2 K^4), ϵ_i is the emissivity of the particle, and A_i is the surface area of particle i. Temperature of the surrounding environment is the volume average of temperature of particles and fluid in an enclosed spherical domain Ω around the particle and is calculated by:

$$T_{se,i} = \varepsilon_f T_{f,\Omega} + \left(1-\varepsilon_f\right)\frac{1}{k_\Omega}\sum_{j=1}^{k_\Omega} T_j \qquad (j \neq i) \tag{6.179}$$

In this equation, $T_{f,\Omega}$ and k_Ω are the fluid temperature and the number of particles, respectively, in the domain Ω. Radius of this spherical domain can be considered to be $1.5d_i$ [123].

6.3.2.5 Heat Generation due to Chemical Reaction or Phase Change

There are situations in which a chemical reaction takes place on the surface or inside particles. This includes the reactions between fluid and solid (such as coal combustion) and fluid-phase catalytic reactions on the active surface of particle (like fluid catalytic cracking). Also, phase change (evaporation, condensation, crystallization, etc.) may occur in/on the particle. The heat associated with this phenomenon for each particle i is:

$$q_{i,c} = A_i\left(-R_c\right)\left(-\Delta H_c\right) \tag{6.180}$$

In this equation, ΔH_c is the heat of chemical or physical phenomena occurring in/on the particle expressed based on unit mass of the particle and R_c is the rate of chemical or physical change in terms of mass rate per unit surface of particle. Multiplication of the surface in this equation appears because the rate R_c is considered to be expressed per unit surface of the particle. Obviously, if the rate is expressed per unit volume of the particle, surface should be replaced by volume of the particle.

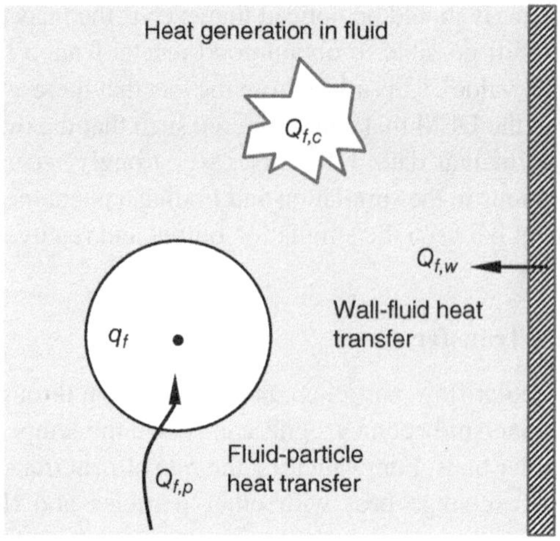

Figure 6.20 Heat exchange/generation mechanisms for the fluid phase

The minus sign for the rate of change in Equation 6.180 is considered based on a system involving chemical reaction. For example, in combustion of coal in a fluidized bed, coal particles shrink due to combustion ($R_c < 0$) and the reaction is exothermic ($\Delta H_c < 0$). In this case, Equation 6.180 produces a positive value for $q_{i,c}$, which indicates that temperature of particles increases due to combustion. But consider a system in which exothermic ($\Delta H_c < 0$) crystallization ($R_c > 0$) takes place. In this situation, we expect that particles become hotter in the course of the process, thus, $q_{i,c}$ should be positive. As a result, we should not use a minus sign for the rate of change. In conclusion, it is very important to inspect and decide about the correct sign of the rate of mass change in Equation 6.180 based on physical and chemical information about the problem in hand.

It should be noticed that for Equation 6.180 in this form, we assumed that all the heat (either generated in the body or on the surface of particle) is totally absorbed by the particle, which is consistent with our primary assumption (lumped formulation) that the inner thermal resistance of particles is negligible.

6.3.3 Rates of Heat Transfer for Fluid

Various mechanisms that contribute to the heat source term in Equation 6.144 are described in this section. As mentioned earlier, studies assessing friction at the wall and on fluids by viscous forces are not considered here. Mechanisms of heat exchange/generation in the fluid phase are shown in Figure 6.20. We present required formulas for calculating these heat sources in this section.

6.3.3.1 Fluid–Particle Heat Exchange

The fluid-to-particle heat transfer rate can be calculated by summation of the heat exchanged between fluid and particle located in a specific fluid computational cell. The details are presented in the previous section.

6.3.3.2 Fluid–Wall Heat Transfer

The convective heat transfer rate between fluid and wall (or other surfaces in the system) can be evaluated from:

$$Q_{f,w} = \frac{h_w A_w}{V_{cell}}\left(T_w - T_f\right) \tag{6.181}$$

In this equation, A_w is the surface area of the wall within or at the border of the fluid cell and h_w is the heat transfer coefficient between fluid and wall. The heat transfer coefficient in this equation should be evaluated from the proper correlation, according to the geometry of the wall or surface. For a cylindrical wall, this heat transfer coefficient can be obtained from [133]:

$$Nu_w = \frac{h_w D}{k_f} = 0.023 Re_w^{0.8} Pr^n \tag{6.182}$$

where D is the column diameter and Re_w is defined as:

$$Re_w = \frac{\rho_f D U}{\mu_f} \tag{6.183}$$

in which U is the superficial fluid velocity. The exponent n in Equation 6.182 is 0.4 for heating the fluid and 0.3 for cooling the fluid.

6.3.3.3 Heat Generation due to Chemical Reaction

If a homogeneous chemical reaction takes place in the fluid phase, its heat should be taken into account in the energy equation. The rate of heat generated by a homogeneous chemical reaction is:

$$Q_{f,c} = \left(-R_A\right)\left(-\Delta H_A\right) \tag{6.184}$$

here, A is a fluid component whose rate of reaction is $R_A (= dC_A/dt)$ and ΔH_A is the heat of reaction expressed per mol (or mass) of A. Note that rate of the homogeneous chemical reaction should be given based on unit volume of fluid.

6.3.4 Sequence of Calculations

We presented the coupling procedure for main governing equations of the CFD-DEM model in Figure 6.21. The heat transfer model described in this section can be coupled with the main governing equations of CFD-DEM and should be solved simultaneously. Therefore, we need to add new modules to it. Figure 6.21 demonstrates the proposed coupling procedure of CFD-DEM coupled with the thermal model of this chapter. This figure shows that three new modules should be added to the original CFD-DEM coupling algorithm: fluid-particle energy coupling, DEM energy equation, and fluid energy equation.

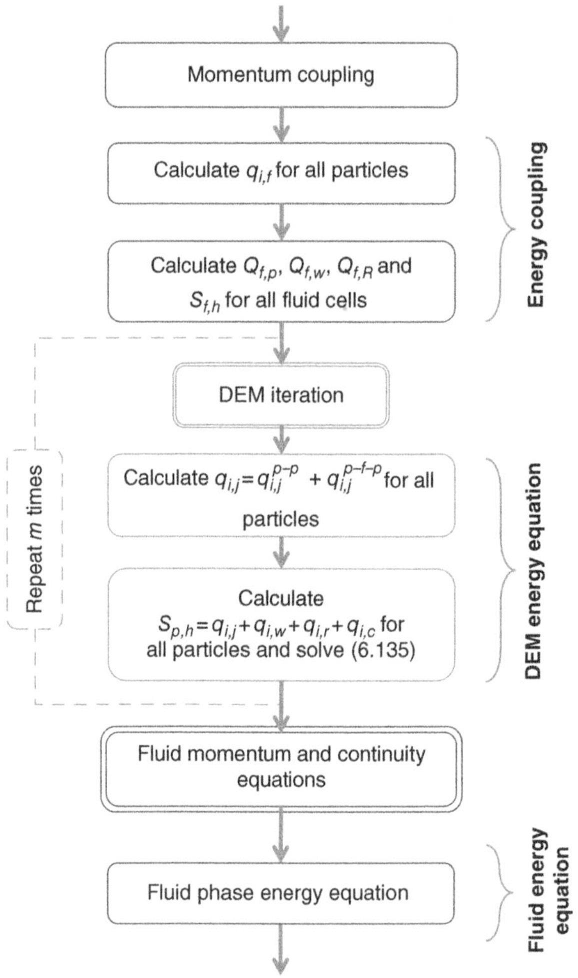

Figure 6.21 Coupling procedure for the thermal CFD-DEM model

In energy coupling, the rate of heat exchanged between particles and fluid is calculated for all particles. Then, the volumetric rate of heat exchanged between fluid and particle in each fluid cell, $Q_{f,p}$, and other source terms, like $Q_{f,w}$ and $Q_{f,R}$, should be evaluated. In the DEM energy equation, all remaining particle scale heat exchange rates that appear in Equation 6.135 are calculated and this equation is integrated for obtaining the new temperature of all particles. In the fluid energy equation step with the source term $S_{p,h}$, which was calculated in the energy coupling step, Equation 6.142 is solved and a new temperature profile of the fluid is obtained.

It should be noted that the coupling procedure is completely arbitrary and can be modified if required. In the proposed coupling procedure of Figure 6.21, the DEM energy equation step is inside the particle solver loop (the loop that is repeated m times in every time step). If heat transfer is a slow process in a given problem, the DEM energy equation step can be moved out of the particle solver loop and the energy equation of particles can be solved only once.

Alternatively, in the case of high rate of heat exchange between phases, $q_{i,f}$, which is calculated in the energy coupling step, can be moved to the particle solver loop and the amount of heat transferred between particles and fluid are calculated based on the updated temperature of particles. In this new strategy, the volumetric source term, $Q_{f,p}$, should be calculated after the completion of particle loop to hold the inter-phase energy conservation.

6.4 Mass Coupling

When we need to calculate concentration of species in a problem, it is necessary to add equations of mass conservation to equations of the model. We encounter this situation when there is exchange of mass between fluid and particles and/or chemical reaction occurs somewhere in the system (in the fluid, in the particle, or between fluid and particle). Such a problem can be considerably complex since in addition to equation of species conservation, we may or may not need to make changes to equations of continuity and momentum of the fluid phase. For example, if there is a homogeneous reaction in the fluid phase or a fluid reaction catalyzed by solids, equations of motion and continuity of the fluid would be the same as those in a system without reaction. On the other hand, in a system that involves chemical reaction between fluid and solid, both these equations should be modified accordingly. Also, size and/or mass of particles can change during the process. In this case, equation of motion of particle should also be modified. Examples of such problems are crystallization, dissolution of particles, and combustion of particles. Therefore, there may be numerous combinations of the previously mentioned processes, which make modeling of the problem of mass transfer and chemical reaction complicated. For instance, in the problem of combustion of coal, we need to modify equation of motion of particle (since particle size changes), equations of continuity, and motion of fluid (since gas is produced in the process as a result of combustion). However, in the process of CO_2 capture by solids, equation of motion of particles remains unchanged (since the size does not change) but equations of continuity and motion of fluid should be modified. On the other hand, in the coating process, the mass added to the particle is determined neither through mass transfer or reaction, but by the rate determined by the spray nozzle. Consequently, it is obvious that we cannot present all these cases by a single unified modeling approach. Nevertheless, we have tried in this section to develop model equations that are commonly encountered in various processes. Thus, if a process is not exactly covered in this chapter, the reader may decide to use combinations of the equations developed here or develop equations that are specific to that process.

6.4.1 Governing Equations

The main variable in a system involving mass transfer is concentration of species in the fluid phase. As such, the main equation to be solved in these systems is the equation of conservation of species for both phases. If the size/mass of particles does not change during the process, the equation of motion of particles would be the same as those given in Chapter 2. Also, if the fluid is not consumed/produced during the process, continuity and momentum equations of the fluid phase remain the same as those described earlier in this chapter. This is the case in solid-catalyzed reactions and when the reaction takes place only in the fluid phase. Examples of a process in which mass of particles as well as amount of fluid stay unchanged are catalytic oxidation of hydrocarbons and fluid catalytic cracking. This condition can also be extended to drying and adsorption/desorption processes. Although there is a change of mass of particles in such processes, this change is usually a small fraction of the initial total mass of the particle and can be ignored during the simulation.

Contrary to these examples, if the size/mass of particles change during the process, the equation of motion of particles should be changed accordingly and if the fluid is consumed/produced, continuity and momentum equations for the fluid should be modified. Coal combustion is an example of a process for which equation of motion of particles, continuity of fluid, and equation of momentum for fluid should be modified. Governing equations for finding concentration of species in the fluid are given here. In this section, we consider the positive direction of mass transfer to be from fluid to particle.

6.4.1.1 Governing Equation for Particles

The size of particle may change during chemical reaction or an interphase mass transfer. In this case, Equation 6.32 cannot be applied anymore and the equation of motion for particle i should be written as:

$$\frac{d\left(m_i\vec{v}_i\right)}{dt} = m_i\frac{d\vec{v}_i}{dt} + \vec{v}_i\frac{dm_i}{dt} = \Sigma\vec{f}_i \tag{6.185}$$

where the summation term in the right-hand side of this equation is the sum of all forces acting on particle i. The rate of mass change of particle, dm_i/dt, in this equation should be obtained from rate of reaction/crystallization/dissolution of solid. We show this rate of mass change of particle as follows:

$$\frac{dm_i}{dt} = R_{Pi} \tag{6.186}$$

in which R_{Pi} is the rate of mass change of particle. Note that R_{Pi} is positive in case of particle enlargement and negative when particle is shrinking and its unit should be given as mass of solid per unit time. Combining Equations 6.185 and 6.186 results in the following modified equation of motion of particle:

$$m_i\frac{d\vec{v}_i}{dt} = -\vec{v}_iR_{Pi} + \Sigma\vec{f}_i \tag{6.187}$$

Likewise, equation of angular momentum for spherical particle should be modified if the mass of particles changes during the process. In this case, Equation 6.33 should be presented as follows:

$$\frac{d\left(I_i\vec{\omega}_i\right)}{dt} = I_i\frac{d\vec{\omega}_i}{dt} + \vec{\omega}_i\frac{dI_i}{dt} = \Sigma\vec{M}_i \tag{6.188}$$

The summation in the right-hand side of this equation stands for the sum of all torques acting on particle. This equation can be rearranged as:

$$I_i \frac{d\vec{\omega}_i}{dt} = -\vec{\omega}_i \frac{dI_i}{dt} + \Sigma \vec{M}_i \tag{6.189}$$

The moment of inertia depends on the shape of particle and its derivative in the above equation should be evaluated accordingly. Moment of inertia for a spherical particle is:

$$I_i = \frac{2}{5} m_i R_i^2 \tag{6.190}$$

We also have:

$$m_i = \frac{4}{3} \pi R_i^3 \rho_i \tag{6.191}$$

Therefore, Equation 6.189, after some algebraic operations, becomes:

$$I_i \frac{d\vec{\omega}_i}{dt} = -\frac{2}{3} R_i^2 R_{Pi} \vec{\omega}_i + \Sigma \vec{M}_i \tag{6.192}$$

It is worth mentioning that the change in size of particle is negligible during short (real-) time simulations. Therefore, the first term of the right hand side of Equations 6.187 and 6.192 can be omitted in simulations that cover only few seconds of the process, but should be kept in place and taken into account when the long term simulation of the real process is performed.

Knowing the rate of change of particle mass, the change in the radius of a sphere in each time step can be obtained from:

$$\Delta R_i = \frac{R_{Pi} \Delta t_p}{4\pi R_i^2 \rho_i} \tag{6.193}$$

where Δt_p is the time step used for integrating the equations of motion of particles. Consider that the radius of particle i is R_i at the start of each iteration. DEM calculations should be carried out with this value and at the end of iteration, the radius of the particle will be changed to $R_i + \Delta R_i$. Calculations in the next iteration should be performed using this new particle radius.

When the mass of particle changes during the process, it is necessary to estimate the rate of mass change during DEM calculations. The mass of a particle can change in the following ways:

- Mass transfer to/from particle (R_{Pi}^{MT}),
- Heterogeneous reaction between fluid and particle ($R_{Pi}^{Het.}$), and
- Homogeneous reaction in the fluid and solid deposition on particle ($R_{Pi}^{Hom.}$).

In other words:

$$R_{Pi} = R_{Pi}^{MT} + R_{Pi}^{Het.} + R_{Pi}^{Hom.} \tag{6.194}$$

We remind the reader that in a particular problem not all these three mechanisms may exist. Therefore, only those rates which exist in the problem should be kept on the right-hand side of Equation 6.194.

6.4.1.2 Governing Equations for Fluid

In addition to equations of momentum and continuity for fluid, the equation of mass balance for species should be considered in systems involving mass transfer and/or reaction. Of course, equations of momentum and continuity need to be modified if mass is generated or consumed in the fluid phase during the process. We present equations of continuity, momentum, and species conservation for the fluid phase in the following.

Equation of Continuity

If mass is added or deducted from the fluid, the following equation should be used in the model:

$$\frac{\partial}{\partial t}\left(\varepsilon_f \rho_f\right) + \nabla.\left(\varepsilon_f \rho_f \vec{u}\right) = S_m \tag{6.195}$$

in which S_m is the rate of production of mass per unit volume of fluid.

Momentum Equation

Equation of momentum conservation should be corrected for addition or deduction of mass to/from fluid due to mass transfer and chemical reaction with particles. We encounter this situation in processes like crystallization or dissolution of particles, to give a few example. Equation of momentum of the fluid phase (e.g., the original model) in this case is:

$$\frac{\partial\left(\rho_f \varepsilon_f \vec{u}\right)}{\partial t} + \nabla.\left(\rho_f \varepsilon_f \vec{u}\vec{u}\right) = -\nabla p - \nabla.\vec{\tau}_f - \vec{F}^{OM} + \rho_f \varepsilon_f \vec{g} + \vec{S}_M \tag{6.196}$$

here, $\vec{S}_M$ is the correction to equation of momentum conservation for taking into account addition or deduction of mass of fluid.

Species Conservation

When the composition of fluid changes in a problem, it is necessary to solve the equation of species conservation for all components in order to keep track of changes of all species during the simulation. This change in the composition may occur due to chemical reaction (either homogeneous or heterogeneous) and/or fluid–particle mass transfer (adsorption, desorption, drying, dissolution, precipitation). The mass balance for species i is:

$$\frac{\partial}{\partial t}\left(\varepsilon_f C_n\right) + \nabla.\left(\varepsilon_f \vec{u} C_n\right) = \nabla.\left(\varepsilon_f D_n \nabla C_n\right) + S_{Y,n} \tag{6.197}$$

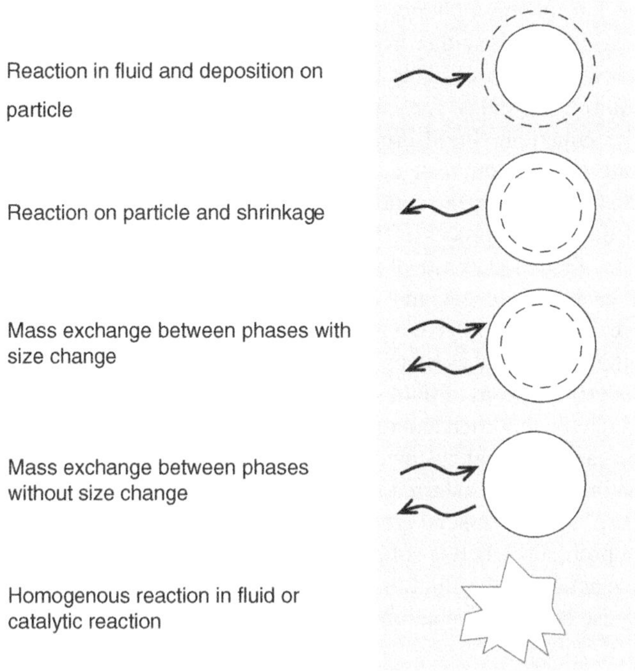

Figure 6.22 Mass exchange mechanisms and reaction in fluid phase. Possible mass transfer between fluid and wall is not considered here

In this equation, $S_{Y,n}$ is the rate of mass production of component n per unit volume of fluid. Effective diffusion coefficient in Equation 6.197 is the sum of molecular and turbulent diffusion coefficients:

$$D_n = D_{n,f} + D_{n,t} \tag{6.198}$$

6.4.1.3 Coupling between Particle and Fluid Equations

In a system that includes mass transfer or reaction, we are interested in concentration profiles of species, in addition to velocities of particles and fluid. There are various sources that give rise to a change of species concentration in the fluid and mass of particle. The most common mass exchange and reaction situations in fluid are illustrated in Figure 6.22. Equations of continuity and momentum remain unchanged in systems involving solid-catalyzed reactions and when homogeneous reactions take place in the fluid phase. However, we need to modify these equations to incorporate consumption or production of mass from/to the fluid phase if one of the following cases exits in the system:

- A reaction that consumes a solid.
- A reaction that produces a solid and deposits it on the surface of particles.
- Enlargement or shrinkage of particles due to mass exchange with the fluid.
- Mass exchange between fluid and solid without changes in particle size.

We showed that the rate of mass change of particle, R_{Pi}, appears in equations of both particles and fluid phases. It can arise either from exchange of mass between fluid and particles or reaction between these two phases. This term couples equations of particles and fluid. Here, we present required formulas for calculating these sources terms in continuity, momentum, and species equations. In all processes involving mass transfer or reaction, we should consider the rate of production of component n per unit volume of fluid, $S_{Y,n}$, for obtaining concentration of species in the fluid. This is regardless of generation or consumption of the total fluid during the process. The rate of production of a species should be determined from the rate of reaction or rate of mass transfer, where applicable. This is specific to every problem for each component and depends on the mechanism of generation or consumption. Therefore, we refrain to give a general formula for $S_{Y,n}$.

In problems where the reaction takes place only in the fluid phase among fluid species or when we encounter catalytic reactions of fluid species, equations of continuity, and momentum conservation do not need to be modified and remain the same as those described earlier in this chapter. This is also the case for problems involving small amount of mass exchange, such as drying. Generally speaking, if the transferred mass during the process is less than 10% of the total mass of either phase, we recommend ignoring corrections to momentum and continuity equations. However, in problems where a solid reacts with fluid, and the solid is produced as a result of a chemical reaction in the fluid or mass exchange between fluid and particles is significant, we need to use corrected equations of continuity and momentum, as given in the previous section.

The production rate of mass of fluid in Equation 6.195 can be evaluated by considering the mass production rate from all particles located in a specific fluid computational cell:

$$S_m = -\frac{1}{V_{cell}}\sum_{i=1}^{k_v} R_{pi} \tag{6.199}$$

Also, the momentum source term in Equation 6.196 is addition to the momentum of fluid due to change of mass of fluid in exchange with particles and is evaluated from:

$$\vec{S}_M = -\frac{1}{V_{cell}}\sum_{i=1}^{k_v} R_{pi}\vec{u} \tag{6.200}$$

In these equation, R_{Pi} is the rate of mass that is consumed from the fluid and goes to particles.

6.4.2 *Rates of Mass Transfer for Particles*

The rate of mass change of particles is determined either by the rate of exchange of mass with fluid (R_{Pi}^{MT}), from the rate of chemical reaction between solid and fluid ($R_{Pi}^{Het.}$) or as a result of homogeneous chemical reaction in the fluid phase ($R_{Pi}^{Hom.}$). These cases are described in the following.

6.4.2.1 Mass Transfer

We can expect a change in mass of particles when they exchange mass with the surrounding fluid. Particle enlargement or shrinkage may also occur due to mass transfer to or from the particle. Examples of exchange of mass at constant particle size are adsorption and desorption and examples of mass transfer with changing particle size are dissolution and crystallization processes. Let's consider A to be the component being exchanged between particle and fluid. The rate of mass change in this case can be obtained from the following equation:

$$R_{Pi}^{MT} = k_{c,i} A_i \left(C_{A,f} - C_{A,i} \right) \tag{6.201}$$

where $k_{c,i}$ is the mass transfer coefficient between particle i and surrounding fluid, $C_{A,f}$ is the concentration of A in the bulk of fluid, and $C_{A,i}$ is the concentration of A at the surface of the particle. The mass transfer coefficient for a single particle in an infinite fluid can be obtained from the correlation of Froessling [134]:

$$Sh_i = \frac{k_{c,i} d_i}{D_{A,f}} = 2 + 0.6\mathrm{Re}_i^{1/2} Sc^{1/3} \tag{6.202}$$

where

$$Sc = \frac{\mu_f}{\rho_f D_{A,f}} \tag{6.203}$$

For a fixed bed, the following correlation, given by Kunii and Levenspiel [135], can be used for evaluating the mass transfer coefficient:

$$Sh_i = 2 + 1.8\mathrm{Re}_i^{1/2} Sc^{1/3} \tag{6.204}$$

In a swarm of particles, the voidage of bed should be taken into account for determining the mass transfer coefficient. For this purpose, either the correlation of Gunn [130] can be used:

$$Sh_i = \left(7 - 10\varepsilon_f + 5\varepsilon_f^2\right)\left(1 + 0.7\mathrm{Re}_i^{0.2} Sc^{1/3}\right) + \left(1.33 - 2.4\varepsilon_f + 1.2\varepsilon_f^2\right)\mathrm{Re}_i^{0.7} Sc^{1/3} \tag{6.205}$$

or it can be obtained from the La Nauze *et al.* correlation [136]:

$$Sh_i = 2\varepsilon_f + 0.69\left(\frac{\mathrm{Re}_i}{\varepsilon_f}\right)^{1/2} Sc^{1/3} \tag{6.206}$$

Correlations introduced here are compared with each other in Figure 6.23. This figure illustrates that various correlations provide different mass transfer coefficients. In order to avoid getting erroneous results in a simulation, it is important to choose the appropriate correct correlation in the problem. We cannot recommend any of the correlations shown in Figure 6.23 as the proper correlation. The appropriate correlation in each problem should be determined based on its agreement with experimental data in a given problem.

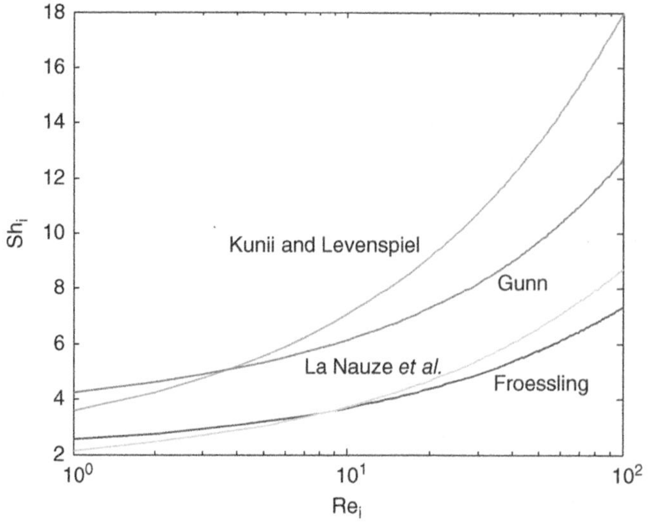

Figure 6.23 Comparison between various correlations for mass transfer coefficient. Sc = 0.7, $\varepsilon_f = 0.7$ obtained from correlations of Froessling [134], Kunii and Levenspiel [135], Gunn [130], and La Nauze *et al.* [136]

6.4.2.2 Heterogeneous Chemical Reaction

Let's consider a chemical reaction between fluid and particle in which the product is in fluid form. An example of such a reaction is combustion of coal particles. A chemical reaction of this kind between fluid and solid can be represented as follows:

$$A_{(f)} + bB_{(s)} \rightarrow cC_{(f)} \tag{6.207}$$

Rate of the reaction between fluid and solid is usually first order with respect to the fluid component and zero order with respect to the solid. Of course, more complex kinetics for fluid–solid reactions can also be considered. But in the following, we assume the common and simplified first-order kinetics with respect to the reactant in fluid. Usually, this rate is expressed based on unit surface of the solid. Therefore, the rate of change in the mass of solid (component B) is then obtained from:

$$R_{Pi}^{Het.} = A_i M_B b(-R_A) = -A_i M_B bkC_{A,f} \tag{6.208}$$

Note that according to reaction (6.207) every mole of A reacts with b moles of B. Therefore, the rate of consumption of B is bR_A, where R_A is the rate of reaction A in molar basis. We should multiply this term by molecular weight of B, M_B, to convert it into its mass rate. It should be noted that the rate of reaction in Equation 6.208 is based on unit surface of solids, thus, the unit for the reaction rate constant, k, is m/s. We put a minus sign in Equation 6.208 due to the fact that the particle is shrinking according to reaction (6.207). Here, the concentration is expressed on a molar basis.

In Equation 6.208 we assumed that the mass transfer rate is such high (compared to the reaction rate) that it does not affect the kinetics of reaction. However, this assumption is not always valid. In a fluid–solid reaction, the reactant A in the fluid phase should first reach the surface of particle and then react with B. The effective rate of reaction (6.207) should be obtained by taking the mass transfer rate into account. The effective reaction rate then, should be determined from [137]:

$$R_{Pi}^{Het.} = -A_i M_B b k_{eff} C_{A,f} \tag{6.209}$$

where

$$k_{eff} = \left(\frac{1}{k} + \frac{1}{k_{c,i}} \right)^{-1} \tag{6.210}$$

As an example of this case, we can consider the combustion of char in a fluidized bed [138].

These equations are applicable in cases where the fluid is in direct contact with the reacting surface of the particle and no product layer is formed. However, there are some cases in which a product layer forms on the particle and thickens as the reaction proceeds. Using K_2CO_3/Al_2O_3 sorbents for CO_2 capture is an example of this situation [139]. Obviously, if a product layer is formed on the surface of particle during the progress of reaction, the reaction rate should be obtained from the shrinking core model [137]:

$$R_{Pi}^{Het.} = -\frac{A_i M_B b C_{A,f}}{\dfrac{1}{k_{c,i}} + \dfrac{R_i (R_i - r_c)}{r_c D_e} + \dfrac{R_i^2}{r_c^2 k}} \tag{6.211}$$

In this equation, r_c is the radius of unreacted core and D_e is the effective diffusion coefficient of the reactant through the product layer.

6.4.2.3 Homogeneous Chemical Reaction

A homogeneous chemical reaction between species in fluid whose product is a solid can be shown as follows:

$$A_{(f)} + bB_{(f)} \rightarrow cC_{(s)} \tag{6.212}$$

Precipitation reactions can be described by this reaction. Assuming that the solid product is evenly dispensed on the surface of all particles in a fluid cell, the rate of mass change of each particle can be obtained from:

$$R_{Pi}^{Hom.} = \frac{M_C c(-R_A)\varepsilon_f V_{cell}}{k_v} \tag{6.213}$$

Note that according to Equation 6.212 every mole of A produces c moles of solid C. Therefore, the rate of production of C is $c(-R_A)$, where $-R_A$ is the rate of reaction A in molar basis.

We should multiply this term by molecular weight of C, M_C, to convert it into mass rate. This rate is multiplied by the volume of fluid in the fluid cell, $\varepsilon_f V_{cell}$, for obtaining the total mass of *C* produced in the cell. Dividing the resultant amount by the number of particles in the fluid cell gives the rate of change of mass for each particle as given in Equation 6.213.

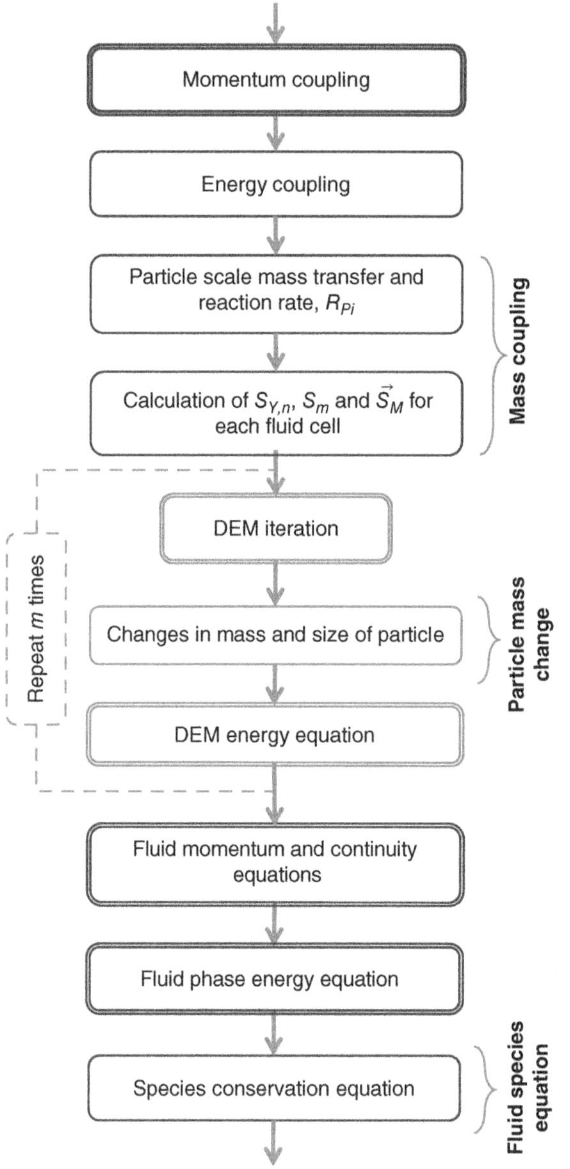

Figure 6.24 Coupling procedure for the CFD-DEM model combined by heat and mass transfer and reactions

6.4.3 Rates of Change in Fluid

We presented possible mechanisms of mass exchange between fluid and particles in the previous section. These mechanisms include both mass transfer between phases and chemical reaction (homogeneous and heterogeneous). The total mass exchange rate between fluid and particles can be calculated by summation of the mass exchanged between fluid and particles located in a specific fluid cell. The details are presented in the previous section.

6.4.4 Sequence of Calculations

The mass coupling procedure for the governing equations described in this section is shown in Figure 6.24. Since the temperature effect is important in most operations involving mass transfer and reaction, we modified the thermal coupling procedure of the CFD-DEM model to obtain a coupling procedure for a multiphase flow with heat and mass transfers. As we showed in the modeling section, we cannot develop a unified model for the mass transfer in the multiphase flow and the governing equations may change in each case. Therefore, the coupling procedure presented here is not general and may change in a specific case. In addition, the exchange rate of mass and heat between phases may also change this coupling procedure.

Based on the sequence shown in Figure 6.24, the particle scale mass transfers and reaction rates are determined using fluid and particle fields profiles at the current time step in the mass coupling stage. Other source terms, like $S_{Y,n}$, S_m, and $\vec{S}_M$, are calculated in each fluid cell. This holds the mass conservation between phases. In the DEM iteration loop, particle scale mass and species balances, if any, are considered and solved. Thereafter, species conservation equations of the fluid phase are solved to obtain the concentration profiles of all components in the next fluid time step.

Nomenclature

Symbol	Unit	Description
A	—	Coefficient in Equations 6.71 and 6.83
A_1, A_2, A_3, A_4	—	Coefficients in Equation 6.105
$A_{f,i}$	m^2	Area of the cell face i
A_i	m^2	Surface area of particle i
A_i^n	—	Coefficient in Equations 6.116 and 6.121
A_i^p	m^2	Projected area of particle i in the direction of motion
A_w	m^2	Surface area of wall
b	—	Correction coefficient
B	—	Coefficient in Equations 6.71 and 6.83
Bi	—	Biot number
c	—	Correction coefficient
c	—	Stoichiometric coefficient
c', c_1, c_2, c_3	—	Constants in Equation 6.149
c_i	—	Particle circularity
$c_{p,i}$	J/kg·K	Heat capacity of particle i
$c_{p,f}$	J/kg·K	Heat capacity of fluid at constant pressure
$c_{v,f}$	J/kg·K	Heat capacity of fluid at constant volume

C	—	Solid phase concentration
C_A	—	Added mass coefficient
$C_{A,f}$	kg/m^3	Concentration of A in fluid
$C_{A,i}$	kg/m^3	Concentration of A on surface of particle i
C_B	—	Basset force coefficient
C_d	—	Drag coefficient
C_d^{ns}	—	Drag coefficient for an isolated non-spherical particle
C_n	kg/m^3	Concentration of component n
CL_i	—	Contact list of particle i
$\langle d \rangle$	m	Mean Sauter diameter
$d_{c,i}$	m	Diameter of porous cube
d_i	m	Diameter of particle i
$d_i^{A^P}$	m	Diameter of the sphere having the same projected area of the particle
$d_{i,k}$	m	Diameter of particle i in group k
d_i^V	m	Diameter of the sphere having the same volume of the particle
ds	m^2	Control surface of the particle
D	m	Column diameter
$D_{A,f}$	m^2/s	Diffusion coefficient of A
D_e	m^2/s	Effective diffusion coefficient
D_n	m^2/s	Diffusion coefficient of component n
$D_{n,f}$	m^2/s	Molecular diffusion coefficient of component n
$D_{n,t}$	m^2/s	Turbulent diffusion coefficient of component n
E_f	J/kg	Specific energy of a fluid
$\dot{E}_f$	W	Rate of energy transported by the fluid within the control volume
E_i	Pa	Young's modulus of particle i
E_{ij}	Pa	Reduced Young's modulus of particles
$E_{ij,0}$	Pa	Young's modulus of particles
$f_{A,i}$	—	Fraction of surface of particle i residing in fluid cell
$\vec{f}_i$	N	Force acting on particle i
$\vec{f}_i^a$	N	Added mass force on particle i
$\vec{f}_i^d$	N	Steady drag force on particle i
$\hat{f}_i^d$	—	Normalized steady drag force on particle i
$\hat{f}_i^d$	—	Mean normalized drag force
$\vec{f}_i^{f-p}$	N	Interaction force between fluid and particle i
$\hat{f}_i^{f-p}$	—	Normalized total force exerted on particle i
$\vec{f}_i^g$	N	Gravitational force on particle i
$\vec{f}_i^l$	N	Lift force on particle i
$\vec{f}_i^{Magnus}$	N	Magnus force on particle i
$\vec{f}_i^{(n)}$	N	Normal force exerted by the surrounding fluid on a particle
$\vec{f}_i^{Staffman}$	N	Saffman force on particle i
$\vec{f}_i^u$	N	Unsteady fluid force on particle i
$\vec{f}_i^{ud}$	N	Unsteady drag force on particle i
$\vec{f}_i^{(t)}$	N	Tangential force exerted by the surrounding fluid on a particle

$\vec{f}_i^{\nabla p}$	N	Pressure gradient force on particle i
$\vec{f}_i^{\nabla \cdot \bar{\tau}_f}$	N	Viscous force due to the fluid shear stress
$\vec{f}_{ij}^{c}$	N	Contact force between particles i and j
$\vec{f}_{ij}^{nc}$	N	Noncontact (long-range) force between particles i and j
$\vec{F}$	N/m^3	Volumetric force acting on particles by the surrounding fluid
$\vec{F}^A$	N/m^3	Volumetric force acting on particles by the surrounding fluid
$\vec{F}^B$	N/m^3	Volumetric force acting on particles by the surrounding fluid
$\vec{F}^{OM}$	N/m^3	Volumetric force acting on particles by the surrounding fluid
F_d	N	Total drag force due to solid phase in the control volume
Fo	—	Fourier number
$\vec{g}$	m/s^2	Gravitational acceleration
h_f	J/kg	Specific enthalpy of a fluid
h_i	W/m$^2\cdot$K	Heat transfer coefficient of particle i
h_w	W/m$^2\cdot$K	Heat transfer coefficient of a wall
H	m	Half of surface to surface distance between particles
I_i	kg$\cdot$m^2	Moment of inertia of particle i
k	Pa$\cdot$s	Dilatational viscosity of a fluid
k	m/s	Reaction rate constant
k	—	Type of particle group
k_0, k_1, k_2, k_3	—	Constants in Equation 6.87
k_1, k_2	—	Constants in Equations 6.100 and 6.107
k_c	—	Number of particles that their centers are located within a cell
$k_{c,i}$	m/s	Mass transfer coefficient between particle i and fluid
k_{eff}	m/s	Effective reaction rate constant
k_f	W/m$\cdot$K	Thermal conductivity of a fluid
k_f^{eff}	W/m$\cdot$K	Effective thermal conductivity of a fluid
$k_{f,t}$	W/m$\cdot$K	Turbulent thermal conductivity of a fluid
k_i	W/m$\cdot$K	Thermal conductivity of particle i
k_v	—	Number of particles in the fluid cell
k_V	—	Number of particles partially located within the cell
k_Ω	—	Number of particles in domain Ω
K_i	N/m^2	Kinetic energy per unit volume
l	m	Characteristic length of flow
L	m	Characteristic length of flow field
m	—	Ratio of time steps in CFD and DEM
m	kg	Mass of particle
$\dot{m}$	kg/s	Mass exchange rate between particle and fluid
m_i	kg	Mass of particle i
m_{ij}	kg	Reduced mass of particles
M_B	kg/kmol	Molecular weight of B
M_C	kg/kmol	Molecular weight of C
$\dot{M}_f$	kg/s	Mass rate of fluid passing through the control volume
$\vec{M}_i$	N$\cdot$m	Torques acting on particle i
$\vec{M}_{ij}^c$	N$\cdot$m	Contact torque acting on particle i by particle j or wall
$\vec{M}_i^d$	N$\cdot$m	Fluid rotational drag on particle i
$\vec{M}_i^{f-p}$	N$\cdot$m	Fluid rotational resistance torque on particle i

$\vec{M}_{ij}$	N·m	Torque acting between particles *i* and *j*
$\dot{M}_p$	kg/s	Mass generated by solid phase in control volume
Mom_f	N	Convective momentum flux of fluid phase in control volume
$\vec{n}$	—	Normal unit vector of control surface
n_c	—	Number of particle group
n_f	—	Number of cell faces
N_k	—	Number of particles in group *k*
Nu_i	—	Nusselt number of particle *i*
p	Pa	Pressure
$\bar{p}$	Pa	Smoothed pressure
p'	Pa	Fluctuating part of pressure
$\tilde{p}$	Pa	Filtered pressure
p_d	Pa	Dynamic pressure
P_i	m	Projected perimeter of particle *i* in the direction of motion
Pr	—	Prandtl number
$q_{i,c}$	J/s	Rate of heat generation in/on particle *i*
$q_{i,f}$	J/s	Rate of convective heat transfer between particle and surrounding fluid
$q_{i,j}$	J/s	Rate of conductive heat transfer between particle *i* and neighboring particle *j*
$q_{i,j}^{p-f-p}$	J/s	Rate of particle–fluid–particle heat conduction
$q_{i,j}^{p-p}$	J/s	Rate of particle–particle heat conduction
$q_{i,r}$	J/s	Rate of radiation heat transfer between particle *i* and environment
$q_{i,w}$	J/s	Rate of particle–wall heat exchanged
$Q_{f,c}$	$J/m^3 \cdot s$	Rate of heat generated by homogeneous chemical reaction
$Q_{f,p}$	$J/m^3 \cdot s$	Rate of heat exchanged between fluid and particles
$Q_{f,R}$	$J/m^3 \cdot s$	Rate of heat generated in fluid due to chemical reaction
$Q_{f,w}$	$J/m^3 \cdot s$	Rate of heat exchanged between fluid and wall
$\dot{Q}_p$	W	Rate of heat exchanged with solid phase in control volume
r	m	Radial coordinate
r_c	m	Contact radius
$r_{ç}$	m	Radius of core
r_c	m	Corrected contact radius
r_{ij}	m	Radius of the lens of fluid between contacting particles *i* and *j*
r_{sf}	m	Radius defined by Equation 6.162
r_{sij}	m	Radius of the contact circle between particles *i* and *j*
R^*	m	Radius defined by Equation 6.164
R_A	$kg/m^2 \cdot s$	Rate of chemical reaction
R_c	$kg/m^2 \cdot s$	Rate of chemical or physical change
R_i	m	Radius of particle *i*
R_{ij}	m	Reduced radius of particles
R_{Pi}	kg/s	Rate of mass change of particle *i*
$R_{Pi}^{Het.}$	kg/s	Rate of heterogeneous reaction between fluid and particle
$R_{Pi}^{Hom.}$	kg/s	Rate homogeneous reaction in fluid
R_{Pi}^{MT}	kg/s	Rate of mass transfer to/from particle
Re_{crit}	—	Reynolds number at the onset of transition flow
Re_i	—	Reynolds number of particle *i*

$\langle \mathrm{Re}_i \rangle$	—	Mean Reynolds number
Re_i^V	—	Reynolds number based on d_i^V
Re_l	—	Reynolds number based on characteristic length l
Re_{turb}	—	Reynolds number at onset of turbulent regime
Re_w	—	Reynolds number of wall
$S_{f,h}$	$J/m^3 \cdot s$	Heat source for fluid
S_i	m^2	Actual surface area of particle i
S_m	$kg/m^3 \cdot s$	Rate of production of mass
$\vec{S}_M$	N/m^3	Rate of momentum production
$S_{p,h}$	J/s	Heat source for particle i
$S_{Y,n}$	$kg/m^3 \cdot s$	Rate of mass production of component n
Sc	—	Schmidt number
Sh_i	—	Sherwood number of particle i
St_{mass}	—	Mass Stokes number
St_{mom}	—	Momentum Stokes number
t	s	Time
t_c	s	Contact time
$T_{0,i}$	K	Initial temperature of particle i
T_f	K	Temperature of fluid
$T_{f,\Omega}$	K	Fluid temperature in domain Ω
T_i	K	Temperature of particle i
$T_{se,i}$	K	Temperature of surrounding environment
T_w	K	Temperature of wall
$\vec{u}$	m/s	Fluid velocity
$\bar{\vec{u}}$	m/s	Smoothed fluid velocity
$\vec{u}'$	m/s	Fluctuating part of fluid velocity
$\tilde{\vec{u}}$	m/s	Filtered fluid velocity
U	m/s	Superficial gas velocity
$\vec{v}_i$	m/s	Velocity of particle i
$\vec{v}_{ij}^n$	m/s	Normal relative velocity between particles i and j
V_{cell}	m^3	Volume of fluid computational cell
V_i	m^3	Volume of particle i
V_{ij}	m^3	Volume of Voronoi polyhedra between particles
$\vec{w}_i$	m/s	Relative velocity
$W_{friction}$	$J/m^3 \cdot s$	Rate of mechanical work done by friction of fluid at wall
$W_{viscous}$	$J/m^3 \cdot s$	Rate of work done on fluid by viscous forces
x_{ij}	m	Center to center distance of particles
x_c	m	Cell centroid positions
x_k	—	Dimensionless parameter
x_p	m	Particle positions
y_k	—	Dimensionless parameter

Greek symbols

α	—	Ratio of the cube diameter to particle diameter
α	—	Thermal diffusivity of particle
α	—	Inertial effect correction factor

α_i	—	Coefficient in Equation 6.127
β	—	Interphase momentum transfer/fluid friction coefficient
β_k	—	Correction factor
χ	—	Empirical coefficient
$\ddot{\delta}$	—	Unit tensor
ΔH_A	J/kg	Heat of chemical reaction
ΔH_c	J/kg	Heat of chemical or physical phenomena
Δt_p	s	Time step of particle
Δt_f	s	Time step of fluid
Δx	m	Size of fluid cell in the x direction
Δy	m	Size of fluid cell in the y direction
ε	m^2/s^3	Dissipation rate of turbulent kinetic energy per unit mass
ε_f	—	Volume fraction of fluid
ε_f^{2D}	—	2D fluid volume fraction
ε_f^{3D}	—	3D fluid volume fraction
ε_i	—	Local porosity around particle
ε_p	—	Volume fraction of solid phase
$\varepsilon_{p,k}$	—	Volume fraction of particles of type k
ϵ_i	—	Emissivity of particle i
ϕ_c	—	Fluid variables stored at the cell position
$\phi_{f,i}$	—	Fluid variables stored at the cell face
ϕ_p	—	Fluid variables stored at the particle position
φ_i	—	Fraction of volume of particle i that is in the fluid cell
λ_K	m	Smallest Kolmogorov length scale
μ_f	Pa · s	Viscosity of fluid
μ_f^t	Pa · s	Fluid turbulent viscosity
$\vec{v}$	m/s	Average velocity of solid phase in the fluid cell
ν_f	m^2/s	Fluid kinematic viscosity
ν_i	—	Poisson ratio
$\ddot{\pi}_f$	N/m^2	Fluid stress tensor
$\ddot{\pi}_p$	N/m^2	Solid phase stress tensor
$\prod_{ener}$	—	Energy coupling parameter
$\prod_{mass}$	—	Mass coupling parameter
$\prod_{mom}$	—	Momentum coupling parameter
ρ_i	kg/m^3	Density of particle i
ρ_f	kg/m^3	Density of fluid
σ	$W/m^2 \cdot K^4$	Stefan–Boltzmann constant
τ_c	s	Average time between particle collisions
τ_f	s	Characteristics time of the flow field
$\ddot{\tau}_f$	N/m^2	Shear stress of fluid
$\overline{\ddot{\tau}}_f$	N/m^2	Time-smoothed shear stress tensor
$\tilde{\ddot{\tau}}_f$	N/m^2	Filtered shear stress tensor
τ_p	s	Particle momentum response time
τ_T	s	Particle thermal response time
$\overline{\ddot{\tau}}_f^t$	N/m^2	Time-smoothed turbulent momentum flux

$\tilde{\bar{\tau}}_f^t$	N/m^2	Filtered turbulent momentum flux
$\bar{\bar{\tau}}_f^v$	N/m^2	Time-smoothed viscous momentum flux
$\tilde{\bar{\tau}}_f^v$	N/m^2	Filtered viscous momentum flux
$\vec{\omega}_i$	rad/s	Angular velocity of particle i
ψ_i	—	Sphericity
$\psi_i^{\parallel}$	—	Lengthwise sphericity
$\psi_i^{\perp}$	—	Crosswise sphericity

References

[1] Crowe, C.T. (2006) Multiphase Flow Handbook, Taylor & Francis.

[2] Crowe, C.T., Schwarzkopf, J.D., Sommerfeld, M., and Tsuji, Y. (2011) Multiphase Flows with Droplets and Particles, 2nd edn, Taylor & Francis.

[3] Vreman, B., Geurts, B., Deen, N.G., Kuipers, J.A.M., and Kuerten, J.G.M. (2009) Two- and four-way coupled Euler–Lagrangian large-eddy simulation of turbulent particle-laden channel flow. *Flow, Turbulence and Combustion*, **82**(1), 47–71.

[4] Dorgan, A.J. (2009) Modeling and Simulation of Bubbles and Particles R. Baj), University of Illinois at Urbana-Champaign.

[5] Tsuji, Y. (2007) Multi-scale modeling of dense phase gas–particle flow. *Chemical Engineering Science*, **62**, 3410–3418.

[6] Deen, N.G., van Sint Annaland, M., and Kuipers, J.A.M. (2004) Multi-scale modeling of dispersed gas–liquid two-phase flow. *Chemical Engineering Science*, **59**(8–9), 1853–1861.

[7] Zhu, H.P., Zhou, Z.Y., Yang, R.Y., and Yu, A.B. (2007) Discrete particle simulation of particulate systems: theoretical developments. *Chemical Engineering Science*, **62**(13), 3378–3396.

[8] Sarkar, S., Kriebitzsch, S.H.L., van der Hoef, M.A., and Kuipers, J.A.M. (2009) Gas–solid interaction force from direct numerical simulation (DNS) of binary systems with extreme diameter ratios. *Particuology*, **7**, 233–237.

[9] Pan, T.W., Joseph, D.D., Baj, R., Glowinski, R., and Sarin, V. (2002) Fluidization of 1204 spheres: simulation and experiment. *Journal of Fluid Mechanics*, **451**, 169-192.

[10] Zhou, H., Flamant, G., Gauthier, D., and Lu, J. (2004) Numerical simulation of the turbulent gas–particle flow in a fluidized bed by an LES-DPM model. *Chemical Engineering Research and Design*, **82**(7), 918–926.

[11] Zhou, H., G. Flamant, and Gauthier, D. (2004) DEM-LES simulation of coal combustion in a bubbling fluidized bed part II: coal combustion at the particle level. *Chemical Engineering Science*, **59**, 4205–4215.

[12] Zhou, H., Flamant, G., and Gauthier, D. (2004) DEM-LES of coal combustion in a bubbling fuidized bed. Part I: gas-particle turbulent flow structure. *Chemical Engineering Science*, **59**, 4193–4203.

[13] Gui, N., Fan, J.R., and Luo, K. (2008) DEM–LES study of 3-D bubbling fluidized bed with immersed tubes. *Chemical Engineering Science*, **63**(14), 3654–3663.

[14] Wu, J. and Aidun, C.K. (2010) A method for direct simulation of flexible fiber suspensions using lattice Boltzmann equation with external boundary force. *International Journal of Multiphase Flow*, **36**, 202–209.

[15] Cook, B.K., Noble, D.R., and Williams, J.R. (2004) A direct simulation method for particle fluid systems. *Engineering Computations*, **21**(2/3/4), 151–168.

[16] Ge, W. and Li, J. (2003) Simulation of particle–fluid systems with macro-scale pseudo-particle modeling. *Powder Technology*, **137**(1–2), 99–108.

[17] Ge, W. and Li, J. (2003) Macro-scale phenomena reproduced in microscopic systems – pseudo-particle modeling of fluidization. *Chemical Engineering Science*, **58**(8), 1565–1585.

[18] Potapov, A.V., Hunt, M.L., and Campbell, C.S. (2001) Liquid–solid flows using smoothed particle hydrodynamics and the discrete element method. *Powder Technology*, **116**(2–3), 204–213.

[19] van Sint Annaland, M., Dijkhuizen, W., Deen, N.G., and Kuipers, J.A.M. (2006) Numerical simulation of behavior of gas bubbles using a 3-D front-tracking method. *AIChE Journal*, **52**(1), 99–110.

[20] van Sint Annaland, M., Deen, N.G., and Kuipers, J.A.M. (2005) Numerical simulation of gas–liquid–solid flows using a combined front tracking and discrete particle method. *Chemical Engineering Science*, **60**(22), 6188–6198.

[21] Cangialosi, F., Liberti, L., Notarnicola, M., and Stencel, J. (2006) Monte Carlo simulation of pneumatic tribocharging in two-phase flow for high-inertia particles. *Powder Technology*, **165**(1), 39–51.
[22] Tsuji, Y., Tanaka, T., and Yonemura, S. (1998) Cluster patterns in circulating fluidized beds predicted by numerical simulation (discrete particle model versus two-fluid model). *Powder Technology*, **95**(3), 254–264.
[23] Bird, R.B., Stewart, W.E., and Lightfoot, E.N. (2002) Transport Phenomena, 2nd edn, John Wiley & Sons, Inc., Hoboken.
[24] Andersson, B., Andersson, R., Hakansson, L., Mortensen, M., Sudiyo, R., and Wachem, B.V. (2012) Computational Fluid Dynamics for Engineers, Cambridge University Press.
[25] Link, J.M., Cuypers, L.A., Deen, N.G., and Kuipers, J.A.M. (2005) Flow regimes in a spout-fluid bed: a combined experimental and simulation study. *Chemical Engineering Science*, **60**, 3425–3442.
[26] Gidaspow, D. (1994) Multiphase Flow and Fluidization: Continuum and Kinetic Theory Description, Academic Press, San Diego, CA.
[27] Bouillard, J.X., Lyczkowski, R.W., and Gidaspow, D. (1989) Porosity distributions in a fluidized bed with an immersed obstacle. *AIChE Journal*, **35**(6), 908–922.
[28] Kafui, K.D., Thornton, C., and Adams, M.J. (2004) Reply to Comments by Feng and Yu on "Discrete particle-continuum fluid modelling of gas–solid fluidised beds" [*Chemical Engineering Science* 59 (2004) 719–722]. *Chemical Engineering Science*, **59**(3), 723–725.
[29] Feng, Y.Q. and Yu, A.B. (2004) Comments on "Discrete particle-continuum fluid modelling of gas–solid fluidised beds" [*Chemical Engineering Science* 57 (2002) 2395–2410]. *Chemical Engineering Science*, **59**(3), 719–722.
[30] Feng, Y.Q. and Yu, A.B. (2004) Assessment of model formulations in the discrete particle simulation of gas-solid flow. *Industrial & Engineering Chemistry Research*, **43**, 8378–8390.
[31] Di Renzo, A. and Di Maio, F.P. (2007) Homogeneous and bubbling fluidization regimes in DEM-CFD simulations: hydrodynamic stability of gas and liquid fluidized beds. *Chemical Engineering Science*, **62**, 116–130.
[32] Feng, Y.Q., Xu, B.H., Zhang, S.J., Yu, A.B., and Zulli, P. (2004) Discrete particle simulation of gas fluidization of particle mixtures. *AIChE Journal*, **50**, 1713–1728.
[33] Zhou, Z.Y., Kuang, S.B., Chu, K.W., and Yu, A.B. (2010) Discrete particle simulation of particle-fluid flow: model formulations and their applicability. *Journal of Fluid Mechanics*, **661**, 482–510.
[34] Hoomans, B.P.B., Kuipers, J.A.M., Briels, W.J., and van Swaaij, W.P.M. (1996) Discrete particle simulation of bubble and slug formation in a two-dimensional gas-fluidised bed: a hard-sphere approach. *Chemical Engineering Science*, **51**(1), 99–118.
[35] Kawaguchi, T., Tanaka, T., and Tsuji, Y. (1998) Numerical simulation of two-dimensional fluidized beds using the discrete element method (comparison between the two- and three-dimensional models). *Powder Technology*, **96**(2), 129–138.
[36] Kawaguchi, T., Sakamoto, M., Tanaka, T., and Tsuji, Y. (2000) Quasi-three-dimensional numerical simulation of spouted beds in cylinder. *Powder Technology*, **109**(1–3), 3–12.
[37] Ouyang, J. and Li, J. (1999) Discrete simulations of heterogeneous structure and dynamic behavior in gas–solid fluidization. *Chemical Engineering Science*, **54**(22), 5427–5440.
[38] Ouyang, J. and Li, J. (1999) Particle-motion-resolved discrete model for simulating gas–solid fluidization. *Chemical Engineering Science*, **54**(13–14), 2077–2083.
[39] Yuu, S., Waki, M., Iwamasa, A., and Umekage, T. (2000) Numerical simulation of the velocity and stress fields for a flowing powder using the smoothed particle method and experimental verification. *Advanced Powder Technology*, **11**(1), 113–128.
[40] Yuu, S., Umekage, T., and Johno, Y. (2000) Numerical simulation of air and particle motions in bubbling fluidized bed of small particles. *Powder Technology*, **110**, 158–168.
[41] Limtrakul, S., Chalermwattanatai, A., Unggurawirote, K., Tsuji, Y., Kawaguchi, T., and Tanthapanichakoon, W. (2003) Discrete particle simulation of solids motion in a gas–solid fluidized bed. *Chemical Engineering Science*, **58**(3–6), 915–921.
[42] Tsuji, Y., Kawaguchi, T., and Tanaka, T. (1993) Discrete particle simulation of two-dimensional fluidized bed. *Powder Technology*, **77**(1), 79–87.
[43] Rong, D., Mikami, T., and Horio, M. (1999) Particle and bubble movements around tubes immersed in fluidized beds – a numerical study. *Chemical Engineering Science*, **54**(23), 5737–5754.
[44] Kaneko, Y., Shiojima, T., and Horio, M. (1999) DEM simulation of fluidized beds for gas-phase olefin polymerization. *Chemical Engineering Science*, **54**, 5809–5821.
[45] Kuwagi, K., Mikami, T., and Horio, M. (2000) Numerical simulation of metallic solid bridging particles in a fluidized bed at high temperature. *Powder Technology*, **109**, 27–40.

[46] Rhodes, M.J., Wang, X.S., Nguyen, M., Stewart, P., and Liffman, K. (2001) Study of mixing in gas-fluidized beds using a DEM model. *Chemical Engineering Science*, **56**(8), 2859–2866.
[47] Rhodes, M.J., Wang, X.S., Nguyen, M., Stewart, P., and Liffman, K. (2001) Onset of cohesive behaviour in gas fluidized beds: a numerical study using DEM simulation. *Chemical Engineering Science*, **56**, 4433–4438.
[48] Rhodes, M.J., Wang, X.S., Nguyen, M., Stewart, P., and Liffman, K. (2001) Use of discrete element method simulation in studying fluidization characteristics: influence of interparticle force. *Chemical Engineering Science*, **56**, 69–76.
[49] Han, T., Levy, A., and Kalman, H. (2003) DEM simulation for attrition of salt during dilute-phase pneumatic conveying. *Powder Technology*, **129**(1–3), 92–100.
[50] Mikami, T., Kamiya, H., and Horio, M. (1998) Numerical simulation of cohesive powder behavior in a fluidized bed. *Chemical Engineering Science*, **53**, 1927–1940.
[51] Xu, B.H. and Yu, A.B. (1998) Comments on the paper "numerical simulation of the gas-solid flow in a fluidized bed by combining discrete particle method with computational fluid dynamics" [*Chemical Engineering Science* 52(1997) 2785-2809]. *Chemical Engineering Science*, **53**, 2646–2647.
[52] Helland, E., Occelli, R., and Tadrist, L. (2000) Numerical study of cluster formation in a gas–particle circulating fluidized bed. *Powder Technology*, **110**(3), 210–221.
[53] Xu, B.H. and Yu, A.B. (1997) Numerical simulation of the gas-solid flow in a fluidized bed by combining discrete particle method with computational fluid dynamics. *Chemical Engineering Science*, **52**, 2785–2809.
[54] Xu, B.H., Yu, A.B., Chew, S.J., and Zulli, P. (2000) Numerical simulation of the gas–solid flow in a bed with lateral gas blasting. *Powder Technology*, **109**(1–3), 13–26.
[55] van Wachem, B.G.M., van der Schaaf, J., Schouten, J.C., Krishna, R., and van den Bleek, C.M. (2001) Experimental validation of Lagrangian–Eulerian simulations of fluidized beds. *Powder Technology*, **116**(2–3), 155–165.
[56] Kafui, K.D., Thornton, C., and Adams, M.J. (2002) Discrete particle-continuum fluid modelling of gas–solid fluidised beds. *Chemical Engineering Science*, **57**, 2395–2410.
[57] Li, J. and Kuipers, J.A.M. (2002) Effect of pressure on gas–solid flow behavior in dense gas-fluidized beds: a discrete particle simulation study. *Powder Technology*, **127**, 173–184.
[58] Li, J. and Kuipers, J.A.M. (2003) Gas-particle interactions in dense gas-fluidized beds. *Chemical Engineering Science*, **58**(3–6), 711–718.
[59] Zhou, H., Flamant, G., Gauthier, D., and Lu, J. (2002) Lagrangian approach for simulating the gas-particle flow structure in a circulating fluidized bed riser. *International Journal of Multiphase Flow*, **28**(11), 1801–1821.
[60] Zhou, H., Abanades, S., Flamant, G., Gauthier, D., and Lu, J. (2002) Simulation of heavy metal vaporization dynamics in a fluidized bed. *Chemical Engineering Science*, **57**(14), 2603–2614.
[61] van Wachem, B.G.M. and Almstedt, A.E. (2003) Methods for multiphase computational fluid dynamics. *Chemical Engineering Journal*, **96**(1–3), 81–98.
[62] Li, J., Langston, P.A., Webb, C., and Dyakowski, T. (2004) Flow of sphero-disc particles in rectangular hoppers – a DEM and experimental comparison in 3D. *Chemical Engineering Science*, **59**(24), 5917–5929.
[63] Bokkers, G.A., Annaland, M.v.S., and Kuipers, J.A.M. (2004) Mixing and segregation in a bidisperse gas–solid fluidised bed: a numerical and experimental study. *Powder Technology*, **140**, 176–186.
[64] Hoomans, B.P.B., Kuipers, J.A.M., and van Swaaij, W.P.M. (2000) Granular dynamics simulation of segregation phenomena in bubbling gas-fluidised beds. *Powder Technology*, **109**, 41–48.
[65] Anderson, T.B. and Jackson, R. (1967) A fluid mechanical description of fluidized beds. *Industrial and Engineering Chemistry Fundamentals*, **6**, 527–539.
[66] Gibilaro, L.G. (2001) Fluidization-Dynamics, the Formulations and Applications of a Predictive Theory for the Fluidized State, Butterworth–Heinmann, Oxford.
[67] Ergun, S. (1952) Fluid flow through packed columns. *Chemical Engineering Progress*, **48**, 89–94.
[68] Wen, C.Y. and Yu, Y.H. (1966) Mechanics of fluidization. *Chemical Engineering Progress Symposium Series*, **62**, 100–111.
[69] Hill, R.J., Koch, D.L., and Ladd, J.C. (2001) Moderate-Reynolds-numbers flows in ordered and random arrays of spheres. *Journal of Fluid Mechanics*, **448**, 243–278.
[70] Hilton, J.E., Mason, L.R., and P.W. Cleary (2010) Dynamics of gas–solid fluidised beds with non-spherical particle geometry. *Chemical Engineering Science*, **65**, 1584–1596.
[71] Alobaid, F. and Epple, B. (2013) Improvement, validation and application of CFD/DEM model to dense gas–solid flow in a fluidized bed. *Particuology*, **11**(5), 514–526.
[72] Wu, C.L., Zhan, J.M., Li, Y.S., Lam, K.S., and Berrouk, A.S. (2009) Accurate void fraction calculation for three-dimensional discrete particle model on unstructured mesh. *Chemical Engineering Science*, **64**(6), 1260–1266.

[73] Peng, Z., Doroodchi, E., Luo, C., and Moghtaderi, B. (2014) Influence of void fraction calculation on fidelity of CFD-DEM simulation of gas-solid bubbling fluidized beds. *AIChE Journal*, **60**(6), 2000–2018.
[74] Wu, C.L., Berrouk, A.S., and Nandakumar, K. (2009) Three-dimensional discrete particle model for gas–solid fluidized beds on unstructured mesh. *Chemical Engineering Journal*, **152**(2–3), 514–529.
[75] Lim, E.W.C., Wang, C.H., and Yu, A.B. (2006) Discrete element simulation for pneumatic conveying of granular material. *AIChE Journal*, **52**, 496–509.
[76] Zhao, J. and Shan, T. (2013) Coupled CFD–DEM simulation of fluid–particle interaction in geomechanics. *Powder Technology*, **239**, 248–258.
[77] Deb, S. and Tafti, D.K. (2013) A novel two-grid formulation for fluid–particle systems using the discrete element method. *Powder Technology*, **246**, 601–616.
[78] Sutkar, V.S., Deen, N.G., Mohan, B., Salikov, V., Antonyuk, S., Heinrich, S., and Kuipers, J.A.M. (2013) Numerical investigations of a pseudo-2D spout fluidized bed with draft plates using a scaled discrete particle model. *Chemical Engineering Science*, **104**, 790–807.
[79] Xiao, H. and Sun, J. (2011) Algorithms in a robust hybrid CFD-DEM solver for particle-laden flows. *Communications in Computational Physics*, **9**(2), 297–324.
[80] Kuang, S.B., Chu, K.W., Yu, A.B., Zou, Z.S., and Feng, Y.Q. (2008) Computational investigation of horizontal slug flow in pneumatic conveying. *Industrial & Engineering Chemistry Research*, **47**(2), 470–480.
[81] Patankar, S.V. (1980) Numerical Heat Transfer and Fluid Flow, Hemisphere Publishing Corporation.
[82] Issa, R.I. (1986) Solution of the implicitly discretised fluid flow equations by operator-splitting. *Journal of Computational Physics*, **62**(1), 40–65.
[83] White, F.M. (1991) Fluid Mechanics, McGraw-Hill, New York.
[84] Cheng, N.S. (2009) Comparison of formulas for drag coefficient and settling velocity of spherical particles. *Powder Technology*, **189**, 395–398.
[85] Di Felice, R. (1994) The voidage function for fluid–particle interaction systems. *International Journal of Multiphase Flow*, **20**, 153–159.
[86] Schiller, L. and Nauman, A. (1935) A drag coefficient correlation. *VDI Zeitung*, **77**, 318–320.
[87] Gibilaro, L.G., Di Felice, R., Waldram, S.P., and Foscolo, P.U. (1985) Generalized friction factor and drag coefficient correlations for fluid-particle interactions. *Chemical Engineering Science*, **40**, 1817–1823.
[88] van der Hoef, M.A., Beetstra, R., and Kuipers, J.A.M. (2005) Lattice Boltzmann simulations of low Reynolds number flow past mono- and bidisperse arrays of spheres: results for the permeability and drag force. *Journal of Fluid Mechanics*, **528**, 233–254.
[89] Benyahia, S., Syamlal, M., and O'Brien, T.J. (2006) Extension of Hill-Koch-Ladd drag correlation overall ranges of Reynolds number and solids volume fraction. *Powder Technology*, **162**, 166–174.
[90] Mazzei, L. and Lettieri, P. (2007) Adrag force closure for uniformly dispersed fluidized suspensions. *Chemical Engineering Science*, **62**, 6129–6142.
[91] Sarkar, S., van der Hoef, M.A., and Kuipers, J.A.M. (2009) Fluid–particle interaction from lattice Boltzmann simulations for flow through polydisperse random arrays of spheres. *Chemical Engineering Science*, **64**, 2683–2691.
[92] Cello, F., Di Renzo, A., and Di Maio, F.P. (2010) A semi-empirical model for the drag force and fluid–particle interaction in polydisperse suspensions. *Chemical Engineering Science*, **65**, 3128–3139.
[93] Beetstra, R., van der Hoef, M.A., and Kuipers, J.A.M. (2007) Erratum to "Drag force of intermediate Reynolds number flow past mono- and bidisperse arrays of spheres" [AIChE Journal 53 (2007) 489–501]. *AIChE Journal*, **53**(11), 3020.
[94] Hill, R.J., Koch, D.L., and Ladd, J.C. (2001) The first effects of fluid inertia on flows in ordered and random arrays of spheres. *Journal of Fluid Mechanics*, **448**, 213–241.
[95] Beetstra, R., van der Hoef, M.A., and Kuipers, J.A.M. (2007) Drag force of intermediate Reynolds number flow past mono- and bidisperse arrays of spheres. *AIChE Journal*, **53**, 489–501.
[96] Hölzer, A. and Sommerfeld, M. (2009) Lattice Boltzmann simulations to determine drag, lift and torque acting on non-spherical particles. *Computers & Fluids*, **38**(3), 572–589.
[97] Zhou, Z.Y., Pinson, D., Zou, R.P., and Yu, A.B. (2011) Discrete particle simulation of gas fluidization of ellipsoidal particles. *Chemical Engineering Science*, **66**, 6128–6145.
[98] Rhodes, M.J. (2008) Introduction to Particle Technology, 2nd edn, John Wiley & Sons, Inc., Hoboken.
[99] Loth, E. (2008) Drag of non-spherical solid particles of regular and irregular shape. *Powder Technology*, **182**(3), 342–353.

[100] Ganser, G.H. (1993) A rational approach to drag prediction of spherical and nonspherical particles. *Powder Technology*, **77**(2), 143–152.
[101] Haider, A. and Levenspiel, O. (1989) Drag coefficient and terminal velocity of spherical and nonspherical particles. *Powder Technology*, **58**(1), 63–70.
[102] Hölzer, A. and Sommerfeld, M. (2008) New simple correlation formula for the drag coefficient of non-spherical particles. *Powder Technology*, **184**(3), 361–365.
[103] Tran-Cong, S., Gay, M., and Michaelides, E.E. (2004) Drag coefficients of irregularly shaped particles. *Powder Technology*, **139**(1), 21–32.
[104] Hilton, J.E. and P.W. Cleary (2011) The influence of particle shape on flow modes in pneumatic conveying. *Chemical Engineering Science*, **66**, 231–240.
[105] Hjelmfelt, A.T., Jr. and Mockros, L.F. (1966) Motion of discrete particles in a turbulent fluid. *Applied Scientific Research*, **16**(1), 149–161.
[106] Odar, F. and Hamilton, W.S. (1964) Forces on a sphere accelerating in a viscous fluid. *Journal of Fluid Mechanics*, **18**(2), 302–315.
[107] Tchen, C.M. (1947) Mean value and correlation problems connected with the motion of small particles suspended in a turbulent fluid. PhD thesis. Delft University of Technology.
[108] Mabrouk, R., Chaouki, J., and Guy, C. (2007) Effective drag coefficient investigation in the acceleration zone of an upward gas–solid flow. *Chemical Engineering Science*, **62**(1–2), 318–327.
[109] Zhu, C., Liu, G., Ye, J., and Xu, H. (2015) Experimental investigation of non-stationary motion of single small spherical particles in an upward flow with different velocities. *Powder Technology*, **273**, 111–117.
[110] Reeks, M.W. and McKee, S. (1984) The dispersive effects of Basset history forces on particle motion in a turbulent flow. *Physics of Fluids*, **27**(7), 1573–1582.
[111] Pannala, S. (2010) Computational Gas-Solids Flows and Reacting Systems: Theory, Methods and Practice, Engineering Science Reference.
[112] Michaelides, E.E. and Roig, A. (2011) A reinterpretation of the Odar and Hamilton data on the unsteady equation of motion of particles. *AIChE Journal*, **57**(11), 2997–3002.
[113] Saffman, P.G. (1965) The lift on a small sphere in a slow shear flow. *Journal of Fluid Mechanics*, **22**(02), 385–401.
[114] Drew, D. (1976) Two-phase flows: constitutive equations for lift and Brownian motion andsomebasicflows. *Archive for Rational Mechanics and Analysis* **62**, 149–164.
[115] Mei, R. (1992) An approximate expression for the shear lift force on a spherical particle at finite reynolds number. *International Journal of Multiphase Flow*, **18**(1), 145–147.
[116] Rubinow, S.I. and Keller, J.B. (1961) The transverse force on a spinning sphere moving in a viscous fluid. *Journal of Fluid Mechanics*, **11**(3), 447–460.
[117] Lun, C.K.K. and Liu, H.S. (1997) Numerical simulation of dilute turbulent gas-solid flows in horizontal channels. *International Journal of Multiphase Flow*, **23**(3), 575–605.
[118] Zhong, W., Xiong, Y., Yuan, Z., and Zhang, M. (2006) DEM simulation of gas–solid flow behaviors in spout-fluid bed. *Chemical Engineering Science*, **61**(5), 1571–1584.
[119] Ren, B., Zhong, W., Jin, B., Yuan, Z., and Lu, Y. (2011) Modeling of gas-particle turbulent flow in spout-fluid bed by computational fluid dynamics with discrete element method. *Chemical Engineering & Technology*, **34**(12), 2059–2068.
[120] Zhang, Y., Jin, B., Zhong, W., Ren, B., and Xiao, R. (2010) DEM simulation of particle mixing in flat-bottom spout-fluid bed. *Chemical Engineering Research and Design*, **88**(5-6), 757–771.
[121] Ren, B., Zhong, W., Jin, B., Yuan, Z., and Lu, Y. (2011) Computational fluid dynamics (CFD)–discrete element method (DEM) simulation of gas–solid turbulent flow in a cylindrical spouted bed with a conical base. *Energy & Fuels*, **25**(9), 4095–4105.
[122] Dennis, S.C.R., Singh, S.N., and Ingham, D.B. (1980) The steady flow due to a rotating sphere at low and moderate Reynolds numbers. *Journal of Fluid Mechanics*, **101**(2), 257–280.
[123] Zhou, Z.Y., Yu, A.B., and Zulli, P. (2009) Particle scale study of heat transfer in packed and bubbling fluidized beds. *AIChE Journal*, **55**(4), 868–884.
[124] Zhou, J.H., Yu, A.B., and Horio, M. (2008) Finite element modeling of the transient heat conduction between colliding particles. *Chemical Engineering Journal*, **139**, 510–516.
[125] Cheng, G.J., Yu, A.B., and Zulli, P. (1999) Evaluation of e!ective thermal conductivity from the structure of a packed bed. *Chemical Engineering Science*, **54**, 4199–4209.

[126] Yang, R.Y., Zou, R.P., and Yu, A.B. (2002) Voronoi tessellation of the packing of fine uniform spheres. *Physical Review E*, **65**(4), 041302.

[127] Zhou, Z.Y., Yu, A.B., and Zulli, P. (2010) A new computational method for studying heat transfer in fluid bed reactors. *Powder Technology*, **197**(1–2), 102–110.

[128] Ranz, W.E. and Marshall, W.R. (1952) Evaporation from drops. *Chemical Engineering Progress*, **48**, 141–146.

[129] Li, J. and Mason, D.J. (2000) A computational investigation of transient heat transfer in pneumatic transport of granular particles. *Powder Technology*, **112**, 273–282.

[130] Gunn, D.J. (1978) Transfer of mass and heat to particles in fixed and fluidized beds. *International Journal of Heat and Mass Transfer*, **21**, 467–476.

[131] Sun, B., Tenneti, S., and Subramaniam, S. (2015) Modeling average gas–solid heat transfer using particle-resolved direct numerical simulation. *International Journal of Heat and Mass Transfer*, **86**, 898–913.

[132] Tavassoli, H., Peters, E.A.J.F., and Kuipers, J.A.M. (2015) Direct numerical simulation of fluid–particle heat transfer in fixed random arrays of non-spherical particles. *Chemical Engineering Science*, **129**, 42–48.

[133] Holman, J.P. (2009) Heat Transfer, 10th edn, McGraw-Hill.

[134] Froessling, N. (1938) Uber die verdunstung fallender tropfen. *Gerlands Beitrage zur Geophysik*, **52**, 170–216.

[135] Kunii, D. and Levenspiel, O. (1991) Fluidization Engineering, Butterworth-Heinemann.

[136] La Nauze, R.D., Jung, K., and Kastl, J. (1984) Mass transfer to large particles in fluidised beds of smaller particles. *Chemical Engineering Science*, **39**(11), 1623–1633.

[137] Levenspiel, O. (1999) Chemical Reaction Engineering, John Wiley & Sons, Inc., Hoboken.

[138] Liu, D., Chen, X., Zhou, W., and Zhao, C. (2011) Simulation of char and propane combustion in a fluidized bed by extending DEM–CFD approach. *Proceedings of the Combustion Institute*, **33**(2), 2701–2708.

[139] Zhang, Z., Liu, D., Zhuang, Y., Meng, Q., and Chen, X. (2014) CFD-DEM modeling of CO_2 capture using alkali metal-based sorbents in a bubbling fluidized bed. *International Journal of Chemical Reactor Engineering*, **12**(1), 441–449.

7

CFD-DEM Applications to Multiphase Flow

Multiphase flows have been the subject of numerous investigations in which momentum, heat, and mass transports occur simultaneously. In addition to experimental investigations, computational fluid dynamics–discrete element method (CFD-DEM) simulations can be also used to properly capture dominated phenomena occurring in these flows. Multiphase flows of interest in this book involve gas or liquid interactions with particles through different forces as well as heat and mass exchanges. Examples of unit operations that include such flows are fluidized beds, spouted beds, pneumatic conveyers, dryers, gas-phase and slurry polymerization reactors, combustors, incinerators, and gasifiers.

This chapter covers some important CFD-DEM applications of multiphase flows with and without energy transport, mass transport, and chemical reactions. It starts with a summary of the most practical applications related to fluidization, spouting, and pneumatic conveying from literature with some figures and facts generated by our in-house CFD-DEM code developed with formulations and implement approach presented in this book. In subsequent sections, non-isothermal and reactive flows are presented shortly in order to cover the essence of these flows. The chapter ends with miscellaneous applications related to fluid–solid interaction systems practiced in chemical industries.

7.1 Fluidization

The hydrodynamics of gas fluidized beds can be affected by many parameters, such as gas velocity, mean particle size, particle size distribution, particle density, operating temperature, and pressure. Depending on the abovementioned parameters, fluidized beds may operate in different regimes such as bubbling, slugging, turbulent, or fast fluidization. Let's consider a

Coupled CFD-DEM Modeling: Formulation, Implementation and Application to Multiphase Flows, First Edition.
Hamid Reza Norouzi, Reza Zarghami, Rahmat Sotudeh-Gharebagh and Navid Mostoufi.

Companion website: www.wiley.com/go/norouzi/CFD-DEM

bubbling fluidized bed. Formation, growth, rise, coalescence, and eruption of bubbles as well as the bed expansion due to the bubble dynamics are the main features of bubbling fluidized beds on meso scales. These macro-scale phenomena/structures are not the only ones existing in the bed. A wake of particles forms below the bubble when it rises in the bed. This wake ascends with the bubble to the bed surface and creates a non-random, convective motion of particles. At the same time, particles in the wake may fall out of the emulsion phase and new particles from the emulsion can enter the wake, which forms the random motion of particles [1]. Moreover, this sort of motion may be also observed in loose regions like bubble cloud or inside the bubbles [2]. These micro-scale (particle scale) phenomena also affect the hydrodynamics in the long-term operation of the bed. Segregation of particles differing in size or density is the result of this random motion of particles and is a well-known phenomenon that causes hydrodynamic changes in the course of fluidization [3]. Thus, the micro-scale phenomena influence macro-scale phenomena and vice versa. Meso-scale structures are intermediate-scale structures that refer to cluster or small bubbles in the fluidized beds. Clusters are mostly observed in circulating fluidized beds or in the freeboard of bubbling fluidized beds. Moreover, the existence of clusters in bubbling fluidized beds has been also proved [4].

It is believed that the CFD-DEM model is robust enough to capture flow structures of a fluidized bed at macro-, meso-, and micro-scales quantitatively and qualitatively. Numerous investigations have been conducted to validate the CFD-DEM model of fluidized beds at different operating conditions, in different fluidization regimes, and with different length scales. A few of these models are referred to in this section for validating various flow structures at different length scales. Bubble behavior and bed expansion in bubbling fluidized beds is considered here as the basis for validation of the CFD-DEM at the macro-scale. In circulating fluidized beds, formation of clusters is considered as the basis for validation of the CFD-DEM at the meso-scale and the particle mixing and segregation at the micro-scale.

7.1.1 *Macro-Scale Phenomena*

The CFD-DEM model can predict formation, rise, growth, coalescence, and bursting of bubbles. Yuu *et al.* [5] performed a soft-sphere CFD-DEM simulation with 100,000 group B particles in the bubbling regime. Results of the simulation were compared with images captured experimentally by a video camera. Direct comparisons showed that the model can reproduce different phenomena like bubble generation, bubble rise, bubble growth, bubble coalescence, and bubble bursting. Ejection of particles at the bed surface due to the bursting of bubble was observed in the simulations. Bokkers *et al.* [6] examined the CFD-DEM model capabilities to predict the size of a single injected bubble into a bed at minimum fluidization. The experiments were conducted in a rectangular fluidized bed with group D particles and the bubble formation process was recorded by a high speed digital camera. The results revealed that with a proper drag force correlation (i.e., the correlation proposed by Hill *et al.* [7], see Chapter 6) their model can predict the bubble size properly. Wang *et al.* [8] studied homogenous and heterogeneous fluidization of group A particles with CFD-DEM. The minimum bubbling velocity is an important parameter that marks the onset of bubbling in fluidization of group A particles. It was found that this model can well predict the minimum fluidization velocity. In addition, they proposed that a sudden change in the standard deviation of pressure drop in the bed is the best way to identify the minimum bubbling velocity when compared to other methods.

Tsuji *et al.* [9] performed three-dimensional, large-scale simulations in bubbling fluidized beds with large cross sections. It was shown that multiple bubbles are simultaneously formed and rise through the bed. In large beds, formation and motion of bubbles are not restricted by the bed walls. The size of bubbles does not change with the bed cross section as it does in beds with small cross sections. The circulation patterns observed in these beds are in qualitative agreement with the experimental observations. These observations were also confirmed in simulations by Olaofe *et al.* [10], which showed that the initial size of a single injected bubble at minimum fluidization velocity does not change with the bed geometry where there is no wall effect. The effect of size distribution [11] and irregularity of particles on the simulation results were also investigated [12]. It was found that the mobility of bubbles is enhanced and bubble formation and growth become faster in a polydispersed system. Moreover, a bed with non-spherical particles experiences a higher pressure drops and has lower minimum fluidization velocity compared to a bed of spherical particles [12].

Operating pressure is another important factor that changes the behavior of macro structures of fluidized beds as well as the micro-scale behavior of particles [13–15]. Many fluidized bed reactors, like polymerization reactors, operate at high pressures. Thus, it is important to know if CFD-DEM can effectively predict the bed behavior at high operating pressures. Increasing the operating pressure reduces the minimum fluidization velocity [13] and increases the initial bed expansion. Distribution of gas among emulsion and bubble phases changes by changing the bed pressure. The local gas volume fraction distribution in a fluidized bed at various pressures, obtained by CFD-DEM simulations, is shown in Figure 7.1 [15]. It can be seen in this figure that there are two peaks in the distribution: a peak at low gas volume fraction, which corresponds to the emulsion phase and a peak at high gas volume fraction, which corresponds to bubbles. Increasing the operating pressure shifts the peak of emulsion to

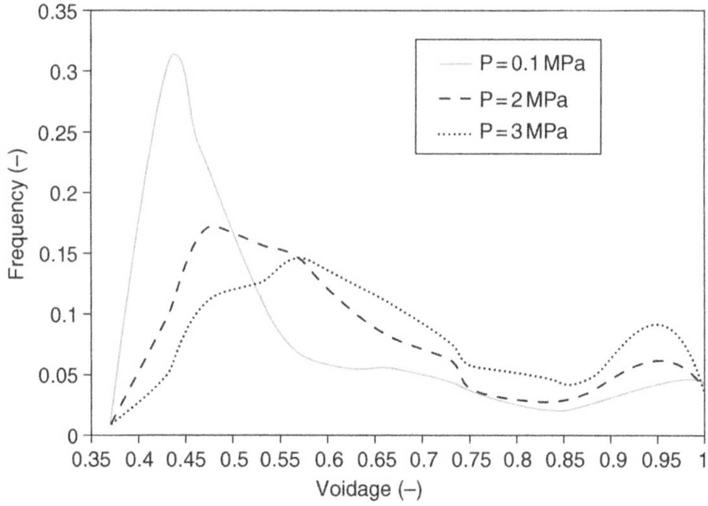

Figure 7.1 Local bed gas volume fraction distribution at different operating pressures. (Source: Mansourpour, Z., Karimi, S., Zarghami, R., Mostoufi, N., Sotudeh-Gharebagh, R., 2010. Insights in hydrodynamics of bubbling fluidized beds at elevated pressure by DEM–CFD approach. *Particuology* **8**, 407–414, with permission)

a higher value and that of the bubbles to a lower value [14–16]. In general, the bed shifts to a more homogeneous state at elevated pressures. In addition, stability of bubbles decreases by increasing the bed pressure, which leads to frequent breakage of bubbles in the bed. Thus, we can expect to observe a reduction in the average bubble diameter as the pressure is increased. The DFD-DEM simulation can correctly predict this trend. Bubbles become smaller when increasing the pressure, which agrees with the theoretical analysis by Yates [17]. In addition, the bubble rise path in the bed changes from a vertical upward movement at atmospheric pressures to a tortuous-like movement at elevated pressures [15].

7.1.2 Meso-Scale Phenomena

Clusters are the main characteristic of circulating fluidized beds. When particles are fluidized at high superficial gas velocity, they form assemblies of particles, called clusters, which travel as ensembles instead of moving as individual dispersed particles. Clusters increase the heterogeneity of the circulating fluidized bed and can affect flow pattern of both solid and gas phases. Despite the importance of clusters and flow structure in circulating fluidized beds, only few attempts have been reported in the literature to simulate and characterize the behaviors of clusters in gas-solid flows [18–22]. Yuu *et al.* [18] observed clusters in their simulations of turbulent fluidized beds. Zhang *et al.* [19] studied the formation of clusters and their behavior in riser and downer of a circulating reactor. They used 40,000 group A particles in their simulations. The simulations were carried out in a small rectangular domain whose dimensions were $0.02 \times 0.14\,m^2$. The effect of van der Waals forces was considered in their simulations. The simulation results showed a non-uniform distribution of solids and clusters. Large and slow clusters were found near the walls while small and fast clusters were located at the center of the bed. This type of distribution is the characteristic of circulating fluidized beds. They confirmed the existence of V-shaped clusters in the riser reactor. Individual particles, dispersed in the free spaces between the clusters, move very fast. They also studied the effect of many parameters, such as initial solid loading, gas velocity, interparticle forces, particle friction, and bed wall friction, on the behavior of clusters.

Zhao *et al.* [21, 22] numerically studied the clustering behavior of group B particles in a long downer and a long riser ($0.1 \times 10\,m^2$). Particles disperse uniformly at the top of the downer and clustering occurs when moving toward the middle. Clusters grow larger and their number reduces as they travel along the downer. Three different zones can be identified in the downer. First, there is an acceleration zone where the velocity of particle is less than the gas velocity and particles accelerate by the resultant of gravity and drag forces. The second acceleration zone is where particles accelerate only by gravity. The third region is the fully developed zone along which the particle velocity remains constant. The long downer considered in simulations of Zhao *et al.* [21, 22] allowed them to observe the fully developed zone. The method of distribution of solids at the inlet, which should be as uniform as possible, is important in simulation. They showed that the inlet design has a significant influence on the flow pattern of both gas and solid phase and may affect the residence time distribution (RTD) of the gas phase considerably [22]. Their simulations in riser and downer show that the RTD of particles in the downer has a narrow distribution, which suggests that particles move nearly with the plug flow pattern while the RTD of solids in the riser has a long tail, showing sever back-mixing in this part [21].

7.1.3 Micro-Scale Phenomena

Micro-scale structures are related to the particle scale phenomena such as particle–particle impact, particle wall contact, and particle motion. Mixing or segregation of particles can reflect the particle scale phenomena. Many investigations have been carried out by CFD-DEM in mixing and segregation of particles in fluidized beds [2, 23–28]. When particles with different sizes or densities (and also shapes) become fluidized, they may segregate. For example, consider a binary mixture of small and large particles which are initially well mixed. If this mixture is fluidized by air, large particles can accumulate at the bottom and small particles near the bed surface [24]. It was found that the segregation of particles mainly occurs when the superficial gas velocity lies between the minimum fluidization velocities of small and large particles. Some experimental data [29] on segregation of binary mixture of particles of different size have been used to validate the CFD-DEM model for segregation and mixing of particles [6, 25, 26, 30]. It was found that this model can predict segregation and mixing of particles properly.

It was well discussed by Feng *et al.* [2] that segregation of particles in a fluidized bed is governed by particle–particle and particle–fluid interactions. When the gas passes through the bed, the drag force acting on small particles relative to its weight is greater than that that on large particles. This difference is the driving force for the segregation of particles. A high particle–particle interaction in dense regions hinders segregation of particles and the motion of particles is restricted by other surrounding particles, thus, no spatial rearrangements of large and small particles can occur. However, in loose regions, like the regions where bubbles exist, particles are less restricted by other particles and rearrangement of small and large particles can occur. Consequently, small particles, which experience a greater relative drag forces, move toward the top of the bed. Different investigations on particle mixing and segregation have been conducted. Feng and Yu [24] studied the flow pattern of solid particles, solid concentration profile, and mixing kinetics of binary mixture of particles with different sizes in gas fluidized beds. Followed by studies on the size segregation of binary mixtures, Norouzi *et al.* [31] studied the effect of fine particle on segregation of a binary mixture. They added 1–5 wt% fine (500 μm) particles to a binary mixture of small (1500 μm) and large (2500 μm) particles. They showed that the segregation tendency increases by addition of fine particles to this mixture. When fine particles are added, they are positioned between larger particles and act as a glidant by decreasing the coordination number and the contact interaction between large particles. This reduces the hindrance effects of particle–particle interactions in dense regions and enhances the mobility of particles. Consequently, with almost the same driving force for segregation (drag force difference) and reduced hindrance effects, the segregation enhances.

Huilin *et al.* [27] studied mixing and segregation of particles with different densities. They showed that the tendency of heavier particles to separate from lighter particles increases as the coefficient of restitution decreases. Tian *et al.* [28] studied the mixing of particles in a gas fluidized bed with two different types of gas distributor: a horizontal distributor with uniform gas supply and an inclined distributor with non-uniform gas supply. They showed that the mixing rate of particles in the fluidized bed with the inclined distributor is higher than the bed equipped with the horizontal distributor due to rapid convective mixing in early stages of fluidization followed by shear mixing.

One way to examine the predictability of CFD-DEM at particle scale is to directly compare the trajectories of particles with the experimental data. The radioactive particle tracking is a technique that provides position of a tracer in the bed during fluidization. Since the CFD-DEM simulation also gives the position of particles, it is possible to directly compare the trajectory of a particle in the simulation with that of the tracer in the experiment. Figure 7.2 shows such typical trajectories in simulation and experiment. It can be seen in this figure that both simulation and experiment trajectory are fairly the same. For example, a particle may travel along the bed height from the distributor to the surface and return back to the distributor zone without any noticeable change in direction (identified with an "A" in the figure) or it may travel a shorter path and then return to its original height (identified with a "B" in the figure). These similarities between the trajectories in experiment and simulation suggest that the CFD-DEM model can reproduce the particle motion pattern in a fluidized bed [32].

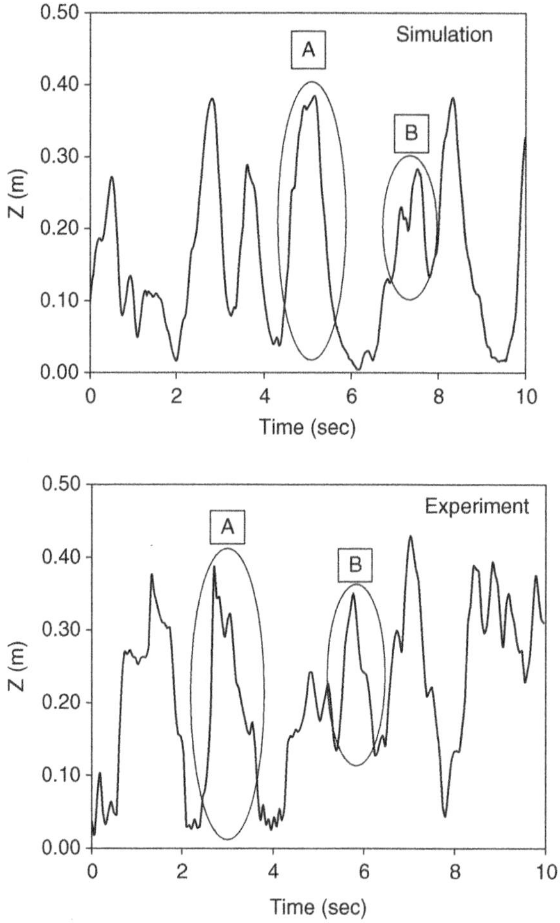

Figure 7.2 Typical trajectory of a particle in physical experiments obtained by radioactive particle tracking technique and that obtained from the CFD-DEM simulation. (Source: Norouzi, H.R., Mostoufi, N., Mansourpour, Z., Sotudeh-Gharebagh, R., Chaouki, J., 2011. Characterization of solids mixing patterns in bubbling fluidized beds. *Chemical Engineering Research and Design* **89**, 817–826, with permission)

7.2 Spouting

Schematic of a conventional conical spouted bed is illustrated in Figure 7.3. A centrally located jet supplies the fluid into the bed. If the fluid velocity is high and the static height of the bed is low enough, the fluid stream will punch through the particles as a spout. The spouted bed is typically structured as a cylindrical vessel with a conical bottom (referred as conical-cylindrical spouted bed) to prevent formation of a stagnant region. The spout fluid pulls solid particles at the spout–annulus interface and then a fountain is formed above the bed. Moreover, there is a transfer of fluid from spout into the annulus. Particles can be continuously fed into the spouted bed through the concentric jet nozzle or into the annulus region and can be withdrawn through the annulus. The beds with high jet-to-column diameter ratio are not spoutable. At high initial bed heights, with increasing the gas velocity through the jet, the bed changes into the fluidized state from its previous fixed state.

In order to have a stable, non-pulsing spouted bed, the jet nozzle diameter to particle diameter should be less than 25 or 30. Moreover, there are the well-known minimum spouting velocity and the maximum spoutable bed height requirements that should be achieved to have a stable spouted condition [33]. Piskova and Mörl [34] experimentally studied transition from the fixed bed to spouting by analyzing the pressure fluctuations in time, frequency, and state space domains. They found different regimes of fixed bed, channel formation, bubbling formation, stable spouting, and slugging flow. Figure 7.4 demonstrate the gas streamlines in a typical conical spouted bed. A gas recirculation region can be observed near to the gas inlet due to the venture effect above the inlet nozzle. The gas flow pattern in the spout zone is close to plug flow and similar to that in a packed bed. A continuous gas transfer exists between the spout and the annulus regions.

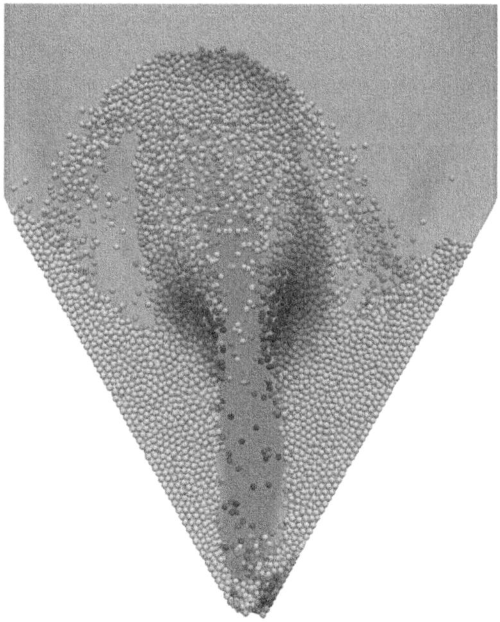

Figure 7.3 A typical conical spouted bed

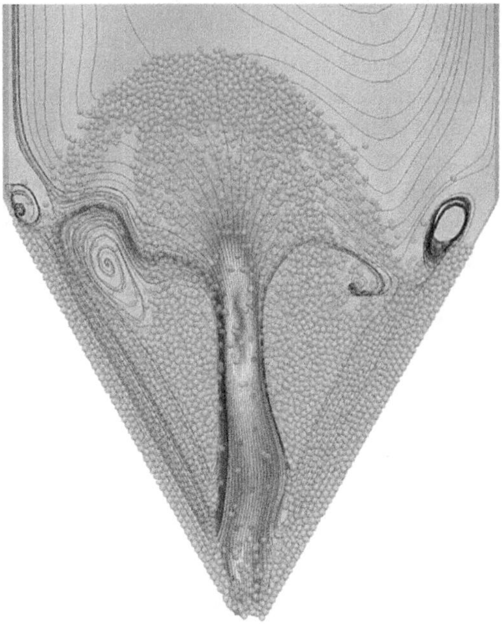

Figure 7.4 Gas streamlines in conical spouted bed

Spouted beds are also constructed as a conical vessel with different cone angles, a spouted fluidized bed with a draft tube, or a conical cylindrical spouted bed with a draft tube. Several investigators have studied the spouted fluidized bed with a draft tube [35–38] and the conical cylindrical spouted bed with a draft tube [39–45]. Operational and design characteristics of a spouted bed are completely different from those of the spouted fluidized bed equipped with the draft tube. One of the most important differences between these two types of spouted beds is the elimination of the maximum spoutable height of the bed requirement and hence the spouted fluidized bed with a draft tube can possess various bed heights. Besides, the minimum spouting velocity in a spouted bed with draft tube is less than that in a spouted bed without draft tube since the gas is limited in the draft tube of the recirculating spouted bed. In fact, the draft tube prevents leakage of the gas from spout region into the annulus. Since there are various operational parameters (such as height of the draft tube) that can be changed and fixed in different cases, the spouted beds with draft tube are more flexible from the operational point of view when compared with typical spouted beds without draft tube. Some researchers also have compared the performance of spouted beds with fluidized bed. For example, Jittanit *et al.* [46] compared the drying performance of rice and wheat seeds with high and medium moisture contents in spouted and fluidized beds. They found that the fluidized bed dryer is better in terms of drying rate and the spouted bed dryer is good in energy consumption.

While the TFM mostly has been used for simulation of spouted beds, nowadays, there is a growing interest in employing the CFD-DEM to investigate the hydrodynamics of spouted beds. The major challenges in CFD-DEM simulations are choosing a suitable turbulent model for the fluid phase in order to determine the central turbulent spout jet and developing a stable spouting [47]. The effect of fluid turbulence in the modeling of fluid phase brings some difficulties in incorporating the turbulent model in the spout-annulus interface. In this context, the laminar flow model has been used in some simulations, while in most of the cases, the k-ε turbulent model is used by researchers for the fluid phase [42, 48–52].

Some researchers carried out 2D and 3D simulations [42, 49, 53, 54] and showed that the CDF-DEM can appropriately predict general flow patterns as well as the mean and fluctuation of particle flux in pseudo-2D spouted beds (a spouted bed with the depth of a few particle diameters). These researchers generally used the particle image velocimetry (PIV) to validate their simulation results. Kawaguchi *et al.* [55] used a quasi-3D modeling for simulation of a cylindrical spouted bed. The fluid motion was calculated in 2D by solving the Navier–Stokes equations and the particle motion was calculated in 3D. Their model could predict the minimum spouting velocity very well, while it could predict the spout diameter and particle velocity profile, qualitatively. This was related to considering the gas flow in 2D in their model. Takeuchi *et al.* [47, 56, 57] implemented a 3D CFD-DEM model in order to simulate flat- and conical-based spouted beds. Link *et al.* [58] developed a regime map by employing the spectral analysis of pressure drop fluctuations and fast image recording and developed a 3D CFD-DEM model to examine the capability of this model for predicting various experimentally observed regimes. They used the porous cubes method for calculating the gas volume fraction in their model. In addition, they investigated several drag closures in their model and found that the frequently used Ergun-Wen and Yu drag correlation [59, 60] produces unsatisfactory results for spouted bed with stable high velocity jets. They recommended that using the Hill–Koch–Ladd drag correlation [7] is preferred in the CFD-DEM simulations of spouted beds. This model has been further compared with experimental measurements on positron emission particle tracking (PEPT) in terms of time averaged particle velocity [61], and on fiberoptic probes in terms of time averaged local porosity and time averaged particle velocity [62]: agreements between model results and experimental measurements were observed.

While most of the reported CFD-DEM simulations are dedicated to spouting behavior of spherical particles, limited investigations have been carried out on particles with irregular shapes. Ren *et al.* [63] used the CFD-DEM to study the hydrodynamics of a 3D conical-cylindrical spouted bed with corn-shaped particles. They used the multi-sphere method to resemble the shape of particles with two arrangements as illustrated in Figure 7.5 and showed that the result of model depends on the number and the arrangement of the spheres in the particle. They found that the particle concentration increases along the radial direction in

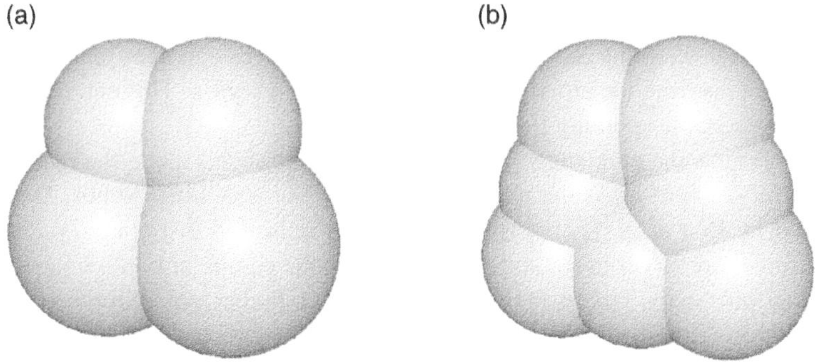

Figure 7.5 Representation of corn-shaped particles with (a) four spheres and (b) seven spheres. (Source: Ren, B., Zhong, W., Chen, Y., Chen, X., Jin, B., Yuan, Z., Lu, Y., 2012. CFD-DEM simulation of spouting of corn-shaped particles. *Particuology* **10**, 562–572, with permission)

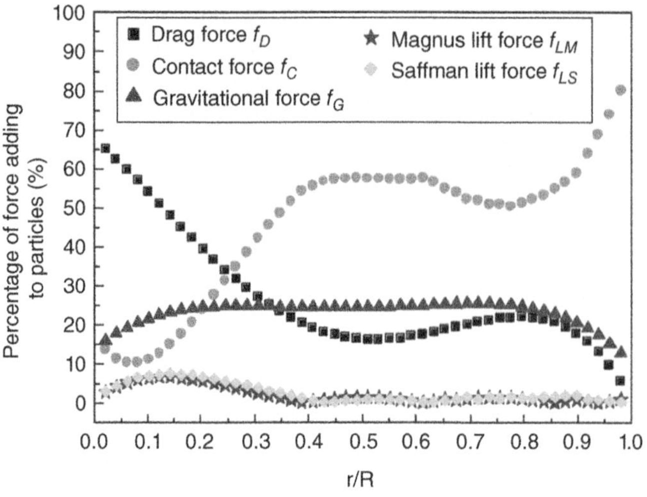

Figure 7.6 Percentage of time averaged forces acting on particles in the spouted bed (*H/D* = 1.6, d_p = 2–3 mm, $u_s = 0.76u_{ms}$). (Source: Zhong, W., Xiong, Y., Yuan, Z., Zhang, M., 2006. DEM simulation of gas–solid flow behaviors in spout-fluid bed. *Chemical Engineering Science* **61**, 1571–1584, with permission)

the spout region and decreases in the fountain, while it is nearly constant in the annulus. In a following study, Ren *et al.* [64] extended their model to the binary particle system and investigated the mixing behavior of mono-shaped (spherical or corn-shaped) mixture and a mixture composed of spherical and corn-shaped particles. Their results showed that the gas velocity and the particle shape have important role in the mixing quality of particles, quantified by the Lacey index. The bed of spherical particles is mixed faster than the bed of corn-shaped particles. Furthermore, particles in the binary system have a better mixing quality with closer sizes and at higher gas velocities.

Ren *et al.* [51] investigated the effect of considering the lift force on the motion of particles in a 3D conical spouted bed. They showed that the lift force is not important in the annulus (where the velocity gradient is not large), whereas it has a minor role in the spout region. Zhong *et al.* [48] showed that Saffman and Magnus lift forces can contribute to entrainment of spherical particles from the annular region into the spout region. This entrainment of particles can develop a stable spouting in the spouted bed.

Figure 7.6 presents the percentage of forces acting on particles along the radial axis of the bed at $u_s = 0.76u_{ms}$, where u_{ms} is the minimum spouting velocity [48]. Almost the same trend was presented by Ren *et al.* [51] at $u_s = 0.68u_{ms}$. As indicated in Figure 7.6, Saffman and Magnus lift forces are larger in the spout region due to the significant difference in gas velocities. The drag force is the largest in the spout region while the contact force is the largest in the annulus region. This shows that in the spout region, the gas–particle interaction is greater than the particle–particle interaction, while the interaction between particles is much greater than gas–particle interactions in the annulus region. The share of gravitational force is greater in the spout and decreases near the wall.

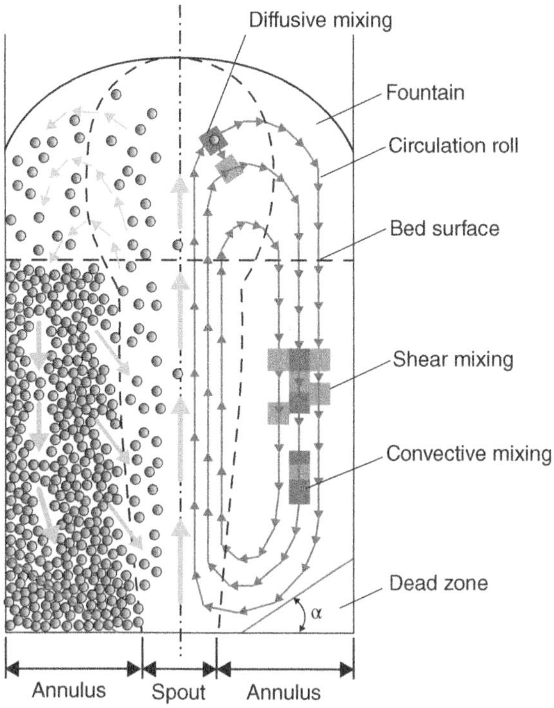

Figure 7.7 Time-smoothed streamlines of particles and mixing mechanism in different regions of a spouted bed. (Source: Zhang, Y., Jin, B., Zhong, W., Ren, B., Xiao, R., 2010. DEM simulation of particle mixing in flat-bottom spout-fluid bed. *Chemical Engineering Research and Design* **88**, 757–771, with permission)

Zhang *et al.* [50] used the CFD-DEM model for simulating a rectangular flat bottom spout-fluid bed. They identified three mixing mechanisms in the bed, which are illustrated in Figure 7.7. These mechanisms are:

- Convective mixing that occurs due to bulk the transport of particles and is caused by the motion of circulation roll,
- Shear mixing that happens due to difference in velocities of the adjacent regions, and
- Diffusive mixing that is caused by the random motion of particles through the gaps between the adjacent circulation rolls.

The convective and shear mixing increase with increasing the fluidizing gas velocity. They also used Saffman and Magnus lift forces in their simulation but did not discuss effects of these forces on the results.

Ren *et al.* [51, 52] used a three-dimensional CFD-DEM model to simulate the turbulent gas-solid flow in a cylindrical spouted bed with a conical base. They used the k-ε turbulent model for gas phase equations of motion. Figure 7.8 shows the spout evolution at various spouting gas velocities. Magnitude of the particle velocity is illustrated with white for maximum and black for minimum particle velocity. The snapshots in Figure 7.8 are generated

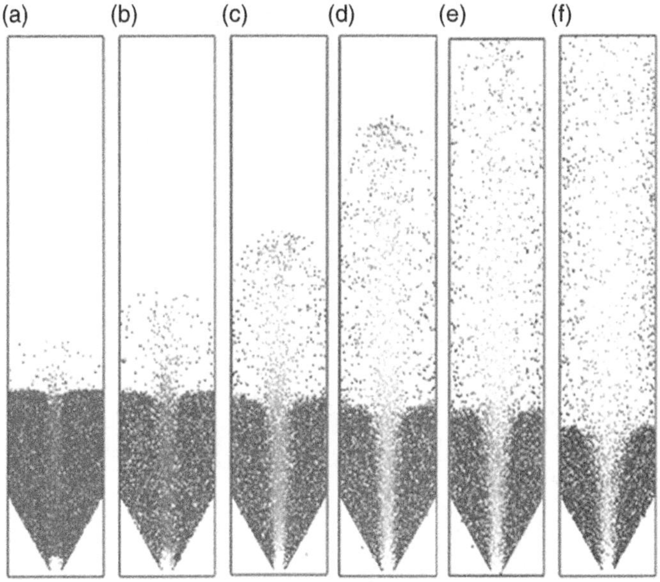

Figure 7.8 Solid flow patterns in a spouted bed at various spouting gas velocities: (a) $u_s = 0.7u_{ms}$, (b) $u_s = 0.8u_{ms}$, (c) $u_s = 0.9u_{ms}$, (d) $u_s = u_{ms}$, (e) $u_s = 1.1u_{ms}$, and (f) $u_s = 1.2u_{ms}$. (Source: Ren, B., Zhong, W., Jin, B., Yuan, Z., Lu, Y., 2011. Modeling of Gas-Particle Turbulent Flow in Spout-Fluid Bed by Computational Fluid Dynamics with Discrete Element Method. *Chemical Engineering & Technology* **34**, 2059–2068, with permission)

by projection of particles in a 2 mm depth cross section at the center of the bed. Three different regions can be observed from the flow patterns in this figure: a central spout region, a fountain zone beyond the peripheral bed level, and an annulus zone between the spout and the column wall. At low gas velocity ($u_s = 0.7u_{ms}$), introducing the spouting gas into the packed bed causes particles to move within the bed. Particles are carried beyond the bed level in a cyclic movement pattern and enter into the fountain region and then rain back onto the annular region. As shown in Figures 7.8b–f, the annular region slightly becomes smaller and the fountain becomes taller with increasing the spouting gas velocity.

Ren *et al.* [52] showed that the particle velocity gradually decreases along the radial direction, while the particles decelerate downward in the cylindrical section and then accelerate in the conical section. Their simulation showed that the particle concentration increases in the spout region and reaches a nearly constant value in the annulus region, then decreases in the fountain region along the radial axis. In addition, the particle collision, particle turbulent intensity, and the drag force are greater in the spout region than in annulus and fountain regions. Although the particle concentration is decreased with an increase in the spouting gas velocity, particles velocity, spout diameter, number of collisions, drag force, and particle turbulent intensity are increased. The time-smoothed vector field of particle velocity at the fully developed spout ($u_s = 1.3u_{ms}$) is illustrated in Figure 7.9. A systematic cycle of particle motion in spout, annulus, and fountain regions can be seen in this figure. While particles continuously are fed into the spout region from the annulus throughout

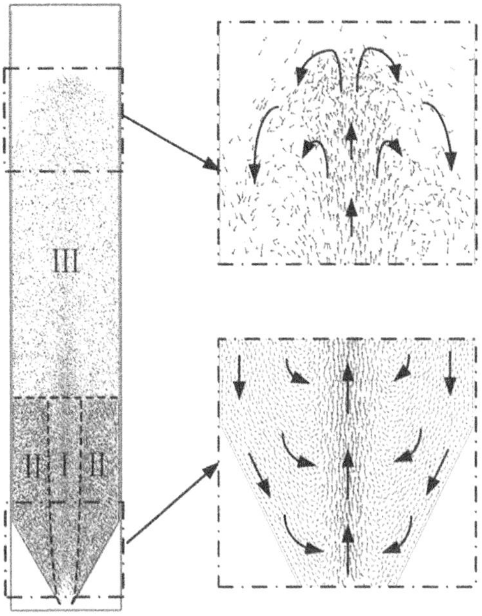

I-Spout region; II-Annulus region; III-Fountain region

Figure 7.9 Time-smoothed vector field of particle velocity of a fully developed spout ($u_s = 1.3u_{ms}$ at t = 3 seconds). (Source: Ren, B., Zhong, W., Jin, B., Yuan, Z., Lu, Y., 2011. Computational Fluid Dynamics (CFD)–Discrete Element Method (DEM) Simulation of Gas–Solid Turbulent Flow in a Cylindrical Spouted Bed with a Conical Base. *Energy & Fuels* **25**, 4095–4105, with permission)

the entire spout region (especially near the conical base) the particles rain back onto the annular region at the top of the fountain region.

Recently, Yang *et al.* [65] used the CFD-DEM model for simulating a 3D double slot, rectangular spouted bed with a vertical partition plate. They used k-ε turbulent model for the gas phase. Their simulation geometry and operating conditions are the same as reported by Chen *et al.* [66] and the geometry is shown in Figure 7.10. This spouted bed has two similar chambers with diverging bases and two parallel slots lying at the bottoms of these two chambers. Flow patterns at the superficial gas velocity of 1 m/s are illustrated in Figure 7.10. After introducing the gas into chambers, two bubbles with few particles inside appear in the lower part of the bed. These bubbles grow larger in the upper region and are followed by two gas routs. Particles are carried upward and injected into the free space after the breakup of bubbles on the bed surface. Finally, at steady particle circulation, three specific regions, that is, spout, annulus, and fountain, can be observed in each chamber. The results show that the insertion of a partition plate leads to the formation of two separated parallel fountains and an increase in the pressure drop. A velocity peak of the fluid phase or solid phase appears in the central region of each individual chamber. Furthermore, a large vertical solid flux appears in the central region of each chamber and shows a distribution strongly dependent to the slot shape. The slot shape influences the spout–annulus interaction boundary, whereas this effect diminishes obviously with increasing the bed height.

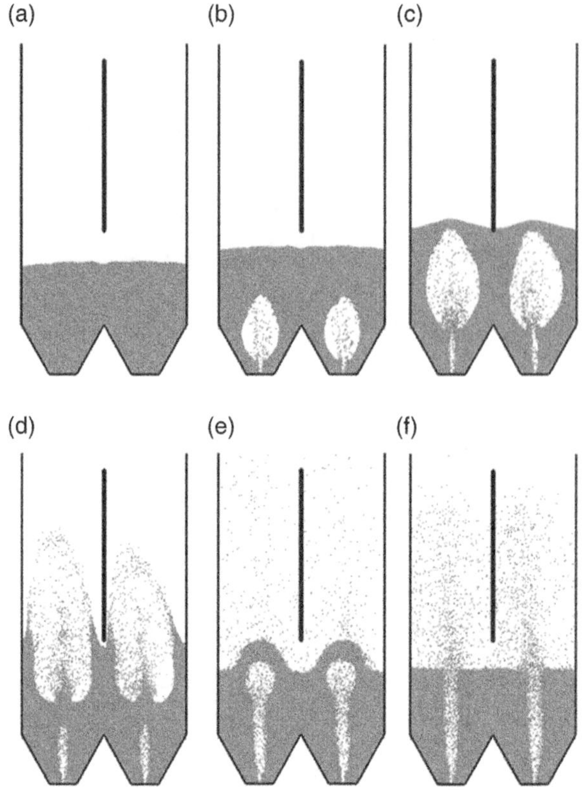

Figure 7.10 Flow pattern in double slot, rectangular spouted bed with a vertical partition plate at a superficial gas velocity 1 m/s, (a) t = 0.1 s; (b) t = 0.2 s; (c) t = 0.3 s; (d) t = 0.4 s; (e) t = 0.5 s; and (f) t = 0.6 s. (Source: Yang, S., Luo, K., Fang, M., Zhang, K., Fan, J., 2014. Parallel CFD–DEM modeling of the hydrodynamics in a lab-scale double slot-rectangular spouted bed with a partition plate. *Chemical Engineering Journal* **236**, 158–170, with permission)

We performed a simulation[1] using the experimental conditions of Liu *et al.* [67] and Zhao *et al.* [49] in a pseudo-2D spouted bed. To have a pseudo-2D behavior, a small bed depth of glass beads with a static height of 100 mm was assumed in their experiment. Particles with a diameter of 2.03 mm and density of 2380 kg/m^3 were used. Velocity of particles was estimated from images of the 2D spouted bed captured by PIV. The drag force was calculated from the Hill–Koch–Ladd correlation [7] and the gas volume fraction was estimated through the equation by Hoomans *et al.* [68]. Figure 7.11 shows velocities of particles and gas at the superficial velocity of 1.58 m/s. The minimum spouting velocity was 0.91 m/s in this case. The spouted bed can be divided into the following three distinct regions:

- the central spout region with low particle concentration, where the gas and particles rise at high velocity,

[1] This simulation was performed by an in-house code not published with this book.

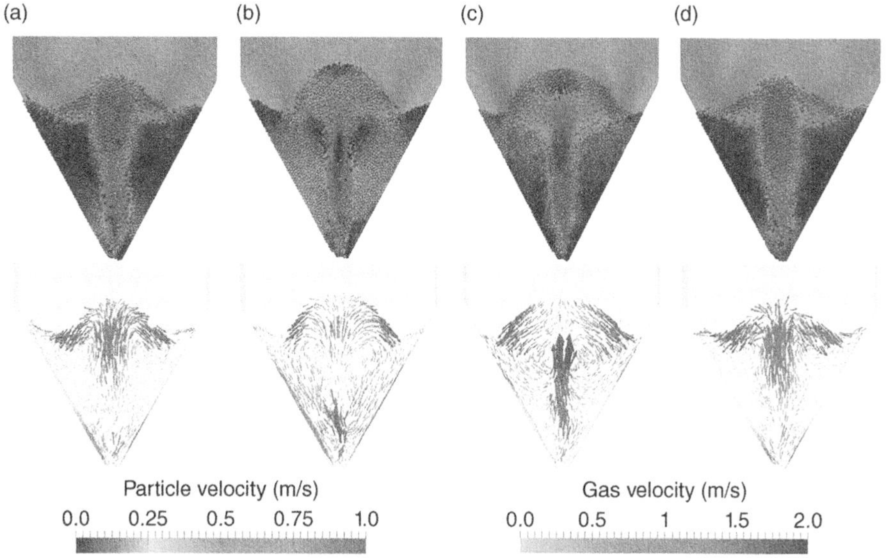

Figure 7.11 Flow patterns in a pseudo-2D spouted bed at the superficial gas velocity of 1.58 m/s (minimum spouting velocity is 0.91 m/s), (a) t = 2.0 s; (b) t = 2.05 s; (c) t = 2.1 s; and (d) t = 2.15 s. Top row shows the gas velocity magnitude contour and particles and bottom row shows the local-averaged velocity vector of particles

- the fountain zone where particles reach to their highest positions and then rain back onto the surface of the annulus, and
- the annulus zone where particles move slowly downward as a dense phase.

Figure 7.12 shows axial velocity of particles along the bed vertical axis at the same conditions of Figure 7.11. The symbols mark experimental results of Zhao *et al.* [49] and the solid line represents our simulation results. It can be seen in this figure that the simulation results are in good agreement with the experimental data in terms of the velocity profile of particles. Particles are accelerated near the gas inlet and reach their maximum velocity near the end of the spouted zone. Deceleration of particles then starts at the upper spout region. Zhao *et al.* [41] showed that in a 3D conical-cylindrical spouted bed, the particles reach their maximum velocity close to the gas inlet region.

7.3 Pneumatic Conveying

Pneumatic conveying is a process that is frequently employed for transporting solid particles from one location to another by the aid of aerodynamic forces in a pipe. Its applications can be found in many industries, including food, chemical, polymer, and pharmaceutical. Depending on the particle loading and the conveying gas velocity, the conveying systems can be categorized into dense and dilute phases.

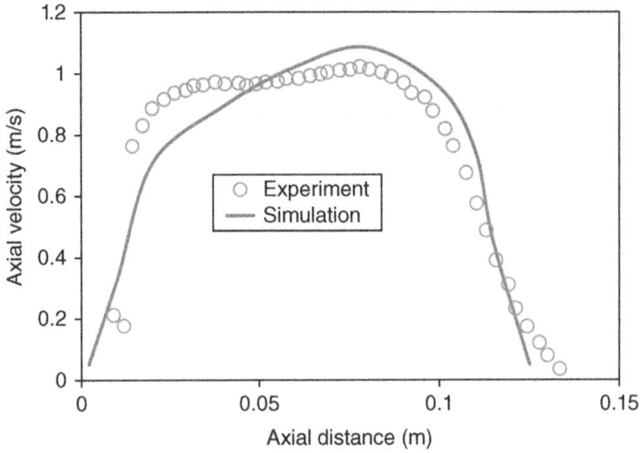

Figure 7.12 Comparison between experimental particle velocity along the bed vertical axis [49] and the simulation performed by the code accompanying this book. The superficial velocity is 1.58 m/s and the minimum spouting velocity is 0.91 m/s

7.3.1 Dilute Phase and Dense Phase Conveying

Dilute phase flow is characterized by high gas velocities (typically higher than 20 m/s), low solid concentrations (less than 1% by volume) and low pressure drops per unit pipe length (less than 5 mbar/m). It is important to note that the interaction of particles is low in the dilute conveying, thus they are dispersed individually. The dense phase flow, however, is characterized by low gas velocities (1–5 m/s), high solid concentrations (greater than 30% by volume), and high pressure drops per unit length of pipe (greater than 20 mbar/m). Particles are not fully dispersed in the dense phase conveying and there is a high interaction between particles, which can form clusters or particle density waves [69, 70]. Since collisional forces are generally low in the dilute phase, this type of conveying is usually preferred for handling of particles that are sensitive to abrasion. As pointed out by Rhodes [70], there are numerous definitions for the transition point between dilute phase and dense phase conveying. This transition is called saltation in horizontal conveying and choking in vertical conveying.

Pressure drop versus gas velocity phase diagram is commonly used to detect the transition between dilute and dense conveying. Pressure drop per unit length for polystyrene particles in the horizontal conveying versus solids loading and air velocities is shown in Figure 7.13. Curves are produced from the correlation proposed by Santos *et al.* [71]. As expected, at zero solid loading, the pressure drop increases monotonously with increasing the air velocity. At a constant solid loading, when the gas velocity is relatively high, all particles are carried in a dilute suspension. This type of conveying in a horizontal pipe is referred to as the dilute phase flow. As the air velocity decreases whilst the solids feed rate is kept constant, the pressure drop reaches its minimum (called saltation point), then increases due to settling of the particles in the lower section of the pipe. At the saltation point, where the gas velocity is not sufficient to suspend all particles, the particles are settled to the bottom of the pipe. The saltation velocity is the boundary between dilute and dense flows in a horizontal pneumatic conveying system.

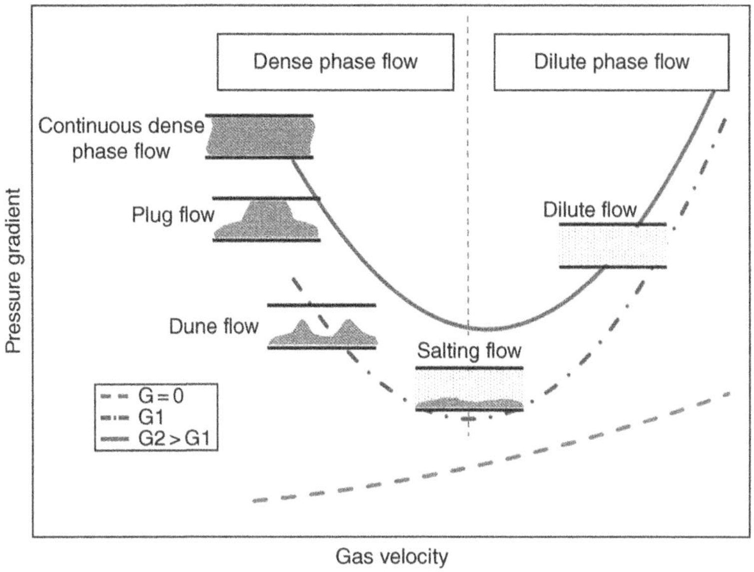

Figure 7.13 Phase diagram for horizontal conveying at constant solids feed rate

In this regime, some particles move in the dense phase along the bottom of the pipe and other particles are transported in the dilute phase. At lower gas velocities, the pressure drop increases with further decrease in the gas velocity where the gas flow area is restricted by the settled bulk of particles. A number of different flow patterns can occur within the dense flow regime, that is, continuous dense phase flow, non-continuous slug flow, and dune flow. The same diagram can be observed for other solid loadings. The phase diagram for vertical conveying is almost similar to that for horizontal conveying.

7.3.2 Horizontal Conveying

The CFD-DEM simulations can capture main features of gas-solid flows in horizontal or inclined pneumatic conveyers [72–86]. Lim *et al.* [77] performed CFD-DEM simulations and observed four different solid flow structures in horizontal conveyers, including homogenous flow, moving dunes flow, scarified flow, and slug flow, which correspond to typical flow regimes in horizontal conveyers identified in experimental observations. Solid concentrations and flow patterns are different in these regimes and depending on both gas and solid flow rates, one of these flow patterns may be observed. Simulation of Kuang *et al.* [81] showed that the speed of slug linearly increases with the gas velocity and is independent on the solid flow rate. They developed a 2D CFD combined with a 3D DEM model, with suitable periodic boundary conditions for both gas and solid phases to investigate the horizontal slug flow. The slug length increases with both solid and gas flow rates [81]. Fluid–particle, particle–wall, and particle–particle interactions control the mechanisms of formation, movement, and sweeping up of the slug. Figure 7.14 shows the horizontal slug flow patterns. The simulation was performed at the gas velocity of 2.09 m/s and with 11,000 particles. As shown in Figure 7.14a,

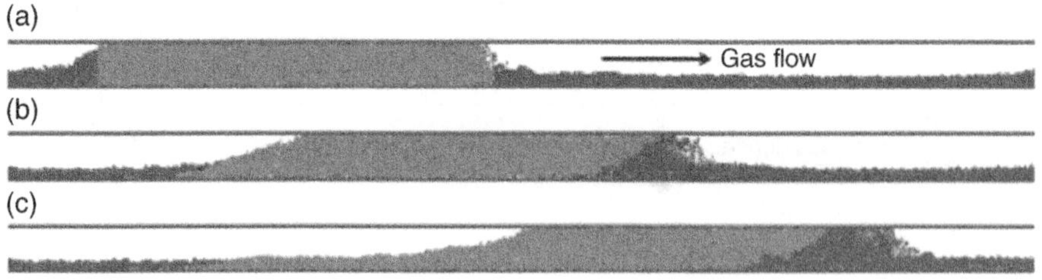

Figure 7.14 The horizontal slug flow patterns predicted by CFD-DEM at simulation times of (a) 14.236 s, (b) 14.498 s, and (c) 14.760 s. (Source: Kuang, S.B., Chu, K.W., Yu, A.B., Zou, Z.S., Feng, Y.Q., 2008. Computational Investigation of Horizontal Slug Flow in Pneumatic Conveying. *Industrial & Engineering Chemistry Research* **47**, 470–480, with permission)

at 12.236 s, particles are divided into the slug and settled layer, shown with dark and light shading, respectively. As the slug moves along the pipe, the particles are swapped in front of the settled layer and leave a new settled layer behind (Figure 7.14a,b).

Another important factor that critically determines the flow behavior of solids in pneumatic conveyers is the shape of solid particles. It is a common practice to consider solid particles as smooth spheres in CFD-DEM simulations. However, Hilton and Cleary [86] showed that the shape of particles, especially particles with high aspect ratio, affects the flow structure and transition points between different flow regimes in horizontal conveyers. They considered four different shapes for particles, namely sphere, prolate ellipsoid, oblate ellipsoid, and cuboid. Formation of slugs in the case of particles with higher aspect ratio is easier because they interlock more readily compared to spherical particles. Moreover, spherical particles can form more stable slugs than ellipsoidal particles with high ellipticity because the porosity of the non-spherical particle is lower and a greater pressure gradient is produced in the gas flow passing through ellipsoidal particles. This high pressure gradient across the slug makes it unstable [86].

The solid flow pattern changes when the pipe becomes inclined. Zhang *et al.* [79] studied the pneumatic conveying in a 45° inclined pipe. Three different flow patterns, including dispersed flow, reversed flow, and half-ring flow, were identified in their simulations. At very high flow rates of air, particles become dispersed throughout the pipe with a maximum velocity at the center (dispersed flow). At moderate air flow rates, two distinct sections can be observed: a dense phase of particles at the bottom and a dilute phase of particles at the top of the pipe (reversed flow). Further reduction of the air flow rate leads to the half-ring flow pattern.

Sakai and Koshizuka [82] applied coarse grain model for the large scale simulation of a 3D slug flow in a horizontal pipeline. In coarse grain model, larger particles are used instead of a group of individual original particles. The slug length and occupied area of the settled layer were compared between the coarse grain model and the original particle model. Their simulations showed that results of the coarse grain model are close to those of the original particle model.

Chu and Yu [80] studied the dense gas–solid flow in a pneumatic conveying bend. They simulated the phenomena like roping, particle segregation, particle velocity reduction, and particle recirculation and compared them with experimental observations. They showed that both particle–particle and particle–fluid interactions are responsible for the typical flow patterns in pneumatic conveying bends. Xiang *et al.* [83] studied segregation of particles with different densities in a horizontal pneumatic conveyer. They found a qualitative agreement between experimental data and their simulation results.

7.3.3 Vertical Conveying

The CFD-DEM model can also capture characteristics of flow patterns in a vertical pneumatic conveying system [77, 87–92]. Two different flow patterns were observed in simulations results: slug flow and dispersed flow. In the slug flow, which occurs at low gas flow rates and high solid flow rates, particles move together as large clusters. In the dispersed flow, occurs at high gas flow rates and low solid flow rates, particles move at high velocity as individuals. In a system of particles with high damping coefficient, particles tend to form large clusters, thus, the slug flow is more probable. Kawaguchi *et al.* [87] used CFD-DEM to investigate the density wave in the dense gas-solid flow in a vertical pipe. They found that particles are not dispersed uniformly in dense flow and form clusters or density waves. Kuang *et al.* [92] developed a 3D CFD-DEM model to study flow regimes in a vertical pneumatic conveying pipe. Their simulation reproduced the transition map from slug flow to dispersed flow with increasing the gas velocity. They showed that the reduction of the pressure drop in the dense phase flow can be attributed to decrease in the fluid drag and pressure gradient forces, while the increase in the pressure drop in the dilute phase flow is due to increases in the fluid drag force and gas wall friction force. They also showed that the particle–particle and particle–wall contact forces increase with increasing the particle–fluid force in the dense phase flow and the change is almost negligible in the dilute phase flow. Figure 7.15 shows the predicted flow pattern for conveying of polyethylene pellets along a vertical pipe, as Kuang *et al.* [92] obtained by simulation. Two distinct regimes, that is, slug flow (Figure 7.15a–c) and dispersed flow (Figure 7.15d,e), can be observed in their simulations that are in agreement with the experimental results of Pan [93]. In slug flow regime, the whole cross section of the pipe at some vertical parts is filled with particles while other parts contain only a few particles. As the slug moves up, some particles are fallen from back of the previous slug, rained into the air gap between slugs and swapped by the next slug. At higher gas velocities, particles are dispersed through the entire pipe (Figure 7.15d) and mostly form clusters (Figure 7.15e).

7.4 Non-Isothermal Flows

Literature on heat transfer modeling by DEM is scarce, but is growing, and available papers do not cover all practical cases in gas-solid flows. Li and Mason [73] were first to use the CFD-DEM approach for investigating the transient heat transfer in pneumatic conveying of particles. This was followed by others on this subject [94–96]. Swasdisevi *et al.* [97] studied hydrodynamics and heat transfer characteristics of gas-particle system in a 2D spouted bed with draft plates. They showed that the heat transfer coefficient and the average temperature of particles in the spout region are greater than those in the downcomer region. Particles are heated for a short time in the spout region and fall into the downcomer. In addition, the main portion of heat transfer between gas and particles occurs in the spout region.

Detailed mechanistic studies on heat transfer has been performed in packed and bubbling gas-solid fluidized beds by Hou *et al.* [98] and Zhou *et al.* [99]. They incorporated the k-ε

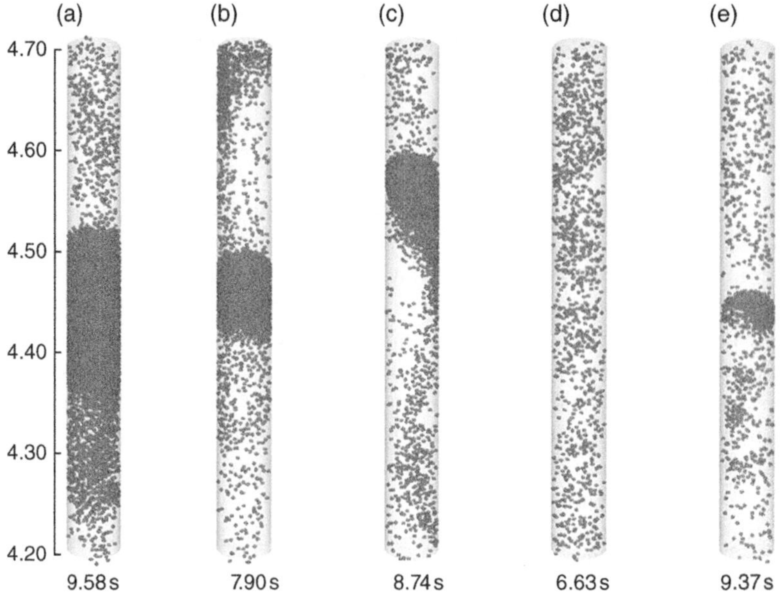

Figure 7.15 Predicted particle flow patterns at different gas velocities (a) 10 m/s, (b) 13 m/s, (c) 18.5 m/s, (d) 30.5 m/s, and (e) 30.5 m/s. (Source: Kuang, S.B., Yu, A.B., Zou, Z.S., 2009. Computational Study of Flow Regimes in Vertical Pneumatic Conveying. *Industrial & Engineering Chemistry Research* **48**, 6846–6858, with permission)

turbulent model to define the turbulent behavior of the gas phase and validated their model with experiments in packed and fluidized beds. They showed that CFD-DEM can satisfactorily predict the effective thermal conductivity of a particle assembly when a real value for Young modulus is selected. In addition, the model can reproduce the experimental data in terms of temperature evolution of hot spheres in a fluidized bed. The thermal CFD-DEM model can give more detailed information about the heat transfer coefficient at the particle scale. At low temperatures, the contribution of radiation is negligible and two mechanisms, namely particle–particle conduction and particle–fluid convection, determine the total heat transfer rate to an individual particle.

It is evident that the convective heat transfer increases with increasing the gas velocity and is not sensitive to the particle thermal conductivity. However, the particle–particle conductive heat transfer is sensitive to the thermal conductivity of particles for packed beds and increases with increasing the thermal conductivity of particles. Nonetheless, this sensitivity is less in fluidized beds because of reduced particle–particle contacts. For particulate systems with high thermal conductivities (such as metals), the effect of particle–particle conduction to the total heat transfer is not negligible even in fluidized beds, while this effect can be neglected (with a reasonable approximation) for particulate systems with low thermal conductivities in fluidized state. Yang *et al.* [100] used this model for investigating particle scale heat transfer in a moving bed and showed that their simulated gas temperature was in good agreement with experimental values. They found that the gas flow rate considerably affects the temperature profile while its effect is negligible in the flow pattern of solids. Convective heat transfer was dominant in the operating conditions of their work and was shown that it increases with increasing the gas flow rate. The solids flow rate, however, has a minor effect on the heat transfer mechanism.

Patil *et al.* [101] performed DEM simulations of a fluidized bed of various particle sizes. They first verified their model by simulation of transient heat transfer in a fixed bed. Next, they used their model and studied the variation of gas temperature during injection of a bubble in a fluidized bed. Bubble size was showed to be affected by both particle size and gas temperature. They used the bubble size in their simulation to estimate bubble heat transfer coefficient and found to be in close agreement with the value theoretical value given by Davidson and Harrison [102] for small particles and deviates from this model when increasing the particle size. These results can be used to determine the heat transfer coefficient in a fluidized bed when the bubble frequency is known.

Horizontally immersed tubes are used to remove or supply heat from/to fluidized beds. Thus, determining the bed to tube heat transfer coefficient is important in designing such beds. The thermal CFD-DEM model was adopted to investigate the bed to tube heat transfer characteristics [103, 104]. Zhao *et al.* [104] obtained profiles of time averaged solid volume fraction, convective heat transfer coefficient between gas and tube and conductive heat transfer coefficient around the tube at various gas velocities. They showed that the solid volume fraction at the top of tube is higher than at bottom and sides. An increase in the gas velocity reduces the solid volume fraction at top and sides while the solid volume fraction remains almost constant at the bottom. At low superficial gas velocities, the maximum convective heat transfer is located at the tube sides and the minimum at the top of the tube where the solid volume fraction is high and the air flow is blocked by tube. Increasing the gas velocity enhances the convective heat transfer between gas and tube. They showed that the conductive heat transfer coefficient is high where the solid concentration is high, which is due to enhanced contacts between particles and the tube surfaces.

Effect of variables such as particle size, solid thermal conductivity, bed temperature, and particle sliding friction on the heat transfer characteristics were also investigated by Zhou *et al.* [105]. Their results suggest that the heat transfer characteristics are not affected by particle sliding friction. They considered effect of radiation since the bed temperature exceeded 400 °C. Malone and Xu [106] studied wall to fluid heat transfer in a liquid-solid fluidized bed with hot walls. They found that different fluid-particle heat transfer correlations used in their thermal CFD-DEM model may lead to quite different results. Di Maio *et al.* [107] studied the characteristics of tube to bed heat transfer in fluidized beds. They examined different particle–particle heat transfer models and concluded that they may lead to different results. These studies show that the thermal CFD-DEM model still suffers from the lack of correlations that can truly predict the particle-fluid and particle–particle heat transfers.

CFD-DEM model can be also applied for investigating less conventional systems. This model was used for studying softening and melting of particles in the cohesive zone of a blast furnace [108]. Softening and melting steps were clearly identified by the model based on particle behavior and bed properties. Hou *et al.* [109] used a 2D model for investigating the thermal behavior of gas-solid flow in the reduction shaft reactor of Corex process for reducing iron ore. Results of this work can be useful for design and optimization of reduction shaft operation. Heat transfer in a generic grate system agitated by bars was simulated by Ricklet *et al.* [110] using the CFD-DEM model. The calculated local minimum, maximum, and average particle surface temperatures in different layers of the bed and showed that these values are in good agreement with experimental data.

7.5 Reactive Flows

In the past two decades, many researchers have used CFD-DEM technique for modeling and simulation of fluid-solid systems involving mass transfer and/or chemical reaction. Rong and Horio [111] were the first to use the CFD-DEM approach for studying char combustion in a bubbling fluidized bed. The 2D fluidized bed was filled with inert sand particles and fluidized by hot air. They considered the effect of particle–particle conduction, particle-fluid convection, radiation, and heat of combustion in their energy balance equations of both solid and gas phases. Number of sand and char particles in their simulation were 2000 and 20, respectively. The simulation was conducted for 5 s real time. Results of their model showed that the maximum temperature of char particles was about 50 °C above the average temperature of bed due to exothermic reactions on the surface of particles. Peters [112] followed the same approach for investigating pyrolysis and combustion of coal or biomass and found a good agreement between experimental data and results of the model.

Zhou *et al.* [113–115] used a 2D CDF-DEM model to investigate combustion of coal particles in a fluidized beds of hot sand particles. They modeled the fluid turbulence with large eddy simulation (LES) turbulence model. A thorough heat transfer model at particle scale was adopted in which particle–particle conduction, particle–gas convection, particle–bed radiation, and heat of combustion were considered. Evolution of size of coal particles during combustion was modeled by the shrinking core model. In this way, by advancement of combustion, the size of particles gradually decreases and the burn out time of coal particles can be obtained. They divided the combustion of coal particles into two distinct stages: a rapid heat up stage and a longer combustion stage. The heating stage mainly occurs in 2 s in which the temperature of coal particles reaches the burning point of coal and the volatiles are released very quickly. It was found that most of the heat that is transferred to coal particles is via convective and radiative mechanisms in this stage. However, in the combustion stage, most of the energy is exchanged through combustion followed by radiation and convection. Thus, the effect of particle–particle conduction is negligible in this stage. They also studied the effect of gas velocity, coal particle diameter, and bed temperature on the combustion characteristics of coal particles. Since the simulations were conducted for just a few seconds, they did not capture the coal particle burnout time, although the variation of coal diameter with combustion was considered in their model.

Geng and Che [116] extended the model of Zhou *et al.* [115] in the following aspects. First, they introduced a source term in the gas continuity equation, which accounts for production of gaseous components (i.e., CO and CO_2) by combustion of coal. Second, they modified the heat balance equation according to the model proposed by Zhou *et al.* [105] for fluidized beds. Third, they took into account the sand inhibitory effects on oxidation of CO to CO_2 in the gas phase. Finally, they considered the shrinking core model consisting of three mechanisms. For large particles, the combustion rate is controlled by bulk diffusion rate and for medium size particles both bulk diffusion and kinetic rates are significant. The mass transfer resistance vanishes as the particle becomes too small and the rate is controlled by the kinetic rate. This model was validated quantitatively against the experimental data obtained in combustion of single graphite particle in a fluidized bed of hot sand particles [117, 118]. The temperature of graphite particles increases from 300 K to a relatively steady temperature of 1400 K, which is about 150 K higher than the bed temperature. The size of graphite continuously decreases against time until it completely burns out at 580 s in the simulation while the time at which the graphite particle disappears is 600 s in physical experiments. Simulated concentration of CO_2 in the off-gas of fluidized bed combustor was also in a reasonable agreement with experiments. Most importantly, the simulation was conducted for several seconds that the graphite burnout point was captured [116].

Liu *et al.* [119] studied the combustion process of char particles in a hot bed of sand particles fluidized with air and propane. They did not consider the effect of particle shrinkage in their model because the simulations time was short (~15 s) and size of char particles did not change noticeably. They showed that presence of propane in the entering gas reduces the char combustion rate. This effect is highlighted when the bed temperature is high or the char is very reactive. These results are consistent with results of Hesketh and Davidson [120] who suggested that higher temperature promotes the combustion of propane instead of char because the combustion rate of char is controlled by diffusion and more oxygen is consumed by propane combustion with increasing bed temperature. They also showed that, depending on the temperature of the bed, the combustible gas can be burnt inside the emulsion phase or inside the rising bubbles at high temperatures or above the bed surface at low temperatures.

Oevermann *et al.* [121] studied the gasification of wood particles in a hot fluidized bed of charcoal particles with a 2D CFD-DEM. They studied effect of change in wood feeding rate on the hydrodynamic changes in the reactor. The results revealed that when the wood feeding rate is increased, frequency and amplitude of concentration signals increase in the exhaust gas. Production of gaseous components through heterogeneous gasification in the reactor intensifies fluidization at high wood feeding rates. Gasification of coal in a fluidized bed was also investigated through CFD-DEM by Snider *et al.* [122]. Both energy and mass conservation equations were considered in their work. They found that the bed is thermally uniform except at the feeding injection points where heat of combustion is released. Snider *et al.* [122] also showed that there is a radial distribution of gaseous components such that this reactor cannot be considered as plug flow reactor. Ku *et al.* [123] simulated a fluidized bed reactor of steam gasification of biomass. Homogeneous and heterogeneous reactions as well as shrinkage of particles were considered in this work. The model was validated through comparison with experimental results, indicating that the proposed CFD-DEM model can capture behavior of both macro-structure (bubbles or slugs) and micro-structure (motion of particles) in the fluidized bed reactor.

Different catalytic or non-catalytic chemical reactions may occur in fluidized beds, like fluid catalyst cracking (FCC) or polymerization of polyolefins. Kaneko *et al.* [16] studied formation of hot spot zones in polyolefin polymerization reactors. They included simple heat transfer model and polymerization kinetics into the CFD-DEM model and showed that an intimate contact between particles and gas phase exists, which leads to an efficient heat transfer between solid and gas phases. In the case of porous plate distributor, no high temperature zone in the fluidized bed was observed while hot spots appear at corners of the bed just above the perforated plate distributor. They related the formation of hot spots to degree of mixing in the reactor. Formation of hot spot is more likely where the degree of diffusive mixing is high. Limtrakul *et al.* [124] studied decomposition of ozone on iron oxide catalyst particles. The reaction was performed in an isothermal spouted bed. They obtained ozone concentration distribution in gas and particle phases as a function of time. A good agreement between simulation and experimental results in similar setup was observed.

Wu *et al.* [125, 126] studied catalytic cracking of heavy oil in riser and downer reactors. They considered the effect of catalyst deactivation, by tracking the history of catalyst particles undergoing reaction, and heat transfer in different locations in the reactor. In this way, they could capture the effect of RTD on the catalyst deactivation and reactor performance. They performed simulations involving 40,000–200,000 particles and showed that non-uniform flow structure and back mixing of catalyst product in the riser reactor, especially at high catalyst-to-oil ratios, result in a wide RTD of catalyst particles that lead to reduced selectivity to gasoline. However, the flow structure in the downer reactor is nearly plug flow, leading to a narrow RTD of catalyst particles and better selectivity to intermediate products.

Karimi *et al.* [127] studied gas-phase polyethylene polymerization with a CFD-DEM model by incorporating mass and energy conservation equations for both solid and gas phases. A comprehensive kinetic model [128] was used to evaluate the rate of copolymerization of ethylene and 1-butene on the Ziegler–Natta catalyst. In this model, a series of elementary reactions, including site activation, propagation, site deactivation, and chain transfer were considered. Distribution of temperature and ethylene in gas and solid phases against time were obtained. They showed that the mixing of particles, as a result of motion of bubbles, is the main reason of approaching a uniform temperature distribution in the bed. One of the most important factors affecting the physico-chemical characteristics of the reactor is its operating pressure. At high pressures, more gas enters the emulsion phase which results in smaller bubbles and a more homogeneous state of the bed. In this condition, a better contact between gas and solid particles is achieved. It was shown that a more uniform distribution of particle temperature with a lower mean value is obtained at elevated pressure. For the gas phase ethylene polymerization, an increase in the operating pressure leads to an increase in the reaction rate and its heat generation. Despite the higher heat generated at elevated pressure, the temperate of particles is lower at such a condition, owing to higher heat transfer rate at elevated pressure.

Zhuang *et al.* [129] modeled catalytic methanol-to-olefins (MTO) reaction in a fluidized bed reactor by CFD-DEM technique. They used a lumped kinetics sub-model in the gas phase and also considered the heat transfer between particles as well as particles and gas in their model. The temperature distribution in the reactor was determined by the exothermic heat of reaction and it was shown that concentration of products is almost uniform throughout the bed due to high rate of mass transfer in the fluidized bed. Modeling of CO_2 capture in a bubbling fluidized bed was carried out by Zhang *et al.* [130]. They used potassium carbonate as the sorbent in their work and considered the shrinking core model in the reaction kinetics. They used the model to show that concentration of CO_2 decreases along the bed height in the axial direction and the concentration is higher at the center of the bed compared to the wall in the radial direction.

7.6 Miscellaneous

Hydrocyclones are devices by which solid particles are separated from a fluid by centrifugal force exerted on particles. Some researchers have studied the gas–solid flow in hydrocyclones and investigated the mechanism of solid particle separations [91, 131, 132]. The main flow characteristics in hydrocyclones were captured by CFD-DEM simulations, which include strand flow pattern of particles, decrease in pressure drop, and tangential velocity due to solid loading. The simulation results were compared with experimental observations in terms of gas pressure drop within the cyclone at different gas velocities, different solid loadings, and different tangential velocity and good agreement was observed between simulation and experimental results. It is also possible to deeply study effect of different parameters (such as fluid drag force, pressure gradient force, and particle–particle and particle–wall contact forces) on the separation of particles from fluid. Based on this valuable information, which can be obtained only through the simulation, we can find the malfunctions in the system and eliminate them by making changes in the geometry of hydrocyclones.

Liquid–solid fluidization is another important two-phase flow case which is mostly used for size classification of particulate systems. Segregation of particles due to their size, shape, or density differences is the basis for separation of particles in liquid–solid fluidized beds. Layer inversion is an important phenomenon that may occur in special cases in fluidization of a binary mixture of particles differing in size and density (larger particle with lower density). Despite important applications of liquid–solid fluidization, a little research has been conducted on these systems by CFD-DEM [133–137]. Malone *et al.* [133] and Di Renzo and Di Maio [134] studied the macroscopic behavior of liquid-solid fluidized beds, including minimum fluidization velocity, homogeneous bed expansion, and transition between particulate and aggregative fluidization regimes. Malone *et al.* [135] and Zhou and Yu [136] reported the capture of the layer inversion phenomenon in liquid-solid fluidized beds by CFD-DEM technique. Di Renzo *et al.* [137] studied the layer inversion phenomena with their new drag correlation expanded for polydispersed systems [138]. They performed a detailed comparison between simulations results and different experimental data in terms of overall bed height, mixing layer interface height, fluid volume fraction, and particle distributions. Their simulations showed that the newly developed correlation for polydispersed systems can reproduce experimental results more precisely.

Three-phase gas–liquid–solid fluidized beds are widely used in chemical, petrochemical, and pharmaceutical applications, such as methanol synthesis process, hydrothreading of tar sand, treatment of lactose, wastewater, and leaching of metals from ores [139]. To model a three phase system, one may model the motion of gas and liquid phases with the Eulerian approach (e.g., two-fluid model) and the motion of particles with the Lagrangian approach. Li *et al.* [140] treated the gas and liquid phases with volume of fluid model wherein gas bubbles were assumed to be void and their motion is governed by liquid–solid motion and surface dynamics. In this model, solids may not present in gas bubbles. Cao *et al.* [141] experimentally studied the gas-liquid–solid circulating fluidized beds and utilized the DEM model to describe solid flow behavior of this system. In their model, gas and liquid phases were assumed to be two interpenetrating continua similar to the two fluid model and solid particles as individual particles similar to the DEM. Particles were allowed to interact with both gas and liquid phases. Different flow phenomena and features of three-phase fluidized beds were studied by this model.

Nomenclature

Symbol	Unit	Description
D	m	Bed diameter
d_p	m	Particle diameter
f_C	N	Contact force
f_D	N	Drag force
f_G	N	Gravitational force

f_{LM}	N	Magnus lift force
f_{LS}	N	Saffman lift force
G	kg/m²·s	Solid mass flux
H	m	Bed height
P	Pa	Pressure
t	s	Time
u_{ms}	m/s	Minimum spouting velocity
u_s	m/s	Spouting velocity
Z	m	Height

References

[1] Stein, M., Ding, Y.L., Seville, J.P.K., and Parker, D.J. (2000) Solid motion in bubbling gas fluidized beds. *Chemical Engineering Science*, **55**, 5291–5300.

[2] Feng, Y.Q., Xu, B.H., Zhang, S.J., Yu, A.B., and Zulli, P. (2004) Discrete particle simulation of gas fluidization of particle mixtures. *AIChE Journal*, **50**, 1713–1728.

[3] Marzocchella, A., Salatino, P., Di Pastena, V., and Lirer, L. (2000) Transient fluidization and segregation of binary mixtures of particles. *AIChE Journal*, **46**, 2175–2182.

[4] Mostoufi, N. and Chaouki, J. (2004) Flow structure of the solids in gas- solid fluidized beds. *Chemical Engineering Science*, **59**, 4217–4227.

[5] Yuu, S., Umekage, T., and Johno, Y. (2000) Numerical simulation of air and particle motions in bubbling fluidized bed of small particles. *Powder Technology*, **110**, 158–168.

[6] Bokkers, G.A., Annaland, M.V.S., and Kuipers, J.A.M. (2004) Mixing and segregation in a bidisperse gas–solid fluidised bed: a numerical and experimental study. *Powder Technology*, **140**, 176–186.

[7] Hill, R.J., Koch, D.L., and Ladd, J.C. (2001) Moderate-Reynolds-numbers flows in ordered and random arrays of spheres. *Journal of Fluid Mechanics*, **448**, 243–278.

[8] Wang, J., van der Hoef, M.A., and Kuipers, J.A.M. (2010) CFD study of the minimum bubbling velocity of Geldart A particles in gas-fluidized beds. *Chemical Engineering Science*, **65**, 3772–3785.

[9] Tsuji, T., Yabumoto, K., and Tanaka, T. (2008) Spontaneous structures in three-dimensional bubbling gas-fluidized bed by parallel DEM–CFD coupling simulation. *Powder Technology*, **184**, 132–140.

[10] Olaofe, O.O., van der Hoef, M.A., and Kuipers, J.A.M. (2011) Bubble formation at a single orifice in a 2D gas-fluidized bed. *Chemical Engineering Science*, **66**, 2764–2773.

[11] Tagami, N., Mujumdar, A., and Horio, M. (2009) DEM simulation of polydisperse systems of particles in a fluidized bed. *Particuology*, **7**, 9–18.

[12] Hilton, J.E., Mason, L.R., and P.W. Cleary (2010) Dynamics of gas–solid fluidised beds with non-spherical particle geometry. *Chemical Engineering Science*, **65**, 1584–1596.

[13] Li, J. and Kuipers, J.A.M. (2002) Effect of pressure on gas–solid flow behavior in dense gas-fluidized beds: a discrete particle simulation study. *Powder Technology*, **127**, 173–184.

[14] Godlieb, W., Deen, N.G., and Kuipers, J.A.M. (2007) A discrete particle simulation study of solids mixing in a pressurized fluidized bed. *The 12th International Conference on Fluidization – New Horizons in Fluidization Engineering*, ECI.

[15] Mansourpour, Z., Karimi, S., Zarghami, R., Mostoufi, N., and Sotudeh-Gharebagh, R. (2010) Insights in hydrodynamics of bubbling fluidized beds at elevated pressure by DEM–CFD approach. *Particuology*, **8**, 407–414.

[16] Kaneko, Y., Shiojima, T., and Horio, M. (1999) DEM simulation of fluidized beds for gas-phase olefin polymerization. *Chemical Engineering Science*, **54**, 5809–5821.

[17] Yates, J.G. (1996) Effects of temperature and pressure on gas-solid fluidization. *Chemical Engineering Science*, **51**, 167–205.

[18] Yuu, S., Nishikawa, H., and Umekage, T. (2001) Numerical simulation of air and particle motions in group-B particle turbulent fluidized bed. *Powder Technology*, **118**, 32–44.

[19] Zhang, M.H., Chu, K.W., Wei, F., and Yu, A.B. (2008) A CFD–DEM study of the cluster behavior in riser and downer reactors. *Powder Technology*, **184**, 151–165.

[20] Suryawanshi, V. and Roy, S. (2009) DEM simulation of gas–solids circulating fluidized beds. *Journal of Chemical Engineering of Japan*, **42**, 130–136.

[21] Zhao, Y., Cheng, Y., Wu, C., Ding, Y., and Jin, Y. (2010) Eulerian–Lagrangian simulation of distinct clustering phenomena and RTDs in riser and downer. *Particuology*, **8**, 44–50.
[22] Zhao, Y., Ding, Y., Wu, C., and Cheng, Y. (2010) Numerical simulation of hydrodynamics in downers using a CFD–DEM coupled approach. *Powder Technology*, **199**, 2–12.
[23] Dahl, S.R. and Hrenya, C.M. (2004) Size segregation in rapid, granular flows with continuous size distributions. *Physics of Fluids*, **16**, 1–13.
[24] Feng, Y.Q. and Yu, A.B. (2006) Discrete particle simulation of size segregation of particle mixtures in a gas fluidized bed. *China Particuology*, **4**, 122–126.
[25] Deen, N.G., Annaland, M.V.S., and Kuipers, J.A.M. (2006) Detailed computational and experimental fluid dynamics of fluidized beds. *Applied Mathematical Modelling*, **30**, 1459–1471.
[26] Beetstra, R., van der Hoef, M.A., and Kuipers, J.A.M. (2007) Numerical study of segregation using a new drag force correlation for polydisperse systems derived from lattice-Boltzmann simulations. *Chemical Engineering Science*, **62**, 246–255.
[27] Huilin, L., Yunhua, Z., Ding, J., Gidaspow, D., and Wei, L. (2007) Investigation of mixing/segregation of mixture particles in gas-solid fluidized beds. *Chemical Engineering Science*, **62**, 301–317.
[28] Tian, F., Zhang, M., Fan, H., Gu, M., Wang, L., and Qi, Y. (2007) Numerical study on microscopic mixing characteristics in fluidized beds via DEM. *Fuel Processing Technology*, **88**, 187–198.
[29] Goldschmidt, M.J.V., Link, J.M., Mellema, S., and Kuipers, J.A.M. (2003) Digital image analysis measurements of bed expansion and segregation dynamics in dense gas-fluidised beds. *Powder Technology*, **138**, 135–159.
[30] Norouzi, H.R., Mostoufi, N., and Sotudeh-Gharebagh, R. (2012) Segregation behaviour of particles in gas solid fluidized beds at elevated pressure. *Journal of Chemical and Petroleum Engineering*, **46**(2), 111–121.
[31] Norouzi, H.R., Mostoufi, N., and Sotudeh-Gharebagh, R. (2012) Effect of fines on segregation of binary mixtures in gas–solid fluidized beds. *Powder Technology*, **225**, 7–20.
[32] Norouzi, H.R., Mostoufi, N., Mansourpour, Z., Sotudeh-Gharebagh, R., and Chaouki, J. (2011) Characterization of solids mixing patterns in bubbling fluidized beds. *Chemical Engineering Research and Design*, **89**, 817–826.
[33] Epstein, N. and Grace, J.R. (2011) *Spouted and Spout-Fluid Beds: Fundamentals and Applications*, Cambridge University Press.
[34] Piskova, E. and Mörl, L. (2007) Fluidization regimes in different spouted bed apparatus constructions. *Chemical Engineering and Processing: Process Intensification*, **46**(8), 695–702.
[35] Milne, B.J., Berruti, F., Behie, L.A., and De Bruijn, T.J.W. (1992) The internally circulating fluidized bed (ICFB): a novel solution to gas bypassing in spouted beds. *The Canadian Journal of Chemical Engineering*, **70**(5), 910–915.
[36] Kim, Y.T., Song, B.H., and Kim, S.D. (1997) Entrainment of solids in an internally circulating fluidized bed with draft tube. *Chemical Engineering Journal*, **66**(2), 105–110.
[37] Ahn, H.-S., Lee, W.-J., KimDg, S.-D., and Song, B.-H. (1999) Solid circulation and gas bypassing in an internally circulating fluidized bed with an orifice-type draft Tube. *Korean Journal of Chemical Engineering*, **16**(5), 618–623.
[38] Kim, S., Kim, Y., Roh, S., and Lee, D. (2002) Solid circulation characteristics in an internally circulating fluidized bed with orifice-type draft tube. *Korean Journal of Chemical Engineering*, **19**(5), 911–916.
[39] Matthew, M.C., Morgan, M.H., and Littman, H. (1988) Study of the hydrodynamics within a draft tube spouted bed system. *The Canadian Journal of Chemical Engineering*, **66**(6), 908–918.
[40] Eng, J.H., Svrcek, W.Y., and Behie, L.A. (1989) Dynamic modeling of a spouted bed reactor with a draft tube. *Industrial & Engineering Chemistry Research*, **28**(12), 1778–1785.
[41] Zhao, X.L., Yao, Q., and Li, S.Q. (2006) Effects of draft tubes on particle velocity profiles in spouted beds. *Chemical Engineering & Technology*, **29**(7), 875–881.
[42] Zhao, X.L., Li, S.Q., Liu, G.Q., Song, Q., and Yao, Q. (2008) Flow patterns of solids in a two-dimensional spouted bed with draft plates: PIV measurement and DEM simulations. *Powder Technology*, **183**(1), 79–87.
[43] Makibar, J., Fernandez-Akarregi, A.R., Díaz, L., Lopez, G., and Olazar, M. (2012) Pilot scale conical spouted bed pyrolysis reactor: draft tube selection and hydrodynamic performance. *Powder Technology*, **219**, 49–58.
[44] Nagashima, H., Suzukawa, K., and Ishikura, T. (2013) Hydrodynamic performance of spouted beds with different types of draft tubes. *Particuology*, **11**(5), 475–482.
[45] Yang, S., Luo, K., Fang, M., and Fan, J. (2013) Discrete element simulation of the hydrodynamics in a 3D spouted bed: influence of tube configuration. *Powder Technology*, **243**, 85–95.
[46] Jittanit, W., Srzednicki, G., and Driscoll, R.H. (2013) Comparison between fluidized bed and spouted bed drying for seeds. *Drying Technology*, **31**(1), 52–56.

[47] Takeuchi, S., Wang, S., and Rhodes, M. (2004) Discrete element simulation of a flat-bottomed spouted bed in the 3-D cylindrical coordinate system. *Chemical Engineering Science*, **59**, 3495–3504.

[48] Zhong, W., Xiong, Y., Yuan, Z., and Zhang, M. (2006) DEM simulation of gas–solid flow behaviors in spout-fluid bed. *Chemical Engineering Science*, **61**(5), 1571–1584.

[49] Zhao, X.L., Li, S.Q., Liu, G.Q., Yao, Q., and Marshall, J.S. (2008) DEM simulation of the particle dynamics in two-dimensional spouted beds. *Powder Technology*, **184**(2), 205–213.

[50] Zhang, Y., Jin, B., Zhong, W., Ren, B., and Xiao, R. (2010) DEM simulation of particle mixing in flat-bottom spout-fluid bed. *Chemical Engineering Research and Design*, **88**(5–6), 757–771.

[51] Ren, B., Zhong, W., Jin, B., Yuan, Z., and Lu, Y. (2011) Modeling of gas-particle turbulent flow in spout-fluid Bed by computational fluid dynamics with discrete element method. *Chemical Engineering & Technology*, **34**(12), 2059–2068.

[52] Ren, B., Zhong, W., Jin, B., Yuan, Z., and Lu, Y. (2011) Computational fluid dynamics (CFD)–discrete element method (DEM) simulation of gas–solid turbulent flow in a cylindrical spouted bed with a conical base. *Energy & Fuels*, **25**(9), 4095–4105.

[53] Link, J., Zeilstra, C., Deen, N., and Kuipers, H. (2004) Validation of a discrete particle model in a 2D spout-fluid bed using non-intrusive optical measuring techniques. *The Canadian Journal of Chemical Engineering*, **82**(1), 30–36.

[54] Rong, L. and Zhan, J. (2010) Improved DEM-CFD model and validation: a conical-base spouted bed simulation study. *Journal of Hydrodynamics, Series B*, **22**(3), 351–360.

[55] Kawaguchi, T., Sakamoto, M., Tanaka, T., and Tsuji, Y. (2000) Quasi-three-dimensional numerical simulation of spouted beds in cylinder. *Powder Technology*, **109**(1–3), 3–12.

[56] Takeuchi, S., Shan Wang, X., and Rhodes, M.J. (2005) Discrete element study of particle circulation in a 3-D spouted bed. *Chemical Engineering Science*, **60**(5), 1267–1276.

[57] Takeuchi, S., Wang, S., and Rhodes, M. (2008) Discrete element method simulation of three-dimensional conical-base spouted beds. *Powder Technology*, **184**(2), 141–150.

[58] Link, J.M., Cuypers, L.A., Deen, N.G., and Kuipers, J.A.M. (2005) Flow regimes in a spout-fluid bed: a combined experimental and simulation study. *Chemical Engineering Science*, **60**, 3425–3442.

[59] Ergun, S. (1952) Fluid flow through packed columns. *Chemical Engineering Progress*, **48**, 89–94.

[60] Wen, C.Y. and Yu, Y.H. (1966) Mechanics of fluidization. *Chemical Engineering Progress Symposium Series*, **62**, 100–111.

[61] Link, J.M., Deen, N.G., Kuipers, J.A.M., Fan, X., Ingram, A., Parker, D.J., Wood, J., and Seville, J.P.K. (2008) PEPT and discrete particle simulation study of spout-fluid bed regimes. *AIChE Journal*, **54**(5), 1189–1202.

[62] Link, J.M., Godlieb, W., Tripp, P., Deen, N.G., Heinrich, S., Kuipers, J.A.M., Schönherr, M., and Peglow, M. (2009) Comparison of fibre optical measurements and discrete element simulations for the study of granulation in a spout fluidized bed. *Powder Technology*, **189**(2), 202–217.

[63] Ren, B., Zhong, W., Chen, Y., Chen, X., Jin, B., Yuan, Z., and Lu, Y. (2012) CFD-DEM simulation of spouting of corn-shaped particles. *Particuology*, **10**(5), 562–572.

[64] Ren, B., Zhong, W., Jin, B., Shao, Y., and Yuan, Z. (2013) Numerical simulation on the mixing behavior of corn-shaped particles in a spouted bed. *Powder Technology*, **234**, 58–66.

[65] Yang, S., Luo, K., Fang, M., Zhang, K., and Fan, J. (2014) Parallel CFD–DEM modeling of the hydrodynamics in a lab-scale double slot-rectangular spouted bed with a partition plate. *Chemical Engineering Journal*, **236**, 158–170.

[66] Chen, Z., Lim, C.J., and Grace, J.R. (2008) Hydrodynamics of slot-rectangular spouted beds: effect of slot configuration on the local flow structure. *The Canadian Journal of Chemical Engineering*, **86**(3), 598–604.

[67] Liu, G.Q., Li, S.Q., Zhao, X.L., and Yao, Q. (2008) Experimental studies of particle flow dynamics in a two-dimensional spouted bed. *Chemical Engineering Science*, **63**(4), 1131–1141.

[68] Hoomans, B.P.B., Kuipers, J.A.M., Briels, W.J., and van Swaaij, W.P.M. (1996) Discrete particle simulation of bubble and slug formation in a two-dimensional gas-fluidised bed: a hard-sphere approach. *Chemical Engineering Science*, **51**(1), 99–118.

[69] Narimatsu, C.P. and Ferreira, M.C. (2001) Vertical pneumatic conveying in dilute and dense-phase flows: experimental study of the influence of particle density and diameter on fluid dynamic behavior. *Brazilian Journal of Chemical Engineering*, **18**, 221–232.

[70] Rhodes, M.J. (2008) *Introduction to Particle Technology*, 2nd edn, John Wiley & Sons, Ltd, Chichester.

[71] Santos, K.G., Murata, V.V., and Barrozo, M.A.S. (2009) Three-dimensional computational fluid dynamics modelling of spouted bed. *The Canadian Journal of Chemical Engineering*, **87**(2), 211–219.

[72] Tsuji, Y., Tanaka, T., and Ishida, T. (1992) Lagrangian numerical simulation of plug flow of cohesionless particles in a horizontal pipe. *Powder Technology*, **71**, 239–250.
[73] Li, J. and Mason, D.J. (2000) A computational investigation of transient heat transfer in pneumatic transport of granular particles. *Powder Technology*, **112**, 273–282.
[74] Xiang, J. and McGlinchey, D. (2004) Numerical simulation of particle motion in dense phase pneumatic conveying. *Granular Matter*, **6**(2–3), 167–172.
[75] Li, J., Webb, C., Pandiella, S.S., Campbell, G.M., Dyakowski, T., Cowell, A., and McGlinchey, D. (2005) Solids deposition in low-velocity slug flow pneumatic conveying. *Chemical Engineering and Processing: Process Intensification*, **44**(2), 167–173.
[76] Fraige, F.Y. and Langston, P.A. (2006) Horizontal pneumatic conveying: a 3d distinct element model. *Granular Matter*, **8**(2), 67–80.
[77] Lim, E.W.C., Wang, C.H., and Yu, A.B. (2006) Discrete element simulation for pneumatic conveying of granular material. *AIChE Journal*, **52**, 496–509.
[78] Lim, E.W.C., Zhang, Y., and Wang, C.-H. (2006) Effects of an electrostatic field in pneumatic conveying of granular materials through inclined and vertical pipes. *Chemical Engineering Science*, **61**(24), 7889–7908.
[79] Zhang, Y., Lim, E.W.C., and Wang, C.H. (2007) Pneumatic transport of granular materials in an inclined conveying pipe: comparison of computational fluid dynamics-discrete element method (CFD-DEM), electrical capacitance tomography (ECT), and particle image velocimetry (PIV) results. *Industrial & Engineering Chemistry Research*, **46**, 6066–6083.
[80] Chu, K.W. and Yu, A.B. (2008) Numerical simulation of the gas-solid flow in three-dimensional pneumatic conveying bends. *Industrial & Engineering Chemistry Research*, **47**, 7058–7071.
[81] Kuang, S.B., Chu, K.W., Yu, A.B., Zou, Z.S., and Feng, Y.Q. (2008) Computational investigation of horizontal slug flow in pneumatic conveying. *Industrial & Engineering Chemistry Research*, **47**(2), 470–480.
[82] Sakai, M. and Koshizuka, S. (2009) Large-scale discrete element modeling in pneumatic conveying. *Chemical Engineering Science*, **64**, 533–539.
[83] Xiang, J., McGlinchey, D., and Latham, J.P. (2010) An investigation of segregation and mixing in dense phase pneumatic conveying. *Granular Matter*, **12**, 345–355.
[84] Zhou, Z.Y., Kuang, S.B., Chu, K.W., and Yu, A.B. (2010) Discrete particle simulation of particle-fluid flow: model formulations and their applicability. *Journal of Fluid Mechanics*, **661**, 482–510.
[85] Brosh, T., Kalman, H., and Levy, A. (2011) DEM simulation of particle attrition in dilute-phase pneumatic conveying. *Granular Matter*, **13**, 175–181.
[86] Hilton, J.E. and P.W. Cleary (2011) The influence of particle shape on flow modes in pneumatic conveying. *Chemical Engineering Science*, **66**, 231–240.
[87] Kawaguchi, T., Tanaka, T., and Tsuji, Y. (2000) Numerical analysis of density wave in dense gas-solid flows in a vertical pipe. *Progress of Theoretical Physics Supplement*, **138**, 696–701.
[88] Watano, S., Saito, S., and Suzuki, T. (2003) Numerical simulation of electrostatic charge in powder pneumatic conveying process. *Powder Technology*, **135–136**, 112–117.
[89] Ouyang, J., Yu, A.B., and Pan, R.H. (2005) Simulation of gas-solid flow in vertical pipe by hard-sphere model. *Particulate Science and Technology*, **23**(1), 47–61.
[90] Xu, M., Ge, W., and Li, J. (2007) A discrete particle model for particle–fluid flow with considerations of sub-grid structures. *Chemical Engineering Science*, **62**(8), 2302–2308.
[91] Chu, K.W. and Yu, A.B. (2008) Numerical simulation of complex particle–fluid flows. *Powder Technology*, **179**, 104–114.
[92] Kuang, S.B., Yu, A.B., and Zou, Z.S. (2009) Computational study of flow regimes in vertical pneumatic conveying. *Industrial & Engineering Chemistry Research*, **48**, 6846–6858.
[93] Pan, R. (1999) Material properties and flow modes in pneumatic conveying. *Powder Technology*, **104**(2), 157–163.
[94] Li, J. and Mason, D.J. (2002) Application of the discrete element modelling in air drying of particulate solids. *Drying Technology*, **20**(2), 255–282.
[95] Li, J., Mason, D.J., and Mujumdar, A.S. (2003) A numerical study of heat transfer mechanisms in gas–solids flows through pipes using a coupled CFD and DEM model. *Drying Technology*, **21**(9), 1839–1866.
[96] Brosh, T. and Levy, A. (2010) Modeling of heat transfer in pneumatic conveyer using a combined DEM-CFD numerical code. *Drying Technology*, **28**(2), 155–164.
[97] Swasdisevi, T., Tanthapanichakoon, W., Charinpanitkul, T., Kawaguchi, T., Tanaka, T., and Tsuji, Y. (2005) Prediction of gas-particle dynamics and heat transfer in a two-dimensional spouted bed. *Advanced Powder Technology*, **16**(3), 275–293.

[98] Hou, Q.F., Zhou, Z.Y., and Yu, A.B. (2009) Computational study of heat transfer in bubbling fluidized beds with Geldart A powder. 7th International Conference on CFD in the Minerals and Process Industries, Melbourne, Australia.
[99] Zhou, Z.Y., Yu, A.B., and Zulli, P. (2009) Particle scale study of heat transfer in packed and bubbling fluidized beds. *AIChE Journal*, **55**(4), 868–884.
[100] Yang, W.J., Zhou, Z.Y., and Yu, A.B. (2015) Particle scale studies of heat transfer in a moving bed. *Powder Technology*, **281**, 99–111.
[101] Patil, A.V., Peters, E.A.J.F., Kolkman, T., and Kuipers, J.A.M. (2014) Modeling bubble heat transfer in gas–solid fluidized beds using DEM. *Chemical Engineering Science*, **105**, 121–131.
[102] Davidson, J. and Harrison, D. (1963) *Fluidised Particles*. Cambridge University Press, Cambridge.
[103] Malone, K.F. and Xu, B.H. (2008) Determination of contact parameters for discrete element method simulations of granular systems. *Particuology*, **6**(6), 521–528.
[104] Zhao, Y., Jiang, M., Liu, Y., and Zheng, J. (2009) Particle-scale simulation of the flow and heat transfer behaviors in fluidized bed with immersed tube. *AIChE Journal*, **55**(12), 3109–3124.
[105] Zhou, Z.Y., Yu, A.B., and Zulli, P. (2010) A new computational method for studying heat transfer in fluid bed reactors. *Powder Technology*, **197**(1–2), 102–110.
[106] Malone, K.F. and Xu, B.H. (2008) Particle-scale simulation of heat transfer in liquid-fluidised beds. *Powder Technology*, **184**(2), 189–204.
[107] Di Maio, F.P., Di Renzo, A., and Trevisan, D. (2009) Comparison of heat transfer models in DEM-CFD simulations of fluidized beds with an immersed probe. *Powder Technology*, **193**(3), 257–265.
[108] Yang, W., Zhou, Z., Pinson, D., and Yu, A. (2015) A new approach for studying softening and melting behavior of particles in a blast furnace cohesive zone. *Metallurgical and Materials Transactions B*, **46**(2), 977–992.
[109] Hou, Q., Li, J., and Yu, A. (2015) CFD-DEM study of heat transfer in the reduction shaft of corex. *Steel Research International*, **86**(6), 626–635.
[110] Rickelt, S., Sudbrock, F., Wirtz, S., and Scherer, V. (2013) Coupled DEM/CFD simulation of heat transfer in a generic grate system agitated by bars. *Powder Technology*, **249**, 360–372.
[111] Rong, D. and Horio, M. (1999) DEM simulation of char combustion in a fluidized bed. 2nd International Conference on CFD in the Minerals and Process Industries, Melbourne, Australia.
[112] Peters, B. (2002) Measurements and application of a discrete particle model (DPM) to simulate combustion of a packed bed of individual fuel particles. *Combustion and Flame*, **131**, 132–146.
[113] Zhou, H., Flamant, G., Gauthier, D., and Flitris, Y. (2003) Simulation of coal combustion in a bubbling fluidized bed by distinct element method. *Chemical Engineering Research and Design*, **81**(9), 1144–1149.
[114] Zhou, H., Flamant, G., and Gauthier, D. (2004) DEM-LES simulation of coal combustion in a bubbling fluidized bed. Part II: coal combustion at the particle level. *Chemical Engineering Science*, **59**(20), 4205–4215.
[115] Zhou, H., Flamant, G., and Gauthier, D. (2004) DEM-LES of coal combustion in a bubbling fluidized bed. Part I: gas-particle turbulent flow structure. *Chemical Engineering Science*, **59**(20), 4193–4203.
[116] Geng, Y. and Che, D. (2011) An extended DEM–CFD model for char combustion in a bubbling fluidized bed combustor of inert sand. *Chemical Engineering Science*, **66**(2), 207–219.
[117] Hayhurst, A.N. and Parmar, M.S. (1998) Does solid carbon burn in oxygen to give the gaseous intermediate CO or produce CO_2 directly? Some experiments in a hot bed of sand fluidized by air. *Chemical Engineering Science*, **53**(3), 427–438.
[118] Hayhurst, A.N. and Parmar, M.S. (2002) Measurement of the mass transfer coefficient and Sherwood number for carbon spheres burning in a bubbling fluidized bed. *Combustion and Flame*, **130**(4), 361–375.
[119] Liu, D., Chen, X., Zhou, W., and Zhao, C. (2011) Simulation of char and propane combustion in a fluidized bed by extending DEM–CFD approach. *Proceedings of the Combustion Institute*, **33**(2), 2701–2708.
[120] Hesketh, R.P. and Davidson, J.F. (1991) The effect of volatiles on the combustion of char in a fluidised bed. *Chemical Engineering Science*, **46**(12), 3101–3113.
[121] Oevermann, M., Gerber, S., and Behrendt, F. (2009) Euler–Lagrange/DEM simulation of wood gasification in a bubbling fluidized bed reactor. *Particuology*, **7**(4), 307–316.
[122] Snider, D.M., Clark, S.M., and O'Rourke, P.J. (2011) Eulerian–Lagrangian method for three-dimensional thermal reacting flow with application to coal gasifiers. *Chemical Engineering Science*, **66**(6), 1285–1295.
[123] Ku, X., Li, T., and Løvås, T. (2015) CFD–DEM simulation of biomass gasification with steam in a fluidized bed reactor. *Chemical Engineering Science*, **122**, 270–283.
[124] Limtrakul, S., Boonsrirat, A., and Vatanatham, T. (2004) DEM modeling and simulation of a catalytic gas–solid fluidized bed reactor: a spouted bed as a case study. *Chemical Engineering Science*, **59**(22–23), 5225–5231.

[125] Wu, C., Cheng, Y., Ding, Y., and Jin, Y. (2010) CFD–DEM simulation of gas–solid reacting flows in fluid catalytic cracking (FCC) process. *Chemical Engineering Science*, **65**(1), 542–549.
[126] Wu, C., Yan, B., Jin, Y., and Cheng, Y. (2010) Modeling and simulation of chemically reacting flows in gas–solid catalytic and non-catalytic processes. *Particuology*, **8**(6), 525–530.
[127] Karimi, S., Mansourpour, Z., Mostoufi, N., and Sotudeh-Gharebagh, R. (2011) CFD-DEM study of temperature and concentration distribution in a polyethylene fluidized bed reactor. *Particulate Science and Technology*, **29**(2), 163–178.
[128] McAuley, K.B., MacGregor, J.F., and Hamielec, A.E. (1990) A kinetic model for industrial gas-phase ethylene copolymerization. *AIChE Journal*, **36**(6), 837–850.
[129] Zhuang, Y.-Q., Chen, X.-M., Luo, Z.-H., and Xiao, J. (2014) CFD–DEM modeling of gas–solid flow and catalytic MTO reaction in a fluidized bed reactor. *Computers & Chemical Engineering*, **60**, 1–16.
[130] Zhang, Z., Liu, D., Zhuang, Y., Meng, Q., and Chen, X. (2014) CFD-DEM modeling of CO_2 capture using alkali metal-based sorbents in a bubbling fluidized bed. *International Journal of Chemical Reactor Engineering*, **12**(1), 441–449.
[131] Chu, K.W., Wang, B., Yu, A.B., and Vince, A. (2009) CFD-DEM modelling of multiphase flow in dense medium cyclones. *Powder Technology*, **193**, 235.
[132] Chu, K.W., Wang, B., Xu, D.L., Chen, Y.X., and Yu, A.B. (2011) CFD–DEM simulation of the gas–solid flow in a cyclone separator. *Chemical Engineering Science*, **66**, 834–847.
[133] Malone, K.F., Xu, B.H., Fairweather, M., and Dixon, C. (2005) liquid fluidisation using a combined continuum and discrete model. 7th World Congress of Chemical Engineering, Glasgow, UK.
[134] Di Renzo, A. and Di Maio, F.P. (2007) Homogeneous and bubbling fluidization regimes in DEM-CFD simulations: hydrodynamic stability of gas and liquid fluidized beds. *Chemical Engineering Science*, **62**, 116–130.
[135] Malone, K.F., Xu, B.H., and Fairweather, M. (2007) Numerical investigation of the layer-inversion phenomenon in binary solid liquid fluidized beds. Fluidization XII.
[136] Zhou, Z.Y. and Yu, A.B. (2009) Simulation of the flow and segregation of particle mixtures in liquid fluidization. Powders and Grains 2009.
[137] Di Renzo, A., Cello, F., and Di Maio, F.P. (2011) Simulation of the layer inversion phenomenon in binary liquid–fluidized beds by DEM–CFD with a drag law for poly disperse systems. *Chemical Engineering Science*, **66**, 2945–2958.
[138] Cello, F., Di Renzo, A., and Di Maio, F.P. (2010) A semi-empirical model for the drag force and fluid–particle interaction in polydisperse suspensions. *Chemical Engineering Science*, **65**, 3128–3139.
[139] Yang, W.C. (2003) *Handbook of Fluidization and Fluid-Particle Systems*, Marcel Dekker, Inc., New York/Basel.
[140] Li, Y., Zhang, J., and Fan, L.S. (1999) Numerical simulation of gas-liquid-solid fluidization systems using a combined CFD-VOF-DPM method: bubble wake behavior. *Chemical Engineering Science*, **54**, 5101–5107.
[141] Cao, C., Liu, M., Wen, J., and Guo, Q. (2009) Experimental measurement and numerical simulation for liquid flow velocity and local phase hold-ups in the riser of a GLSCFB. *Chemical Engineering and Processing: Process Intensification*, **48**, 288–295.

8

Interparticle Forces and External Fields

In this chapter, interparticle forces (IPFs) and external fields are introduced and their implementation in a computational fluid dynamics–discrete element method (CFD-DEM) code is discussed. The most common IPFs are van der Waals, liquid bridging, electrostatic, and magnetic forces. Presence of these forces can affect the packing properties of particles, their fluidization behavior [1, 2], and their granular flow in general [3]. These IPFs form temporal and weak attractive bounds among particles, causing particles to form assemblies called clusters, agglomerates, or cakes.

External fields, such as electric and magnetic fields, are sometimes used to improve flowability of adhesive particles, to enhance gas–solid contact efficiency, or to ease separation of particles from the fluid. These fields are applied externally and exert force on particles in the flow. The same effect can be expected from applying vibration and acoustic fields. When the wall of a vessel vibrates, energy is transmitted to particles through the wall. Sound waves also transfer energy to particles via the fluid. These extra forces may overcome adhesion force among particles and ease their movements.

We present procedures for calculating IPFs (including van der Waals, liquid bridge and electrostatic between charged particles) and forces arising from external fields (including electrical and magnetic) in the equation of motion of particles. Applying acoustic field as well as describing the effect of imposing vibration of walls is also described in this chapter. Although these latter cases cannot be directly named as IPFs, these fields add additional kinetic energy to particles in a granular flow. As such, their effects can be considered similar to IPFs since they exert additional forces on particles as discussed in this chapter.

Coupled CFD-DEM Modeling: Formulation, Implementation and Application to Multiphase Flows, First Edition.
Hamid Reza Norouzi, Reza Zarghami, Rahmat Sotudeh-Gharebagh and Navid Mostoufi.

Companion website: www.wiley.com/go/norouzi/CFD-DEM

8.1 Governing Equations

Particles in the multiphase flows interact with each other through physical and non-physical contacts. Physical contact refers to a condition where the surfaces of two particles touch each other while non-physical contact refers to a condition where particles do not necessarily touch but still interact with each other. In Figure 8.1, we have illustrated particles that are in physical and non-physical contact. The collision force between two particles with physical contact is calculated according to the force-displacement laws presented in Chapter 2. The interaction force between two particles with a non-physical contact is called the interparticle force (also known as long-range force) and exists to a certain radius around the particle according to long-range interaction laws. Also, particles may be in physical contact while still an IPF exists between particles. For example, solid and liquid bridging are other types of interparticle interactions when particles are in contact.

Since the pair-wise interaction between particles can be considered in the soft-sphere approach, these forces can be directly incorporated into the equation of motion of particles. Thus, the soft-sphere CFD-DEM model makes it possible to study the behavior of granular and multiphase flows in which IPFs play an important role. Based on this approach, the equation of translational motion of the particle, as discussed in Chapter 2, is as follows:

$$m_i \frac{d\vec{v}_i}{dt} = m_i \frac{d^2\vec{x}_i}{dt^2} = \sum_{j \in CL_i} \vec{f}_{ij}^{\,p-p} + \vec{f}_i^{\,f-p} + \vec{f}_i^{\,ext} \tag{8.1}$$

where m_i, $\vec{x}_i$, and $\vec{v}_i$ are mass, position vector, and linear velocity vector of particle i. The first term on the right-hand side of Equation 8.1 ($\vec{f}_{ij}^{\,p-p}$) is the sum of particle–particle interaction forces acting on particle i. Collisional forces due to surface contact of particles are among these forces that were detailed in Chapter 2. Other particle–particle forces can be IPFs, like electrostatic or van der Waals forces. The second term ($\vec{f}_i^{\,f-p}$) represents the total fluid-particle

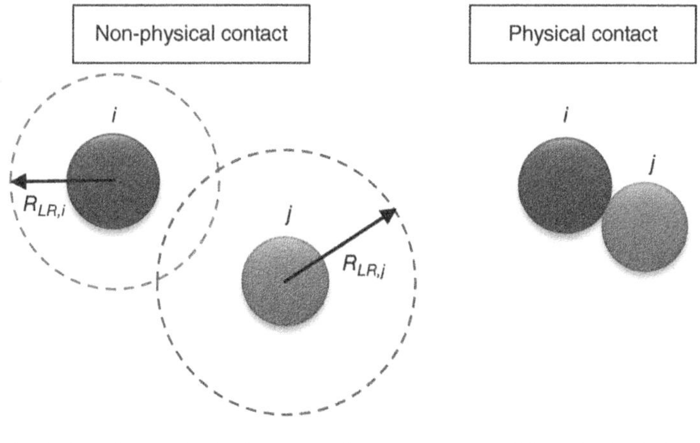

Figure 8.1 Physical and non-physical contacts between two particles. A physical contact happens when the surfaces of two particles touch each other. It is used to calculate collisional interactions between particles. A non-physical contact happens when long-range interactions between particles exist. R_{LR} is the maximum distance at which the interparticle force is still significant

interaction forces, described in Chapter 6. This term is important in multiphase flows in which fluid effects are significant. The third term in Equation 8.1 ($\vec{f}_i^{ext}$) represents all external forces acting on particle i due to uniform or non-uniform external fields. For example, we can name the gravitational force due to the existence of the uniform gravitational field of earth.

The summation on particle–particle forces is performed on all particles in the contact list of particle i, CL_i. It should be noted that for each type of particle–particle contact, a separate contact list may be used. For example, the contact list of particle i for collisional force is different from that for the van der Waals force since the distances at which these forces act may differ. Therefore, we rephrase the particle–particle interaction forces term in Equation 8.1 as follows:

$$\sum_{j \in CL_i} \vec{f}_{ij}^{p-p} = \sum_{j \in CL_{c,i}} \vec{f}_{ij}^{c} + \sum_{j \in CL_{IP,i}} \vec{f}_{ij}^{IP} \tag{8.2}$$

On the right-hand side of this equation, the first term is the summation over all particles that have a collisional interaction with particle i and the second term is the summation over all particles having interparticle interactions with particle i. $\vec{f}_{ij}^{IP}$ is the IPF existing between particles i and j and $CL_{IP,i}$ is the contact list of the particle i and contains particles that have a interparticle interaction with particle i. It is clear that if there is no particle in the contact list of particle i, the particle–particle interaction term would be zero.

The critical step in the implementation of IPFs in a DEM code is the construction of the contact list. The Verlet list contact search algorithm is commonly used for this purpose [4]. In this method, a cut-off radius is considered for IPFs. All particles around the target particle that lie within a circle with the radius $R_{cut-off}$ are added to the Verlet list of the target particle as shown in Figure 8.2. The main feature of this method is that particles in the Verlet list of each particle will stay in this list at least for a few time steps. Thus, it is not required to update the Verlet list in each time step. This method can be combined by cell-based contact search algorithms to improve its computational performance threefold [5]. For each particle in the system,

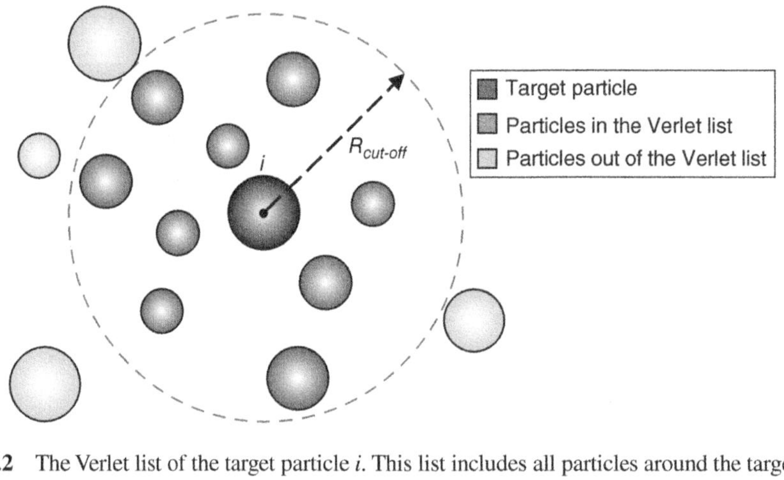

Figure 8.2 The Verlet list of the target particle i. This list includes all particles around the target particle whose center to center distance (or surface to surface distance, depending on the type of interparticle force) is less than $R_{cut\text{-}off}$

a new bounding sphere is defined whose center and radius are the center of the particle and $R_{cut-off}$. Then, by applying any of the contact search methods (described in Chapter 3) to all these new bounding spheres, a contact list is obtained for each particle. This contact list is the Verlet list of the particle.

8.1.1 Sequence of Calculations

Figure 8.3 shows the execution sequence of the main parts of a DEM code that handles IPFs. The flowchart of DEM calculations is the same as what described in Chapter 3 and in this figure we have added only the new parts necessary for implementing the IPFs into the calculations.

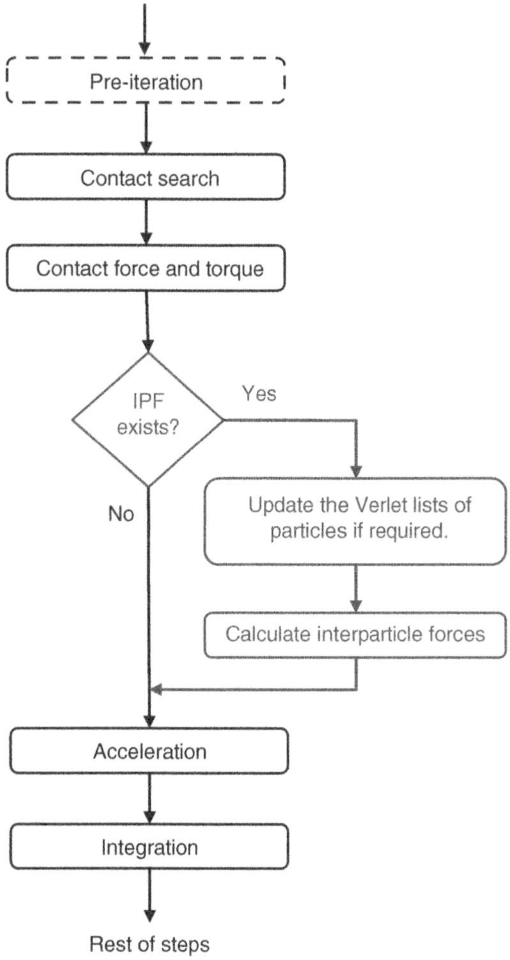

Figure 8.3 Calculation sequence of the main steps of a DEM code including interparticle forces. Note that the calculation sequence for solving fluid phase equations and interphase coupling remain intact, hence not shown here

Since the inclusion of IPFs considered in this book does not affect the overall algorithm of the execution sequence of the fluid phase (solution of Navier–Stokes and continuity equations) and coupling part (porosity and drag), we have omitted these sections in this figure. Based on the flowchart shown in Figure 8.3, after finding the contact pairs and calculating the collisional forces and torques, the program should check if any kind of IPF should be included in the simulation. In this case, the execution control is directed to evaluation of IPFs. The program first checks if the Verlet lists of particles need to be updated. If so, the program updates it using a convenient contact search algorithm. Thereafter, according to the type of the IPF and the Verlet lists of particles, the program calculates the IPF for each pair of particles and adds it to the total particle–particle force acting on particles in that pair.

8.2 Interparticle Forces

8.2.1 van der Waals Force

Attraction forces between two fine particles are attributed mainly to London-van der Waals forces. Origin of the van der Waals force is existence of momentary electric and magnetic fields occurring spontaneously on the particle surface. Electric charges on the surface fluctuate due to thermal motion and quantum mechanical uncertainties in position and momentum of molecules on the surface of particles. In general, the van der Waals force is the time averaged sum of interactions between moving charges of molecules on the surface of particles. Therefore, this force is always present, although is usually neglected when handling large particles.

According to Hamaker [6], attraction forces between two particles decay as the separation distance between two particles increases. Consider two spherical particles with diameters d_i and d_j and the surface to surface separation distance l as shown in Figure 8.4. The interaction force among these particles is described as follows:

$$\begin{aligned}\vec{f}_{ij}^{vdW} &= -\frac{A_H}{12d_i}(2x+y+1) \\ &\times\left(\frac{-y}{\left(x^2+xy+x\right)^2}-\frac{y}{\left(x^2+xy+x+y\right)^2}+\frac{2}{x^2+xy+x}-\frac{2}{x^2+xy+x+y}\right)\vec{n}_{ij}\end{aligned} \tag{8.3}$$

where $x = l/d_i$, $y = d_j/d_i$, and $\vec{n}_{ij}$ is the unit vector that points from center of particle i to center of particle j. A_H is the Hamaker constant, which depends on properties of materials and surface condition of the particle. This constant varies from 1.0×10^{-21} to 1.0×10^{-18} J and lies between 1.0×10^{-20} and 1.0×10^{-19} J in most cases. In this equation, the attraction force between two same size spheres ($y=1$) reduces to:

$$\vec{f}_{ij}^{vdW} = -\frac{A_H}{6d_i}\left(\frac{2(x+1)}{x^2+2x}-\frac{x+1}{\left(x^2+2x\right)^2}-\frac{2}{x+1}-\frac{1}{(x+1)^3}\right)\vec{n}_{ij} \tag{8.4}$$

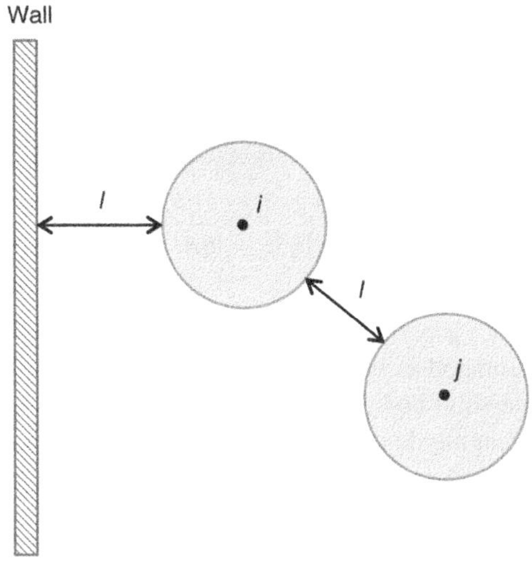

Figure 8.4 Relative distance between two particles' surfaces and particle's surface and wall for calculating van der Waals force

As an extreme case, when $x \ll 1$, Equation 8.4 becomes:

$$\vec{f}_{ij}^{\,vdW} = \frac{A_H}{24 d_i} \frac{1}{x^2} \vec{n}_{ij} \tag{8.5}$$

The van der Waals force acting between a sphere and a flat wall ($y = \infty$) is obtained from (see Figure 8.4):

$$\vec{f}_{iw}^{\,vdW} = -\frac{A_H}{12 d_i} \left(\frac{2}{x} - \frac{1}{x^2} - \frac{2}{x+1} - \frac{1}{(x+1)^2} \right) \vec{n}_{iw} \tag{8.6}$$

in which $\vec{n}_{iw}$ which unit vector pointing from center of particle to wall. Likewise, when $x \ll 1$, the van der Waals force between a flat wall and a sphere is approximated from:

$$\vec{f}_{iw}^{\,vdW} = \frac{A_H}{12 d_i} \frac{1}{x^2} \vec{n}_{iw} \tag{8.7}$$

When $x < 0.05$, the approximate Equations 8.5 and 8.7 can be used instead of exact equations with an error less than 5%. The van der Waals force increases significantly as the two particles approach each other and it becomes infinity when the separation distance l approaches zero. To avoid this problem, a cut-off value is defined for the separation distance, $h_{cut-off}$, in which the attraction force is maximum. Different values have been considered for $h_{cut-off}$ in literature ranging between 0.2 and 0.4 nm [7–15].

In the presence of surface asperities, calculation of van der Waals force should not be based on the radius of particles but on the radius of the curvature of the surface elevation where the surfaces are in contact. Thus, the van der Waals force attributed to the adhesion of the adjacent particles can be formulated as [16]:

$$\vec{f}_{ij}^{vdW} = \frac{A_H}{6l^2} R_{asp} \left(1 + \frac{A_H}{6\pi l^3 H_r} \right) \vec{n}_{ij} \tag{8.8}$$

The mean radius of surface asperities, R_{asp}, depends on the natural geometry of the adherent surfaces and it can be estimated as half of its radius for a spherical particle. Also, H_r is the hardness of undeformable solid bodies. In the case that Equation 8.8 prevails, regardless of the size of particles, the magnitude of van der Waals force at the contact point (or asperities of particles) is a finite value. A typical value of 0.1 μm can be considered for the asperity in this case [17].

8.2.1.1 Evaluation of the Hamaker Constant

The Hamaker constant depends on the physical properties of particles and the interacting medium. In the interaction of the two identical material 1 with an intervening medium 2, the Hamaker constant can be obtained from [6]:

$$A_H = \frac{3}{4} k_B T \left(\frac{\epsilon_1 - \epsilon_2}{\epsilon_1 + \epsilon_2} \right)^2 + \frac{3\hbar v_e}{16\sqrt{2}} \frac{\left(n_1^2 - n_2^2\right)^2}{\left(n_1^2 + n_2^2\right)^{3/2}} \tag{8.9}$$

where ϵ and n are permittivity and refractive index of materials and medium, respectively, T is the absolute temperature, $\hbar$ is the reduced Planck constant ($1.0545718 \times 10^{-34}$ J·s), ν_e is the UV absorption frequency of the medium, and k_B is the Boltzmann constant ($1.3806488 \times 10^{-23}$ J/K). These values for some selected materials are given in Table 8.1.

Relations given here for evaluating the van der Waals force are limited to particles in a vacuum. In the case where a fluid exists between particles, it is necessary to define the Hamaker constant according to the following equation [19]:

$$A_H = \left(\sqrt{A_{H,i}} - \sqrt{A_{H,f}} \right) \left(\sqrt{A_{H,j}} - \sqrt{A_{H,f}} \right) \tag{8.10}$$

where $A_{H,i}$, $A_{H,j}$, and $A_{H,f}$ are the Hamaker constants for particles i and j and fluid, respectively.

To have an idea about the order of magnitude of the van der Waals force between particles in different fluids, van der Waals forces between two amorphous silica particles ($\rho_i = 2500$ kg/m^3 and $d_i = 60$ μm) and two porous alumina particles ($\rho_i = 1900$ kg/m^3 and $d_i = 60$ μm) in air and in water at 20 °C are evaluated and shown in Figure 8.5. These particles are assumed to be perfectly smooth with no asperities (which is not the case in reality). The force is calculated from Equation 8.6 and is made dimensionless with respect to the weight of particle. The Hamaker constant was calculated from Equations 8.9 and 8.10 for air and water as the fluid,

Table 8.1 Permittivity ϵ, refractive index n, and main absorption frequency ν_e in UV for various solids, liquids, and polymers at 20 °C [18]

	ϵ	n	ν_e ($\times 10^{15}$ Hz)
Alumina (Al_2O_3)	9.3–11.5	1.75	3.2
Diamond	5.3	2.40	2.7
Calcium carbonate ($CaCO_3$)	8.2	1.59	3.0
Flurite (CaF_2)	6.7	1.43	3.8
Muscovite mica ($KAl_2Si_3AlO_{10}(OH)_2$)	5.4	1.58	3.1
Potassium chloride (KCl)	4.4	1.48	2.5
Sodium chloride (NaCl)	5.9	1.53	2.5
Silicon nitrate, amorphous (Si_3N_4)	7.4	1.99	2.5
Quartz (SiO_2)	4.3–4.8	1.54	3.2
Silica, amorphous (SiO_2)	3.82	1.46	3.2
Titania (TiO_2)	114	2.46	1.2
Zinc oxide (ZnO)	11.8	1.91	1.4
Acetone	20.7	1.359	2.9
Chloroform	4.81	1.446	3.0
n-Hexane	1.89	1.38	4.1
n-Octane	1.97	1.41	3.0
n-Hexadecane	2.05	1.43	2.9
Ethanol	25.3	1.361	3.0
1-Propanol	20.8	1.385	3.1
1-Butanol	17.8	1.399	3.1
1-Octanol	10.3	1.43	3.1
Toluene	2.38	1.497	2.7
Water	78.5	1.333	3.6
Polyethylene	2.26–2.32	1.48–1.51	2.6
Polystyrene	2.49–2.61	1.59	2.3
Polyvinyl chloride	4.55	1.52–1.55	2.9
Polytetrafluoroethylene	2.1	1.35	4.1
Polymethyl methacrylate	3.12	1.5	2.7
Polydimethyl cyloxane	2.6–2.8	1.4	2.8
Nylon 6	3.8	1.53	2.7

respectively. This figure demonstrates that the dimensionless van der Waals force sharply decreases when the two particles move away from each other such that it becomes negligible when surfaces distance of these particles becomes greater than 1 μm. Figure 8.5 also illustrates that the van der Waals force in water is weaker than that in air with about 1 order of magnitude.

8.2.2 Liquid Bridge Force

Many gas–solid flows involve particles with free liquid on their surface, which makes the liquid bridge force dominant in comparison to other forces. This type of gas–solid flow can be found in coating processes, agglomerating flows, driers, pneumatic conveying, and chemical reactors. Calculation of liquid bridge force is more complicated than the van der Waals force.

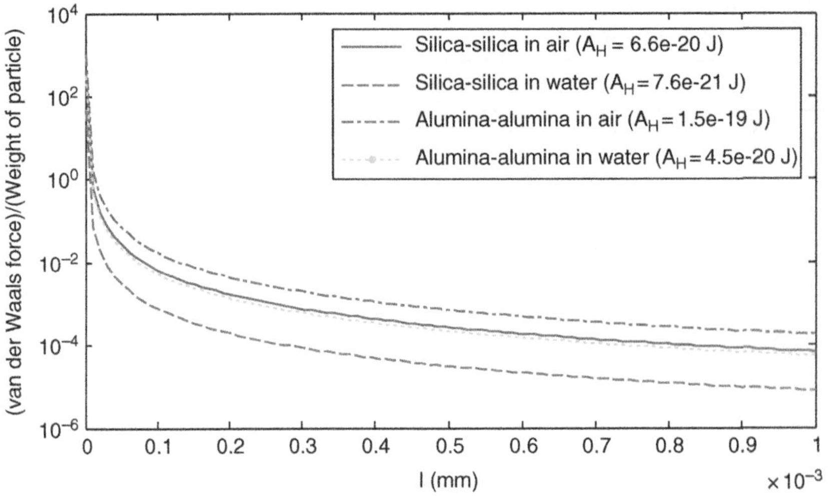

Figure 8.5 Dimensionless van der Waals force against the distance between two 60 μm particles in air and water

Real behavior of the liquid film on a particle with a rough surface is rather complex and needs an accurate model to account for all parameters. However, some hypotheses have been made to simplify it and at the same time to obtain a reasonable approximation of the liquid bridge force with less computational cost. Mikami *et al.* [20] were among the first who implemented the liquid bridge force into the CFD-DEM model for investigating the hydrodynamics of fluidized beds in the presence and absence of humidity. Assuming that the surface of particle is smooth (thus, no liquid can be trapped between pores and cracks of the particle) and uniformly surrounded by the liquid film, all the liquid on the surface of particle contributes to the liquid bridging. When two such particles come into contact, their liquid films join at the contact point forming a liquid bridge that creates, in most cases, an attraction force between particles [19]. The liquid bridge formed between two particles and its related parameters is shown in Figure 8.6.

The total liquid bridge force, $\vec{f}_{ij}^{lb}$, between two colliding particles is the sum of capillary and viscous forces:

$$\vec{f}_{ij}^{lb} = \vec{f}_{ij}^{cap} + \vec{f}_{ij}^{vis} \tag{8.11}$$

where $\vec{f}_{ij}^{cap}$ and $\vec{f}_{ij}^{vis}$ are the capillary and viscous forces due to the liquid bridging, respectively.

8.2.2.1 Capillary Force

The capillary force acting on particles in the normal direction is the sum of surface tension and capillary pressure effects due to the curvature of the liquid bridge:

$$\vec{f}_{ij}^{cap} = \left[2\pi R\sigma \sin\phi \sin(\phi+\theta) - \Delta p \pi R^2 \sin^2\phi\right]\vec{n}_{ij} \tag{8.12}$$

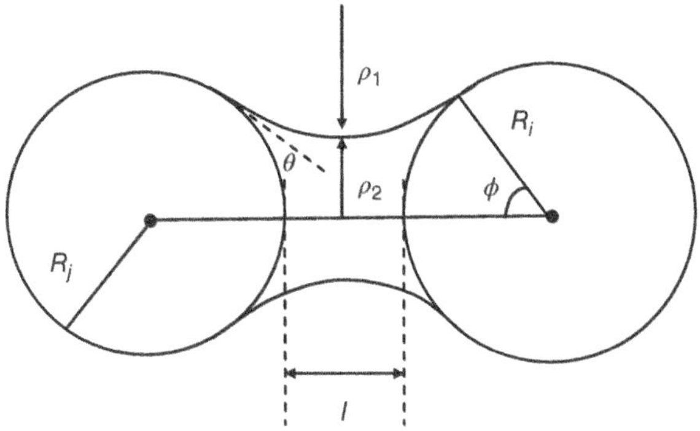

Figure 8.6 Schematic of a liquid bridge between two particles

where ϕ and θ are the half-filling and solid–liquid–gas contact angles in radians, respectively, σ is the surface tension and $\vec{n}_{ij}$ is the unit vector, which points from the center of particle i to the center of particle j. The capillary pressure difference due to mean curvature of liquid bridge, Γ, is given as:

$$\Delta p = 2\sigma\Gamma \tag{8.13}$$

The mean curvature of liquid bridge is obtained through solving the Young–Laplace equation. By solving this equation, an implicit equation is obtained that relates the capillary force to half-filling angle, contact angle, and dimensionless curvature. However, it is more convenient (from the computational point of view) to have an equation that explicitly calculates the capillary force. The first suggestion is an approximation that considers a circular shape for the liquid bridge that results in the following formula [21]:

$$\vec{f}_{ij}^{cap} = \pi\,\sigma\,\rho_2\left(1+\frac{\rho_2}{\rho_1}\right)\vec{n}_{ij} \tag{8.14}$$

with

$$\rho_1 = \frac{l/2 + R(1-\cos\phi)}{\cos(\theta+\phi)} \tag{8.15}$$

$$\rho_2 = R\sin\phi - (1-\sin(\theta+\phi))\rho_1 \tag{8.16}$$

The error of this estimation is less than 10%, which is acceptable from a practical point of view. Yet, the half-fill angle is required to calculate the capillary force based on Equation 8.14, which is not suitable for the DEM.

Mikami *et al.* [20] solved the Young–Laplace equation to obtain explicit relations for capillary force and critical separation distance, which are functions of liquid volume and separation distance. These equations, obtained by regression analysis, are suitable for implementing into a discrete element code. They read as:

Particle–particle capillary force:

$$\vec{f}_{ij}^{cap} = \pi R\sigma\left(\exp\left(A\hat{l} + B\right) + C\right)\vec{n}_{ij} \tag{8.17}$$

where

$$A = -1.1\hat{V}^{\ 0.53} \tag{8.18}$$

$$B = \left(-0.34\ln\hat{V} - 0.96\right)\theta^2 - 0.019\ln\hat{V} + 0.48 \tag{8.19}$$

$$C = 0.0042\ln\hat{V} + 0.078 \tag{8.20}$$

Particle–wall capillary force:

$$\vec{f}_{iw}^{cap} = \pi R\sigma\left(\exp\left(A\hat{l} + B\right) + C\right)\vec{n}_{iw} \tag{8.21}$$

where $\vec{n}_{iw}$ is the unit vector that points from center of particle i perpendicular to the wall and:

$$A = -1.9\hat{V}^{-0.51} \tag{8.22}$$

$$B = \left(-0.016\ln\hat{V} - 0.76\right)\theta^2 - 0.12\ln\hat{V} + 0.1.2 \tag{8.23}$$

$$C = 0.013\ln\hat{V} + 0.18 \tag{8.24}$$

In these equations, $\hat{V}$ is the dimensionless liquid bridge volume defined as:

$$\hat{V} = \frac{V_L}{R^3} \tag{8.25}$$

where V_L is the liquid bridge volume and R is the harmonic mean radius, which is twice the effective radius of the two spheres:

$$R = 2\left(\frac{1}{R_i} + \frac{1}{R_j}\right)^{-1} \tag{8.26}$$

Also, the dimensionless distance is defined as follows:

$$\hat{l} = \frac{l}{R} \tag{8.27}$$

In another attempt, Maugis [22] assumed that the liquid bridge is cylindrical and approximated the capillary force by the following equation:

$$\vec{f}_{ij}^{cap} = 2\pi R\sigma\, X_V cos\theta.\vec{n}_{ij} \tag{8.28}$$

where X_V is the volume factor and is given by:

$$X_V = 1 - \frac{1}{\sqrt{1 + 2V_L / \pi R l^2}} \tag{8.29}$$

8.2.2.2 Viscous Force

Relative motion of particles can induce a tension in the liquid bridge among particles that leads to a viscous force [23]. This force is calculated by the following equation based on the Reynolds' lubrication theory [22, 24]:

$$\vec{f}_{ij}^{vis} = \frac{3}{2}\pi\mu_L R^2 \frac{X_V^2}{l} \vec{v}_{ij}^n \tag{8.30}$$

where μ_L and $\vec{v}_{ij}^n$ are the liquid viscosity and relative normal velocity of particles, respectively. The viscous force becomes infinity when the separation distance approaches zero. To avoid this problem, like the limitation used for van der Waals force, a cut-off (or minimum) surface separation distance is considered, which depends on roughness of the particle surface. Seville *et al.* [25] compared capillary and viscous forces of liquid bridge between two particles (992 μm) in the presence of water. They showed that in comparison to the capillary force, the viscous force is significant at small separation distances. However, its significance reduces as the separation distance increases and the separation rate decreases.

Pitois *et al.* [24] examined Equations 8.28 and 8.30 for evaluating capillary and viscous forces, respectively, against experimental data at various separation velocities. It was shown that this model can reproduce the experimental data accurately. Furthermore, the viscous force is dominant at short separation distances while the liquid bridge force is dominated by capillary force at longer separation distances. These results were also confirmed by Seville *et al.* [25]. Ennis *et al.* [23] showed that the viscosity effect is negligible when the capillary number is less than 10^{-3}. Capillary number is the ratio of viscous force to surface tension force as given by the following expression:

$$Ca = \frac{\mu_L \left|\vec{v}_{ij}^n\right|}{\sigma} \tag{8.31}$$

8.2.2.3 Rupture Distance

The liquid bridge between two colliding particles remains stable until they reach a critical separation distance l_{rup} at which the liquid bridge ruptures and the attraction force due to the liquid bridge vanishes. Lian *et al.* [26] proposed the following relationship for evaluating the rupture distance:

$$\frac{l_{rup}}{R} = \left(1 + \frac{\theta}{2}\right)\left(\frac{V_L}{R^3}\right)^{1/3} \tag{8.32}$$

This relationship shows that the rupture distance increases when the volume of liquid bridge increases. However, this is valid when the effect of liquid viscosity is negligible. Pitois *et al.* [27] showed that experimental rupture distance is longer than that obtained from Equation 8.32 in the case of viscous liquids. Moreover, the rupture distance is proportional to the square root of the relative velocity. Accordingly, they later proposed the following relation for the rupture distance:

$$\frac{l_{rup}}{R} = \left(1 + \frac{\theta}{2}\right)\left(1 + Ca^{1/2}\right)\left(\frac{V_L}{R^3}\right)^{1/3} \tag{8.33}$$

When the viscosity of liquid or the relative velocity is low, the capillary number approaches zero, and Equation 8.33 reduces to Equation 8.32.

Mikami *et al.* [20] also obtained the rupture distance by regression analysis as follows:

Particle–particle:

$$\hat{l}_{rup} = \left(0.62\theta + 0.99\right)\hat{V}^{0.34} \tag{8.34}$$

Particle-wall:

$$\hat{l}_{rup} = \left(0.22\theta + 0.95\right)\hat{V}^{0.32} \tag{8.35}$$

here, $\hat{l}_{rup}$ is the dimensionless rupture distance given as:

$$\hat{l}_{rup} = \frac{l_{rup}}{R} \tag{8.36}$$

It is worth noting that a liquid transfer exists between the two colliding particles and the liquid is redistributed amongst particles after their separation. Shi and McCarthy [28] investigated this redistribution numerically and showed that the liquid transfer ratio depends on the size of particles and the contact angle. We will not repeat the details of this model here and refer the reader to Shi and McCarthy [28] for more information.

8.2.2.4 Bridge Volume

There are many formulas for estimating the liquid bridge volume. Pietsch and Rumpf [29] obtained the following formula for calculating the volume of the liquid bridge between two particles:

$$\begin{aligned} V_L = 2\pi \Big\{ & \left[\rho_1^2 + \left(\rho_1 + \rho_2\right)^2\right]\rho_1 \cos\left(\phi + \theta\right) - \frac{1}{3}\rho_1^3 \cos^3\left(\phi + \theta\right) \\ & - \rho_1^2\left(\rho_1 + \rho_2\right)\cos\left(\phi + \theta\right)\sin\left(\phi + \theta\right)\left(\frac{\pi}{2} - \phi - \theta\right) \\ & - \frac{1}{24}\left(2 + \cos\phi\right)\left(1 - \cos\phi\right)^2 \Big\} \end{aligned} \tag{8.37}$$

Assuming that the contact angle is zero and the shape of the interface is circular, the volume of the liquid bridge can be estimated by Kuwagi *et al.* [30]:

$$V_L = 2\pi\left\{\left(C^2 + r_0^2\right)a - C\left(a\sqrt{r_0^2 - a^2} + r_0^2\alpha\right) - \frac{a^3 - b^2(3-b)}{3}\right\} \tag{8.38}$$

where

$$\alpha = \frac{\pi}{2} - \phi \tag{8.39}$$

$$C = \left(R_i + \frac{l}{2}\right)\tan\phi \tag{8.40}$$

$$r_0 = \frac{2R_i + l}{2\cos\phi} - R_i \tag{8.41}$$

$$a = R_i\left(1 - \cos\phi\right) + \frac{l}{2} \tag{8.42}$$

$$b = R_i\left(1 - \cos\phi\right) \tag{8.43}$$

A much simpler explicit expression to relate liquid volume to the half filling angle, with a less than 4% error, is [31]:

$$V_L = 0.96R^3\sin^4\phi\left(1 + 3\hat{l}\right)\left(1 + 1.1\sin\theta\right) \tag{8.44}$$

Rabinovich *et al.* [32] also proposed the following formula:

$$V_L = \pi R^2\phi^2 l + \frac{1}{2}\pi R^3\phi^4 \tag{8.45}$$

A numerical example of evaluating the liquid bridge force is shown in Figure 8.7 for three common liquids. Among these liquids, glycerin has the highest viscosity and n-hexane has the lowest viscosity. The model of Mikami *et al.* [20] was used for calculating the liquid bridge force in this figure. The particles were considered to be 1 mm glass beads ($\rho_i = 2500\,\text{kg/m}^3$). The liquid bridge force is made dimensionless in this figure by dividing it to the weight of one particle. For the sake of simplicity, the liquid bridge volume (V_L) was assumed to be 3% of the volume of a particle, the relative normal impact velocity of particles ($\left|\vec{v}_{ij}^{\,n}\right|$) was considered to be 1 m/s and the contact angle (θ) was set to zero. It can be seen in this figure that the water bridge force is greater than the glycerin bridge force and less than the n-hexane bridge force. At short separation distances the viscous force is dominant, which makes the order of magnitude of the glycerin bridge force to become greater than the other two liquids due to its high viscosity. However, by increasing the distance between particles, the viscous force decreases sharply and the capillary force becomes important. This is the reason that, when close to the

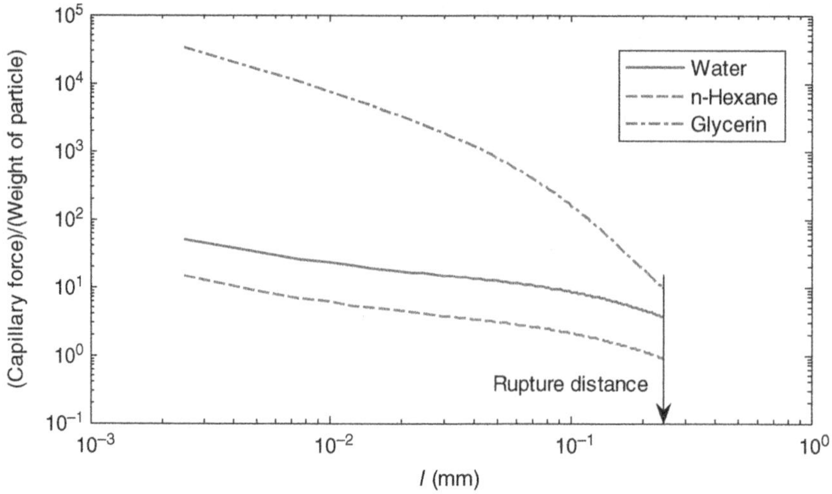

Figure 8.7 Dimensionless liquid bridge force against the distance between two 1 mm glass beads for various liquids: water ($\rho_f = 1000\,\text{kg/m}^3$, $\mu_f = 0.001\,\text{Pa}\cdot\text{s}$, $\sigma = 0.073\,\text{N/m}$), glycerin ($\rho_f = 1260\,\text{kg/m}^3$, $\mu_f = 0.95\,\text{Pa}\cdot\text{s}$, $\sigma = 64\,\text{N/m}$), and n-hexane ($\rho_f = 655\,\text{kg/m}^3$, $\mu_f = 0.000297\,\text{Pa}\cdot\text{s}$, $\sigma = 18\,\text{N/m}$)

rupture distance, the difference between the forces becomes smaller. In case of water and n-hexane, the capillary force is complexly dominant in this condition while the viscosity of glycerin is so high that the viscous force still plays its role even at the rupture distance. Figure 8.7 also demonstrates that, according to Equation 8.34, the rupture distance for all three liquids is the same, provided that the contact angle was assumed to be the same for these liquids.

8.2.3 Electrostatic Force

When two objects with different materials collide, an electric charge is transferred from one to another. In a system with a large number of particles, there are many particle–particle and particle–wall collisions. Thus, electrification of particles and walls is a common phenomenon in such systems. The generated charges on the surface of particles result in the electrostatic force between particles. The electrostatic forces influence the hydrodynamics of gas-solid fluidized beds and pneumatic transport systems considerably. These forces may also play an important role in polymerization reactors by influencing the reactor performance through particle agglomeration, particle segregation, wall sheeting, and non-uniform temperature distribution [33].

Two charged particles are shown in Figure 8.8. The electrostatic force between these particles is calculated based on the Coulomb's law:

$$\vec{f}_{ij}^{\,elec} = \frac{-q_i q_j}{4\pi\epsilon_f x_{ij}^2}\vec{n}_{ij} \tag{8.46}$$

where q_i and q_j are the charges of particles, $x_{ij} = \left|x_i - x_j\right|$ and ϵ_f is the absolute permittivity of the fluid between particles i and j (the value for vacuum is 8.854×10^{-12} F/m).

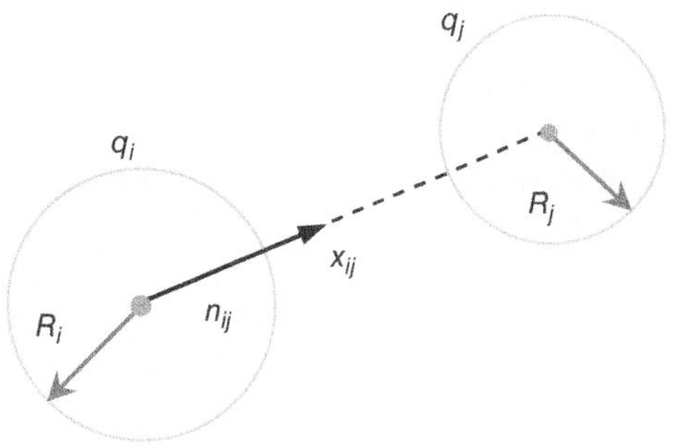

Figure 8.8 Two charged particles that exert electric force on each other. Geometric parameters required for calculation of electric force are shown in this figure

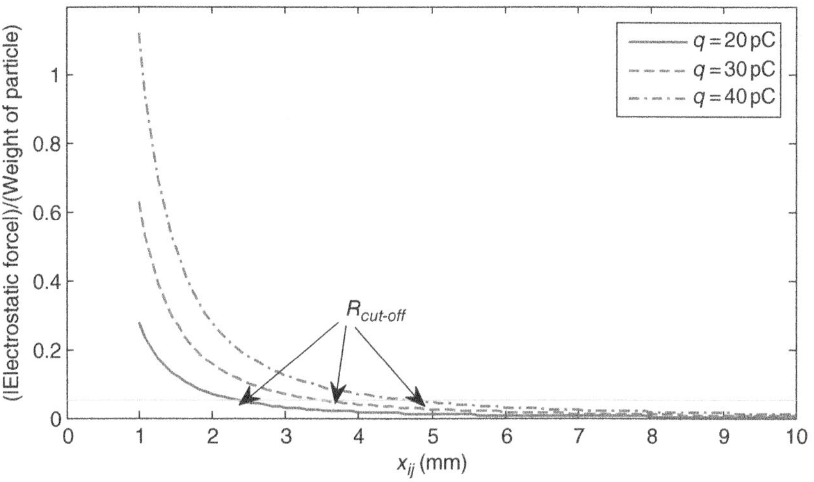

Figure 8.9 Dimensionless electrostatic force against the distance between two 1 mm particles at various particle charges

An example of variation of electrostatic force between two particles in a gaseous medium as a function of the distance between the particles is shown in Figure 8.9. The particles were considered to be 1 mm glass beads ($\rho_i = 2500\,\text{kg/m}^3$). Charge of each particle was considered to be 20, 30, and 40×10^{-12} C. This figure demonstrates that the electrostatic force is significant when the particles are close (up to about 2 mm in this case). However, it rapidly decreases when the particles move away from each other. As mentioned earlier, we consider a Verlet distance ($R_{cut\text{-}off}$) at which the IPF is significance. If we consider this significant value to be 5% of the weight of the particle, we can determine the Verlet list for each particle in this example. These distances are shown in Figure 8.9 for different charge of particles. It can be seen that this distance increases with increasing the charge of the particle.

8.2.3.1 Tribocharging

When two objects with different work functions collide (even if they are electrically neutral), electrons are transferred from the object with a lower work function to the one with higher work function. This is called triboelectric charging. If the transferred charge is Δq, charge of particles i and j after the collision will be $q_i + \Delta q$ and $q_j - \Delta q$, respectively. In case of charge transfer between particle i and a wall, charge of the wall would become $q_w - \Delta q$.

Matsusaka *et al.* [34] extensively reviewed the electrification of particles, different mechanism that lead to the generation of charge on the surface of particles and the theoretical models that can be used for evaluating charge generation in particle–wall contacts. The charge generated in a particle–wall contact can be described by the condenser model. In this model, the contact region between wall and particle is considered as a capacitor. Assuming that the contact time is long enough for charge transfer, the amount of charge transferred between the wall and particle in a collision is given by Matsusaka *et al.* [35]:

$$\Delta q = k_c C V \tag{8.47}$$

where k_c is the charging efficiency, C is the capacitance, and V is the total potential difference between wall and particle. The capacitance is obtained from [35]:

$$C = \frac{\epsilon_f S}{Z_0} \tag{8.48}$$

where Z_0 is the critical gap including the geometrical factors between contacting objects and S is the contact area. Value of the critical gap can be from a few nanometers up to hundreds of nanometers [36]. The total potential difference is:

$$V = V_c - V' \tag{8.49}$$

where V_c is the contact potential difference and V' is the total induced potential difference. The contact potential difference between surfaces of particle and wall with different work functions is:

$$V_c = -\frac{(\varphi_i - \varphi_w)}{e} \tag{8.50}$$

in which φ_i and φ_w are effective work functions of particle and wall, respectively, e is the elementary charge. The total induced potential is:

$$V' = V_{ind} + V_{im} - V_{ext} \tag{8.51}$$

The potential difference arising from external electric fields, V_{ext}, is not considered here and we discuss the other two potentials in this section.

The induced potential charge difference caused by the initial charge of particle in the vicinity of a conductive wall, V_{ind}, is the induced potential charge difference caused by the initial charge of particle [37]:

$$V_{ind} = \frac{Z_0}{4\pi\epsilon_f R_i^2} q_i \tag{8.52}$$

This equation is the potential induced among the particle i and the wall. The potential induced by other surrounding charged particles is ignored in this equation, which is valid in dilute systems and low initial charge of particles. It is relatively difficult to obtain an equation for the induced potential caused by all surrounding particles. Nevertheless, there are relationships for evaluating this potential in some simple configurations when considering simplifying assumptions [34].

The image effect potential, V_{im} is the induced potential charge difference near the conductive walls caused by image charge [34]:

$$V_{im} = k_{im}\, q_i = \frac{8Z_0}{\pi \epsilon_f R_i^2} q_i \tag{8.53}$$

The image charge does not exist for insulating walls.

Equation 8.52 is valid for collision of an insulating particle and a grounded metal surface for which the electric charge does not accumulate due to its conductivity. If the wall is insulating, the potential charge difference should be evaluated from [38]:

$$V_{ind} = \frac{Z_0}{\epsilon_f}\left(\frac{q_i}{4\pi R_i^2} - \frac{\sigma_w}{2}\right) \tag{8.54}$$

where σ_w is the surface charge density of the wall at the contact point.

Similar to the previous discussion, the amount of charge transferred between two charged particles of insulating materials during their collision can be obtained from Equation 8.47. In this case, however, the induced charge potential difference should be calculated from [39]:

$$V_{ind} = \frac{Z_0}{4\pi \epsilon_f}\left(\frac{q_i}{R_i^2} - \frac{q_j}{R_j^2}\right) \tag{8.55}$$

Also, it should be noted that the potential difference between surfaces of two particles with different work functions can be obtained from Equation 8.50 with the difference that the work functions of the two particles should be used in this equation.

The electrification model mentioned here has been incorporated into our in-house CFD-DEM code and the electrification of neutral particles in a tribocharger with square cross-section was investigated. Simulation and operating conditions of this tribocharger are given in Table 8.2. The walls of tribocharger were non-conductive. The results are shown in Figure 8.10. Particles are shaded based on the electric charge. It can be seen in Figure 8.10a that neutral particles enter the tribocharger on the left and gain electric charge along the charger due to frequent contact with the walls. Distributions of charge of particles and number of contacts during the process are also shown in Figure 8.10b,c. These figures demonstrate that the statistical mode of number of contacts is 5 and the statistical mode of particle charge is about 8.5 pC.

Table 8.2 Simulation and operating condition of the tribocharger

Tribocharger (non-conductive)	
Dimensions	8 cm × 8 cm × 50 cm
Wall work function	7.52×10^{-19} J
Particles	
Mean diameter	375 μm
Particle work function	7.232×10^{-19} J
Density	2500 kg/m^3
Flow rate	0.105 g/s
Gas (air)	
Velocity	10 m/s

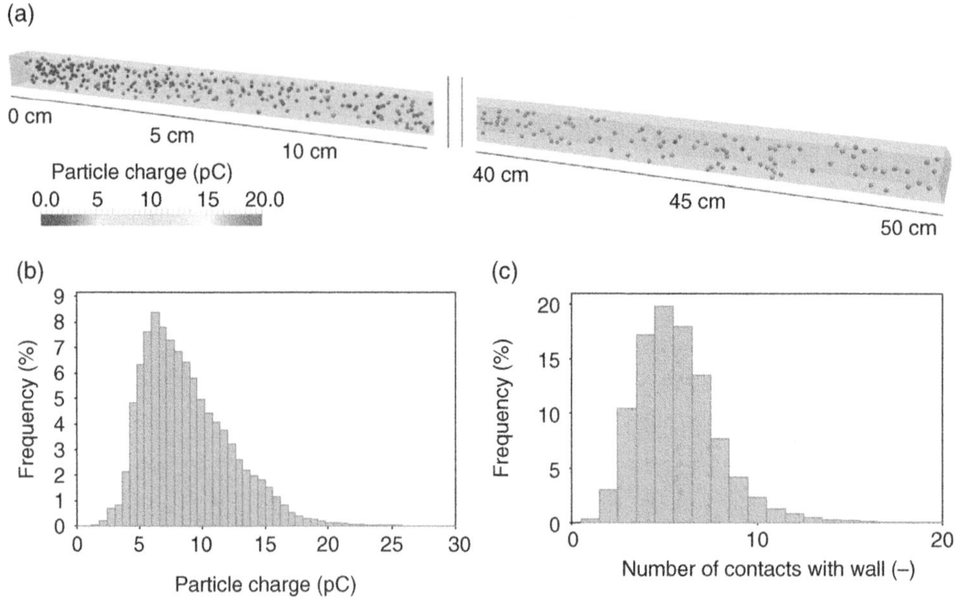

Figure 8.10 (a) Snapshot of the tribocharger. Particles are shaded based on their surface charge and their sizes are doubled in the figure for visibility, (b) distribution of particle charge in the exit, and (c) distribution of number of particle–wall contacts

8.3 External Fields

8.3.1 Electric Field

When a particle is placed in an electric field, an electric dipole is induced in it, as shown in Figure 8.11. This electric dipole results in charge separation in the non-conductive particle and the particle becomes polarized. This effect leads to an electric force between particles. The charge separation does not result in accumulation of electrical charge among particles and the swarm

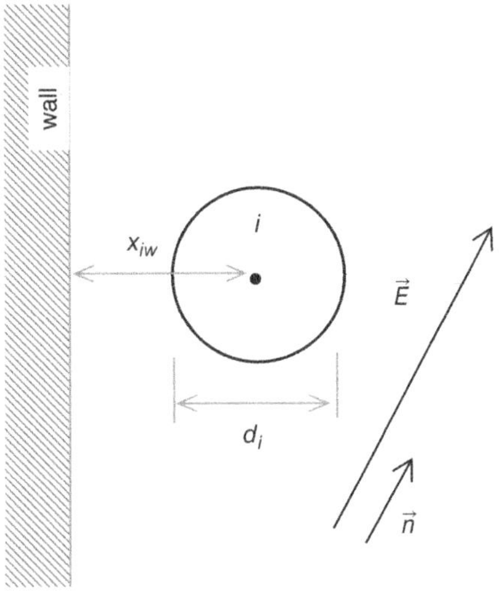

Figure 8.11 A particle placed in an electric field. Geometric parameters required for calculation of electric field induced interparticle force are shown in this figure

of particles remains neutral. Assuming that the charging process is instantaneous, the charge on the particle can bȩ expressed as [40]:

$$q = 16\pi\epsilon_f (2K+1) d_i^2 |\vec{E}| \tag{8.56}$$

in which

$$K = \frac{\epsilon_p - \epsilon_f}{\epsilon_p + 2\epsilon_f} \tag{8.57}$$

where ϵ_p and ϵ_f are permittivity of particle and fluid, respectively. The parameter K is the Clausius–Mossotti function that determines the strength of effective polarization in the particle. The Coulomb's electrostatic force exerted on this particle due to the electric field is:

$$\vec{f}_i^{elec} = q\vec{E} \tag{8.58}$$

In the vicinity of a wall, an additional force is exerted on the polarized particle due to the image effect. The collected electrical forces acting on this particle then is the sum of Coulomb's electrostatic force and the image force between the charged particle and the collection plate [41]:

$$\vec{f}_i^{elec} = \left(q|\vec{E}| + \frac{q^2}{16\pi\epsilon_f x_{iw}^2} \right) \vec{n} \tag{8.59}$$

where $\vec{n}$ is the unit vector in the direction of electric field and x_{iw} is the distance between the particle center and the wall.

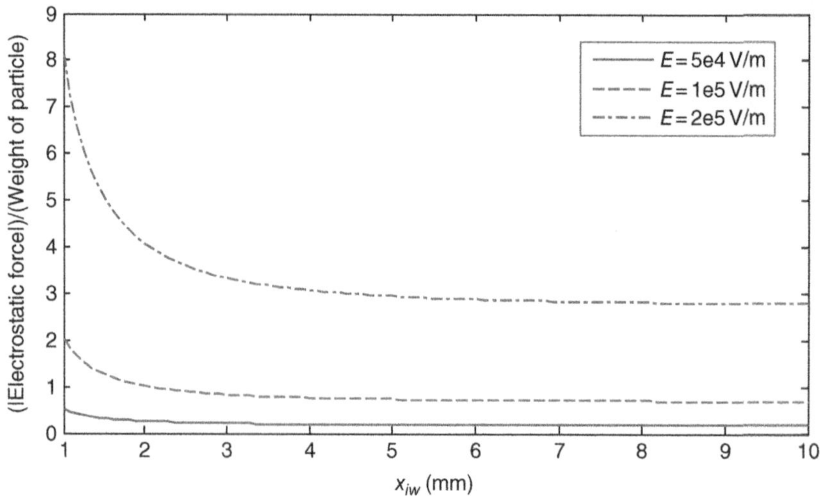

Figure 8.12 Dimensionless electrostatic force exerted on a 1 mm glass bead placed in an electric field. Horizontal axis shows the distance of particle from a dielectric wall

A sample calculation of the electrostatic force exerted on a 1 mm glass beads ($\rho_i = 2500$ kg/m^3 and $\epsilon_p = 3.82$) in various strengths of electrical fields is shown in Figure 8.12. The horizontal axis in this figure is the distance between the center of the particle and the wall. When close to the wall, the image effect is significant and the image force changes with the distance between the particle and the wall. However, it can be seen in this figure that the electrostatic force reaches a constant value in distances far from the wall (roughly 3 mm in this case) since the image effect becomes nil. Nevertheless, Coulomb's force due to the electric field would not vanish and remains unchanged with the distance between the particle and the wall. This figure also shows that the electrostatic force increases by increasing the field strength.

Kleijn van Willigen *et al.* [42] adopted a similar approach for evaluating $\vec{f}_{ij}^{elec}$. Considering the dipoles as geometric points, the IPF between two particles induced by the electric field can be obtained in a 2D space from [43]:

$$\vec{f}_{ij}^{elec} = C\frac{3\pi}{16}\epsilon_f d_i^2 K^2 E^2 \left(\frac{d_i}{x_{ij}}\right)^4 \left[\left(3\cos^2\theta_{ij} - 1\right)\vec{n}_r + \left(sin2\theta_{ij}\right)\vec{n}_\theta\right] \tag{8.60}$$

In this equation, C is the multi-pole correction factor, E is the strength of the electric field, θ_{ij} is the angle between the center-to-center line of particles, and the electric field and $\vec{n}_r$ and $\vec{n}_\theta$ are the unit vectors in r and θ directions, respectively. The total electrical IPF acting on the particle i due to induction by the external electric field is:

$$\vec{f}_i^{elec} = \sum_{j \in CL_{i,elec}} \vec{f}_{ij}^{elec} \tag{8.61}$$

where $CL_{i,elec}$ contains the list of particles that have significant electrical filed force interaction with particle i.

8.3.2 Magnetic Field

Applying an external magnetic field to magnetizable particles can improve the fluidization quality of particles [44], enhance cake filtration by boosting the filtration rate [45] and improving the transport properties of magnetorheological fluid [46]. For instance, magnetized fluidized beds can involve a number of advantages of both fixed beds and ordinary fluidized beds such as suppressing or delaying formation of bubbles in the fluidized bed as a result of imposing the magnetic field. In this way, the contact efficiency between gas and solid is improved and the problem of solids back-mixing is reduced.

At low to moderate intensity of magnetic field, orientation and arrangement of magnetizable particles takes place according to the direction of the field, but effect of magnetization of particles can be ignored. However, when the magnetic field is strong, magnetization of particles becomes significant and interparticle magnetic force caused by the interaction with neighboring particles also should be taken into account. Therefore, the total magnetic force exerted on a particle is caused by external magnetic field ($\vec{f}_i^{mag,ext}$) as well as magnetized particles ($\vec{f}_i^{mag,\text{int}}$):

$$\vec{f}_i^{mag} = \vec{f}_i^{mag,ext} + \vec{f}_i^{mag,\text{int}} \tag{8.62}$$

The magnetic force acting on a single particle in the presence of a magnetic field is [41]:

$$\vec{f}_i^{mag,ext} = V_i \chi \mu_0 \vec{H}.\nabla\vec{H} \tag{8.63}$$

in which V_i is the volume of the particle, μ_0 is the magnetic permeability of vacuum ($4\pi\times10^{-7}$ V s A^{-1} m^{-1}), and $\vec{H}$ is the magnetic field intensity. χ is the susceptibility of the mixture (fluid and particle) and is related to the particle susceptibility through the following formula:

$$\frac{\chi}{\chi+3} = \left(1-\varepsilon_f\right)\frac{\chi_p}{\chi_p+3} \tag{8.64}$$

where χ_p is the magnetic susceptibility of particles. In this equation, ε_f is the local porosity of the system, which can be considered as the porosity of the fluid computational in which the particle resides.

The interparticle magnetic force can be attractive or repulsive, depending on the relative position of particles with respect to their dipole moment. The force is attractive if the movement of the two particles is along their dipole moment axis [47]:

$$\vec{f}_{attractive}^{mag,ext} = -\frac{6\left(\chi_p V_i B\right)^2}{\mu_0}\left[\frac{2\pi^2 x_{ij}^5 + \chi_p V_i \pi x_{ij}^2}{\left(2\pi x_{ij}^3 - \chi_p V_i\right)^3}\right]\vec{n}_{ij} \tag{8.65}$$

This force is repulsive if moving perpendicular to this axis:

$$\vec{f}_{repulsive}^{mag,ext} = \frac{12\left(\chi_p V_i B\right)^2}{\mu_0}\left[\frac{4\pi^2 x_{ij}^5 - \chi_p V_i \pi x_{ij}^2}{\left(4\pi x_{ij}^3 + \chi_p V_i\right)^3}\right]\vec{n}_{ij} \tag{8.66}$$

Table 8.3 Magnetic susceptibility of various materials [48]

Material	Susceptibility	Material	Susceptibility
Graphite (perpendicular to atomic planes)	-5.95×10^{-4}	Carbon	-2.04×10^{-4}
Bismuth	-1.64×10^{-4}	Antimony	-6.7×10^{-5}
Indium	-5.1×10^{-5}	Thallium	-3.7×10^{-5}
Gold	-3.4×10^{-5}	Mercury	-2.8×10^{-5}
Beryllium	-2.4×10^{-5}	Silver	-2.4×10^{-5}
Gallium	-2.3×10^{-5}	Tin (α-gray)	-2.3×10^{-5}
Diamond	-2.18×10^{-5}	Phosphorous (white)	-2.0×10^{-5}
Selenium	-1.9×10^{-5}	Phosphorous (red)	-1.85×10^{-5}
Alumina (Al_2O_3)	-1.81×10^{-5}	Silica	-1.63×10^{-5}
Lead	-1.58×10^{-5}	Zinc	-1.57×10^{-5}
Pyrex glass	-1.388×10^{-5}	Sulfur (α)	-1.26×10^{-5}
Sulfur (β)	-1.14×10^{-5}	Magnesia (MgO)	-1.14×10^{-5}
Copper	-9.63×10^{-6}	Water	-9.05×10^{-6}
Silicon nitrate (Si_3N_4)	-9.0×10^{-6}	Graphite (parallel to atomic planes)	-8.5×10^{-6}
Zirconia (ZrO_2)	-8.3×10^{-6}	Germanium	-7.1×10^{-6}
Silicon	-4.2×10^{-6}	Air	3.6×10^{-7}
Tin (β-white)	2.4×10^{-6}	Rubidium	3.8×10^{-6}
Cesium	5.2×10^{-6}	Potassium	5.8×10^{-6}
Sodium	8.5×10^{-6}	Magnesium	1.17×10^{-5}
Yttria (Y_2O_3)	1.24×10^{-5}	Aluminum	2.07×10^{-5}
Calcium	2.17×10^{-5}	Tungsten	7.72×10^{-5}
Zirconium	1.09×10^{-4}	Yttrium	1.19×10^{-4}
Molybdenum	1.23×10^{-4}	Rhodium	1.69×10^{-4}
Tantalum	1.78×10^{-4}	Titanium	1.82×10^{-4}
Niobium	2.37×10^{-4}	Platinum	2.79×10^{-4}
Chromium	3.20×10^{-4}	Vanadium	3.84×10^{-4}
Hematite (α-Fe_2O_3)	1.46×10^{-3}	Lepidocrocide (γ-FeOOH)	2.12×10^{-3}
Geothite (α-FeOOH)	2.65×10^{-3}	Magnetite (Fe_3O_4)	7×10^{1}
Cobalt	2.5×10^{2}	Nickel	6×10^{2}
Stainless steel	$(4–11)\times10^{2}$	Mumetal[a]	1×10^{5}
Iron	2×10^{5}	Supermalloy[b]	1×10^{6}

[a] (16% Fe, 5% Cu, 2% Cr, 77% Ni).
[b] (16% Fe, 5% Mo, 79% Ni).

In these two formulas, B is the magnitude of the external magnetic flux density. Table 8.3 shows the magnetic susceptibility of different material.

Numerical evaluation of magnetic force between two ferromagnetic particles as a function of the distance between particles is illustrated in Figure 8.13. The magnetic force in this figure was evaluated between two 1 mm particles containing 20% ferrite ($\rho_i = 1090\,\text{kg/m}^3$ and $\chi_p = 2.75$ [47]). Only magnetic repulsive and attractive forces based on Equations 8.67 and 8.68 are considered in the presence of a uniform magnetic field of $B = 0.01\,\text{T}$. This figure shows that the magnetic force is significant in distances less than 3 mm between particles and sharply decreases by increasing the distance between particles.

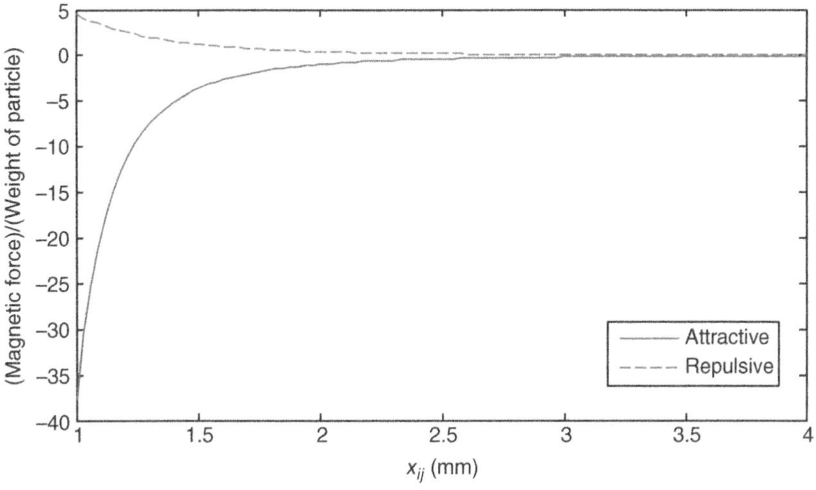

Figure 8.13 Dimensionless repulsive and attractive magnetic forces as a function of the distance between two 1 mm ferromagnetic particles in the presence of 0.01 T uniform magnetic field

Equations 8.65 and 8.66 are valid in only two directions (moving along or perpendicular to the dipole moment axis) and we need to adopt a general approach for evaluating the interparticle magnetic force. For this purpose, consider two particles of equal size as shown in Figure 8.14. Although the magnetic field may not be uniform throughout the system, let's assume that it is uniform in the short distance between these particles, which are considered ideal dipoles under the influence of this uniform magnetic field. The radial component of the magnetic force between particles is:

$$f_{ij,r}^{mag,\mathrm{int}} = -\frac{\mu_0}{4\pi}\left\{-\frac{6M^2}{x_{ij}^3}\frac{\partial\gamma}{\partial r}\cos(\theta-\gamma)\sin(\theta-\gamma)+\left[1-3\cos^2(\theta-\gamma)\right]\left(-\frac{3M^2}{x_{ij}^4}+\frac{2M}{x_{ij}^3}\frac{\partial M}{\partial r}\right)\right\} \tag{8.67}$$

where

$$\frac{\partial\gamma}{\partial r} = -\frac{9\chi V_i\sin 2\theta}{8\pi x_{ij}^4\left\{\left[1+\frac{\chi V_i}{8\pi x_{ij}^3}(3\cos 2\theta-1)\right]^2+\left(\frac{3\chi V_i}{8\pi x_{ij}^3}\sin 2\theta\right)^2\right\}} \tag{8.68}$$

$$\begin{aligned}\frac{\partial M}{\partial r} = &-\frac{\chi V_i B}{\mu_0\left\{\cos\gamma-\frac{\chi V_i}{8\pi x_{ij}^3}\left[\cos\gamma+3\cos(2\theta-\gamma)\right]\right\}^2}\\ &\times\left\{\left[\left(1-\frac{\chi V_i}{8\pi x_{ij}^3}\right)\sin\gamma+\frac{3\chi V_i}{8\pi x_{ij}^3}\sin(2\theta-\gamma)\right]\frac{\partial\gamma}{\partial r}-\frac{3\chi V_i}{8\pi x_{ij}^4}\left[\cos\gamma+\cos(2\theta-\gamma)\right]\right\}\end{aligned} \tag{8.69}$$

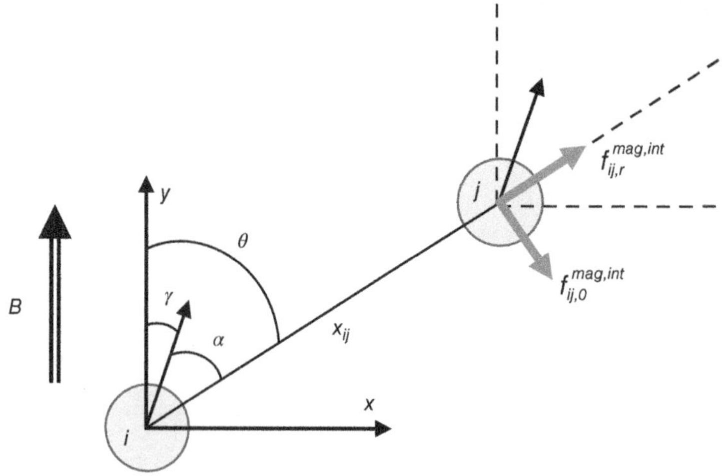

Figure 8.14 Repulsive and attractive magnetic force between two ideal dipoles under the influence of a uniform external magnetic field

The angular component of the magnetic force between particles is:

$$f_{ij,\theta}^{mag,\text{int}} = -\frac{\mu_0}{4\pi x_{ij}^4}\left\{6M^2\cos(\theta-\gamma)\sin(\theta-\gamma)\left(1-\frac{\partial\gamma}{\partial\theta}\right)+2M\left[1-3\cos^2(\theta-\gamma)\frac{\partial M}{\partial\theta}\right]\right\} \tag{8.70}$$

where

$$\frac{\partial\gamma}{\partial\theta} = \frac{2\left(\frac{3\chi V_i}{8\pi x_{ij}^3}\sin 2\theta\right)^2 + \frac{3\chi V_i}{4\pi x_{ij}^3}\cos 2\theta\left[1+\frac{\chi V_i}{8\pi x_{ij}^3}(3\cos 2\theta-1)\right]}{\left[1+\frac{\chi V_i}{8\pi x_{ij}^3}(3\cos 2\theta-1)\right]^2+\left(\frac{3\chi V_i}{8\pi x_{ij}^3}\sin 2\theta\right)^2} \tag{8.71}$$

$$\frac{\partial M}{\partial\theta} = \frac{\chi V_i B}{\mu_0\left[\left(\left(1-\frac{\chi V_i}{8\pi x_{ij}^3}\right)\cos\gamma-\frac{3\chi V_i}{8\pi x_{ij}^3}\cos(2\theta-\gamma)\right)\right]^2} \times\left\{\left[\left(1-\frac{\chi V_i}{8\pi x_{ij}^3}\right)\sin\gamma+\frac{3\chi V_i}{8\pi x_{ij}^3}\sin(2\theta-\gamma)\right]\frac{\partial\gamma}{\partial\theta}-\frac{3\chi V_i}{4\pi x_{ij}^3}\sin(2\theta-\gamma)\right\} \tag{8.72}$$

In these equations, M is the magnetic dipole moment. Once the radial and angular components of the magnetic forces are evaluated from Equations 8.67 and 8.70, respectively, the magnetic force exerted on particle i from magnetized particle j is given by:

$$\vec{f}_{ij}^{mag,\text{int}} = f_{ij,r}^{mag,\text{int}}\left(\vec{n}_r\right)_{ij} + f_{ij,\theta}^{mag,\text{int}}\left(\vec{n}_\theta\right)_{ij} \tag{8.73}$$

in which $\vec{n}_r$ and $\vec{n}_\theta$ are unit vectors in radial and angular directions, respectively, which vary with respect to relative position of particles i and j. The total magnetic force acting on particle i due to magnetization of surrounding particles is then obtained from:

$$\vec{f}_i^{mag,\text{int}} = \sum_{j \in CL_{i,mag}} \vec{f}_{ij}^{mag,\text{int}} \tag{8.74}$$

where $CL_{i,mag}$ contains the list of particles that have significant magnetic force with particle i.

Similar to the force, the torque acting on a magnetic particle due to the magnetic field comprises of the torque directly due to the imposed external magnetic field ($\vec{B}^{ext}$) and the torque due to magnetic dipole moment induced by all other particles ($\vec{B}^{\text{int}}$) [49]:

$$\vec{M}_i^{mag} = \vec{M}_i \times \left(\vec{B}^{ext} + \vec{B}^{\text{int}}\right) \tag{8.75}$$

where

$$\vec{B}^{\text{int}} = \sum_{j \in CL_{i,mag}} \frac{\mu_0}{4\pi x_{ij}^5}\left[3\left(\vec{M}_j . \vec{x}_{ij}\right)\vec{x}_{ij} - x_{ij}^2 \vec{M}_j\right] \tag{8.76}$$

8.3.3 Vibration Field

Vibration is used in gas-solid fluidized beds containing fine particles, which are difficult to operate due to high cohesive force, to improve the quality of fluidization by breaking up agglomerates. This is usually done by vertically vibrating the vessel and distributor, simultaneously. This technique is also used in hoppers and packing of particles. Vibration affects IPFs through intensifying the kinetic energy of particles. Therefore, we describe the effect of vibration in this section.

In order to incorporate the vibration induced forces into a DEM model, we should introduce the movement of walls (including the distributor) to the model. Assuming a sinusoidal vibration, displacement, and velocity of walls ($\vec{x}_w$ and $\vec{v}_w$, respectively) can be expressed by the following equations:

$$\vec{x}_w = \vec{x}_{w,0} + \vec{A}\sin\left(2\pi ft\right) \tag{8.77}$$

$$\vec{v}_w = 2\pi f\vec{A}\cos\left(2\pi ft\right) \tag{8.78}$$

where $\vec{A}$ and f are the amplitude and frequency of vibration, respectively, and $\vec{x}_{w,0}$ is the position of the wall at rest. The DEM steps remain the same as described in Chapter 3 and position and velocity of the wall should be calculated by Equations 8.77 and 8.78.

8.3.4 Acoustic Field

Acoustic fields are used in many processes to affect the flowability of particles. Forming agglomerates [50], breaking agglomerates [51], and cleaning the surface from particles [52] are examples of applications of acoustic and ultrasonic waves. Sound waves are also used for improving the fluidization quality of ultrafine and cohesive particles [53]. In fact, sound wave penetrates the fluidized bed and breaks up the large agglomerates formed by cohesive force, thus, a stabilized fluidization can be achieved.

The sound wave can be expressed by a periodic function. The sound velocity of an acoustic field with a frequency f can be given by:

$$\vec{U}_{sound} = \vec{U}\sin(2\pi ft) \tag{8.79}$$

here, U is the absolute magnitude of sound velocity that can be calculated from [54]:

$$SLP = 20\log\left(\frac{|\vec{U}|}{\sqrt{2}U_{ref}}\right) \tag{8.80}$$

where U_{ref} is the reference velocity at reference pressure. For air, the reference pressure is typically chosen to be the value at the threshold of human hearing, 2×10^{-5} Pa and the value of U_{ref} is 4.83×10^{-8} m/s.

SPL in Equation 8.80 is the sound pressure level. A sound wave attenuates through a gas-solid medium and follows exponential attenuation [55]:

$$\frac{SPL}{SPL_0} = l^{-2}\exp(-2\alpha l) \tag{8.81}$$

where SPL_0 is the inlet sound wave pressure (in decibel) and SPL is the sound wave pressure at distance l in the medium with an attenuation coefficient α. Taking the attenuation into account, the sound velocity at distance l in the gas-solid mixture is:

$$\vec{U}_{sound}(l,t) = \vec{U}\exp(-\alpha l)\sin(2\pi ft - kl) \tag{8.82}$$

here, k is the wave number:

$$k = \frac{2\pi f}{c} \tag{8.83}$$

and the attenuation coefficient can be obtained from [56]:

$$\alpha = \frac{9\pi}{2\lambda}(1-\varepsilon_f)\left(\frac{2\delta}{d_i}\right)^2\left(1+\frac{d_i}{2\delta}\right) \tag{8.84}$$

in which ε_f is the local porosity of the system, which can be considered as the porosity of the fluid computational cell in which the particle resides, and:

$$\delta = \sqrt{\frac{\mu\lambda}{\pi\rho_f c}} \tag{8.85}$$

$$\lambda = \frac{c}{f} \tag{8.86}$$

where c is the speed of sound in the gas and μ is the gas viscosity.

The acoustic force acting on a particle in a gas-solid mixture arises from the relative velocity between the velocity of particle and the oscillatory component of gas (sound velocity). Therefore, the acoustic force exerted on particle i can be evaluated from [15]:

$$\vec{f}_i^{ac} = \frac{\beta_{gs}V_i}{1-\varepsilon_f}\left(\vec{U}_{sound} - \vec{v}_i\right) \tag{8.87}$$

In this equation, the slip velocity, $\vec{U}_{sound} - \vec{v}_i$, is the velocity difference between particle and gas, imposed due to the acoustic field and β_{gs} is the interphase momentum transfer coefficient, which can be evaluated from:

$$\beta_{gs} = \begin{cases} 150\dfrac{\left(1-\varepsilon_f\right)^2 \mu_f}{\varepsilon_f^2 d_i^2} + 1.75\dfrac{\rho_f\left(1-\varepsilon_f\right)\left|\vec{U}-\vec{v}_i\right|}{\varepsilon d_i}, & \varepsilon_f < 0.8 \\[2ex] \dfrac{3}{4}C_d\dfrac{\rho_f\left(1-\varepsilon_f\right)\left|\vec{U}-\vec{v}_i\right|}{d_i}\varepsilon_f^{-2.65}, & \varepsilon_f \geq 0.8 \end{cases} \tag{8.88}$$

where C_d is the drag coefficient for an isolated particle (see Chapter 6 for more details).

8.4 Applications

Some researchers considered the van der Waals or liquid bridge force to be proportional to the buoyant force and included it into their CFD-DEM simulations as the IPF [9, 12, 57–59]. The following equation was used to represent different values of IPF:

$$f_{coh} = KV_i\left(\rho_i - \rho_f\right)g \tag{8.89}$$

where K is a constant that determines the magnitude of IPF. Rhodes *et al.* [8, 9] and Pandit *et al.* [57] studied the effect of IPFs on the hydrodynamics of fluidized beds of Geldart group B and D powders. Presence of IPFs among particles of group B powder changes the fluidization behavior of these powders. As the IPF increases, the average bubble size of bubbles decreases. Three different regimes can be observed when the gas velocity increases in the bed.

In the first regime, particles are fixed and pressure drop across the bed changes linearly with velocity. In the second regime, which corresponds to the homogeneous fluidization condition, the particles are completely suspended and pressure drop remains constant against gas velocity. In the third regime, bubbles are initiated in the bed and pressure drop fluctuates noticeably as a result of bubble movement. Such a behavior is similar to what we observe in the fluidization of group A powders. According to Pandit *et al.* [57], the minimum fluidization velocity is not sensitive to the presence of IPF while the minimum bubbling velocity increases with increasing the strength of this force. The maximum bed expansion before the onset of bubbling fluidization increases with increasing the IPF.

Pandit *et al.* [58, 59] studied the fluidization behavior of group A powder with and without IPFs. They found the same behavior as explained previously for group A powders in the absence of IPF. Introducing the IPF to the particles intensifies the homogeneous fluidization (increasing the ratio of minimum bubbling velocity to minimum fluidization velocity). The mechanism of bubble formation of group B powders in the absence and presence of IPFs is different. The presence of cohesive IPFs changes the size and shape of bubbles, delays the bubble formation, and higher gas velocity is required for bubble formation [59].

In a more phenomenological approach, some researchers used the Hamaker expression [6] to represent the cohesive van der Waals forces [60–64]. Xu [60] studied the fluidization behavior of mixtures of cohesive and non-cohesive particles. They showed that adding non-cohesive particles can improve the movability of cohesive particles and in this way the fluidization of very cohesive particles (which cannot be achieved in a normal situation) becomes possible. Ye *et al.* [61, 62] and Wang *et al.* [64] studied the fluidization behavior of group A powders with different strengths of van der Waals forces. The gross circulations of particles were observed in the homogeneous fluidization and bubbles in the bubbling fluidization were fast bubbles in the presence of weak van der Waals forces. They analyzed velocity fluctuations of particles and found that in the homogeneous fluidization, particle–particle contact, particle–fluid interaction, and van der Waals forces are main sources of velocity fluctuations while the effect of particle–fluid interactions become dominant in the bubbling fluidization. Furthermore, presence of van der Waals force can help stabilization of homogeneous fluidization [61]. Also, inelastic collisions between particles, sliding friction, and weak van der Waals force have insignificant effects on the hydrodynamic behavior of group A powders [64].

Jia *et al.* [65] investigated the packing properties of cohesive powder by the DEM simulation in which van der Waals force was considered as the cohesive force. It was shown that in the case of a mono-sized particle, the packing is looser when smaller particles are used. In the case of a bimodal powders, the particle size ratio does not influence the compactness of packing when this ratio is small and the packing becomes more compact when the particle size ratio increases. Simulations showed that if a powder has a Gaussian size distribution, the packing becomes more compact by increasing the average size. Kinetics of aggregation of nanoparticles in colloidal mixtures was investigated by Macpherson *et al.* [66] through DEM simulation. They showed that increasing the Hamaker constant (thus, increasing the van der Waals force) slows down the aggregation with a power law fall. Flow and mixing of cohesive particles (with van der Waals force) in a vertical bladed mixer was studied by Chandratilleke *et al.* [67]. They showed that the bed remains above the rotating blades without mixing if the cohesion between particles and walls is strong. Hou *et al.* [68] modeled the flow of cohesive powders in a screw feeder. In their model, magnitude of the cohesive force was obtained from the van der Waals force for fine particles and the capillary force in the case of wet particles. They correlated the solid flow rate to the magnitude of cohesive force and the rotational speed of the screw. Impact of ultrafine glass particles on the partition wall of a multichannel cyclone was investigated by Jasevicius *et al.* [69]. Ostanin *et al.* [70] developed a DEM model for multiscale modeling of cross-linked carbon nanotubes. They considered nanotubes as bonded cylinder segments on which van der Waals adhesion, viscous friction, and contact bonding are exerted. They showed that how cross-linking can transform from a soft bundle into a stiff one.

As mentioned earlier, when a free film of liquid exists on the surface of particles in a gas-solid flow, the liquid bridge is formed between contacting particles [20, 30, 71–76]. Mikami *et al.* [20] studied the fluidization behavior of wet and dry particles in a two dimensional fluidized bed. Cohesive force and critical rupture distance were included in their simulations. Formation of agglomerates was observed when particles were wet. They showed that the fluidization behavior of wet particles is completely different from that of dry particles and minimum fluidization velocity, bed voidage, and pressure fluctuations in a wet fluidized bed are higher than those in a dry fluidized bed. Xu *et al.* [71] included cohesive and viscose forces in the liquid bridge force model. They studied formation and breakage of wet agglomerates, mixing of wet solid particles, and force distribution in the bed. They suggested that the viscous force must be considered in the liquid bridge force because it greatly influences the behavior of fluidized bed. Kuwagi *et al.* [30] considered the effect of liquid lubrication and combined the fluid shear resistance with the Coulomb friction law to account for the lubrication effects. By introducing the liquid lubrication effects, particles move more actively in the whole bed and become less adhesive to the wall, while the pressure fluctuations do not change noticeably. Jain *et al.* [75] used a dimensionless number, the granular capillary number, which is the ratio of capillary force to drag force and characterized fluidization and mixing of wet particles with this dimensionless number. Zhu *et al.* [76] studied the flow pattern and particle mixing of wet particles in a three dimensional spouted bed. At low volumes of liquid bridge, the separation of particles increases. As a result, the drag force dominates the liquid bridge force and particles move in the form of clusters in the spout region. However, at high liquid volumes, the liquid bridge force is much greater than the drag force that causes formation of agglomerates at the spout-annulus interface.

The effect of liquid bridge force has been investigated in other types of granular flows. For example, Muguruma *et al.* [72] studied the flow behavior of mono-sized wet particles in a centrifugal tumbling granulator. Hsiau and Yang [74] studied the self-diffusion and mixing of wet powders with three types of wetting agents in a vibrator granulator bed. Performance of a filter dryer was modeled by Sahni *et al.* [77]. They incorporated a simplified heat transfer model into their calculations in order to determine the optimum drying conditions. DEM simulations were used for studying the excavation and pushing of soil by a bulldozer blade [78]. The water in soil forms liquid bridge among particles and increases the cohesive force among them, which is the main reason for the resistance of soil against the blade. Liu *et al.* [79] investigated the mixing of wet particles in a rotary drum. They found that the mixing slows down when the capillary force increases. Mechanical properties of piles of wet particles were obtained using the DEM model by Zhao *et al.* [80]. They showed that shape of particles and magnitude of the liquid bridge force have significant effects on the shape of pile and also have major influence on the force exerted on the base surface.

When the temperatures of particles increase beyond the softening temperature, the surface of particles becomes sticky. When these sticky particles collide, they may form larger agglomerates due to the formation of permanent solid bridges. Such agglomeration occurs in a polymerization reactor and iron oxide reduction process in the hot spot zones and causes some operational difficulties such as undesired change in the hydrodynamics of fluidized bed (de-fluidization, gas channeling, and bypassing), process shut down, and low quality of products. Some researchers have carried out experiments to understand this phenomenon at high temperature and develop the method for avoiding it [81, 82]. However, the CFD-DEM simulation is a viable alternative to such tedious and expensive experiments. Kuwagi *et al.* [83] studied the agglomeration of iron particles at high temperature based on the CFD-DEM with uniform temperature distribution of particles. They considered the surface diffusion mechanism for neck growth and could capture the agglomeration process of particles. Kuwagi and Horio [84] studied formation and breakage of agglomerates of fine particles. They showed that agglomerates are formed in the bubble wake region while they break in the upper region of bubbles.

Mansourpour *et al.* [85–87] studied the agglomeration of polyethylene particles in a bubbling fluidized bed in which they considered the viscous neck growth mechanism for polyethylene particles. Temperature distribution of both solid and gas phases were obtained by a typical thermal CFD-DEM model. Translational and rotational motions of agglomerates were modeled according to the multi-sphere method proposed by Kruggel-Emden *et al.* [88]. In their simulations, the inlet gas temperature (383 K) was considered to be above the softening temperature of polyethylene particles (368 K). They showed that the completely fluidized bed at the beginning of the simulation gradually is transformed into a de-fluidized bed when the agglomeration process advances. When the bed is completely fluidized, distinct bubbles form at the distributor zone, grow up through the bed and burst at the bed surface. Then, small agglomerates comprised of two or three particles form in the dense regions as well as some zones close to the distributor. Increasing the number of agglomerates in the bed promotes the possibility of particle–agglomerate and agglomerate–agglomerate contacts. Also, by growth of agglomerates, their movement becomes more restricted. Consequently, the bed expansion reduces and so does the bubble size. In this situation, the gas passing through the bed does not exert enough drag force on the agglomerates to compensate their weight. Therefore, the majority of the gas passes through channels formed among agglomerates. At long enough time, massive agglomerates accumulate at the bottom of the bed and a de-fluidized layer is formed at the top of the distributor. Nevertheless, in this situation particles and agglomerates in the middle or the bed surface are still fluidized and the height at which bubbles are formed shifts to the top of the de-fluidized layer. This situation continues until all particles form large agglomerates and the whole bed becomes de-fluidized.

Hassani *et al.* [89] used CFD-DEM for investigating the effect of electrostatic forces between particles on the hydrodynamics of fluidized beds. They showed that size of bubbles and solids diffusivity decrease when electric charge of bubbles increases. Hogue *et al.* [90] electrified spherical particles through their movement over surfaces of different material and measured charge to mass of particles. They then modeled this process by DEM. The electric charge generated on particles through their impact on metallic and polymeric surfaces was determined experimentally by Yoshida *et al.* [91]. They developed a model based on the proposed mechanism of this process and validated it by the DEM simulation. Watano [92] described the mechanism and control of electrification of particles in pneumatic transport of powders both experimentally and numerically and used CFD-DEM for describing the movement of particles in 3D at various operating conditions.

Electrostatic field has been used in pneumatic transport systems to counter the effect of electrification of particles on the flow. Lim *et al.* [93] used a CFD-DEM model to investigate the effect of electrostatic field on pneumatic conveying of powders in inclined and vertical pipes. They found reversed flow of particles in the presence of a mild electrostatic field in the dense region close to the bottom and forward flow in a more dilute region in the space above. However, at sufficiently high field strengths, complete backflow of solids in the inclined pipe was observed which requires higher gas velocity to sustain a net positive flow along the pipe. They also reported that it would take longer to reach the steady state condition when the electrostatic field strength is increased. Packing of fine particles in an electrical field in an electrostatic precipitator was studied numerically by Yang *et al.* [94]. They showed that increasing either particle size or strength of the electrical field increases packing density until reaching the density of random loose packing obtained by gravitational settling. Especially, they found that the correlation between packing density and the ratio of cohesive force to packing driven force can be applied to the packing in electrostatic separators if the gravity is replaced with the electrical field force.

Electric field can improve the quality of fluidization. Mixing degree increases when applying electric field into the fluidized bed [95]. Kleijn van Willigen *et al.* [42] showed with DEM simulations that gas bubbles become smaller when either horizontal or vertical electric field is applied to a fluidized bed. However, they observed in their simulations that at very high field strength, the bed becomes de-fluidized. They concluded that moderate strength electric fields cause better distribution of gas in the bed at the distributor level and decreases the rate of coalescence, which results in smaller bubbles at the top.

Zhenghua *et al.* [44] simulated the flow of magnetizable particles in a magnetically assisted gas-solid fluidized bed by CFD-DEM using an external constant gradient magnetic field along bed height. They showed that by increasing the strength of the magnetic field, particles form straight chains along the direction of the field and these strings become fluidized. At very high strength field, particles strongly stick to each other and gas channels can be observed in between the magnetic particle chains. It was concluded that applying a moderate strength magnetic field provides high fluctuations of particles and distributes the gas more evenly in the bed. Hydrodynamic characterization of liquid-solid fluidized beds in the presence of uniform transverse magnetic field was carried out by Wang *et al.* [96, 97] by means of CFD-DEM simulations. They observed the same behavior of particles reported by Zhenghua *et al.* [44] in a gas–solid fluidized bed concerning formation of straight chains along the direction of the field and channeling of the fluid. Their simulation also indicated that fluctuation of particles diminishes as the field strength is increased. Deng *et al.* [49] investigated the mixing of cohesive particles under magnetically assisted impaction condition. Their simulation showed that the mixing becomes faster when smaller magnets are used and when the surface energy is decreased. They found that the mixing process reaches the steady state faster when the collision energy between magnets and non-magnets exceeds the cohesive energy. It was shown by CFD-DEM simulation that filtration is enhanced when a magnetic field is applied to the cake [45]. The results confirmed that, at low field strength, the change in permeability of the filter cake due to magnetic dipole–dipole interactions improves filtration performance, which was already suggested by the experiments [98].

Effect of vibration on the fluidization behavior of cohesive powders has been studied by CFD-DEM simulations [13, 14, 99]. It was shown that vibration enhances the fluidization of cohesive particles by creating large tensile stresses that break the whole assembly of the bed into the agglomerates. The vibration energy is transported to particles near the walls via particle–wall collisions and is propagated in the whole bed which causes the particles to move vigorously. When the effect of vibration is incorporated with the effect of gas velocity, the bubbles are initiated and a better mixing of particles is achieved with higher amplitudes, frequencies, and gas velocities. Asmar *et al.* [100] simulated vibrated fluidized bed of cohesive powders by CFD-DEM and proposed some benchmark mathematical tests for validation of the model. These tests were used to confirm correct implementation of several force algorithms in isolation, free motion, single contact, and multiple contact situations. Effect of vertical and horizontal vibration on the hydrodynamic of a fluidized bed of Geldart group B particles was investigated by Tatemoto *et al.* [101]. They showed that when vertical vibration is applied, large bubbles appear in the bed caused by vibration gap. The vibration gap is the gap between particle bed and wall caused by vibration. In the case of horizontal vibration, they found it difficult to fluidize the bed at high vibration strength and low frequency due to the fact that vibration gaps act as channels and bypass the gas. They later used their model for simulation of motion of cohesive particles in a vibrated fluidized bed [13] and showed that as the vibration gap increases, the effect of vibration on the cohesive particle motion becomes larger. There are three regimes along bed height in vertical vibration [102]: a low particle concentration regime with a vibration gap near the bottom, a high concentration regime in the middle, and a transition regime at the bed surface. Effect of vibration on segregation of a binary mixture in a vibrating fluidized bed was investigated by Sun *et al.* [103]. It was shown that vibration enhances the segregation of spherical particles of the same size with different densities.

Shuai *et al.* [15] studied the effects of sound fields on the fluidization of cohesive particles. Their simulation results showed that the fluidization of cohesive powders is enhanced in the presence of sound fields. The granular temperature is increased with increasing the level of sound pressure. The fluidization is more enhanced at high sound pressure levels and an optimum range of frequency between 100 and 120 Hz exists at which the cohesive force is minimized.

Nomenclature

Symbol	Unit	Description
a	m	Distance defined by Equation 8.42
A	—	Constant in Equations 8.17 and 8.21
$\vec{A}$	m	Amplitude of vibration
A_H	J	Hamaker constant
$A_{H,i}$	J	Hamaker constant of particle i
$A_{H,f}$	J	Hamaker constant of fluid
b	m	Distance defined by Equation 8.43
B	—	Constant in Equations 8.17 and 8.21
B	T	Magnetic flux density
$\vec{B}^{ext}$	T	External magnetic flux density
$\vec{B}^{\text{int}}$	T	Magnetic flux density due to dipole moment of other particles
c	m/s	Speed of sound
C	—	Constant in Equations 8.17 and 8.21

C	F	Capacitance
C	—	Multi-pole correction factor
Ca	—	Capillary number
C_D	—	Drag coefficient
CL_i	—	Contact list of particle i
$CL_{IP,i}$	—	Contact list of particle i containing particles with interparticle interaction
d_i	m	Diameter of particle i
e	C	Elementary charge
$\vec{E}$	V/m	Electric field
f	Hz	Frequency
f_{coh}	N	Cohesive force
$\vec{f}_i^{ac}$	N	Acoustic force acting on particle i
$\vec{f}_{ij}^{c}$	N	Collision force between particle i and particle j
$\vec{f}_{ij}^{cap}$	N	Capillary force between particles i and j
$\vec{f}_{iw}^{cap}$	N	Capillary force between particle and wall
$\vec{f}_i^{elec}$	N	Electrostatic force acting on particle i
$\vec{f}_{ij}^{elec}$	N	Electrostatic force between particles i and j
$\vec{f}_i^{ext}$	N	External forces acting on particle i
$\vec{f}_i^{f-p}$	N	Fluid–particle interaction forces acting on particle i
$\vec{f}_{ij}^{lb}$	N	Liquid bridge force between particles i and j
$\vec{f}_{ij}^{IP}$	N	Interparticle force between particles i and j
$\vec{f}_i^{mag}$	N	Magnetic force exerted on particle i
$f_i^{mag,ext}$	N	Magnetic force exerted on particle i caused by an external field
$\vec{f}_{attractive}^{mag,ext}$	N	Attractive magnetic force
$\vec{f}_{repulsive}^{mag,ext}$	N	Repulsive magnetic force
$f_i^{mag,int}$	N	Magnetic force exerted on particle i caused by magnetized particles
$f_{ij,r}^{mag,int}$	N	Radial component of magnetic force due to magnetized particles
$f_{ij,\theta}^{mag,int}$	N	Tangential component of magnetic force due to magnetized particles
$\vec{f}_{ij}^{p-p}$	N	Particle–particle interaction forces acting on particle i
$\vec{f}_{ij}^{vdW}$	N	van der Waals force between particles i and j
$\vec{f}_{iw}^{vdW}$	N	van der Waals force between particle and wall
$\vec{f}_{ij}^{vis}$	N	Viscous force between particles i and j
$\vec{g}$	m/s^2	Gravitational acceleration
$\hbar$	$J \cdot s$	Reduced Planck constant
$\vec{H}$	T	Magnetic field intensity
H_r	Pa	Hardness of particle
k	m^{-1}	Wave number
k_B	J/K	Boltzmann constant
k_c	—	Charging efficiency
k_{im}	F^{-1}	Image effect constant
K	—	Clausius–Mossotti function

K	—	Constant in Equation 8.89
l	m	Surface to surface distance between particles
l	m	Distance into mixture
$\hat{l}$	—	Dimensionless distance
l_{rup}	m	Rupture distance
$\hat{l}_{rup}$	—	Dimensionless rupture distance
m_i	kg	Mass of particle *i*
M	$A \cdot m^2$	Magnetic dipole moment
$\vec{M}_i$	$A \cdot m^2$	Magnetic dipole moment of particle *i*
$\vec{M}_i^{mag}$	$N \cdot m$	Torque due to magnetic force acting on particle *i*
n	—	Refractive index
$\vec{n}$	—	Unit vector in the direction of electric field
$\vec{n}_{ij}$	—	Center to center unit vector
$\vec{n}_{iw}$	—	Unit vector pointing from center of particle to wall
$\vec{n}_r$	—	Unit vector in *r* direction
$\vec{n}_\theta$	—	Unit vector in θ direction
p	Pa	Pressure
q_i	C	Electric charge of particle *i*
q_w	C	Electric charge of wall
r_0	m	Radius defined by Equation 8.41
R	m	Effective radius of particle
R_{asp}	m	Radius of surface asperities
$R_{cut-off}$	m	Cut-off radius for interparticle forces
R_{LR}	m	Maximum radius at which long range force is significant
S	m^2	Contact area
SPL	dB	Sound pressure level
SPL_0	dB	Sound pressure level at inlet
t	s	Time
T	K	Temperature
U	m/s	Magnitude of velocity of sound
U_{ref}	m/s	Reference velocity of sound
$\vec{U}_{sound}$	m/s	Velocity of sound
$\vec{v}_i$	m/s	Velocity of particle *i*
$\vec{v}_{ij}^n$	m/s	Normal relative velocity between particles *i* and *j*
$\vec{v}_w$	m/s	Velocity of wall
V	V	Electric potential
V'	V	Total induced potential
$\hat{V}$	—	Dimensionless liquid bridge volume
V_c	V	Contact potential
V_{ext}	V	Potential difference due to external electric field
V_i	m^3	Particle volume
V_{im}	V	Image effect potential
V_{ind}	V	Induced potential
V_L	m^3	Liquid bridge volume
V_p	m^3	Particle volume
x	—	Dimensionless surface to surface distance

x_i	m	Position of particle i
x_{ij}	m	Center to center distance of particles
x_{iw}	m	Distance between particle and wall
$\vec{x}_w$	m	Position of wall
$\vec{x}_{w,0}$	m	Position of wall at rest
X_V	—	Volume factor
y	—	Dimensionless diameter of particle
Z_0	m	Critical gap
Greek symbols		
α	rad	Supplementary half-filling angle
α	m^{-1}	Attenuation coefficient
β_{gs}	$kg/m^3 \cdot s$	Interphase momentum transfer coefficient
γ	rad	Dipole angular orientation
Γ	m^{-1}	Curvature of liquid bridge
δ	m	Thickness of Stokes boundary layer
Δq	C	Change in electric charge
ϵ	F/m	Permittivity
ϵ_f	F/m	Absolute permittivity of fluid
ϵ_p	F/m	Absolute permittivity of particle
θ	rad	Solid–liquid–gas contact angle
θ	rad	Angle between magnetic field and center to center line of particles
θ_{ij}	rad	Angle between center-to-center line of particles and electric field
λ	m	Wavelength
μ	$Pa \cdot s$	Viscosity
μ_0	$V \cdot s/A \cdot m$	Magnetic permeability of vacuum
μ_f	$Pa \cdot s$	Viscosity of fluid
ρ_1, ρ_2	m	Radii of principal curvature of liquid bridge
ρ_i	kg/m^3	Density of particle i
ρ_f	kg/m^3	Density of fluid
σ	N/m	Surface tension
σ_w	C/m^2	Surface charge density
ν_e	Hz	UV absorption frequency
ϕ	rad	Half-filling angle
φ_i	J	Work function of particle i
φ_w	J	Work function of wall
χ	—	Magnetic susceptibility
χ_p	—	Magnetic susceptibility of particle

References

[1] Shabanian, J. and Chaouki, J. (2014) Local characterization of a gas–solid fluidized bed in the presence of thermally induced interparticle forces. *Chemical Engineering Science*, **119**, 261–273.

[2] Shabanian, J. and Chaouki, J. (2015) Hydrodynamics of a gas–solid fluidized bed with thermally induced interparticle forces. *Chemical Engineering Journal*, **259**, 135–152.

[3] Guo, Y. and Sinclair Curtis, J. (2015) Discrete element method simulations for complex granular flows. *Annual Review of Fluid Mechanics*, **47**, 21–46.

[4] Verlet, L. (1967) Computer experiments on classical fluids. I. Thermodynamical properties of Lennard-Jones molecules. *Physical Review E*, **159**, 98–103.

[5] Yao, Z., Wang, J., Liu, G., and Cheng, M. (2004) Improved neighbor list algorithm in molecular simulations using cell decomposition and data sorting method. *Computer Physics Communications*, **161**, 27–35.
[6] Hamaker, H.C. (1937) The London-van der Waals attraction between spherical particles. *Physica IV*, **10**, 1058–1080.
[7] Yang, R.Y., Zou, R.P., and Yu, A.B. (2000) Computer simulation of the packing of fine particles. *Physical Review E*, **62**, 3900–3908.
[8] Rhodes, M.J., Wang, X.S., Nguyen, M., Stewart, P., and Liffman, K. (2001) Onset of cohesive behaviour in gas fluidized beds: a numerical study using DEM simulation. *Chemical Engineering Science*, **56**(14), 4433–4438.
[9] Rhodes, M.J., Wang, X.S., Nguyen, M., Stewart, P., and Liffman, K. (2001) Use of discrete element method simulation in studying fluidization characteristics: influence of interparticle force. *Chemical Engineering Science*, **56**(1), 69–76.
[10] Wang, J., van der Hoef, M.A., and Kuipers, J.A.M. (2010) CFD study of the minimum bubbling velocity of Geldart A particles in gas-fluidized beds. *Chemical Engineering Science*, **65**(12), 3772–3785.
[11] Yang, R.Y., Zou, R.P., and Yu, A.B. (2003) Effect of material properties on the packing of fine particles. *Journal of Applied Physics*, **94**(5), 3025–3034.
[12] Pandit, J.K., Wang, X.S., and Rhodes, M.J. (2005) Study of Geldart's group A behaviour using the discrete element method simulation. *Powder Technology*, **160**(1), 7–14.
[13] Tatemoto, Y., Mawatari, Y., and Noda, K. (2005) Numerical simulation of cohesive particle motion in vibrated fluidized bed. *Chemical Engineering Science*, **60**(18), 5010–5021.
[14] Limtrakul, S., Rotjanavijit, W., and Vatanatham, T. (2007) Lagrangian modeling and simulation of effect of vibration on cohesive particle movement in a fluidized bed. *Chemical Engineering Science*, **62**(1–2), 232–245.
[15] Shuai, W., Xiang, L., Huilin, L., Guodong, L., Jiaxing, W., and Pengfei, X. (2011) Simulation of cohesive particle motion in a sound-assisted fluidized bed. *Powder Technology*, **207**(1–3), 65–77.
[16] Parsegian, V.A. (2006) *Van der Waals Forces: A Handbook for Biologists, Chemists, Engineers, and Physicists*, Cambridge University Press, Cambridge.
[17] Shabanian, J., Jafari, R., and Chaouki, J. (2012) Fluidization of ultrafine powders. *International Review of Chemical Engineering*, **4**, 16–50.
[18] Butt, H.J., Graf, K., and Kappl, M. (2006) *Physics and Chemistry of Interfaces*, Wiley-VCH Verlag GmbH, Darmstadt.
[19] Li, S., Marshall, J.S., Liu, G., and Yao, Q. (2011) Adhesive particulate flow: the discrete-element method and its application in energy and environmental engineering. *Progress in Energy and Combustion Science*, **37**, 633–668.
[20] Mikami, T., Kamiya, H., and Horio, M. (1998) Numerical simulation of cohesive powder behavior in a fluidized bed. *Chemical Engineering Science*, **53**(10), 1927–1940.
[21] Hotta, K., Takeda, K., and Iinoya, K. (1974) The capillary binding force of a liquid bridge. *Powder Technology*, **10**(4–5), 231–242.
[22] Maugis, D. (1987) Adherence of elastomers: fracture mechanics aspects. *Journal of Adhesion Science and Technology*, **1**, 105–134.
[23] Ennis, B.J., Li, J., Gabriel I, T., and Robert, P. (1990) The influence of viscosity on the strength of an axially strained pendular liquid bridge. *Chemical Engineering Science*, **45**(10), 3071–3088.
[24] Pitois, O., Moucheront, P., and Chateau, X. (2000) Liquid bridge between two moving spheres: an experimental study of viscosity effects. *Journal of Colloid and Interface Science*, **231**(1), 26–31.
[25] Seville, J.P.K., Willett, C.D., and Knight, P.C. (2000) Interparticle forces in fluidisation: a review. *Powder Technology*, **113**(3), 261–268.
[26] Lian, G., Thornton, C., and Adams, M.J. (1993) A theoretical study of the liquid bridge forces between two rigid spherical bodies. *Journal of Colloid and Interface Science*, **161**(1), 138–147.
[27] Pitois, O., Moucheront, P., and Chateau, X. (2001) Rupture energy of a pendular liquid bridge. *The European Physical Journal B*, **23**, 79–86.
[28] Shi, D. and McCarthy, J.J. (2008) Numerical simulation of liquid transfer between particles. *Powder Technology*, **184**(1), 64–75.
[29] Pietsch, W.B. and Rumpf, H. (1967) Haftkraft, Kapillardruck, FlWussigkeitsvolumen und Grenzwinkel einer/ WussigkeitsbrWucke zwischen zwei Kugeln. *Chemie Ingenieur Technik*, **39**, 885–893.
[30] Kuwagi, K., Takano, K., and Horio, M. (2000) The effect of tangential lubrication by bridge liquid on the behavior of agglomerating fluidized beds. *Powder Technology*, **113**, 287–298.

[31] Weigert, T. and Ripperger, S. (1999) Calculation of the liquid bridge volume and bulk saturation from the half-filling angle. *Particle & Particle Systems Characterization*, **16**(5), 238–242.
[32] Rabinovich, Y.I., Esayanur, M.S., and Moudgil, B.M. (2005) Capillary forces between two spheres with a fixed volume liquid bridge: theory and experiment. *Langmuir*, **21**, 10992–10997.
[33] Bi, X.T. (2011) Electrostatic phenomena in fluidized systems: present status of understanding, and research needs. International Conference on Circulating Fluidized Beds and Fluidization Technology – CFB-10, Sunriver, OR.
[34] Matsusaka, S., Maruyama, H., Matsuyama, T., and Ghadiri, M. (2010) Triboelectric charging of powders: a review. *Chemical Engineering Science*, **65**(22), 5781–5807.
[35] Matsusaka, S., Ghadiri, M., and Masuda, H. (2000) Electrification of an elastic sphere by repeated impacts on a metal plate. *Journal of Physics D: Applied Physics*, **33**, 2311–2319.
[36] Lowell, J. and Roseinnes, A.C. (1980) Contact electrification. *Advances in Physics*, **29**, 947–1023.
[37] Matsuyama, T. and Yamamoto, H. (1995) Characterizing the electrostatic charging of polymer particles by impact charging experiments. *Advanced Powder Technology*, **6**(3), 211–220.
[38] Norouzi, Y. (2016) Modeling of particle electrification by discrete element method in a tribocharger. MSc thesis. University of Tehran.
[39] Pei, C., Wu, C.Y., England, D., Byard, S., Berchtold, H., and Adams, M. (2013) Numerical analysis of contact electrification using DEM–CFD. *Powder Technology*, **248**, 34–43.
[40] White, H.J. (1951) Particle charging in electrostatic precipitation. *Transactions of the American Institute of Electrical Engineers*, **70**(2), 1186–1191.
[41] Jackson, J.D. (1999) *Classical Electrodynamics*, 3rd edn, John Wiley & Sons, Inc., New York.
[42] Kleijn van Willigen, F., van Ommen, J.R., van Turnhout, J., and van den Bleek, C.M. (2003) Bubble size reduction in a fluidized bed by electric fields. *International Journal of Chemical Reactor Engineering*, **1**, A21.
[43] Parthasarathy, M. and Klingenberg, D.J. (1996) Electrorheology: mechanisms and models. *Materials Science and Engineering: R: Reports*, **17**(2), 57–103.
[44] Zhenghua, H., Xiang, L., Huilin, L., Guodong, L., Yurong, H., Shuai, W., and Pengfei, X. (2010) Numerical simulation of particle motion in a gradient magnetically assisted fluidized bed. *Powder Technology*, **203**(3), 555–564.
[45] Eichholz, C., Nirschl, H., Chen, F., and Hatton, T.A. (2012) DEM-simulation of the magnetic field enhanced cake filtration. *AIChE Journal*, **58**(12), 3633–3644.
[46] Kargulewicz, M., Iordanoff, I., Marrero, V., and Tichy, J. (2012) Modeling of magnetorheological fluids by the discrete element method. *Journal of Tribology*, **134**, 031706-1–031706-9.
[47] Jovanovic, G.N., Sornchamni, T., Atwater, J.E., Akse, J.R., and Wheeler Jr, R.R. (2004) Magnetically assisted liquid–solid fluidization in normal and microgravity conditions: experiment and theory. *Powder Technology*, **148**(2–3), 80–91.
[48] Schenck, J.F. (1996) The role of magnetic susceptibility in magnetic resonance imaging: MRI magnetic compatibility of the first and second kinds. *Medical Physics*, **23**(6), 815–850.
[49] Deng, X., Scicolone, J.V., and Davé, R.N. (2013) Discrete element method simulation of cohesive particles mixing under magnetically assisted impaction. *Powder Technology*, **243**, 96–109.
[50] Riera-Franco de Sarabia, E., Elvira-Segura, L., González-Gómez, I., Rodríguez-Maroto, J.J., Muñoz-Bueno, R., and Dorronsoro-Areal, J.L. (2003) Investigation of the influence of humidity on the ultrasonic agglomeration of submicron particles in diesel exhausts. *Ultrasonics*, **41**(4), 277–281.
[51] Ding, P. and Pacek, A.W. (2008) De-agglomeration of goethite nano-particles using ultrasonic comminution device. *Powder Technology*, **187**(1), 1–10.
[52] Brereton, G.J. and Bruno, B.A. (1994) Particle removal by focused ultrasound. *Journal of Sound and Vibration*, **173**(5), 683–698.
[53] Guo, Q., Liu, H., Shen, W., Yan, X., and Jia, R. (2006) Influence of sound wave characteristics on fluidization behaviors of ultrafine particles. *Chemical Engineering Journal*, **119**(1), 1–9.
[54] Morse, P.M. and Ingard, K.U. (1968) *Theoretical Acoustics*, McGraw-Hill, New York.
[55] Wu, D., Qian, Z.W., and Shao, D. (1993) Sound attenuation in a coarse granular medium. *Journal of Sound and Vibration*, **162**(3), 529–535.
[56] Valverde, J.M. (2013) Acoustic streaming in gas-fluidized beds of small particles. *Soft Matter*, **9**(37), 8792–8814.
[57] Pandit, J.K., Wang, X.S., and Rhodes, M. (2003) Influence of interparticle cohesive force on fluidized bed behaviour by DEM simulation. 3rd International Conference on CFD in the Minerals and Process Industries, Melbourne, Australia.

[58] Pandit, J.K., Wang, X.S., and Rhodes, M.J. (2006) On Geldart group A behaviour in fluidized beds with and without cohesive interparticle forces: a DEM study. *Powder Technology*, **164**(3), 130–138.
[59] Pandit, J.K., Wang, X.S., and Rhodes, M.J. (2007) A DEM study of bubble formation in group B fluidized beds with and without cohesive inter-particle forces. *Chemical Engineering Science*, **62**(1–2), 159–166.
[60] Xu, B.H. (2003) Modelling of the gas fluidization of a mixture of cohesive and cohesionless particles by a combined continuum and discrete model. *KONA*, **21**, 100–108.
[61] Ye, M., van der Hoef, M.A., and Kuipers, J.A.M. (2004) A numerical study of fluidization behavior of Geldart A particles using a discrete particle model. *Powder Technology*, **139**(2), 129–139.
[62] Ye, M., van der Hoef, M.A., and Kuipers, J.A.M. (2005) The effects of particle and gas properties on the fluidization of Geldart A particles. *Chemical Engineering Science*, **60**(16), 4567–4580.
[63] Moreno-Atanasio, R., Xu, B.H., and Ghadiri, M. (2007) Computer simulation of the effect of contact stiffness and adhesion on the fluidization behaviour of powders. *Chemical Engineering Science*, **62**(1–2), 184–194.
[64] Wang, J., van der Hoef, M.A., and Kuipers, J.A.M. (2010) The role of particle-particle interactions in bubbling gas-fluidized beds of Geldart A particles: a discrete particle study. 6th International Symposium on Multiphase Flow, Heat Mass Transfer and Energy Conversion, Xi'an, China.
[65] Jia, T., Zhang, Y.W., Chen, J.K., and He, Y.L. (2012) Dynamic simulation of granular packing of fine cohesive particles with different size distributions. *Powder Technology*, **218**, 76–85.
[66] Macpherson, S.A., Webber, G.B., and Moreno-Atanasio, R. (2012) Aggregation of nanoparticles in high ionic strength suspensions: effect of Hamaker constant and particle concentration. *Advanced Powder Technology*, **23**, 478–484.
[67] Chandratilleke, R., Yu, A.B., Bridgwater, J., and Shinohara, K. (2014) Flow and mixing of cohesive particles in a vertical bladed mixer. *Industrial and Engineering Chemistry Research*, **53**, 4119–4130.
[68] Hou, Q.F., Dong, K.J., and Yu, A.B. (2014) DEM study of the flow of cohesive particles in a screw feeder. *Powder Technology*, **256**, 529–539.
[69] Jasevicius, R., Baltrenas, P., Kacianauskas, R., and Grubliauskas, R. (2014) DEM simulation of the impact of ultrafine glass particles on the partition wall of the multichannel cyclone. *Particulate Science and Technology*, **32**, 576–587.
[70] Ostanin, I., Ballarini, R., and Dumitrica, T. (2015) Distinct element method for multiscale modeling of cross-linked carbon nanotube bundles: from soft to strong nanomaterials. *Journal of Materials Research*, **30**, 19–25.
[71] Xu, B.H., Yu, A.B., Chew, S.J., and Zulli, P. (1999) A study of the effect of liquid-induced forces on gas-solid flow by a combined continuum and discrete model. 2nd International Conference on CFD in the Minerals and Process Industries, Melbourne, Australia.
[72] Muguruma, Y., Tanaka, T., and Tsuji, Y. (2000) Numerical simulation of particulate flow with liquid bridge between particles (simulation of centrifugal tumbling granulator). *Powder Technology*, **109**(1–3), 49–57.
[73] Yang, S.C. and Hsiau, S.S. (2001) The simulation of powders with liquid bridges in a 2D vibrated bed. *Chemical Engineering Science*, **56**, 6837–6849.
[74] Hsiau, S.S. and Yang, S.C. (2003) Numerical simulation of self-diffusion and mixing in a vibrated granular bed with the cohesive effect of liquid bridges. *Chemical Engineering Science*, **58**, 339–351.
[75] Jain, K., Shi, D., and McCarthy, J.J. (2004) Discrete characterization of cohesion in gas–solid flows. *Powder Technology*, **146**, 160–167.
[76] Zhu, R.R., Zhu, W.B., Xing, L.C., and Sun, Q.Q. (2011) DEM simulation on particle mixing in dry and wet particles spouted bed. *Powder Technology*, **210**, 73–81.
[77] Sahni, E., Hallisey, J., Morgan, B., Strong, J., and Chaudhuri, B. (2012) Quantifying drying performance of a filter dryer: experiments and simulations. *Advanced Powder Technology*, **23**(2), 239–249.
[78] Tsuji, T., Nakagawa, Y., Matsumoto, N., Kadono, Y., Takayama, T., and Tanaka, T. (2012) 3-D DEM simulation of cohesive soil-pushing behavior by bulldozer blade. *Journal of Terramechanics*, **49**(1), 37–47.
[79] Liu, P.Y., Yang, R.Y., and Yu, A.B. (2013) DEM study of the transverse mixing of wet particles in rotating drums. *Chemical Engineering Science*, **86**, 99–107.
[80] Zhao, L.L., Zhao, Y.M., Liu, C.S., and Li, J. (2014) Discrete element simulation of mechanical properties of wet granular pile. *Acta Physica Sinica*, **63**, 034501.
[81] Arastoopour, H., Huang, C.S., and Weil, S.A. (1988) Fluidization behavior of particles under agglomerating conditions. *Chemical Engineering Science*, **43**(11), 3063–3075.
[82] Hariri, M.H., Arastoopour, H., and Rehmat, A. (1993) Agglomeration of polyolefin particles in a fluidized bed with a central jet part I. Experimental study. *Powder Technology*, **74**(3), 231–238.
[83] Kuwagi, K., Mikami, T., and Horio, M. (2000) Numerical simulation of metallic solid bridging particles in a fluidized bed at high temperature. *Powder Technology*, **109**(1–3), 27–40.

[84] Kuwagi, K. and Horio, M. (2002) A numerical study on agglomerate formation in a fluidized bed of fine cohesive particles. *Chemical Engineering Science*, **57**(22–23), 4737–4744.
[85] Mansourpour, Z., Mostoufi, N., and Sotudeh-Gharebagh, R. (2014) Investigating agglomeration phenomena in an air-polyethylene fluidized bed using DEM–CFD approach. *Chemical Engineering Research and Design*, **92**(1), 102–118.
[86] Mansourpour, Z., Mostoufi, N., and Sotudeh-Gharebagh, R. (2014) A numerical study on agglomeration in high temperature fluidized beds. *Journal of Chemical and Petroleum Engineering*, **48**, 15–25.
[87] Mansourpour, Z., Mostoufi, N., and Sotudeh-Gharebagh, R. (2013) A mechanistic study of agglomeration in fluidized beds at elevated pressures. *Canadian Journal of Chemical Engineering*, **91**, 560–569.
[88] Kruggel-Emden, H., Wirtz, S., and Scherer, V. (2008) A study on tangential force laws applicable to the discrete element method (DEM) for materials with viscoelastic or plastic behavior. *Chemical Engineering Science*, **63**(6), 1523–1541.
[89] Hassani, M.A., Zarghami, R., Norouzi, H.R., and Mostoufi, N. (2013) Numerical investigation of effect of electrostatic forces on the hydrodynamics of gas–solid fluidized beds. *Powder Technology*, **246**, 16–25.
[90] Hogue, M.D., Calle, C.I., Curry, D.R., and Weitzman, P.S. (2009) Discrete element modeling (DEM) of triboelectrically charged particles: revised experiments. *Journal of Electrostatics*, **67**(4), 691–694.
[91] Yoshida, M., Shimosaka, A., Shirakawa, Y., Hidaka, J., Matsuyama, T., and Yamamoto, H. (2003) Estimation of electrostatic charge distribution of flowing toner particles in contact with metals. *Powder Technology*, **135–136**, 23–34.
[92] Watano, S. (2006) Mechanism and control of electrification in pneumatic conveying of powders. *Chemical Engineering Science*, **61**(7), 2271–2278.
[93] Lim, E.W.C., Zhang, Y., and Wang, C.-H. (2006) Effects of an electrostatic field in pneumatic conveying of granular materials through inclined and vertical pipes. *Chemical Engineering Science*, **61**(24), 7889–7908.
[94] Yang, S., Dong, K., Zou, R., Yu, A., and Guo, J. (2013) Packing of fine particles in an electrical field. *Granular Matter*, **15**(4), 467–476.
[95] Lu, L.-S. and Hsiau, S.-S. (2005) Mixing in vibrated granular beds with the effect of electrostatic force. *Powder Technology*, **160**(3), 170–179.
[96] Wang, S., Shen, Y., Ma, Y., Gao, J., Lan, X., Dong, Q., and Cheng, Q. (2013) Study of hydrodynamic characteristics of particles in liquid–solid fluidized bed with uniform transverse magnetic field. *Powder Technology*, **245**, 314–323.
[97] Wang, S., Sun, Z., Li, X., Gao, J., Lan, X., and Dong, Q. (2013) Simulation of flow behavior of particles in liquid–solid fluidized bed with uniform magnetic field. *Powder Technology*, **237**, 314–325.
[98] Eichholz, C., Stolarski, M., Goertz, V., and Nirschl, H. (2008) Magnetic field enhanced cake filtration of superparamagnetic PVAc-particles. *Chemical Engineering Science*, **63**(12), 3193–3200.
[99] Moon, S.J., Kevrekidis, I.G., and Sundaresan, S. (2006) Particle simulation of vibrated gas-fluidized beds of cohesive fine powders. *Industrial and Engineering Chemistry Research*, **45**, 6966–6977.
[100] Asmar, B.N., Langston, P.A., Matchett, A.J., and Walters, J.K. (2002) Validation tests on a distinct element model of vibrating cohesive particle systems. *Computers & Chemical Engineering*, **26**(6), 785–802.
[101] Tatemoto, Y., Mawatari, Y., Yasukawa, T., and Noda, K. (2004) Numerical simulation of particle motion in vibrated fluidized bed. *Chemical Engineering Science*, **59**(2), 437–447.
[102] Xiang, L., Shuyan, W., Huilin, L., Goudong, L., Juhui, C., and Yikun, L. (2010) Numerical simulation of particle motion in vibrated fluidized beds. *Powder Technology*, **197**(1–2), 25–35.
[103] Sun, L., Zhao, F., Zhang, Q., Li, D., and Lu, H. (2014) Numerical simulation of particle segregation in vibration fluidized beds. *Chemical Engineering & Technology*, **37**(12), 2109–2115.

Index

Coupled CFD-DEM Modeling: Formulation, Implementation and Application to Multiphase Flows, First Edition.
Hamid Reza Norouzi, Reza Zarghami, Rahmat Sotudeh-Gharebagh and Navid Mostoufi.

Companion website: www.wiley.com/go/norouzi/CFD-DEM